Space Sciences Series of ISSI

Volume 82

More information about this series at http://www.springer.com/series/6592

Teodolina Lopez · Anny Cazenave ·
Mioara Mandea · Jérôme Benveniste
Editors

Geohazards and Risks Studied from Earth Observations

Previously published in *Surveys in Geophysics*,
Volume 41, Issue 6, 2020

 Springer

Editors
Teodolina Lopez
GET
IRT Saint Exupéry—Fondation STAE
Toulouse, France

Mioara Mandea
Centre National d'Etudes Spatiales
Paris, France

Anny Cazenave
International Space Science Institute (ISSI)
Bern, Switzerland

Jérôme Benveniste
European Space Agency
ESA-ESRIN
Frascati, Roma, Italy

ISSN 1385-7525
Space Sciences Series of ISSI
ISBN 978-3-030-87991-4

Cover image: Puzzle representing the most common and the most observed geohazards from Earth
observations. From upper left to right bottom: Holuhraun eruption (2014-09-06) captured by Landsat 8
(USGS); North beach erosion (2017-10-06); Typhoon Hagibis heading towards Japan (2019-10-10)
observed by Copernicus Sentinel-3 (Copernicus data/ESA); Interferogram of the Napa valley earthquake
(2014-08-24, Mw = 6) from Sentinel-1A (Copernicus data (2014)/ESA/PPO.labs/Norut/COMET-
SEOM); The Kerala flood (2018-08-22) captured by Sentinel-2 (Copernicus data/ESA); The Camp Fire
(2018-11-08) observed from Landsat 8 (USGS). The background is a passover on Earth (2015) picture
from ESA astronaut Samantha Cristoforettiin in the ISS (ESA).

This Springer imprint is published by the registered company Springer Nature Switzerland AG
The registered company address is: Gewerbestrasse 11, 6330 Cham, Switzerland

Contents

Guest Editorial: International Space Science Institute (ISSI) Workshop on Geohazards and Risks Studied from Earth Observations 1
T. Lopez, A. Cazenave, M. Mandea, and J. Benveniste

Earth Observations for Coastal Hazards Monitoring and International Services: A European Perspective 7
Jérôme Benveniste, Mioara Mandea, Angélique Melet, and Pierric Ferrier

Space-Based Earth Observations for Disaster Risk Management 31
G. Le Cozannet, M. Kervyn, S. Russo, C. Ifejika Speranza, P. Ferrier, M. Foumelis, T. Lopez, and H. Modaressi

Early Warning from Space for a Few Key Tipping Points in Physical, Biological, and Social-Ecological Systems 59
Didier Swingedouw, Chinwe Ifejika Speranza, Annett Bartsch, Gael Durand, Cedric Jamet, Gregory Beaugrand, and Alessandra Conversi

Geoscientists in the Sky: Unmanned Aerial Vehicles Responding to Geohazards 107
R. Antoine, T. Lopez, M. Tanguy, C. Lissak, L. Gailler, P. Labazuy, and C. Fauchard

Earth Observation for the Assessment of Earthquake Hazard, Risk and Disaster Management 145
J. R. Elliott

Earth Observation for Crustal Tectonics and Earthquake Hazards 177
J. R. Elliott, M. de Michele, and H. K. Gupta

Remote Sensing for Assessing Landslides and Associated Hazards 213
Candide Lissak, Annett Bartsch, Marcello De Michele,
Christopher Gomez, Olivier Maquaire, Daniel Raucoules,
and Thomas Roulland

Fire Danger Observed from Space 259
M. Lucrecia Pettinari and Emilio Chuvieco

**On the Use of Satellite Remote Sensing to Detect Floods
and Droughts at Large Scales** 283
T. Lopez, A. Al Bitar, S. Biancamaria, A. Güntner, and A. Jäggi

**Earth Observations for Monitoring Marine Coastal Hazards
and Their Drivers** 311
A. Melet, P. Teatini, G. Le Cozannet, C. Jamet, A. Conversi, J. Benveniste,
and R. Almar

**Contributions of Space Missions to Better Tsunami Science:
Observations, Models and Warnings** 357
H. Hébert, G. Occhipinti, F. Schindelé, A. Gailler, B. Pinel-Puysségur,
H. K. Gupta, L. Rolland, P. Lognonné, F. Lavigne, E. Meilianda,
S. Chapkanski, F. Crespon, A. Paris, P. Heinrich, A. Monnier, A. Jamelot,
and D. Reymond

Air Pollution and Sea Pollution Seen from Space 405
Camille Viatte, Cathy Clerbaux, Christophe Maes, Pierre Daniel,
René Garello, Sarah Safieddine, and Fabrice Ardhuin

**Geomagnetic Field Processes and Their Implications
for Space Weather** 433
Mioara Mandea and Aude Chambodut

Surveys in Geophysics (2020) 41:1179–1183
https://doi.org/10.1007/s10712-020-09617-1

Guest Editorial: International Space Science Institute (ISSI) Workshop on Geohazards and Risks Studied from Earth Observations

T. Lopez[1,2] · A. Cazenave[2,3] · M. Mandea[4] · J. Benveniste[5]

Accepted: 23 September 2020 / Published online: 13 October 2020
© Springer Nature B.V. 2020

The Earth's environment and humankind have always suffered from natural hazards such as earthquakes, volcanic eruptions, tsunamis, occasional meteor impacts, wildfires and hydro-meteorological extremes (e.g., cyclones, floods, storm surges, landslides triggered by heavy rainfall or floods, heat waves and droughts). With the development of human societies, a new kind of hazards has appeared; these are referred to as man-made hazards. These new hazards include global warming and greenhouse gas emissions, air and water pollution, plastic wastes throughout the ecosystem and technological disasters such as oil spills, chemical or nuclear explosions (Alexander 2018). Among the man-made hazards, technological disasters may cause huge damages in terms of environmental degradation and human casualties, but due to their scarcity they are less susceptible to interact with natural hazards, as opposed to pollution and, above all, to greenhouse gas releases. Indeed, the development of anthropogenic hazards can enhance or interact with already existing natural risks, thus leading to more complex effects, which in turn are more difficult to predict. Hydro-meteorological hazards such as cyclones, storms, heat waves, floods and droughts are intensifying as time goes on and some of them are becoming more frequent (e.g., heat waves and floods) (Stocker et al. 2013; Yamazaki et al. 2018 and references therein), and increasing the vulnerability of populations. However, the societal and economic impacts of natural and man-made hazards will differ depending on the geographical region, southern countries paying generally a heavier price in terms of human casualties and goods lost (Hyndman and Hyndman 2016).

With the development of a large variety of advanced sensors aboard satellites and the growing amount of available data, space-based Earth Observations (EOs) are increasingly being used to better support disaster monitoring, mitigation, adaptation and risk

✉ T. Lopez
  lopez.teodolina@gmail.com

1 Present Address: IRT Saint Exupéry, Institut de Recherche Technologique Antoine de Saint-Exupéry – Fondation STAE, GET, 14 Avenue Édouard Belin, 31400 Toulouse, France

2 ISSI – International Space Science Institute, Hallerstrasse 6, 3012 Bern, Switzerland

3 LEGOS – Laboratoire d'Études en Géophysique et Océanographie Spatiales, 18 Avenue Édouard Belin, 31401 Toulouse Cedex 9, France

4 CNES – Centre National d'Études Spatiales, 2 Place Maurice Quentin, 75039 Paris, France

5 ESA – European Space Agency, Largo Galileo Galilei, 1, 00044 Frascati, Italy

 Springer

management. The space-based observing systems have several advantages compared to in situ networks. Since they are not affected by the hazards occurring at the surface of the Earth, they collect consistent data over different spatiotemporal scales and give us access to dangerous and/or remote areas. The EO datasets provide human society with the benefits of a synoptic view of natural hazards and their associated risks.

The Copernicus Programme of the European Union now provides routine space-based and in situ observations of the Earth and the environment, in particular with the operational Sentinel missions developed and operated by the European Space Agency (ESA). These missions offer invaluable observations of natural and man-made hazards and disasters. The scientific community can then better understand the underlying processes and their complex interactions and further improve forecasts and projections. As the above-mentioned hazards cannot be treated in isolation, since several of them are interconnected (e.g., land use changes have an impact on hydrology and coastal zone dynamics; pollution affects fresh and coastal zone water quality, etc.), a Workshop entitled "Natural and man-made hazards monitoring by the Earth Observation missions: current status and scientific gaps" was held at the International Space Science Institute (ISSI) in Bern, Switzerland (from 15 to 18 April 2019), in order to facilitate cross-disciplinary discussions. The Workshop brought together scientists from different horizons working on different aspects of geohazards and their impacts on society and the environment. More specifically, the objectives of the Workshop consisted of:

(1) Presenting the various natural and man-made hazards for which remote sensing data from the Copernicus Sentinels and other Earth Observation missions can be used in synergy, and with UAV (Unmanned Aerial Vehicles)-based and in situ data,
(2) Discussing the interactions between different natural and man-made hazards,
(3) Providing added-value information on the physical processes causing these phenomena, and
(4) Addressing how the space-based data can be assimilated into predictive models in order to improve our forecasting abilities.

The first section of this Special Issue presents both the role and the benefits of EOs for hazards monitoring, post-disaster management and forecasting. The paper by Benveniste et al. (2020) provides an informative overview of European space missions that routinely monitor geohazards. The authors also describe relevant scientific and societal applications of the collected data and provide information on the downstream International Services such as the Copernicus ocean and land monitoring services that freely distribute remote sensing data products and information to a broad variety of users from scientists to managers and decision-makers.

With the increasing amount of EO data available and their improved coverage over large areas, Le Cozannet et al. (2020) investigate how space-based EOs best support disaster risk management. More precisely, they consider how and where EOs can be an asset for disaster prevention, preparedness and crisis management, as well as for post-crisis management. This article emphasizes the advantages of cross-disciplinary studies between the environmental and social sciences in order to improve disaster risk reduction. In the context of global warming, the monitoring of the evolution of key tipping points that will cascade global warming and other natural hazards, could benefit from the use of more EO datasets. Indeed, these datasets will stimulate the detection of early warning signals of such tipping elements. This is the issue discussed in the article by Swingedouw et al. (2020), whose focus is on the overturning circulation of the Atlantic Ocean and on the subpolar gyre system, marine and terrestrial ecosystems, permafrost, and runaway melting of the Greenland and Antarctica ice sheets. The last article of this section on EOs reviews the use of UAVs deployed in the geohazards context (Antoine et al. 2020). The tremendous increase in the amount of UAV-based data available gives us access to very high-spatial resolution datasets which leads to the opening of new research areas. This article reviews the latest scientific and technical advances made in different domains such as mass movements of the Earth surface (e.g., landslides), volcanic eruptions, flood events and earthquakes. The benefits of combining EO datasets from space-borne sensors with UAVs' datasets in order to better monitor and forecast hazards are discussed.

The second section emphasizes the hazards that impact the Earth and its environment, such as fires, mass movements of the Earth, e.g., landslides, and earthquakes. As opposed to the number of flood and drought hazards which increased during the 1900s, fatality rates due to earthquakes remained relatively constant during the last century. In that context, Elliott (2020) proposes a review of how the assimilation of EO data into models helps to quantify the crustal deformation and associated risks during the seismic event and the subsequent longer-time scale inter-seismic deformation. Specific examples are presented in Elliott et al. (2020) to illustrate the use of EO, airborne and ground-based datasets to measure the deformation associated with the failure of the Earth's crust occurring at faults. This article also provides a study of the interest of EOs for smaller magnitude earthquakes that are associated with artificial water reservoirs.

Ground deformations also occur in other contexts such as coastal landslides, mountain debris flows, rockfalls and mudflows in periglacial environments. Due to their occurrence in such different environments, Lissak et al. (2020) present the main remote sensing tools currently being used for hazard mapping, for making inventories, and for monitoring surface deformations. They also focus on the assets becoming available for Earth mass redistribution studies and disaster risk prevention via multi-platform remote sensing using spaceborne, airborne and in situ datasets. Fires have an increasing impact on the natural environment and human and animal populations, as demonstrated by the recent disasters in Australia, Northern Europe and the West Coast of the USA. The article by Pettinari and Chuvieco (2020) describes the different causes of fires and their impacts on the environment. Due to the increased vulnerability of populations suffering from fires, EOs provide invaluable information with worldwide and frequent coverage which identifies the different factors triggering the fires and giving access to near real time information on them.

The last section of this Special Issue deals with some of the hazards that affect the hydrosphere (both continental and oceanic), the atmosphere and the upper atmosphere where space weather effects predominate. Recent advances on the monitoring of floods and droughts are rapidly progressing. This evolution is achieved because of moderate-to-low-resolution datasets obtained from the space gravimetry missions, GRACE and GRACE-FO, the SMOS and SMAP missions dedicated to measuring soil moisture, and the various nadir radar altimetry missions that measure surface water levels. Lopez et al. (2020) review the innovative applications of the datasets collected by these sensors, in particular to improve our understanding of the physical processes involved as well as the forecasting of floods and droughts. Due to their social, economic and environmental values to human society, coastal zones experience high stresses and suffer from higher vulnerability to different natural and anthropogenic hazards such as storm surges, marine heat waves, coastal flooding, sea level rise, erosion and shoreline retreat, acidification and destruction of ecosystems. In that context, Melet et al. (2020) investigate what are the benefits of EOs to monitor coastal hazards and their drivers via the development of monitoring programs and early warning forecasting systems. One of the most massive natural hazards that also causes tremendous destruction in coastal areas is tsunamis. The major tsunamis that followed the Sumatra earthquake (in Indonesia, Mw = 9.1) in 2004 and the Tohoku earthquake (in Japan, Mw = 9.1) in 2011 have raised the public's awareness of this specific disaster. Hébert et al. (2020) prepared a review of the different challenges encountered in tsunami science and of how EO datasets improve the development of tsunami warning systems at the global scale. Another major anthropogenic hazard that impacts human health and the natural environment is pollution, in particular air and sea pollution. Viatte et al. (2020) review the benefits of EOs for monitoring different forms of pollution and discuss how their combination with in situ data increases our understanding of the complex interactions of all types of pollution within the biosphere. Finally, space weather hazards are also discussed in this Special Issue. Space weather deals with the complex interactions between the Sun, the solar wind and the Earth's magnetosphere, ionosphere and atmosphere. The manifestation of strong space weather events has a number of physical effects in the near-Earth environment: the acceleration of charged particles in space forming the Van Allen belts, extraordinary auroral displays in both polar regions, the intensification of electric currents in space and of induced currents on the ground, and global magnetic disturbances at the Earth's surface. In Mandea and Chambodut (2020), the most recent developments in space-based measurements of the external and internal magnetic fields of the Earth and their interpretation are provided together with a discussion of related technological issues, such as the malfunctioning of satellites, impacts on radio communications and navigation systems, as well as impacts on electrical power distribution grid systems.

The organizers of this Workshop are most grateful to ISSI for hosting and for the sponsorship of this event, to the reviewers for their hard work, which led to improvements in the quality of the published articles, to the Editor in Chief, Professor Michael Rycroft, for his advice and for editorial improvements to the English language of some papers, and to the staff of Springer Nature for publishing this Special Issue efficiently.

References

Alexander D (2018) Natural disasters. Routledge, London

Antoine R, Fauchard C, Gaillet L et al (2020) Geoscientists in the sky: unmanned aerial vehicles for geohazards response. Surv Geophys 41(6):1285–1321. https://doi.org/10.1007/s10712-020-09611-7

Benveniste J, Mandea M, Melet A et al (2020) Earth observations for coastal hazards monitoring and international services: a European perspective. Surv Geophys 41(6):1185–1208. https://doi.org/10.1007/s10712-020-09612-6

Elliott JR (2020) Earth observation for the assessment of earthquake hazard, risk and disaster management. Surv Geophys 41(6):1323–1354. https://doi.org/10.1007/s10712-020-09606-4

Elliott JR, de Michele M, Gupta HK (2020) Earth observation for crustal tectonics and earthquake hazards. Surv Geophys 41(6):1355–1389. https://doi.org/10.1007/s10712-020-09608-2

Hébert H, Gailler A, Gupta H et al (2020) Contribution of space missions to better tsunami science: observations, models and warnings. Surv Geophys 41(6):1535–1581. https://doi.org/10.1007/s10712-020-09616-2

Hyndman D, Hyndman D (2016) Natural hazards and disasters. Cengage Learning, Boston

Le Cozannet G, Kervyn M, Russo S et al (2020) Space-based earth observations for disaster risk management. Surv Geophys 41(6):1209–1235. https://doi.org/10.1007/s10712-020-09586-5

Lissak C, Bartsch A, De Michele M et al (2020) Remote sensing for assessing landslides and associated hazards. Surv Geophys 41(6):1391–1435. https://doi.org/10.1007/s10712-020-09609-1

Lopez T, Al Bitar A, Biancamaria S et al (2020) On the use of satellite remote sensing to detect floods and droughts at large scales. Surv Geophys 41(6):1461–1487. https://doi.org/10.1007/s10712-020-09618-0

Mandea M, Chambodut A (2020) Geomagnetic field processes and their implications for space weather. Surv Geophys 41(6):1611–1627. https://doi.org/10.1007/s10712-020-09598-1

Melet A, Teatini P, Le Cozannet G et al (2020) Earth observations for monitoring marine coastal hazards and their drivers. Surv Geophys 41(6):1489–1534. https://doi.org/10.1007/s10712-020-09594-5

Pettinari ML, Chuvieco E (2020) Fire dangers observed from space. Surv Geophys 41(6):1437–1459. https://doi.org/10.1007/s10712-020-09610-8

Stocker TF, Qin D, Plattner G-K et al (2013) IPCC 2013: climate change 2013—the physical science basis. Contribution of working group I to the fifth assessment report of the intergovernmental panel on climate change. Cambridge University Press, Cambridge, New York

Swingedouw D, Speranza CI, Bartsch A et al (2020) Early warning from space for a few key tipping points in physical, biological, and social-ecological systems. Surv Geophys 41(6):1237–1284. https://doi.org/10.1007/s10712-020-09604-6

Viatte C, Clerbaux C, Maes C et al (2020) Air pollution and sea pollution seen from space. Surv Geophys 41(6):1583–1609. https://doi.org/10.1007/s10712-020-09599-0

Yamazaki D, Watanabe S, Hirabayashi Y (2018) Global flood risk modeling and projections of climate change impacts. In: Schumann GJ-P, Bates PD, Apel H, Aronica GT (eds) Global flood hazard: applications in modeling, mapping, and forecasting, AGU geophysical monograph series 233. Wiley, London. https://doi.org/10.1002/9781119217886.ch11

Publisher's Note Springer Nature remains neutral with regard to jurisdictional claims in published maps and institutional affiliations.

Surveys in Geophysics (2020) 41:1185–1208
https://doi.org/10.1007/s10712-020-09612-6

Earth Observations for Coastal Hazards Monitoring and International Services: A European Perspective

Jérôme Benveniste[1] · Mioara Mandea[2] · Angélique Melet[3] · Pierric Ferrier[4]

Received: 11 April 2020 / Accepted: 6 August 2020 / Published online: 30 September 2020
© Springer Nature B.V. 2020

Abstract

This article aims to provide a tour of satellite missions for Coastal Hazards Monitoring, of relevant applications, as well as the downstream International Services such as the Copernicus Ocean and Land Monitoring Services. Earth observation (EO) satellite remote sensing provides global, repetitive and long-term observations with increasing resolution with every new generation of sensors. They permit the monitoring of small-scale signals like the ones impacting the coastal zone. EO missions are showcased in this article. Transforming the data products based on the satellite mission ground segment (usually called geophysical products, geophysical data records or so-called Level 2 products) into information useable by managers and decision-makers is done by downstream international services. This is an essential step to increase the uptake of satellite data for the benefit of society. Here, the type of services provided by, e.g., the European Copernicus Programme, is described along with examples of applications, such as monitoring storm surges.

Keywords Earth observation satellite missions · Satellite data products · International services · Coastal zone · Coastal hazards · Copernicus

1 Introduction

Coastal zones are observed by Earth observation (EO) satellite missions since the onset of the space era. Data products provided by the satellite mission ground segment are used by scientists to monitor coastal zone parameters. Scientists develop alternate novel EO data processing algorithms to enhance the data quality (precision, resolution) for a better monitoring of the coastal zone, both on the marine side and the land side.

The definition of the coastal zone, its population and urbanisation, its social and economic impact, its biological value (marine flora and fauna), its development and evolution,

---

✉ Jérôme Benveniste
  Jerome.Benveniste@esa.int

[1] European Space Agency (ESA-ESRIN), Largo Galileo Galilei, 1, 00044 Frascati, Rome, Italy

[2] CNES - Centre National d'Etudes Spatiales, 2 Place Maurice Quentin, 75039 Paris, France

[3] Mercator Ocean International, 8 rue Hermès, 31520 Ramonville-Saint-Agne, France

[4] CNES- Centre National d'Etudes Spatiales, 18 Ave. E. Belin, 31400 Toulouse, France

and its exploitation (offshore oil and gas, maritime and coastal tourism, port and transport activities, fisheries, aquaculture, ocean-extractable energy, marine safety and surveillance) are reported in Melet et al. (2020). The important role of the coastal zone and therefore the detrimental impact of coastal hazards, be they natural or anthropogenic, is also stressed in Melet et al. (2020).

Monitoring the coastal zone is consequently a fundamental first step to protect the well-being of the large number of people living nearby and even of the rest of the world depending on its ecosystem. Satellite missions provide a cost-effective way to supply repeated, dense and global observations of the essential variables representing the dynamics, the health and trends of the coastal zone, as well as events leading to coastal hazards (Benveniste et al. 2019). For a better exploitation of the measurements provided by satellites, International Services have been set up to convert these data into exploitable information for managers and decision-makers, but they are used as well by scientists, to analyse the impacts of coastal hazards, develop predictive models and mitigation strategies and improve resilience.

The scientific community has developed the precursor processing algorithms aiming at global, regional and local monitoring that have been implemented in operational services. The use of the data operationally supplied by the International Services and by the scientific community ensures a continuous loop of data quality and dedicated algorithms to improve the operational services.

2 Earth Observation Satellite Missions and Applications

2.1 ESA-Led Earth Observation Satellite Missions

The Earth Observation Missions launched by the European Space Agency (ESA) are focused on three main lines of activities: Earth System science (the Earth Explorer Programme), operational global monitoring of the environment and security (the Copernicus Programme) and meteorology (Fig. 1). Since the latest launch of Aeolus on 22 August 2018, ESA has 25 EO satellites in development and 15 satellites in operation. The next mission to be launched is the radar altimetry Sentinel-6 Mission (Sect. 2.1.2). The Earth Explorer programme (Sect. 2.1.1) and Copernicus Sentinel missions (Sect. 2.1.2) are developed hereafter.

2.1.1 Earth System Science

The Earth Explorer Programme provides innovative specific scientific data requested via Open Calls to design and launch satellite missions to support Earth's system science (Fig. 2). The main satellite mission used for the coastal zone is the Synthetic Aperture and Interferometric Altimeter mission CryoSat, launched in 2010 and still in excellent working order 10 years after. The Gravity and Ocean Circulation Experiment (GOCE) Mission (2009–2013) collected data to produce the most accurate and highest resolution static gravity field and geoid, used as a reference for oceanography. The Soil Moisture and Ocean Salinity (SMOS) Mission, launched in 2009, still supplies not only sea surface salinity but also wind speed during extreme events. The Aeolus Mission provides 3-dimensional wind profiles that may turn out very useful for predicting coastal hazards linked to weather.

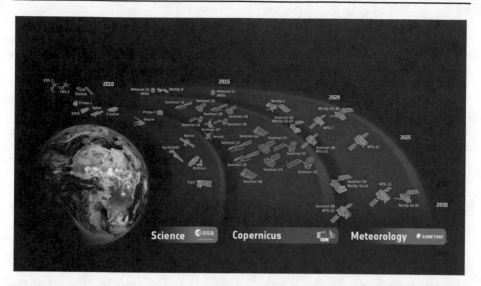

Fig. 1 The ESA-developed Earth Observation Missions; some additional missions are being proposed beyond 2025 in the Science and Copernicus Programmes. *Credit:* ESA

Fig. 2 Satellites of the ESA Earth Explorer Programme. *Credit:* ESA

2.1.2 Copernicus Sentinels Missions

The Copernicus Programme fosters a user-driven system for Global Monitoring of the Environment and Security (formerly known as GMES). Copernicus, which is led and mainly financed by the European Union, is an operational "system of systems" that integrates space, in situ data and numerical models. Beyond the family of Sentinel satellites, it hosts six services (Sect. 3) and offers a full and open data policy, over very long term, systematically and operationally. ESA is the development and procurement agency and the coordinator of the Copernicus Space Component offering 6 Sentinel Missions (Fig. 3), which have redundant and backup satellites, and the associated Ground Segment, which is jointly operated by Eumetsat (the European Organisation for the Exploitation of Meteorological Satellites). ESA is also responsible for the definition of the future evolution of the overall Copernicus Space Component architecture.

Copernicus is the largest producer of EO data in the world, pushing forward an unprecedented Big Data revolution in EO. For instance, today all landmasses are observed every 5 days at 10 m resolution with Sentinel-2 (optical) and every 12 days at 20 m resolution with Sentinel-1 (radar). Copernicus enables the measurement of Earth's vital signs from space, providing an Earth's system view, which is key to understand how the ocean, atmosphere, land operate and interact as part of a unique system. Also its life component, which is critical and at the centre of these feedback mechanisms, can be seen, for example through chlorophyll, both on the ocean and land part of the coastal zone.

Operating the Sentinel satellites is very complex because of the many parameters that have to be taken into account and reconciled. This is extra challenging by the fact that Copernicus is an operational constellation and because the Sentinel satellites are state-of-the-art technology, which leads to both a big data volume challenge and an innovation challenge. The data volume challenge tackles 14 terabytes/day, which will lead to an exponential increase of ESA's Data Archive. Acquiring this amount of data and managing it requires technical and scientific expertise regarding data storage and processing. The data continuity challenge imposed on an operational system means that continuity should be ensured between different generations of Sentinel missions, but also when Copernicus takes over a previous scientific mission into an operational context, such as the ocean altimetry Sentinel-6/Michael Freilich mission, the next generation of the Topex/Jason series started in 1992, realised thanks to the cooperation between ESA, NASA (National Aeronautics and Space Administration of USA), the European Commission, EUMETSAT and NOAA (National Oceanographic and Atmospheric Administration of USA). The

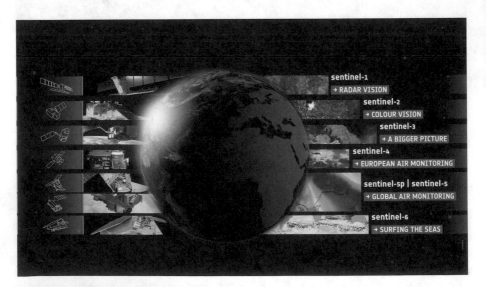

Fig. 3 The family of Copernicus Sentinels. *Credit:* ESA

Springer

data sharing challenge means that the full, open, timely and free data policy imposes that data shall be shared with various stakeholders, which requires that ESA implements the Collaborative Ground Segment in cooperation with the Member States. The data quality challenge stems from the fact that Copernicus is a system on which policy-makers rely to get accurate information, which imposes that data quality must be constantly guaranteed. Another challenge is timeliness, which is always critical but all the more since the Sentinels provide data for a service like the Emergency Management Service (e.g., to reduce the impact of forest fires, floods, storm surge, etc.): timeliness is a matter of saving lives. Sentinel satellites are enabling innovation thanks to their scope and operational character, allowing, for example, to monitor the effects of earthquakes, landslides, storm surges and other natural phenomena (as reported in the other articles of this Special Issue) in a way that was not feasible before in terms of revisiting time, resolution and accuracy. Copernicus has also a great value and potential for scientific purposes. Some Sentinels offer capabilities that are unique, for instance the interleave mode of Sentinel-6/Michael Freilich that will enable to link the traditional altimetry with the new generation of Synthetic Aperture Radar (SAR) altimeters on Sentinel-3, thus extending accurately the sea-level trend (Sect. 2.2.3).

2.1.2.1 Sentinel-1 (A/B/C/D) Synthetic Aperture Radar (SAR) Imaging The Sentinel-1 SAR is an all-weather microwave active sensor for day/night radar imaging applications and interferometry on a 12-day repeat orbit. The Sentinel-1 Mission was designed to ensure C-band (5.405 GHz) SAR data continuity, started in 1991 with the European Remote Sensing satellite ERS-1 and pursued by ERS-2 and Envisat, the large Environmental satellite with ten instruments onboard launched in 2002. Sentinel-1, as well as other Sentinels, was designed to support the three priority Services that have been identified by user consultation working groups gathered by the European Union: The Copernicus Marine Environment Monitoring Service (CMEMS), The Copernicus Land Monitoring Service (CLMS) and the Copernicus Emergency Management Service (Copernicus EMS). Topical applications of Sentinel-1 data covered by these services are: (1) monitoring sea ice zones and the Arctic environment; (2) surveillance of marine environment; (3) monitoring land surface motion risks; (4) mapping of land surfaces: forest, water and soil, agriculture and (5) mapping in support of humanitarian aid in crisis situations. These services have been functional since the launch of Sentinel-1A on 3 April 2014. Sentinel-1B was launched on 25 April 2016 to complete the 2-satellite constellation. Sentinel-1C and -1D will be launched in 2022 or 2023 depending on the health of Sentinel-1A and Sentinel-1B. Thanks to the open data policy, Sentinel data are provided also to national user services and to the entire scientific user community.

2.1.2.2 Sentinel-2 (A/B/C/D) Multispectral Imaging The second Copernicus Sentinel Mission is an optical multispectral operational imaging mission for the high-resolution and high-revisit-frequency monitoring of vegetation, soil and water cover for land, inland waterways and coastal areas that was designed to provide continuity and enhancements of the SPOT-5 and Landsat-7 missions (e.g., wide-swath, better revisit time of 5 days instead of 26 and 16 days, respectively). The 13 spectral bands, from the visible and the "red-edge" (near and shortwave infrared) at different spatial resolutions ranging from 10 to 60 m, ensure global land and coastal zone monitoring with unprecedented enhancements. The Sentinel-2A satellite was launched on 23 June 2015, followed by Sentinel-2B on 7 March 2017. The assembly of Sentinel-2C and Sentinel-2D will be completed by 2021 and launched depending on the health of Sentinel-2A and Sentinel-2B.

2.1.2.3 Sentinel-3 (A/B/C/D) Global Ocean and Land Monitoring The third Copernicus Sentinel Mission carries four instruments to measure ocean colour, vegetation, sea and land surface temperature over a wide-swath and all the variables derived from radar altimetry, that is surface height (sea, river, lake or ice sheet level), wave height and wind speed. The Sentinel-3 Mission, composed of 4 satellites operating in pairs, responds to the requirements for operational and near-real-time global monitoring over a period of 20 years. The Sentinel-3A satellite was launched on 16 February 2016, followed by Sentinel-3B on 25 April 2018. The assembly of Sentinel-3C and Sentinel-3D will be completed by 2021 and launched depending on the health of Sentinel-3A and Sentinel-3B, in 2023 for S-3C, with S-3D a year later.

The optical payload consists of the Ocean and Land Colour Instrument (OLCI) and Sea and Land Surface Temperature Radiometer (SLSTR) instruments. They provide a quasi-simultaneous overlapping view of the Earth that can be used in synergy. The synergy products generated in the Payload Data Ground Segment are a co-registration of images from OLCI and SLSTR to provide continuity to SPOT-VEGETATION products.

OLCI is a medium-resolution push-broom imaging spectrometer inherited from the Medium Resolution Imaging Spectrometer (MERIS) on Envisat, with a slightly modified observation geometry: the field of view is tilted towards the west (~ 12° away from the sun), minimising the sun-glint effect over the ocean and offering a wider effective swath of 1270 km, with 21 spectral bands. The sampling resolution is 1200 m over the open ocean and 300 m over the coastal zone and on land. OLCI ensures every 2 days a global land and coastal zone monitoring with unprecedented enhancements.

Global daily surface temperature data are supplied by the Sentinel-3 Sea and Land Surface Temperature Radiometer (SLSTR) instrument to maintain continuity with the Along-Track Scanning Radiometer series of instruments onboard ERS-1/2 and Envisat (ATSR and Advanced ATSR). SLSTR measures energy radiating from Earth's surface in nine spectral bands and two viewing angles with an absolute accuracy of better than 0.3 K and a temporal stability of 0.1 K/decade. The swath is 1400 km for the nadir view and 740 km for the oblique view. The resolution is 1 km for the infrared-fire spectral channels and 500 m for the visible/near-infrared/mid-wave/thermal-infrared spectral channels.

The radar altimeter (RA) onboard Sentinel-3 is mainly inherited from the CryoSat Earth Explorer Mission with characteristics imported from the Envisat RA-2 and the Jason-3/Poseidon radar altimeters. The Sentinel-3 radar altimeter (SRAL) is a dual-frequency Synthetic Aperture Radar (delay-Doppler) altimeter. SRAL operates either in the low-resolution mode (LRM) or in the high-resolution (SAR) mode. The SAR mode has an enhanced along-track (azimuth) resolution of the order of 300 m when processed with the classical unfocused SAR algorithm. The data can also be processed with the fully focused SAR (FF-SAR) algorithm to produce a range estimate every 0.5 m along-track thanks to an extended synthesis duration. The two usual calibration modes for the control of the internal instrument impulse response and for the determination of the transfer function of the receiver chain are naturally still on. The radar tracks the surface at nadir in two different modes: a closed-loop mode for autonomous positioning of the range window and an open-loop mode to position the range window based on a priori knowledge of terrain altitude derived from a Digital Elevation Model (DEM) compressed and stored onboard (that can be modified and re-uploaded in flight).

To support the wet tropospheric correction of the range measurement a Microwave Radiometer (MWR) instrument, derived from Envisat MWR with the same 23.8 GHz and 36.5 GHz channels, is onboard. A Precise Orbit Determination (POD) package is needed to control the other end of the range measurement, which includes a Global Navigation Satellite Systems (GNSS) instrument, a Doppler Orbit determination and Radio-positioning Integrated on Satellite (DORIS) instrument (on Sentinel-3A only) and a Laser Retro-Reflector (LRR).

2.1.2.4 Sentinel-6 (A/B) The Sentinel-6 Mission is an extension of the altimetry reference series of missions started in 1992 with Topex/Poseidon, followed by the three Jason missions. It is designed, built, launched, commissioned and operated in a multi-agency partnership comprising ESA, The European Union, EUMETSAT, NASA-Jet Propulsion Laboratory, NOAA and CNES, to deliver operational ocean and land observation services. It was initially coined Jason-CS, for its design based on CryoSat and its objective of Continuity of Service on the reference orbit. After Michael Freilich retired from his duties as Director of the NASA Earth Science Division in 2019, a position that he had held since late 2006, the stakeholders of the Sentinel-6 mission decided to renamed Sentinel-6 in his honour for his dedication, creativity, operational vision and his conducting of the National Academy of Sciences first-ever Earth Science and Applications from Space Decadal Survey, among other remarkable achievement. Sentinel-6 is now called "Sentinel-6/Michael Freilich".

The main objective of the Sentinel-6/Michael Freilich Mission is to measure sea surface topography, as well as inland water level when flying over the continents, with high accuracy, stability and reliability to support ocean-forecasting systems, large-scale ocean circulation and extend and enhance the sea-level Essential Climate Variable (ECV), as defined by the Global Climate Observing System (GCOS 2016). The spacecraft will carry the altimetry package composed of the usual support instrument and a next-generation high-resolution Synthetic Aperture Radar (delay-Doppler) altimeter (Poseidon-4) that can operate simultaneously the LRM and SAR modes described in the Sentinel-3 section. This is possible thanks to the new interleaved mode, based on the open burst Ku-band pulse transmission performing a near continuous transmission of Ku-band pulses. It will allow simultaneous processing of the measurements to obtain high-resolution synthetic aperture along-track (HR or SAR mode) and low-resolution along-track (LR mode or LRM) estimates of the range. The usual calibration modes are of course used, but for the purpose of minimising long-term instrumental drifts, which would impact adversely the sea-level trends, as many as 20 calibration pulses per second are generated.

Sentinel-6 will also carry a GNSS radio occultation instrument to derive atmospheric profiles of pressure and temperature as a function of height as a secondary mission objective.

2.1.2.5 Contributing Missions The Copernicus Programme is supported by contributing missions from other space agencies (Fig. 4). These missions were developed for other purposes but are making important data available for Copernicus. For optical low, medium, high and very high resolution, missions such as the Disaster Monitoring Constellation (DMC, an international programme led by Surrey Satellite Technology Ltd, UK), Pléiades (France-Italy), RapidEye (Germany), Deimos-2 (Spain), SPOT 6/7, Spot-5 High Resolution Geometric (HRG), SPOT-5 High Resolution Stereoscopic (HRS) (France) and Proba-V (Belgium, ESA) are making their data available to the Copernicus Programme. Similar agreements exist for Synthetic Aperture Radar missions such as COSMO-SkyMed (Italy), Radarsat (Canada), TerraSAR-X (Germany) and Tandem-X (Germany). As for altimetry missions, the Jason mission series (France-USA), the SARAL/AltiKa (Franco-Indian Ka-

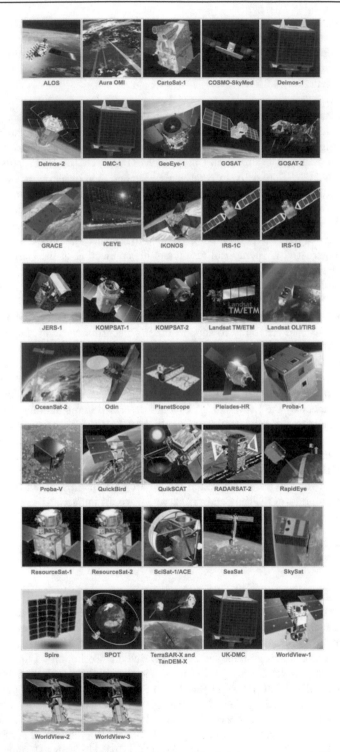

Fig. 4 Missions from other space agencies than ESA contributing to the Copernicus Programme. *Credit:* ESA

band Altimeter) mission, as well as ERS1/2 (ESA), Envisat (ESA) and the ESA Earth Explorer CryoSat, are also contributing to the Copernicus Programme. The reader interested in more details is referred to the https://spacedata.copernicus.eu/web/cscda/data-offer/missi ons web pages or to specific national agencies.

2.2 Earth Observation Applications

2.2.1 Sea Surface Temperature

Sea surface temperature (SST) is a major indicator of climate change, an important ECV, an input for assimilative ocean models, and is also used for retrieval of sea ice parameters. Maps of SST show thermal front location and intensity, used by researchers to improve our understanding and qualification of the vertical structure of the water mass, the internal wave propagation, ocean primary production and seasonal blooms (Fig. 5). Upwellings, which bring deep, oxygen- and nutrients-rich waters up to the surface, are monitored though their cold temperature signature. Foundation SST data are assimilated in physical models with the objective to improve ocean forecasts and reanalyses delivered in the Copernicus Marine Environment Monitoring Service (CMEMS) (Le Traon 2019). For skin or sub-skin SST assimilation or representation of the diurnal cycle of SST observed by EO in models, refer to, e.g., Jansen et al. (2019).

2.2.2 Ocean Colour

The Sentinel-3 Ocean and Land Colour Instrument (OLCI) provides reflectance or waterleaving radiance measurements used to estimate chlorophyll concentration, a proxy of phytoplankton abundance, which is a key variable for the study of the marine ecosystem and plays a major role in the ocean carbon cycle and in the air-sea exchange of energy. The ESA Ocean Colour Climate Change Initiative Project (OC_CCI) has merged data from past Ocean Colour Missions to provide the first 17-year (1997–2013) high-quality, inter-mission

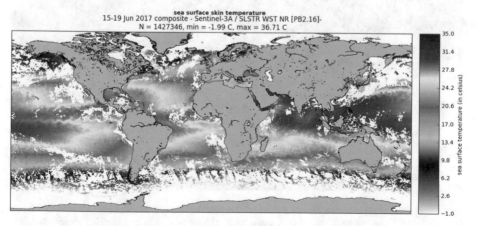

Fig. 5 Sea surface skin temperature from Sentinel-3A SLSTR, a composite of all measurements between 15 and 19 June 2017. The image contains modified Copernicus Sentinel data (2017), processed by ESA, CC BY-SA 3.0 IGO

bias-corrected and error-characterised data record of ocean colour (Sathyendranath et al. 2019) with the aim to study inter-annual variations and long-term trends in the state of primary production and study how the marine ecosystem responds to climate variability and climate change.

Repeated and frequent monitoring of harmful algae and cyanobacteria blooms is crucial to protect populations in the coastal zone (humans and marine mammals, fish and shellfish, birds) from serious threats to their health. Harmful algal bloom monitoring systems rely on operational missions such as Sentinel-3 to rapidly acquire the data needed to analyse and forecast the biological state of the ocean and inland waters, to mitigate the risk and reduce the impact on seafood and tourism industries. Sentinel-3 OLCI data permit the initial identification of the chlorophyll pigment of the bloom and therefore its species and toxicity (Fig. 6).

2.2.3 Sea-Level Variations and Trends

The sea surface height is measured by radar altimeters. The sea-level anomalies are computed as the differences in sea surface height compared to a mean along-track profile or a mean sea surface. Using local sea-level variations measured by a radar altimeter and comparing them to a reference level, major ocean currents can be computed and mapped. In addition to the importance of sea-level variation measurements for science purposes such as oceanography and climate research, the Sentinel-3A&B and soon the Sentinel-6/ Michael Freilich radar altimeters, while contributing to the extension of the sea-level ECV, also measure significant wave height and provide estimates of surface wind speeds over the ocean, which is important information for ocean and wave forecasting systems operated by the Copernicus Marine Environment Monitoring Service.

Fig. 6 A significant algae bloom covering most of the southern part of the North Sea is revealed by the Sentinel-3A OCLI sensor on 27 May 2017. Harmful algal blooms caused by excessive growth of marine algae have occurred in the North Sea and the English Channel area in recent years, with satellite data being used to track their growth, as well as forecasting their evolution on a seasonal timescale. Contains modified Copernicus Sentinel data (2017), processed by ESA, CC BY-SA 3.0 IGO

The ESA Sea Level Climate Change Initiative Project (SL_CCI) developed an algorithm "standard" by selecting through a series of round-robin exercises the most suitable algorithms for the altimetry data processing, with climate requirements from GCOS (Legeais et al. 2018). One of the SL_CCI products obtained from multi-mission SL_CCI fundamental climate data records based on the along-track radar altimeter data that are gridded to compute the trend in each 25-km pixel is the global map of sea-level trends (Fig. 7). Regional differences vary between − 10 and + 10 mm/year, while the global average is around 3 mm/year.

Global mean sea level has not only been rising at a rate of ~3 mm/year on global average over the past 26 years, but it has also accelerated by ~0.1 mm/year^{-2} over that period (WCRP Global Sea Level Budget Group (The) 2018, Fig. 8). Global mean sea level could rise by another metre in the next 80 years (with a 17% chance of a rise larger than 1.1 m in 2100 compared to the 1986–2005 baseline in the high-emission Representative Concentration Pathway RCP8.5, Oppenheimer et al. 2019). Sea level could rise substantially more (up to ~+30%) than the global mean in some parts of the ocean, as sea level is not rising uniformly around the globe. The acceleration of sea-level rise, driven mainly by an enhanced mass loss from the Greenland and Antarctic ice sheets, has the potential to double the total sea-level rise by 2100 as compared to projections that assume a constant rate (Nerem et al. 2018). Global mean sea-level rise could be higher in 2100 in case of widespread and rapid ice sheets mass loss, a contribution whose amplitude remains uncertain in 2100 and beyond (DeConto and Pollard 2016; Wong et al. 2017; Kopp et al. 2017; Le Bars et al. 2017; Kulp and Strauss 2019; Shepherd et al. 2020).

The contemporary global sea-level maps produced from altimetric data are delivered on a 25-km grid. Therefore, they are not representative of the ocean near the coast. It is not yet known if the sea level is rising in the same manner in the open ocean and at the coast. It is very challenging to acquire accurate sea surface heights very near the coast due to the contamination of the radar echoes by the nearby land and often very

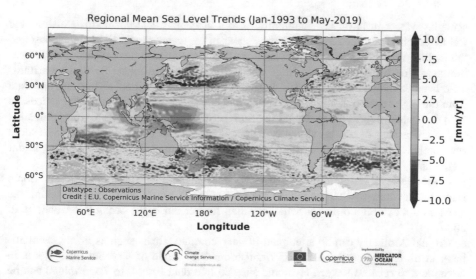

Fig. 7 Global map of regional sea-level trends over 1993–2019; a Copernicus product, with support from ESA SL_CCI. There are marked regional differences that vary between − 10 and + 10 mm/year, mainly in the western tropical Pacific and in major ocean currents

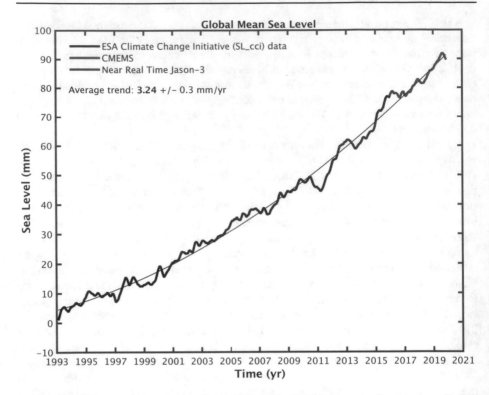

Fig. 8 Global sea-level trend over the period 1993–2019 obtained from the Sea_Level_CCI Project, the Copernicus Marine Service and the near-real-time data from Jason-3 from the CNES data centre for Altimetry and Doris products AVISO. *Credits:* ESA, LEGOS, CLS, CNES, and CMEMS—Assembled at LEGOS

specular waters in bays, not mentioning errors in geophysical corrections such as the wet tropospheric correction due to strong shear in the water vapour field at the coast and the inability for the onboard water vapour radiometer to make clean measurements near the coast due to strong signals from land entering the side lobes. Special algorithms have been developed to circumvent these difficulties for traditional altimeters, and data from the high-resolution altimeters on Sentinel-3A and Sentinel-3B are being processed with dedicated algorithms to extend the time series in the coastal zone. The refinement of the processing of coastal altimetry is on-going so as to get closer to the coast more accurately, developing such methods as the Fully Focused Synthetic Aperture Radar algorithm.

Is the sea level rising at the coast in the same manner as in the open ocean? Some Jason tracks show that this is indeed the case (Fig. 9), while other tracks reaching the coast do not exhibit a different behaviour than further out in the ocean (Gouzenes et al. 2020).

Coastal Altimetry can be attempted in very complex zones such as fjords. Abulaiti-jiang et al. (2015) used the SARin (interferometric) mode of CryoSat over Fjords in Greenland to retrieve the sea level and showed that data from 0 to 7 km inland can be re-allocated to the coastal water using the off-nadir range correction. Idžanovíc et al. (2017) used also the SARin mode of CryoSat-2 near the Norwegian coast and compared

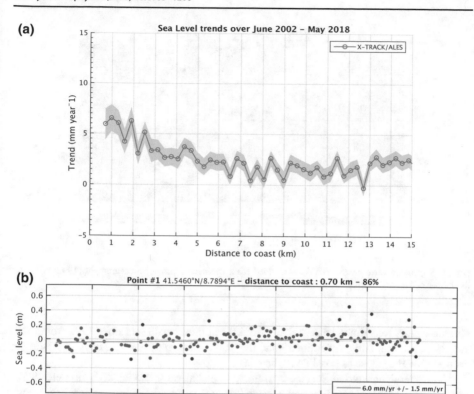

Fig. 9 a Sea-level trend along-track near the coast for the Jason track 085, near Senetosa, Corsica. The trend increases with the proximity of the coast. **b** Sea-level evolution with time at a point on the Jason track 085 near Senetosa, Corsica, at 700 m from coast. Red points were discarded, leaving 86% of valid data for this point. The slope of the linear regression is 6.0 mm/year (± 1.5 mm/year). *Source:* ESA Sea_Level_CCI Project

the corresponding sea-level data with tide gauge records to show that the SARin data have potential for coastal sea-level studies, with a smaller SAR footprint enabling a monitoring closer to the coast than conventional altimeters.

Coastal sea-level rise is still poorly known (e.g., Cipollini et al. 2018; Birol et al. 2017). There are reasons to suspect that coastal sea-level change is not just an extension of open ocean sea-level change but may differ at the coast. High-resolution ocean circulation models show that sea level can be different in the coastal zone compared to the open ocean (consequence of trapped coastal waves, shelf currents, baroclinic instabilities, etc.) (Williams and Hughes 2013, Hughes et al. 2019). Some processes (e.g., waves, river runoff in deltas and estuaries, coastal currents) only occur at the coast and can also impact coastal sea level (Woodworth et al. 2019).

2.2.4 Wave Height

Significant wave height (SWH) is derived from the radar altimeter measurement. Figure 10 shows an example of satellite-derived climatology (1992–2013) for the month of January estimated in the ESA GlobWave Project. The Sea State CCI Project will derive

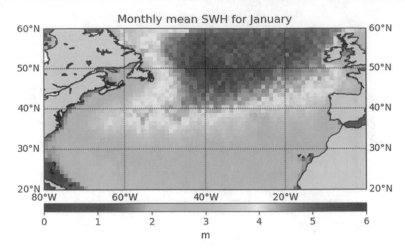

Fig. 10 Significant wave height (SWH) as derived from the radar altimeter measurement (1992–2013) for the month of January. *Source:* GlobWave, SatOC

climate-quality long time series of SWH from the altimetry data sets since 1992 from ERS-1, ERS-2, Envisat, AltiKa, Sentinel-3 and the TOPEX/Jason series. The challenge is to get accurate SWH as close as possible to the coast.

The Copernicus Marine Service is running a global wave model with altimeter SWH and Sentinel-1A&B SAR directional wave spectra data assimilation, which has crucial

Fig. 11 Significant wave height (warm colours are high) when Hurricane Florence was impacting the US Carolinas and Virginia coasts in September 2018. *Source:* Copernicus, Météo-France, L. Aouf

Surveys in Geophysics (2020) 41:1185–1208

implication for safeguarding ships in extreme weather. As an example, while Hurricane Florence was impacting the US Carolinas and Virginia coasts in September 2018, The Copernicus Marine Service was producing the map in Fig. 11.

2.2.5 Hurricane Monitoring

Hurricanes are one of the forces of nature that can be tracked only by satellites, providing up-to-date imagery so that authorities know when to take precautionary measures. Satellites deliver information on a storm extent, wind speed and path, and on key features such as cloud thickness, temperature and water and ice content. The Copernicus Sentinel-3A satellite saw the temperature at the top of Hurricane Harvey on 25 August 2017 at 04:06 GMT as the storm approached Texas, USA (Fig. 12). The brightness temperature of the clouds at the top of the storm, some 12–15 km above the ocean, ranges from about minus 80 °C near the eye of the storm to about 20 °C at the edges.

The UK Met Office has started examining the impact of assimilating SMOS wind speeds on numerical weather prediction (NWP) analyses and forecasts of tropical storms. Hurricane Marie (2014) located in the East Pacific was chosen as an initial NWP case study since it was a very intense storm (maximum sustained winds of 260 km/h and minimum pressure of 918 hPa; a Category 5 hurricane) and there were several SMOS intercepts around peak intensity. A series of short assimilation experiments were performed from 19 to 31 August 2014, covering the main lifecycle of Marie. A control experiment was run in order to provide a reference, and this incorporated all model and observational upgrades made to the global model since 2014. The set of observations assimilated included all conventional and satellite observations, with ocean surface wind vectors observations assimilated from the Advanced Scatterometer (ASCAT) onboard MetOp-A/B/C (a series of three polar-orbiting meteorological satellites developed by ESA and operated by EUMETSAT, launched in 2006, 2012 and 2018) and WindSat (a joint NOAA/Department of Defense/

Fig. 12 Temperature at the top of Harvey as the storm approaches Texas, based on Sentinel-3A data on 25 August 2017[©] BY-SA 3.0 IGO

NASA instrument demonstration project designed for a 3-year lifetime, launched in 2003 onboard the Coriolis satellite and still operating). A set of trial experiments was run to include the addition of SMOS wind speeds above a minimum wind speed threshold of 15 m/s. After quality control to remove possible radiofrequency interference contaminations, the SMOS observations were thinned to a distance of 80 km to mitigate the impact of spatial error correlations, since the assimilation assumes errors to be uncorrelated. The initialisation and forecasts of Hurricane Marie's intensity and track were validated for each of the experiments using brightness temperature data. Figure 13 shows the SMOS wind field at the time when crossing the path of Hurricane Marie.

2.2.6 Storm Surge Monitoring

A storm surge is an unexpectedly high sea level caused by an extreme weather event.

The flooding resulting from a storm depends on the sum of the mean sea level, the tide, the storm surge (wind, low pressure) and the wave height (wave setup and wave energy, influenced by the bathymetry). Storm surges have a dramatic impact on populations and infrastructures. The North Sea flood in 1953 was the worst natural disaster in modern Europe with 2100 fatalities; Hurricane Katrina in 2005 had a toll of 1833 deaths and 81 billion dollars of damage with severe devastation in New Orleans and surroundings; in 2007 the Cyclone Sidr was one of Bangladesh's worst disasters with at least 3500 deaths estimated; in 2008 the Cyclone Nargis caused at least 138,000 fatalities in Myanmar's worst natural disaster.

The role of altimetry to support storm surge predictions has been investigated as part of the ESA eSurge and the eSurge Venice Projects, which ran from 2011 to 2015 (http://www.storm-surge.info). In Venice there is a regional amplification due to the Sirocco wind directed along the main axis of the Adriatic Sea. The stronger amplification is in the north-western part, especially in the city of Venice where the well-known "Acqua Alta"

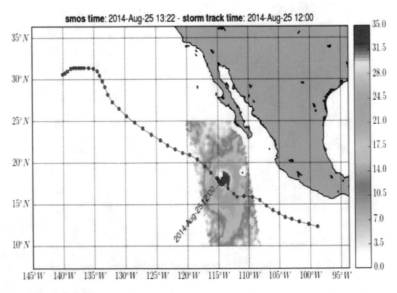

Fig. 13 Soil Moisture and Ocean Salinity (SMOS) satellite wind field in m/s at the time when crossing the path of Hurricane Marie on 25 August 2014. *Credits*: ESA

phenomenon (High Water) takes place. The challenge is to forecast the height of water at the time of a surge event and the extent to which it will inundate the city. In total, 22 storm surge events recorded in Venice from 2004 to 2012 were selected for assimilation experiments in Shallow water HYdrodynamic Finite Element Model (SHIFEM) by Bajo et al. (2017). For each event, all data sets (altimetry, scatterometry, model and tide gauge) were

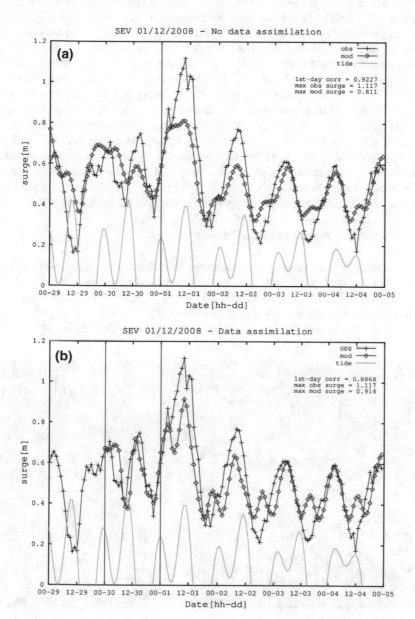

Fig. 14 Assimilation experiments in the Shallow water HYdrodynamic Finite Element Model (SHIFEM), in Venice: **a** 24 h before, the predictive model badly reproduces the moderate surge that happened on 1 December 2008 at noon; **b** the assimilation of altimetry data clearly improves the prediction. *Credits:* ESA, CNR, e-Surge-Venice Project

prepared. The evolution of the storm surge event of 1 December 2008 was up to 156 cm as observed by the Venice tide gauge. The model badly reproduces the moderate surge 1 day before (Fig. 14a) resulting in a strong underestimation of the maximum surge. The altimeter assimilation (Fig. 14b) clearly improves the estimation of the peak (De Biasio et al. 2017). Altimetry contributes to reduce model errors due to initial conditions at the start of simulations and due to the seiche effects. Scatterometry also contributes to reduce errors due to inaccurate wind field from numerical weather prediction models (e.g., from the European Centre for Medium-range Weather Forecasts—ECMWF) as the accuracy can be diminished in the coastal zone. Another experiment was run in the Danish North Sea coast and in Kattegat by Madsen et al. (2015) for the Bodil/Xaver storm surge that hit the North Sea coast on 5 December 2013 and created a 1000-year event on the northern coast of Zealand (Denmark). Madsen et al. (2015) report that altimetry is a powerful tool for improving storm surge modelling and forecasting through data assimilation, which improves the initial conditions prior to the forecast. Even the standard open ocean processing (i.e. non-coastal) of altimetry is useful for calibrating models and ensemble pruning, but coastal altimetry products adds the extra estimates near the coast that is most efficient to drive the model, resulting in practical benefits of better forecasts, longer warning period and fewer false alarms. There is nevertheless an issue with the sparse sampling (inter-track distance), which limits to information that can be provided to the model's much denser grid.

2.2.7 Sediment Transport and Coastal Erosion

Remote sensing tools are essential for monitoring turbidity and suspended particulate matter to study water quality, eutrophication and sediment, organic matter, bacteria and pollutants transport. Turbidly bursts can occur with excess of melt water, fires and agricultural and industrial stresses. They are also used to validate sediment transport numerical models and dynamics in estuarine systems forced by tidal range, river discharge, water depth and

Fig. 15 Sentinel-2 image showing the extent of the flooding (in brown) in the Sistan and Baluchestan province on 13 January 2020. Sediment and mud from the Bahu Kalat River, Iran, and Dasht River, Pakistan, flow into Gwadar Bay. *Credits*: ESA, the image contains modified Copernicus Sentinel data (2020), processed by ESA, CC BY-SA 3.0 IGO

wind speed (Sherwood et al. 2004). Figure 15 is a brilliant illustration of a recent Sentinel-2 image where sediment and mud from the Bahu Kalat River, Iran, and Dasht River, Pakistan, flow into Gwadar Bay after a torrential rain event.

Sea-level rise and extreme events have a dramatic impact on coastal-dwelling societies (Nicholls and Cazenave 2010; Cazenave and Le Cozannet 2014; Vousdoukas et al. 2018). ESA has setup a Coastal Change consortium to detect and track coastal erosion and accretion by analysing 25 years of ESA archive data to quantify historical coastal change and volumetric change of sediment transport. The work is relying on radar data from Sentinel-1 and optical data from Sentinel-2 to monitoring coastal change from space for the mitigation of the negative effects of erosion and to support decision-makers fostering coastal management plans to increase resilience through development-oriented adaptation.

Exploiting the wave field from Sentinel-1 and Sentinel-2 images allow scientists to derive bathymetry in shallow coastal waters, an important variable for coastal ocean circulation, tide propagation, wave setup and for monitoring seabed changes to estimate sediment transport. The waterline can also be extracted from these images (using the right combination of spectral bands), taking into account the tide locally, and by analysing long-term archives of microwave and optical images derive the evolution of coastal erosion or accretion.

Marine debris can also be tracked by analysing the reflectance in Sentinel-2 images, see Maximenko et al. (2019) and Biermann et al. (2020).

3 Copernicus Services

The European Copernicus services are game-changers regarding operational systems. The services setup today are Land, Marine Environment, Atmosphere, Emergency Management, Security and Climate Change. For operational coastal hazard monitoring we focus on the Marine, Land and Climate Services only, even though an Emergency Management Service does also contribute to the mitigation of Coastal Hazards. Other services are operated around the world by numerous nations and agencies, such as the NOAA office for Coastal Management (https://coast.noaa.gov), providing the Coastal Change Hazards Portal (https://coast.noaa.gov/digitalcoast/tools/hazards-portal.html). Here, we showcase the European Copernicus Marine and Land Monitoring Services.

3.1 Copernicus Marine Environment Monitoring Service

CMEMS is operational since 2015 and provides regular and systematic core reference information on the physical state, variability and dynamics of the ocean (ocean circulation, waves, sea ice) and marine ecosystems for the global ocean and the European regional seas. This capacity encompasses the description of the current ocean situation, the prediction of the ocean state a few days to weeks ahead (forecast) and the provision of consistent retrospective data records for recent decades (reprocessed observations and re-analyses). As a European core service, CMEMS focuses on activities best performed at European level to support the downstream development of expert services with essential products and services, with a clear delineation with coastal monitoring services, operated by Member States or private groups.

All CMEMS products are quality-assessed and freely accessible through a single Internet interface (http://marine.copernicus.eu/getting-started/). They can be downloaded using different authenticated download mechanisms (File Transfer Protocol, direct get file, sub-setter and soon opendap). Regarding EO, CMEMS provides added-value satellite products for different ocean-related variables (sea level, ocean colour, sea surface temperature, sea ice, winds and waves). CMEMS gathers level-2 EO products (Geophysical Data Records) of Sentinels and other contributing missions from the Copernicus satellite component through ESA and EUMETSAT. CMEMS then provides added-value, high-level EO products that are useable for downstream applications related for instance to safety at sea, water quality or pollution monitoring. CMEMS EO-derived products include level-3 products (e.g., along-track observations put on a reference track) and level-4 products corresponding to gridded and/or multi-sensor products. In addition, CMEMS generates tailored and validated EO data sets that can be assimilated in modelling systems to produce reanalyses and forecasts.

CMEMS is a user-driven service and serves a wide range of users (more than 22,000 registered users as of February 2020) and applications. User requests and needs are regularly analysed to drive evolutions of products and services. Observation-based (EO and in situ) and model-based (forecasts and reanalyses) CMEMS products support various marine applications. With regard to the coastal environment, CMEMS also provides information driven by European Union policies and directives relevant to the coastal ocean (e.g., integrated maritime policy, maritime spatial planning, marine strategy framework directive, water framework directive, common fishery policy) and by societal and blue economy needs. More than 150 examples of use of CMEMS products can be found on the CMEMS web site (http://marine.copernicus.eu/markets/) for various sectors of the blue economy, broadly categorised into safety and disaster, coastal monitoring, marine navigation, water quality, sea ice monitoring, natural resources and energy, marine food, marine conservation and policies, society and education, science and climate.

CMEMS products and services continuously evolve to better answer user needs, including in the coastal zone. This includes monitoring and forecasting the ocean at finer scales, improving the EO monitoring of coastal zones and providing a more seamless and better forcing for very-high-resolution coastal models (Le Traon 2019).

3.2 Copernicus Land Monitoring Service

The Copernicus Land Monitoring Service (CLMS) is Europe's eyes on the terrestrial environment. The CLMS has been operational since 2012 and systematically produces a series of qualified bio-geophysical products on the status and evolution of the land surface, at global scale and at mid to low spatial resolution, complemented by the constitution of long-term time series. The products are used to monitor the vegetation, the water cycle, the energy budget and the terrestrial cryosphere as well as urban planning, forest management, water management, agriculture and food security, nature conservation and restoration, rural development, ecosystem accounting and mitigation/adaptation to climate change.

CLMS works on four main components:

- Global—a service that provides a series of bio-geophysical products on the status and evolution of the land surface at global scale but with a mid- and low spatial resolution, produced every 10 days;
- Pan-European—a service that provides information about the land cover and use, as well as their changes and characteristics;

- A local component focuses on the different hot spots, area characterised by environmental challenges and problems;
- An imagery and reference data component that provides the possibility to use in synergy satellite data together with in situ data.

The global component (Copernicus Global Land Service, CGLS) reliably provides a set of biophysical variables, which describe the state and the evolution of the vegetation, the energy budget, the water cycle and the cryosphere over the land surface at global scale, including the coastal zone. In detail, the following themes are covered:

- Vegetation: To monitor the changes on continental biomes, the CGLS provides a set of biophysical variables describing the state, the dynamism and the disturbances of the terrestrial vegetation (normalised difference vegetation index, and its derived condition and productivity indices, leaf area index, fraction of vegetation cover, fraction of radiation absorbed for the photosynthesis, maps of burnt areas, etc.).
- Energy: To quantify the main responses of the continents to solar illumination, the CGLS provides biophysical variables assessing the energy budget at the land surface (reflectance and albedo, land surface temperature, vegetation cover, soil moisture).
- Water: To contribute to a better understanding of Earth's water cycle, the CGLS helps to monitor the water surface temperature and water quality, the surface extent covered by water on a permanent basis or occasional occurrence such as from severe floods, the water level of lakes and rivers.
- Cryosphere: The key cryosphere parameters monitored within the CGLS are the areal snow extent, snow water equivalent and lake ice extent.

The service coined Ground-Based Observations for Validation (GBOV) of Copernicus Global Land Products provides multiple years of high-quality in situ measurements to validate the seven core land products (Top-of-canopy reflectances, Surface albedo, Fraction of Absorbed Photosynthetically Active Radiation (fAPAR), Leaf Area Index (LAI), Fraction of green Vegetation Cover (fCover), Land Surface Temperature and Soil Moisture). To be of accurate use in environmental monitoring and decision-making, it is essential to ensure that satellite-derived products are of high quality, accuracy and consistency. The in situ data are essential for the validation processes and the continuous quality control of satellite-derived products, a prerequisite for quality downstream applications.

3.3 Copernicus Climate Change Service

The objective of the Copernicus Climate Change Service (C3S) is to provide an operational service to produce and deliver climate data record products for the essential climate variables to support society by providing authoritative information about global climate. While the processing algorithms and standards were developed within the ESA Climate Change Initiative for the ECVs from space, the on-going operational production is done in the C3S, under the responsibility of ECMWF on behalf of the European Commission. The Services produces more than 30 ECVs and additional or alternative ECV products are planned, including from non-European sources such as from NOAA or from the Global Precipitation Climatology Project (University of Maryland, USA). C3S is also providing ECVs from in situ data, as well as global atmospheric and oceanic reanalyses (ERA5 and

ORA5). C3S is also producing multi-model seasonal forecasts and publishes reference sets of climate projections (see https://climate.copernicus.eu).

3.4 Access to Copernicus Data, Information and Services

Any citizen or organisation around the world can access the Copernicus Services on a free, full and open access basis via the web site (https://copernicus.eu). The Copernicus Services support many EU policies and directives, including the Habitats Directive, the Water Framework Directive, the Marine Strategy Framework Directive, the Flood Directive, Maritime Spatial Planning, Bathing Waters, Integrated Coastal Zone Management and the Global Framework for Climate Services, facilitating the maritime spatial planning to balance the competing claims of economic sectors on space and resources in marine areas, as well as supplying a comprehensive coastal zone monitoring solution to address the complex and dynamic situations of coastal environments, including the monitoring and prediction of coastal hazards.

The Copernicus Data and Information Access Services (DIAS) is a common European Commission DG-GROW and ESA approach to EO data exploitation with Copernicus at its core, initiated in June 2018. The objective is to create and enable a European EO data ecosystem for research and business to close the European "gap" between the data generation and the exploitation layer. Four DIAS have been established through ESA and one through Eumetsat.

The Copernicus Programme and the scientific research effort supported by space and in situ agencies strive to support the balancing of the competing interests of human development with the need to ensure healthy and resilient coastal ecosystems in the ocean, coastal lagoon and wetlands, and land.

4 Conclusion

The user-driven Copernicus Programme, having inherited from the ESA preoperational missions ERS-1, ERS-2, Envisat, the Earth Explorer Programme (SMOS, CryoSat) and the precursor ESA initiative on the Global Monitoring of the Environment and Security Service Element Programme (GSE), is an invaluable asset to supply data for coastal hazards monitoring. The Copernicus Services convert data into information both in reanalyses covering the past decades, nowcast and forecast, which is essential for coastal zone planner to establish development and management plans to mitigate the impacts and improve resilience against coastal hazards. In the article of this Special Issue entitled "Earth Observations for Monitoring Marine Coastal Hazards and Their Drivers" by Melet et al., further applications of Earth Observation by satellites, such as coastland flooding and coastline dynamics, maritime security, marine pollution, water quality and marine ecology shifts, are developed in further details.

The Copernicus Programme provides accurate, high-quality data and information. It also relies on in situ observations from ground, seafaring and airborne sensors, as well as geospatial ancillary or reference data. The in situ component is mainly used to identify data access gaps and bottlenecks, support the provision of cross-cutting data, manage partnerships with data providers to improve access and use conditions, and to broker innovative solutions with services, providers or national authorities. More information is available on https://insitu.copernicus.eu/.

Surveys in Geophysics (2020) 41:1185–1208

The Copernicus Programme will ensure the future supply of Earth Observation by satellite products with the plan to launch the "C" and "D" Sentinels near the end of the lifetime of the "A" and "B" satellites. Looking to the more distant future, six high-priority candidate missions are being studied to address the European Union policy and gaps in Copernicus user needs, and to expand the current capabilities of the Copernicus space component.

Acknowledgements This article is an outcome of the International Space Science Institute workshop on "Natural and man-made hazards monitoring by the Earth Observation missions: current status and scientific gaps" held at the ISSI-Bern, Switzerland, on 15–18 April 2019. Warm thanks are extended to the International Space Science Institute and to Dr Teodolina Lopez and Dr Anny Cazenave for the organisation. We are also grateful to the reviewers for their appreciated contributions to improve the manuscript.

References

Abulaitijiang A, Andersen OB, Stenseng L (2015) Coastal sea level from inland CryoSat-2 interferometric SAR altimetry. Geophys Res Lett 42:1841–1847. https://doi.org/10.1002/2015GL063131

Bajo M, De Biasio F, Umgiesser G, Vignudelli S, Zecchetto S (2017) Impact of using scatterometer and altimeter data on storm surge forecasting. Ocean Model 113:85–94. https://doi.org/10.1016/j.ocemo d.2017.03.014

Benveniste J et al (2019) Requirements for a coastal hazards observing system. Front Mar Sci 6:348. https://doi.org/10.3389/fmars.2019.00348

Biermann L, Clewley D, Martinez-Vicente V et al (2020) Finding plastic patches in coastal waters using optical satellite data. Sci Rep 10:5364. https://doi.org/10.1038/s41598-020-62298-z

Birol F, Fuller NX, Lyard F et al (2017) Coastal applications from nadir altimetry: example of the X-TRACK regional products. Adv Space Res 59:936–953. https://doi.org/10.1016/j.asr.2016.11.005

Cazenave A, Le Cozannet G (2014) Sea level rise and coastal impacts. Earth's Fut 2(2):15–34. https://doi.org/10.1002/2013EF000188

Cipollini P, Benveniste J, Birol F et al (2018) Satellite altimetry in coastal regions. In: Stammer D, Cazenave A (eds) Satellite altimetry over the oceans and land surfaces. CRC Press, Taylor and Francis Group, Boca Raton, London, New York, pp 343–373. https://doi.org/10.1201/9781315151779-11

De Biasio F, Bajo M, Vignudelli S, Umgiesser G, Zecchetto S (2017) Improvements of storm surge forecasting in the Gulf of Venice exploiting the potential of satellite data: the ESA DUE eSurge-Venice project. Eur J Remote Sens 50:428–441. https://doi.org/10.1080/22797254.2017.1350558

DeConto RM, Pollard D (2016) Contribution of Antarctica to past and future sea-level rise. Nature 531:591–597

GCOS (2016) The global observing system for climate: implementation needs. Technical report GCOS-200, World Meteorological Organization

Gouzenes Y, Léger F, Cazenave A, Birol F, Bonnefond P, Passaro M, Nino F, Almar R, Laurain O, Schwatke C, Legeais J-F, Benveniste J (2020) Coastal sea level rise at Senetosa (Corsica) during the Jason altimetry missions. Ocean Sci Discuss. https://doi.org/10.5194/os-2020-3

Hughes CW, Thompson K, Minobe S, Fukumori I, Griffies S, Huthnance J, Spence P (2019) Sea level and the role of coastal-trapped waves in mediating the interaction between the coast and open ocean. Surv Geophys. https://doi.org/10.1007/s10712-019-09535-x

Idžanović M, Ophaug V, Andersen OB (2017) Coastal sea level from CryoSat-2 SARIn altimetry in Norway. Adv Space Res 62:1344–1357. https://doi.org/10.1016/j.asr.2017.07.043

Jansen E, Pimentel S, Tse W-H, Denaxa D, Korres G, Mirouze I, Storto A (2019) Using Canonical Correlation Analysis to produce dynamically-based highly-efficient statistical observation operators. Ocean Sci Discuss. https://doi.org/10.5194/os-2018-166

Kopp RE et al (2017) Evolving understanding of antarctic ice-sheet physics and ambiguity in probabilistic sea-level projections. Earth's Fut 5:1217–1233. https://doi.org/10.1002/2017EF000663

Kulp SA, Strauss BH (2019) New elevation data triple estimates of global vulnerability to sea-level rise and coastal flooding. Nat Commun 10:4844. https://doi.org/10.1038/s41467-019-12808-z

Le Bars D, Drijfhout S, de Vries H (2017) A high-end sea level rise probabilistic projection including rapid Antarctic ice sheet mass loss. Environ Res Lett 12:044013. https://doi.org/10.1088/1748-9326/aa6512

Le Traon P-Y et al (2019) From observation to information and users: the Copernicus Marine Service perspective. Frontiers in Marine Science, Oceanobs'19 Community White Paper. https://doi.org/10.3389/fmars.2019.00234

Legeais J-F, Ablain M, Zawadzki L, Zuo H, Johannessen JA, Scharffenberg MG, Fenoglio-Marc L, Fernandes MJ, Andersen OB, Rudenko S, Cipollini P, Quartly GD, Passaro M, Cazenave A, Benveniste J (2018) An improved and homogeneous altimeter sea level record from the ESA Climate Change Initiative. Earth Syst Sci Data 10:281–301. https://doi.org/10.5194/essd-10-281-2018

Madsen KS, Høyer JL, Fu W, Donlon C (2015) Blending of satellite and tide gauge sea level observations and its assimilation in a storm surge model of the North Sea and Baltic Sea. J Geophys Res 120(9):6405–6418. https://doi.org/10.1002/2015JC011070

Maximenko N et al (2019) Toward the integrated marine debris observing system. Front Mar Sci 6:447. https://doi.org/10.3389/fmars.2019.00447

Melet A, Teatini P, Le Cozannet G, Jamet C, Conversi A, Benveniste J, Almar R (2020) Earth observations for monitoring marine coastal hazards and their drivers. Surv Geophys. https://doi.org/10.1007/s1071 2-020-09594-5

Nerem RS, Beckley BD, Fasullo JT, Hamlington BD, Masters D, Mitchum GT (2018) Climate-change-driven accelerated sea-level rise detected in the altimeter era. PNAS 115(9):2022–2025. https://doi.org/10.1073/pnas.1717312115

Nicholls RJ, Cazenave A (2010) Sea level change and the impacts in coastal zones. Science 328:1517–1520. https://doi.org/10.1126/science.1185782

Oppenheimer M, Glavovic BC, Hinkel J, van de Wal R, Magnan AK, Abd-Elgawad A, Cai R, Cifuentes-Jara M, DeConto RM, Ghosh T, Hay J, Isla F, Marzeion B, Meyssignac B, Sebesvari Z (2019) Sea level rise and implications for low-lying islands, coasts and communities. In: Pörtner H-O, Roberts DC, Masson-Delmotte V, Zhai P, Tignor M, Poloczanska E, Mintenbeck K, Alegría A, Nicolai M, Okem A, Petzold J, Rama B, Weyer NM (eds) IPCC special report on the ocean and cryosphere in a changing climate. Available at https://www.ipcc.ch/srocc/

Sathyendranath S, Brewin RJW, Brockmann C, Brotas V, Calton B, Chuprin A, Cipollini P, Couto A, Dingle J, Doerffer R, Donlon C, Dowell M, Farman A, Grant MG, Groom S, Horseman A, Jackson T, Krasemann H, Lavender S, Martinez-Vicente V, Mazeran C, Mélin F, Moore T, Müller D, Regner P, Roy S, Steele C, Steinmetz F, Swinton J, Taberner M, Thompson A, Valente A, Zühlke M, Brando V, Feng H, Feldman G, Franz B, Frouin R, Gould R, Hooker S, Kahru M, Kratzer S, Mitchell B, Muller-Karger F, Sosik H, Voss K, Werdell J, Platt T (2019) An ocean-colour time series for use in climate studies: the experience of the ocean-colour climate change initiative (OC-CCI). Sensors 19(19):4285. https://doi.org/10.3390/s19194285

Shepherd A, Ivins E, Rignot E et al (2020) Mass balance of the Greenland Ice Sheet from 1992 to 2018. Nature 579:233–239. https://doi.org/10.1038/s41586-019-1855-2

Sherwood CR, Book JW, Carniel S, Cavaleri L, Chiggiato J, Das H, Doyle JD, Harris CK, Niedoroda AW, Perkins H (2004) Sediment dynamics in the adriatic sea investigated with coupled models. Oceanography 17:58–69. https://doi.org/10.5670/oceanog.2004.04

Short AD (2020) The Australian coast: review and overview. In: Australian coastal systems. Coastal Research Library, vol 32, Springer, Cham. https://doi.org/10.1007/978-3-030-14294-0_34

Vousdoukas M, Mentaschi L, Voukouvalas E, Bianchi A, Dottori F, Feyen L (2018) Climatic and socioeconomic controls of future coastal flood risk in Europe. Nat Clim Change 8:776–780. https://doi.org/10.1038/s41558-018-0260-4

WCRP Global Sea Level Budget Group (The) (2018) Global sea level budget, 1993-present. Earth Syst Sci Data 10:1551–1590. https://doi.org/10.5194/essd-10-1551-2018

Williams J, Hughes CW (2013) The coherence of small island sea level with the wider ocean: a model study. Ocean Sci 9:111–119. https://doi.org/10.5194/os-9-111-2013

Wong TE, Bakker AM, Keller K (2017) Impacts of Antarctic fast dynamics on sea-level projections and coastal flood defense. Clim Change 144:347–364. https://doi.org/10.1007/s10584-017-2039-4

Woodworth P, Melet A, Marcos M, Ray RD, Wöppelmann G, Sasaki YN, Cirano M, Hibbert A, Huthnance JM, Montserrat S, Merrifield MA (2019) Forcing factors affecting sea level changes at the coast. Surv Geophys 40:1351–1397. https://doi.org/10.1007/s10712-019-09531-1

Surveys in Geophysics (2020) 41:1209–1235
https://doi.org/10.1007/s10712-020-09586-5

Space-Based Earth Observations for Disaster Risk Management

G. Le Cozannet[1] · M. Kervyn[2] · S. Russo[3] · C. Ifejika Speranza[4] · P. Ferrier[5] ·
M. Foumelis[1] · T. Lopez[6,7] · H. Modaressi[1,8]

Received: 26 November 2019 / Accepted: 24 February 2020 / Published online: 10 March 2020
© The Author(s) 2020

Abstract

As space-based Earth observations are delivering a growing amount and variety of data, the potential of this information to better support disaster risk management is coming into increased scrutiny. Disaster risk management actions are commonly divided into the different steps of the disaster management cycle, which include: prevention, to minimize future losses; preparedness and crisis management, often focused on saving lives; and post-crisis management aiming at re-establishing services supporting human activities. Based on a literature review and examples of studies in the area of coastal, hydro-meteorological and geohazards, this review examines how space-based Earth observations have addressed the needs for information in the area of disaster risk management so far. We show that efforts have essentially focused on hazard assessments or supporting crisis management, whereas a number of needs still remain partly fulfilled for vulnerability and exposure mapping, as well as adaptation planning. A promising way forward to maximize the impact of Earth observations includes multi-risk approaches, which mutualize the collection of time-evolving vulnerability and exposure data across different hazards. Opportunities exist as programmes such as the Copernicus Sentinels are now delivering Earth observations of an unprecedented quality, quantity and repetitiveness, as well as initiatives from the disaster risk science communities such as the development of observatories. We argue that, as a complement to this, more systematic efforts to (1) build capacity and (2) evaluate where space-based Earth observations can support disaster risk management would be useful to maximize its societal benefits.

Keywords Space-based Earth observations · Satellite remote sensing · Disaster risk management · Prevention · Preparedness · Crisis management · Recovery · Resilience · Adaptation

---

✉ G. Le Cozannet
  g.lecozannet@brgm.fr

Extended author information available on the last page of the article

1 Introduction

As early as the 1960s, data from the TIROS-1 satellite delivered meteorological forecasts for the first time. This major breakthrough in Earth Observation opened a new era for disaster risk management, whereby meteorological hazards can be better monitored, understood and ultimately anticipated (Manna 1985). The remarkable integration of space-based observations in the computations of meteorologists raises an obvious question: to what extent can this success be replicated for other coastal, hydro-meteorological and geohazards? Several initiatives and programmes have addressed this issue for a number of hazards such as soil and coastal erosion, coastal, groundwater and inland flooding, subsidence (Lopez et al. this issue; Melet et al. this issue), landslides (Lissak et al. this issue), earthquakes and volcanic eruptions (Elliot et al. this issue), tsunamis (Hébert et al. this issue) or wildfires (Pettinari et al. this issue). These include the IGOS (Integrated Global Observing Strategy) established in 1998, the Group on Earth Observations (GEO) and its Geohazard Supersites and Natural Laboratories (GSNL) since 2003, the European "Copernicus" programme (formerly known as GMES, Global Monitoring for Environment and Security) and its Sentinel missions that have been deployed since 2014, the Japanese supported "Sentinel-Asia", or the "Natural Laboratories" in the USA (Aschbacher et al. 2014; Aschbacher and Milagro-Perez 2012; Kaku 2019; Koike et al. 2010; Plag et al. 2010; Salichon et al. 2007). The common vision of these various initiatives and projects involves constellations of satellites for monitoring key observable determinants of risks, informing users through a global data and modelling infrastructure, and ultimately benefiting population at risks from disasters. While there is evidence that satellite-based remote sensing applications are indeed increasingly used for disaster risk management (Tralli et al. 2005), the simple fact that these projects and initiatives have existed for decades demonstrates that making this vision real remains a major challenge (Denis et al. 2016). Hence, the following question remains timely: how can space-based Earth observations best support disaster risks management?

The terminology of disaster risk management is complex, not only due to the numerous different hazards to be considered, but also because disaster risk management is a crosscutting issue that intersects several different policies, which all have their own languages. These policies include economic development, land-use planning, building codes and regulation as well as climate change adaptation (see Romieu et al. (2010) for a discussion of coastal risk prevention and adaptation). Here we rely on the terminology of the United Nation Office for Disaster Risks Reduction (UNDRR, https://www.undrr.org/publicatio n/2009-unisdr-terminology-disaster-risk-reduction), which defines disaster risk management as "the organization, planning and application of measures preparing for, responding to and recovering from disasters". Note that the International Panel on Climate Change uses a rather similar definition (IPCC 2018), whereby disaster risk management is the process "for designing, implementing, and evaluating strategies, policies, and measures to improve the understanding of disaster risk, foster disaster risk reduction and transfer, and promote continuous improvement in disaster preparedness, response, and recovery practices (…)". The latter definition gives more emphasis on the governance and institutional dimensions of disaster risk management, institutions being understood in their broad sense here of "habitualized behavior" and rules and norms that govern society" (Adger 2000). In both cases, disaster risk management refers to all actions and decisions that aim at minimizing losses from disasters.

Earth observations refer to the "gathering of information about planet Earth's physical, chemical and biological systems" (Group on Earth Observations; https://www.earthobservatio ns.org/index.php). Therefore, it is not limited to satellite remote sensing but also includes all in situ and aerial monitoring instruments and networks such as GNSS or seismometer networks, unmanned aerial vehicles or field observations (Antoine et al. this issue; Balsamo et al. 2018; Salichon et al. 2007). We focus here on space-based observations because space agencies have played a prominent role in organizing the community of Earth observations over the last decades (Bally 2012; Desnos et al. 2014, 2016). Furthermore, space observations are explicitly recognized as important contributions in the disaster risk reductions strategies of both the Hyogo and Sendai frameworks for the United Nations (2005, 2015). For example, the Sendai Framework includes priorities for actions such as "understanding disaster risks", "investing in disaster reduction for resilience" and "enhancing disaster preparedness", which all can be supported by improved Earth observations. However, in many cases, neither space-based nor in situ and aerial observations are directly supporting disaster risk management: instead, they are realizing an intermediate layer of analysis, which in turn informs users regarding risk (Salichon et al. 2007). For example, in the area of Earthquake risks, field and satellite observations are first included within a risk model, whose results are transferred to those informing disaster risk management (Sedan et al. 2013). Hence, users of Earth Observations are usually not directly those in charge of disaster risk management, but rather they are service providers. The existence of these multiple layers may explain why the full potential of Earth Observations has not been exploited yet.

Previous reports and papers have extensively discussed the technical and scientific challenges required to maximize the benefits from space-based observations in the area of disaster risk management (Bello and Aina 2014; Guo 2010; Joyce et al. 2009). Here, we evaluate to what extent space-based Earth observations being delivered actually meet the decision frameworks of disaster risk management (Plag et al. 2010; Taubenbock et al. 2008). Our starting point differs from the common approach towards users of Earth observation as we do not start from *formalized requirements of users*, but from *decisions and workflows within disaster risk management*. Some users have already expressed their requirements to the Earth Observation sector (Smolka and Siebert 2013). However, adopting an approach centred on decisions and workflows may help maximize the benefits of Earth Observation. The perspective taken in this paper is not new and has been investigated both at the theoretical and practical levels in a number of research domains (Hinkel et al. 2019; Steen et al. 2007). However, despite the recommendations of the IGOS Geohazards initiative (Plag et al. 2010), this perspective remains not well established in the area of space-based Earth observations for disaster risk management, as previous papers that adopted this perspective have focused only on specific hazards, such as tropical cyclones (Hoque et al. 2017).

This paper fills this gap by first presenting the disaster risk management cycle, which holds as the reference framework for classifying disaster risk management actions (Sect. 2), then by assessing how space-based Earth Observation fits within disaster risk management workflows as in prevention in Sect. 3 and preparedness, crisis management and post-disaster response in Sect. 4. Finally, in Sect. 5, we review the current ways forward and challenges. The overall outline of this paper is summarized in Fig. 1, which highlights how different layers within the end-to-end chain of observation providers and users interact, and where this is reviewed in the present paper.

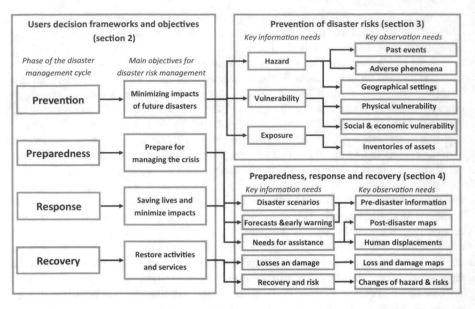

Fig. 1 Space-based observation needs and challenges according to the different user decision frameworks and objectives relevant to disaster risk management

2 Users Decision Frameworks and Objectives

2.1 The Disaster Risk Management Cycle

Disaster risk management includes a wide range of actions, whose primary aim is to reduce the impacts of disasters. These actions are commonly divided according to the following steps of the disaster management cycle (Fig. 2):

(1) prevention, which includes all actions aiming at minimizing future losses;
(2) preparedness, which aims at being prepared to a threat and managing disasters (Denis et al. 2016; Klomp 2016; Voigt et al. 2007);
(3) crisis and post-crisis management, aiming at saving lives, minimizing impacts, and finally re-establishing services supporting human activities (Myint et al. 2008; Wang and Xie 2018).

2.2 Decision Levels Relevant to Disaster Risk Management

Disaster risk management stakeholders such as governments, municipalities, industries, large businesses, disaster management organizations, civil protection can act at different levels ranging from operational management to strategic planning. Therefore, there is a wide diversity of actions relevant to each component of the disaster risk management cycle. This can therefore be illustrated for actions belonging to prevention, which include, for example:

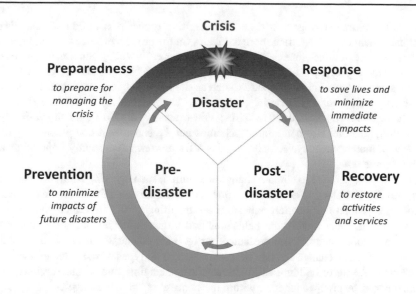

Fig. 2 Disaster management cycle

- regional to local authorities defining land-use policies in a way that minimizes exposure to hazards;
- regional to local authorities defining and enforcing specific mitigation measures, such as reducing the vulnerability of buildings for flooding or earthquakes by means of structural interventions;
- states and transnational organizations establishing regulations defining the principles and rules for operational disaster risk prevention;
- states, ministries and adaptation funds ranking the severity of hazards and risks among countries or regions, in order to optimize investments and allocate resources for prevention;
- the private or public reinsurance industry assessing the vulnerability of finance mechanisms covering post-disaster costs (Smolka and Siebert 2013);
- international to national organizations (e.g., Delta Commission in the Netherlands after the 1953 flood, or the United Nation Office for Disaster Risk Reduction) strengthening the governance of risks, as recommended by the Sendai Framework for disaster risk reduction.

All the actions listed above belong to prevention and can therefore be placed in the disaster risk management cycle. Whatever the level considered, the disaster risk management cycle remains a common reference for classifying actions aiming at reducing the impacts of disasters.

2.3 Criticism to the Disaster Risk Management Cycle

The disaster risk management cycle framework has been criticized for being a too simplistic classification of disaster risk management actions, which ignores synergies among, e.g., post-crisis and prevention actions (Lavell et al. 2012). As a response to this criticism, recent improvements have recognized that risks evolve as the environment, human pressures and climate change over time. The resulting frameworks build upon the disaster risk management cycle by putting more emphasis on the process of reviewing and updating risk management actions (Jones et al. 2014; Stammer et al. 2019). Such new frameworks are recognized especially useful to address the specific problem of climate change, which involves a constant evolution of hazards, such as the more frequent and intense heatwaves, drought and storm surges (Alexander et al. 2006; Cazenave and Le Cozannet 2014; Meehl and Tebaldi 2004; Russo et al. 2013, 2014; Vousdoukas et al. 2018). As shown in Fig. 3, the resulting "disaster risk continuum" schemes can be considered complementary to the disaster risk management cycle, as it provides a framework to constantly improve disaster risk management.

Far from being limited to climate change adaptation, such evolving frameworks are useful as well for all risks that change over time. For example, even for earthquakes, changes in exposure, ground deformation pattern, the retrofitting of the structures and the slight damage after an event modify the behaviour and vulnerability of the built environment. Hence, frameworks such as those presented in Fig. 3 are potentially useful well beyond the sole issue of climate change and can be generalized to all types of risks that evolve due to urbanisation, ageing of building and infrastructure, changing human interventions, which collectively can be grouped together within the concept of "global change".

2.4 Relevance to the Earth-Observation Sector

The previous subsection shows that the disaster risk management cycle is not the unique framework to classify decisions and actions pertaining to risk management. However, despite the known limitations discussed above, the disaster risk management cycle remains a common reference for many stakeholders concerned with disaster risk management. For the Space-based Earth Observation sector, the value of the disaster risk management framework is to identify where satellite data can be useful within the existing workflows

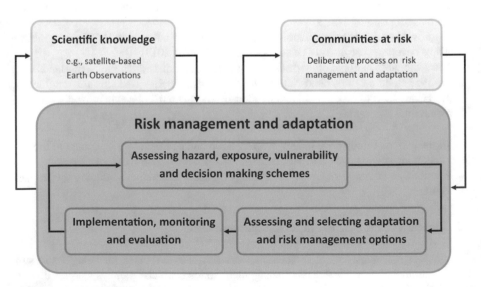

Fig. 3 Iterative "disaster risk continuum" approach; adapted from Jones et al. (2014) and Stammer et al. (2019)

of users. Figure 1 lists the main objectives of users during each phase of the disaster risk management cycle, and links user's potential actions and decisions with the related services and observation needs. These actions and decisions can be grouped in two categories (Fig. 1):

- Actions and decisions relevant to the prevention of disaster risks (Sect. 3): here, the main objective of users such as disaster risk prevention agencies is to minimize the impacts of future disasters, both in terms of human lives and costs. This overarching objective can be met in three ways: by reducing the hazard, where possible (Sect. 3.1), the vulnerability of exposed assets (Sect. 3.2) or the exposure itself (Sect. 3.3). To illustrate this concept, the risks induced by landslides can be reduced by various measures such as revegetating the slopes (reducing the hazard), reinforcing the infrastructure such as roads or building exposed to landslides (reducing vulnerability) or relocating exposed building (reducing exposure).
- Actions and decisions relevant to preparedness, response and recovery of disaster risks (Sect. 4): here, the main objective of users such as civil protection agencies is to prepare for managing the crisis, which requires pre-disaster information (Sect. 3), to forecast disasters to deliver appropriate alerts (Sect. 4.1), to save lives and provide immediate assistance (Sect. 4.2), minimize the impacts and to progressively restore activities and services during and after the crisis, which requires damage mapping (Sect. 4.3) and pre-disaster information (Sect. 3). For example, these users will test their procedures against pre-defined disaster risk scenarios before crisis, which requires pre-disaster knowledge on the hazard, exposure and vulnerability. During the crisis, other information such as displacement and disaster damage maps is required. This illustrates that workflows within disaster risk management involve different procedures, constraints and needs depending on the positioning of each stakeholder with respect to the disaster management cycle. Specifically, the time constraint is critical during the preparedness, response and recovery phases, whereas prevention is less constrained by time but often more constrained by the limited amount of resources (Salichon et al. 2007).

The remainder of this paper explores further how Earth observations can support the prevention of disaster risks (Sect. 3), preparedness, response and recovery (Sect. 4) and ways forward to improve the current practices (Sect. 5).

3 Earth Observation in Support to Prevention of Disaster Risks

Stakeholders concerned with prevention primarily aim at minimizing future damage, whether affecting human lives, economic activities or environmental assets. Achieving this aim implies reducing one or several determinants of the risk, that is, the hazard itself, exposure to these hazards, or the vulnerability of the exposed assets.

3.1 Reducing the Hazard

The hazard is commonly defined as a threat with potentially adverse consequences for humans, the economy or the environment. Quantitatively, the hazard is the combination of the probability of occurrence and the intensity of an adverse phenomenon in a given location (UNDRR terminology; https://www.undrr.org/publication/2009-unisdr-terminology-disaster-risk-reduc

tion). Prevention actions focused on the hazard include removing the hazards (for example by levelling landslide-prone slopes), mitigating the hazard (for example by restoring vegetation on dunes or unstable slopes) or changing the hazard (for example by building dikes and limiting flooding overflow), which sometimes leads to creating new hazards such as breaching of dikes (Ludy and Kondolf 2012). Reducing the hazard is not always possible: for example, earthquake or pyroclastic flow hazards generally cannot be reduced.

Removing, mitigating or changing the hazard typically requires hazard maps. Hazard maps may either consider a single hazard (e.g., river flooding), or combine multiple hazards (e.g., river flooding and landslides) (Gill and Malamud 2016). For example, volcanic hazard maps typically superimpose or combine multiple hazards such as lava and pyroclastic flows, ash falls, lahars, earthquakes, landslides (Felpeto et al. 2007; Neri et al. 2013; Thierry et al. 2008). Despite decades of research in quantitative hazard assessment and the related physical processes, a lack of full understanding of the hazards remains a critical knowledge gap today. Even for earthquakes where the knowledge of vulnerability is key for assessing potential damage, unknowns in the seismic activity (e.g., magnitude, occurrence, focal depth) can still remain the most important source of uncertainty in quantitative risk assessment (Rohmer et al. 2014). This raises the need for more hazard observations, whether in situ or space-based.

Space-based observations relevant for hazard assessment include observations of past events, of physical processes associated to the hazard, or of characteristics of the location of interest (i.e., settings). Observations of past events can directly or indirectly inform about the frequency or intensity of a hazard. For example, direct observations of subsidence caused by mining activities or groundwater pumping can be used to better estimate where, when and how much vertical ground motions may affect the urban or peri-urban environment (Foumelis et al. 2016; Raucoules et al. 2003, 2013). Indirect observations are extremely common as well, for example assessing drought intensity through observations of their impacts on vegetation (Kogan 1997). Space observations can also be used for monitoring physical processes associated to the hazard, such as thermal activity associated with volcanic eruptions (Vaughan et al. 2008) accumulation of strain along faults that may rupture and generate earthquakes (Barbot et al. 2013; de Michele et al. 2011; Hussain et al. 2016; Parcharidis et al. 2009), or rainfall causing flood hazard (Tote et al. 2015). Finally, typical characteristics of the location of interest that can be derived from Earth Observations include high-resolution digital surface models, which can be used for example to estimate the potential intensity and frequency of flooding, lahar or pyroclastic flows in volcanic areas (Kervyn et al. 2008; Neri et al. 2013). In the latter case, space-based Earth Observations do not appear prominently in the final product delivered to users, because they are integrated within physical models or geographic information system. However, the quality of these observations (e.g., resolution, precision and accuracy) remains essential for producing trustworthy information on the hazard processes.

For adaptation planning, characterizing changes of the hazards over multi-decadal timescales can be relevant (see Sect. 2.3), as exemplified for long-term sea-level rise aggravating coastal erosion and flooding (Melet et al. this issue). While precise needs for information have not completely yet been formalized, some scientists have proposed to focus on the detection of early signals of acceleration of changes in order to plan adaptation in a timely way (Haasnoot et al. 2018; Stephens et al. 2018).

3.2 Reducing the Vulnerability

The concept of vulnerability has been defined in many different ways by multiple scientific communities (Birkmann et al. 2013; Brooks et al. 2005; Kienberger et al.

2013; Romieu et al. 2010). In a disaster risk management context, vulnerability is defined as the physical, social and economic conditions that influence the susceptibility of humans or assets to the impacts of hazards (see https://www.undrr.org/publicatio n/2009-unisdr-terminology-disaster-risk-reduction).

Structural measures aiming at reducing the physical vulnerability of buildings are typically extremely efficient in the context of prevention of earthquake risks, because most of the damage and casualties are due to the collapse of buildings. Users concerned with reducing the physical dimension of vulnerability typically require information that allows them to design vulnerability or fragility curves, which give the probability of damage for an event of a given intensity for various types of assets (e.g., wood, concrete or masonry buildings) and different types of hazards. Fragility curves have long been established for earthquake risk assessment (Rossetto and Elnashai 2003), but they are increasingly used as well for landslides, flooding, tsunamis, tephra load and other volcanic hazards (Fuchs et al. 2007; Garcin et al. 2008; Jenkins et al. 2014; Jongman et al. 2012; Papathoma-Kohle et al. 2012; Spence et al. 2005). The fact that the fragility curves approach is not equally adopted across all natural hazards is partly due to the characteristics of hazards themselves (Douglas 2007). For example, coastal cliffs erosion or pyroclastic flows commonly cause total collapse of buildings (Dawson et al. 2009; Gehl et al. 2013). However, this is also due to the focus of scientific communities, who tend to put more emphasis on the physical processes and hazards than on the characterization of physical vulnerability (Douglas 2007; Geiss and Taubenbock 2013).

The physical determinants of vulnerability can be obtained by direct observations (e.g., building's shapes, materials, presence of chimneys) (Ehrlich and Tenerelli 2013), or by indirect observations, linking observable features of the built environment (e.g., shape and width of streets, size of buildings, colour of roofs) with the actual determinants of vulnerability (Geiss et al. 2014; Muck et al. 2013; Taubenbock et al. 2008). For example, the current practice of operational physical vulnerability assessment consists in collecting in situ proxies during field surveys, eventually complemented by a visual examination of remote sensing images (Sedan et al. 2013). The approach based on direct observations is limited by the fact that there remain numerous proxies that cannot be observed yet with remote sensing, and which are important drivers of the vulnerability, such as the internal structures of buildings. For example, the uncertainty in the average vulnerability of earthquakes cannot be reduced by more than 50% by direct methods (Le Cozannet et al. 2018). Combining both indirect and direct proxies delivers results that are accurate enough for rapid screening of the vulnerability (Geiss et al. 2014).

Social vulnerability, that is, the propensity for human well-being to be adversely affected, interacts with economic vulnerability that is, the propensity for financial damage or disruption of productive capacity. Observing social or economic vulnerability involves assessing the extent to which populations are able or unable to cope with exposure to hazards. Earth observation data provide such information indirectly. For example, nighttime lights are a straightforward indicator of urban services, which needs to be considered in vulnerability assessments (de Sherbinin et al. 2015). However, other features that can be identified almost automatically in remote sensing images can help retrieving proxies of economic and social vulnerability. For example, Jean et al. (2016) combined machine learning (convolutional neural network) with high-resolution satellite imagery to predict the spatial distribution in economic well-being (poverty and wealth) in five African countries, hence explaining up to 75% of local variations of economic outcomes (Jean et al. 2016).

3.3 Reducing the Exposure

The exposure is defined by UNDRR as "people, infrastructure, housing, production capacities and other tangible human assets located in hazard-prone areas". A first aim of disaster risk reduction can be to avoid further exposure to natural hazards, for example by avoiding building in coastal wetlands exposed to storms or in areas subject to pyroclastic flows. A second aim can be to reduce exposure through relocation of assets.

Satellite-based Earth Observation has long demonstrated capacities to map elements at risk. Usually, the main elements retrieved from Earth observations include the distribution of human population, buildings and strategic infrastructure, and roads, but the mapping of crops and natural ecosystems can be considered as well (De Bono and Mora 2014; Van Westen 2013). The spatial distribution and amount of most of these features can be readily extracted from moderate to very high spatial resolution satellite image through supervised classification algorithms, using pixel or object-based approaches (Jozdani et al. 2019). Challenges remain in the classification of different types of elements based on their material (e.g. dirt road versus tarmac road; houses with different roof types) or the discrimination of individual elements in densely built area (e.g., segmentation of individual houses in densely built slumps; Mossoux et al. 2018). Today, satellite-based Earth observations are becoming increasingly integrated with other sources of information within national geo-data repository or volunteered geographic information systems such as OpenStreetMap (Mahabir et al. 2018). As a consequence, despite being an essential component of state-of-the art data repository used for exposure mapping, satellite-based Earth observations are becoming less apparent to users.

Population maps and density are generally obtained from national census, and available per administrative units, going from small census tracts to coarser municipality or district units. Censuses are expensive, labour intensive and time-consuming and, as such, are often only organized once every 10 years or less. Many developing countries currently lack accurate population data at a fine geographical scale. Remote sensing can be a cost-effective alternative for mapping of population distribution. While it is not expected to result in the same accuracy as a traditional census, it can be a useful surrogate in absence of up-to-date census data. Two different remote sensing approaches have been developed for population mapping:

- top-down approaches using interpolation methods based on land cover maps and other spatial indicators to disaggregate census data to a finer spatial scale (e.g., WorldPop, LandScan, World Population Estimate, Global Human Settlement Layer-Population (GHS-POP), Global Rural Urban Mapping Project (GRUMP)) (Lung et al. 2013);
- bottom-up approaches aiming at defining a statistical relationship between remote sensing derived variables (nighttime light, number or size of dwelling units, settlement area, type of buildings) and population numbers (Mossoux et al. 2018; Wu et al. 2005) (see Fig. 4).

For example, Lung et al. (2013) showed that population distribution at regional scale derived from object-based image analysis of very high-resolution satellite imagery

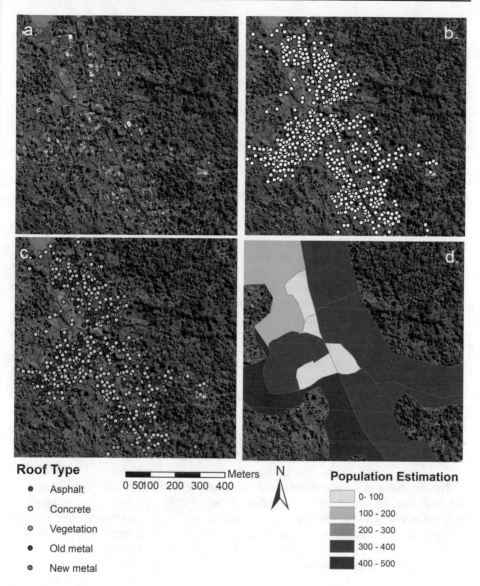

Fig. 4 Example of exposure mapping in a remote area (Mitsoudje, Comoros, a volcanic archipelago off the East coast of Africa) using high-resolution Pléiades images (**a**) to identify (**b**), classify roof type of houses (**c**) and quantify the number of residents based on average number of residents per type of house extracted from field survey (**d**) (adapted from Mossoux et al. 2018)

compare well with information contained in census data in Kenya. As a further example, satellite derived observations of night lights have been related to the regional dynamics of population in China (Ma 2018). At coarser scales, imagery showing urban extent has been combined with nighttime light emissions, to assess population distribution at continental to global scales (Tan et al. 2018; Wu et al. 2005; Yu et al. 2018).

Exposure mapping is typically an area where automated processing of space-based Earth observations can substantially lower costs of information acquisition, because it is relevant across all hazards. This potential benefit of thinking beyond one single risk is one of the motivation for so-called "multi-risk" approaches, which attempt to address all components of risks across threats affecting a territory (Gallina et al. 2016; Grunthal et al. 2006; Marzocchi et al. 2012).

4 Earth Observation in Support to Crisis Management and Disaster Response

Needs relevant to preparedness, response and recovery include (Fig. 1):

- disaster scenarios to simulate the impacts of crisis events and test emergency procedures: such scenarios primarily require information about vulnerability and exposure (see Sects. 3.2 and 3.3), in order to compute the potential impacts of adverse phenomena such as earthquakes or volcanic eruption (Gehl et al. 2013; Marrero et al. 2012; Sedan et al. 2013; Zuccaro et al. 2008). Information about the hazard is important as well in order to communicate the likelihood of each scenario to the bodies in charge of disaster management (see Sect. 3.1).
- forecasts and early warning systems to plan for specific protection measures as a disaster is becoming more likely: this phase also requires pre-disaster information, but also forecasts and early warning systems, that require near-real-time observations of the ongoing event (e.g., cyclonic tracks, escalating volcanic unrest) in order to assess which scenarios are becoming more likely and which protection measures need to be taken (see Sect. 4.1).
- information for delivering immediate assistance (see Sect. 4.2)
- damage mapping and information about the changes of hazards and risks to assess reconstruction costs and needs (see Sect. 4.3).

The list above shows that many of the satellite-based Earth Observation needs for hazard, exposure and vulnerability information discussed in Sect. 3 are relevant to preparedness and recovery. In the remaining of this section, we focus on the three aspects of crisis management that require more specific information and procedures: forecasts and early warning (Sect. 4.1), crisis management to deliver immediate assistance (Sect. 4.2) and damage mapping to assess reconstruction costs and needs (Sect. 4.3).

4.1 Forecasts and Early Warning Systems

The expected benefits from efficient early warning systems or forecasts are first to save lives, second to protect assets at risk. The meteorological forecasts are the most obvious example of satellite-based Earth Observations supporting early warning, for example in the case of storms or cyclones. During volcanic eruptions, early warning and forecasts can benefit from thermal hotspot, SO_2 degassing and ground deformation satellite-based Earth observations, which in combination with in situ seismic monitoring have proven useful for forecasting and following ongoing crisis (Ernst et al. 2008; Smets et al. 2014; Surono et al. 2012). Overall, the potential of satellite-based Earth observations to inform forecasts of disasters is increasingly being explored for nearly 30 years (e.g., Tralli et al. 2005; Tronin 2010; Freund 2013).

New research may even offer more benefits in this area: for example, current early warning systems for Earthquakes take advantage that the first seismic waves propagate quicker than the most destructive ones, which allows immediate warning measures to save lives and avoid cascade events (Salichon et al. 2017). However, other non-seismic signals may suggest that an earthquake is about to happen, such groundwater temperature and/or chemistry changes (Claesson et al. 2004; Rigo 2010; King and Chia 2018 and references therein) and radon emissions (e.g., King 1986; Einarsson et al. 2008; Awais et al. 2017 and references therein). While such changes can be monitored in situ, satellite-based Earth Observation may also help monitoring these non-seismic signal changes in the ionospheric Total Electron Content (TEC) (e.g., Zakharenkova et al. 2008; Harrison et al. 2010; Kon et al. 2011; Zhang et al. 2014 and references therein), the surface temperature (e.g., Tronin 2000; Tramutoli et al. 2013; Barkat et al. 2018 and references therein), surface latent heat flux (Dey et al. 2004; Cervone et al. 2006; Qin et al. 2014) and Earthquake-cloud formation (e.g., Harrison et al. 2014 and references therein). While these partial examples confirm the variety of non-seismic signal that can be monitored, research is still required to improve the earthquake forecasting. In fact, several processes can create non-seismic signals that can be interpreted as early signals of an upcoming Earthquake.

4.2 Immediate Assistance

During and after a disaster, information is required to evaluate impacts of the disaster to humans and deliver immediate assistance (Boccardo and Tonolo 2015; Denis et al. 2016; Voigt et al. 2007, 2016). During this phase of crisis management, the priority is to save lives. Therefore, the information required relates to the likelihood of casualties and population displacements caused by the event. As an example of what Earth Observations can provide during this phase, nighttime lights from satellite images have been demonstrated useful to assess population size and dynamics during a 2010 humanitarian crisis caused by political conflict in Côte d'Ivoire (Bharti et al. 2015). They found increased brightness in neighbouring regions of Liberia where refugees had flown to, which they related to the number of refugees recorded by the United Nations simultaneously.

Traditionally, such needs are filled by conventional communication and aerial or in situ monitoring, but there is evidence that the space sector is increasingly informing operational crisis management and post-disaster response as well. For example, the decision to extend exclusion zones during the 2010 Merapi eruption was not only based on information from conventional geophysical data, but also from satellite images (Surono et al. 2012). Furthermore, a formal user feedback gathered after the Tohoku Earthquake and tsunami in 2011 in Japan suggests that space technologies delivered information that brought benefit to the definition, planning, implementation, monitoring and assessment of disaster relief operations (Kaku et al. 2015).

Several initiatives aim at delivering information in near real time to support crisis management and disaster response, including the European Copernicus Emergency Management Service (EMS) or Sentinel Asia (Kaku 2019; Lorenzo-Alonso et al. 2019). Here, international cooperation plays a key role because it allows each agency to benefit from third parties data potentially acquired earlier. The international charter for space and major disasters (https://disasterscharter.org/en/web/guest/home) is a worldwide collaboration, through which satellite data are made available for the benefit of crisis and post-disaster management. By combining Earth observations from different space agencies, the Charter allows resources and expertise to be coordinated for rapid response to major disaster situations, thereby helping civil protection authorities and the international humanitarian community at no cost for the user. It mobilizes agencies around the world and benefit from their know-how and their satellites through a single access point. As the Charter immediate response is key to success, this access point operates 24 h a day, 7 days a week.

The Charter has been founded in 1999 and declared operational as of November 2000, with 3 participating agencies: ESA, CNES and CSA. In two decades, it has been activated more than 600 times worldwide. The international charter now comprises 17 space agencies worldwide. The Charter is triggered for a large variety of disasters, either natural or related to human activity: earthquakes, fires, floods, ice jams, landslides, tsunamis, ocean storms, volcanic eruptions, oil spills and industrial accidents. It is functioning on the best efforts basis. Each agency is committed to provide space data resources and value-added processing as needed in the framework defined by the charter regulations. The so-called "Authorized Users" (AU) have a permanent access to the Charter access point in order to declare an emergency. There are currently 66 AU spread over 66 countries and this number is constantly increasing.

Once activated, the charter provides the local authorities in charge of the disaster management on site, with detailed value-added maps outlining the level of infrastructure destruction. Depending on the type of disaster and its location on Earth, the adequate sensors (medium and/or high-resolution optical images or all-weather capability radar images) of the flying-by satellites are used. Today, the current procedures allow the delivery of satellite products 48 h after the event, so that the request of users for a 6-8 h delivery cannot be met yet in many cases (Denis et al. 2016; Voigt et al. 2016). Here, constellations of satellites allowing for more frequent data acquisition could become a game changer over the coming decades (Denis et al. 2016). Already, many private companies (e.g., Planet Labs, ICEYE, UrtheCast) aim at providing daily or even hourly Earth Observation data coverage using both optical and Synthetic Aperture Radar (SAR) spaceborne sensors. It should be underlined that this involvement of the private sector comes as an encouraging step to the uptake of the market for such operational observations systems.

4.3 Damage Mapping

Once human lives have been saved, recovery actions can take place. These actions are essentially intended to restore infrastructure and economic and environmental services. Here, an assessment of reconstruction needs and the effects of post-disaster response is required. Two types of products based on satellite-based Earth observations can be schematically distinguished for these post-disaster phases:

- Rapid damage mapping for an assessment of the needs for ongoing assistance (e.g., shelters for people whose buildings have collapsed)
- Re-assessment of hazards and risks in order to reconstruct and restore services in a way that minimizes the impacts of future events.

While the two types of actions essentially require the same types of data, rapid mapping involves more severe time constraints but lower demands for accuracy than the full re-assessment of hazards and risks.

Damage mapping refers first to the identification of buildings or assets that have been entirely destroyed or cannot deliver essential services anymore. For example, satellite-based Earth observations have proven useful for mapping the destruction of settlements of ethnic minorities during armed conflicts (Bhattacharjee 2007; Levin et al. 2018). Changes to the land use or to the environment can be used as indicators for crisis-induced changes in population vulnerability, as exemplified in Northern Iraq where crops changed to grassland during the conflict (Mubareka and Ehrlich 2010). Furthermore, assessing the damage to the vegetation can be considered a goal per se, as in the case of the mining accident of Norilsk that caused degradation of the boreal forest (Zubareva et al. 2003) or to assess potential impacts for food security after volcanic ash-fallout events in densely populated areas (De Rose et al. 2011; De Schutter et al. 2015).

Damage mapping also refers to the quantification of vulnerability changes. After disasters, the vulnerability needs to be reassessed, because more fragile buildings, infrastructure or assets may represent a threat if another adverse event occurs. For example, differential subsidence or earthquakes may slightly damage the structure of buildings without causing their full collapse, and, ultimately, make them more vulnerable to a future seismic shock (Ehrlich and Tenerelli 2013). Here, following the work of Negulescu et al. (2014), Earth observations, whether in situ or space based, can be used to detect fragile structures, the result serving as information basis for identifying future vulnerable structures and areas in need of reconstruction.

After a disaster event, the hazard can change drastically. For example, volcanic eruption, lahars or landslides may change the elevation, slopes and landslide susceptibility (de Belizal et al. 2013; Joyce et al. 2009; Kerle et al. 2003), tsunamis and cyclones can modify shorelines and the vegetation (Paris et al. 2009), potentially favouring the propagation of surges inland if another coastal flood occurs. Hence, hazard ideally needs to be reassessed after disasters before planning for reconstruction. This especially update of available Digital Elevation Model and land cover maps, including for identifying changes in the exposure of elements at risk.

While the distinctions of actions according to the disaster management circle are convenient to understand where Earth observation can best support disaster risk reductions, there are linkages between pre- and post-disaster disaster management actions. Observation of damage after a disaster is commonly used to support crisis management and recovery (see Sect. 5). However, it can be used as well for designing and improving vulnerability curves, which can then be used to design scenarios useful for prevention and preparedness (Papathoma-Kohle et al. 2012).

5 Challenges, Opportunities and Ways Forward

5.1 Challenges

Overall, this review presents evidence that Earth Observation is already used in many ways to support disaster risk management. Current applications are sometimes not directly visible by the end-users, as space-based observations are used within complex approaches and workflows. Hence, what is finally visible to users is the most salient information (e.g., a hazard map) and not on the key datasets that allow to produce it (e.g., a very high-resolution digital elevation model).

We argue that the potential of Earth Observation for prevention is still underused. In fact, crisis management has received much attention from the Earth Observation community so far, as exemplified by the development of the International Charter and the Copernicus Emergency Management Service (EMS). On the other hand, prevention and some post-disaster phases aiming at increasing resilience have been less considered.

Studies focused on other phases of the disaster management cycle are not well balanced either, with usually a much stronger focus towards hazard assessment and less emphasis on vulnerability and exposure mapping. Surprisingly, this statement holds even for earthquake risk assessment, although quantifying vulnerability is a key in this area. A review of seismic risk assessment studies based on space-based Earth observations showed a strong emphasis on observations of the hazards and of post-disaster damage, but much less efforts for vulnerability mapping (Geiss and Taubenbock 2013).

Despite some promising research results (Ong et al. 2019; te Brake et al. 2013; Capes and Teeuw 2017), some phenomena with important impacts for the society still remain incompletely observed (e.g. shrinking and swelling of clays, groundwater induced flooding), and many needs of adaptation practitioners still need to be addressed. For example, the increasing occurrence and magnitude of heatwaves and droughts which is already being registered in the last decades (Russo et al. 2014; Russo and Sterl 2011) represents a potential important threat for human lives and well-being, for regional economies, and for natural ecosystems (Meehl and Tebaldi 2004). The challenge of better anticipating more frequent and intense heatwaves and droughts, or generally extreme natural phenomena, comes along with Earth Observation needs, which have not been systematically assessed yet. For example, adaptation to sea-level rise requires improved monitoring of shoreline changes and sediment transport (Benveniste et al. 2019), but the most urgent needs for such information comes from current risk management whereas adaptation at timescales relevant to sea-level rise (i.e., decades and centuries) is usually considered less urgent. For these sectors where no specific requirements have been formalized yet, we argue that a systematic review of user workflows and decisions is required to maximize the societal benefits of space-based Earth observations.

A final challenge will be to develop a sustainable market and adequate/tailored business models for the satellite-based Earth Observations supporting disaster risk management. Current estimates suggest that the financial investments that are planned today are relatively small, although funds for future Earth Observation missions, serving both scientific and commercial needs, have been increased. For example, the insurance market for the Earth Observation techniques supporting natural risks has been estimated at about 75 million euros per year in 2030 for Europe (Source: https://www.cvt-allenvi.fr/etudes/technologies-dobservation-environnementale-pour-lagriculture-et-les-risques-naturels/). For comparison, a single satellite typically represents a cost of 100 to 300 million Euros, which still remains much smaller than the economic damages from single disasters: for example, the 2010 Xynthia coastal storm in France and the costliest 2011 Tohoku Earthquake in Japan represented more than 2.5 billion Euros and 300 billion USD, respectively, notwithstanding casualties (Genovese and Przyluski 2013; Daniell et al. 2011). As long as the planned investments of the private sector remain so small in Earth observations, public investment will remain key to support the development of the sector.

5.2 The New Context of Satellite-Based Observations

There are today many opportunities to progress in using satellite-based Earth observations for disaster risk management. These opportunities are allowed by initiatives of the space sector, which has been developing an unprecedented Earth Observation infrastructure, in particular the Copernicus Sentinel constellation of satellites (Aschbacher et al. 2014). Although the Sentinel missions were initially envisaged to address the operational monitoring needs of Copernicus EMS, they have become a game changer not only for science but also to the space applications industry. Indeed, ensuring the systematic flow of Earth Observation data and the long-term sustainability of services is a prerequisite for commercial business models. This is apart from the actual contribution to science in terms of understanding and quantifying physical phenomena. Thus, further developments and innovative ideas should be expected in various domains, among which higher resolution soil moisture measurements, mapping of clay swelling, correcting tide gauge measurements and even monitoring of ground motion at nation-wide scale. Opportunities are not limited to hazard monitoring: as observations resources are becoming available through the Copernicus Land Service and other open historical archives (Koks et al. 2019), there is now an opportunity to better understand and how the vulnerability and exposure have evolved in the past, reached their current status and may evolve in the future (Duvat et al. 2017).

However, to allow the above-mentioned achievements dedicated dissemination and utilization strategies for such record amount of Earth Observation data should be considered. This is exactly where the support of space agencies arises, introducing concepts of moving algorithms and codes close to data, contrary to past practices that will soon become obsolete due to limited local storage and processing capacities. The Sentinels come along with web-based data dissemination mechanism (including collaborative ground segments etc.) like the Copernicus Open Access Hub (https://scihub.coper nicus.eu), the Sentinel Product Exploitation Platform (PEPS) of CNES (https.//peps. cnes.fr/rocket/#/home) as well as the development of processing platforms such as the ESA Geohazards Exploitation Platform (GEP, https://geohazards-tep.eu/), with numerous hosted processing services for geohazards applications (Foumelis et al. 2019). This constitutes a radical change in working procedures of Earth observations scientists and engineers, as less time is required to data management and processing, so that more efforts can be dedicated to the integration of different datasets, the interpretation of the observed phenomena, and the development of new algorithms to gather hazard-relevant information, as illustrated recently for shoreline changes (Mentaschi et al. 2018).

The conceptual model for such platform-based systems is relatively simple, having as a basis the infrastructure where Earth Observation data are stored and accessed from, followed by an upper layer of processing segments, represented in principal by several cloud service providers. Finally, at a higher end, dedicated web-based interfaces are the portals for user interaction with data and processing tools. Each of these layers is being developed separately by international and national initiatives, each at different level of maturity. The Data and Information Access Services (DIAS) platforms, funded by the European Commission (EC), are such example aiming to facilitate centralised and standardize access, manipulation and processing to Copernicus Sentinel data (https:// www.copernicus.eu/en/access-data/dias).

Even though the evolution of these platforms cannot yet be easily projected into the future, since they are still heavily supported by space agencies, they count already several success stories in publishing scientific results (Galve et al. 2017; Papageorgiou et al. 2019), supporting geohazards related initiatives (ESA EO4SD project, https://www.eo4sd-drr.eu/news/government-indonesia%E2%80%99s-users-eo-products-disaster-risk-reduction) and providing rapid and valuable information in response to geohazards events (example of the 2019 M5 earthquake in mainland France, https://www.esa.int/Applications/Observing_the_Earth/Copernicus/Sentinel-1/French_earthquake_fault_mapped). Here again, it is clear that the role of the private sector will also be critical to ensure sustainability and support further enhanced processing and e-collaboration capabilities of platform-based solutions.

In this context, initiatives such as the Group on Earth Observations (GEO), the European Plate Observing System (EPOS) and the Geohazards Lab (http://ceos.org/ourwork/workinggroups/disasters/geohazards-lab) of the Committee of the Earth Observation Satellites (CEOS) are more useful than ever to foster the organization of the Earth Observation community concerned with disaster risks management.

5.3 Initiatives from the Disaster-Risk Management Community

Other opportunities result from initiatives from the community concerned with disaster risk reduction. Several scientific or operational observatories are focused on a particular system or territory. For example, this includes seismic networks, volcano or coastal observatories. Space agencies have long identified the prominent role of these observatories and linked with them, for example through initiatives such as the "Supersites" or the "Natural Laboratories", through which they deliver all available data to stimulate scientific research for critical areas. These sites include, for example, the earthquake-prone San Andrea and Marmara region in California and Turkey, or the Virunga in the democratic republic of Congo, Vesuvius, Campi Flegrei and Etna volcanic areas in Italy (see a list of existing and candidate Supersites at https://geo-gsnl.org/). For many of these Supersites, there is not a single hazard threatening communities and human assets, but multiple natural and anthropogenic risks, which can be more efficiently observed if one takes advantage of synergies among observations relevant to these multiple risks. An obvious element here is the characterization of the built environment, which has long been considered in multi-risk approaches developed by the scientific community concerned with disaster risk reduction (Grunthal et al. 2006; Kappes et al. 2012; Thierry et al. 2008). As shown in Sect. 3, characteristics of the built environment can hardly be observed directly, but there are opportunities in valuing indirect proxies, such as local knowledge of building practices or nightlights for assessing post-disaster recovery. Overall, these efforts of the communities concerned by disaster risk reduction are supported by research in the area of geospatial information management (Wilkinson et al. 2016) that now allow for interoperable information systems and web services to share data relevant to disaster risks (Douglas et al. 2008; Le Cozannet et al. 2014; Tellez-Arenas et al. 2018). Finally, some major frameworks such as the Sendai Framework (https://www.unisdr.org/files/43291_sendaiframeworkfordrren.pdf) highlight on key priorities for reducing disaster risks, including improving the governance of risk management, supporting operational disaster risk management and long-term strategic planning. For the Earth Observation community concerned with disaster risks, the detail of these priorities is relevant to consider because they link to actions presenting the best potential to reduce risks.

5.4 Need for Mutual Exchange of Knowledge

From the perspective of space agencies and space-based Earth Observation providers, the challenge of improving the use of Earth observations techniques in everyday disaster risk management practices goes through proper training and increase of awareness of the capacity of Earth observations missions and techniques. We argue that a multi-directional mutual exchange of knowledge is required, which involves liaising the disaster risk reduction and the Earth observations research and practitioner communities. The usefulness of Earth observations can be demonstrated with successful showcases, but understanding workflows of practitioners and training them to the most appropriate tools plays also a significant role. Several space agencies have been investing towards the direction of training by organizing regular face-to-face as well as massive online training courses addressed mainly to academic students and members of scientific institutes, while other initiatives focus also on non-Earth observations experts, such as the Research and User Support (RUS) service of Copernicus (http://rus-copernicus.eu). Furthermore, capacity building activities, as organized by space agencies (e.g., ESA) together with other relevant funding organisms (e.g., World Bank and Asian Development Bank) are contributing likewise towards improved awareness. Tailored education of the population for realizing the potential positive impact of Earth observations to the disaster risk management is considered nowadays crucial for the successful acceptance of these technologies and proper utilization of the Earth observations assets. We argue that an equally ambitious agenda is required to inform the Earth observations data and service providers with needs, workflows, and practices of technicians, engineers and researchers working on disaster risk reduction.

6 Conclusions

This paper argues that user-centric approaches linking specific decisions schemes with information needs are useful to maximize the benefits of satellite-based Earth observations. This statement is obtained from a review of satellite-based Earth Observation inputs into the different phases of the disaster risk management cycle (Sect. 2), including prevention (Sect. 3), preparedness, crisis and post-crisis management (Sect. 4).This review shows that while satellite-based Earth Observation is useful in a number of cases, there are still opportunities to develop, in particular to support prevention of risks, to better monitor the vulnerability and exposure and to detect early changes in hazards caused by climate change.

Several opportunities exist, originating from both the space sector and the disaster risk scientific and management communities (Sect. 4). A straightforward way forward to take advantage of these opportunities will consist in linking these top-down and bottom-up approaches, for instance following successful experiences in hazard and risk observatories, "supersites" and natural laboratories. This requires appropriate capacity building and mutual exchange of knowledge of all communities concerned with disaster risk reduction. For the research community, this creates opportunities for trans-disciplinary research linking Earth observations and environmental science with decision making and social science.

Acknowledgements This paper arose from the international workshop on "Natural and man-made hazards monitoring by the Earth Observation missions: current status and scientific gaps" held at the International Space Science Institute (ISSI), Bern, Switzerland, on April 15-18 2019. GLC and MF are supported by the Geohazards Lab and ERA4CS INSeaPTION project (Grant 690462). The authors are grateful to the guest editors Anny Cazenave, Miora Mandea, Jérôme Benveniste, Stephen Belcher and Teodolina Lopez for inviting this paper in this special issue. We thank a number of colleagues who have participated to various geohazards initiatives, especially Daniel Raucoules, Marcello de Michele, Marc Paganini, Sophie Mossoux, Jérôme Salichon, Steven Hosford, John Labrecque and Claudie Carnec. We thank John Douglas and an anonymous reviewer for their constructive review that helped improving the paper significantly.

References

Adger WN (2000) Social and ecological resilience: are they related? Prog Hum Geogr 24(3):347–364

Alexander LV, Zhang X, Peterson TC, Caesar J, Gleason B, Tank A, Haylock M, Collins D, Trewin B, Rahimzadeh F, Tagipour A, Kumar KR, Revadekar J, Griffiths G, Vincent L, Stephenson DB, Burn J, Aguilar E, Brunet M, Taylor M, New M, Zhai P, Rusticucci M, Vazquez-Aguirre JL (2006) Global observed changes in daily climate extremes of temperature and precipitation. J Geophys Res Atmos 111:D05109

Antoine R, Lissak C, Fauchard C, Tanguy M, Smet B, Gomez C (this issue) UAVs for geohazards. Surv Geophys (in review)

Aschbacher J, Milagro-Perez MP (2012) The European Earth monitoring (GMES) programme: status and perspectives. Remote Sens Environ 120:3–8

Aschbacher J, Beer T, Ciccolella A, Filippazzo G, Milagro M, Tassa A (2014) COPERNICUS Moving from development to operations. ESA Bull Eur Space Agency 157:30–37

Awais M, Barkat A, Ali A et al (2017) Satellite thermal IR and atmospheric radon anomalies associated with the Haripur earthquake (Oct 2010; Mw 5.2), Pakistan. Adv Space Res 60:2333–2344. https://doi.org/10.1016/j.asr.2017.08.034

Bally P (2012) (edt) Scientific and technical memorandum of the international forum on satellite EO and geohazards, 21–23 May 2012. Santorini, Greece, https://doi.org/10.5270/esa-geo-hzrd-2012. http://esamultimedia.esa.int/docs/EarthObservation/Geohazards/esa-geo-hzrd-2012.pdf

Balsamo G, Agusti-Parareda A, Albergel C, Arduini G, Beljaars A, Bidlot J, Bousserez N, Boussetta S, Brown A, Buizza R, Buontempo C, Chevallier F, Choulga M, Cloke H, Cronin MF, Dahoui M, De Rosnay P, Dirmeyer PA, Drusch M, Dutra E, Ek MB, Gentine P, Hewitt H, Keeley SPE, Kerr Y, Kumar S, Lupu C, Mahfouf JF, McNorton J, Mecklenburg S, Mogensen K, Munoz-Sabater J, Orth R, Rabier F, Reichle R, Ruston B, Pappenberger F, Sandu I, Seneviratne SI, Tietsche S, Trigo IF, Uijlenhoet R, Wedi N, Woolway RI, Zeng XB (2018) Satellite and in situ observations for advancing global earth surface modelling: a review. Remote Sens 10(12):72

Barbot S, Agram P, De Michele M (2013) Change of apparent segmentation of the San Andreas fault around Parkfield from space geodetic observations across multiple periods. J Geophys Res Solid Earth 118(12):6311–6327

Barkat A, Ali A, Rehman K et al (2018) Thermal IR satellite data application for earthquake research in Pakistan. J Geodyn 116:13–22. https://doi.org/10.1016/j.jog.2018.01.008

Bello OM, Aina YA (2014) Satellite remote sensing as a tool in disaster management and sustainable development: towards a synergistic approach. In: 3rd international geography symposium, Geomed2013, vol 120, pp 365–373

Benveniste J, Cazenave A, Vignudelli S, Fenoglio-Marc L, Shah R, Almar R, Andersen O, Birol F, Bonnefond P, Bouffard J, Calafat F, Cardellach E, Cipollini P, Le Cozannet G, Dufau C, Fernandes MJ, Frappart F, Garrison J, Gommenginger C, Han GQ, Hoyer JL, Kourafalou V, Leuliette E, Li ZJ, Loisel H, Madsen KS, Marcos M, Melet A, Meyssignac B, Pascual A, Passaro M, Ribo S, Scharroo R, Song YT, Speich S, Wilkin J, Woodworth P, Woppelmann G (2019) Requirements for a coastal hazards observing system. Front Mar Sci 6:348

Bharti N, Lu X, Bengtsson L, Wetter E, Tatem AJ (2015) Remotely measuring populations during a crisis by overlaying two data sources. Int Health 7(2):90–98

Bhattacharjee Y (2007) Human rights—Myanmar's secret history exposed in satellite images. Science 318(5847):29

Birkmann J, Cardona OD, Carreno ML, Barbat AH, Pelling M, Schneiderbauer S, Kienberger S, Keiler M, Alexander D, Zeil P, Welle T (2013) Framing vulnerability, risk and societal responses: the MOVE framework. Nat Hazards 67(2):193–211

Boccardo P, Tonolo FG (2015) Remote sensing role in emergency mapping for disaster response. Engineering geology for society and territory, Vol 5: urban geology, sustainable planning and landscape exploitation, pp 17–24

Brooks N, Adger WN, Kelly PM (2005) The determinants of vulnerability and adaptive capacity at the national level and the implications for adaptation. Glob Environ Change Hum Policy Dimens 15(2):151–163

Capes R, Teeuw R (2017) On safe ground? Analysis of European urban geohazards using satellite radar interferometry. Int J Appl Earth Obs Geoinf 58:74–85. https://doi.org/10.1016/j.jag.2017.01.010

Cazenave A, Le Cozannet G (2014) Sea level rise and its coastal impacts. Earths Future 2(2):15–34

Cervone G, Maekawa S, Singh RP et al (2006) Surface latent heat flux and nighttime LF anomalies prior to the $M_w = 8.3$ Tokachi-Oki earthquake. Nat Hazards Earth Syst Sci 6:109–114

Claesson L, Skelton A, Graham C et al (2004) Hydrogeochemical changes before and after a major earthquake. Geology 32:641–644. https://doi.org/10.1130/G20542.1

Daniell JE, Khazai B, Wenzel F, Vervaeck A (2011) The CATDAT damaging earthquakes database. Nat Hazards Earth Syst Sci 11(8):2235

Dawson RJ, Dickson ME, Nicholls RJ, Hall JW, Walkden MJ, Stansby PK, Mokrech M, Richards J, Zhou J, Milligan J, Jordan A (2009) Integrated analysis of risks of coastal flooding and cliff erosion under scenarios of long term change. Clim Change 95(1–2):249–288

de Belizal E, Lavigne F, Hadmoko DS, Degeai JP, Dipayana GA, Mutaqin BW, Marfai MA, Coquet M, Le Mauff B, Robin AK, Vidal C, Cholik N, Aisyah N (2013) Rain-triggered lahars following the 2010 eruption of Merapi volcano, Indonesia: a major risk. J Volcanol Geoth Res 261:330–347

De Bono A, Mora MG (2014) A global exposure model for disaster risk assessment. Int J Disaster Risk Reduct 10:442–451

de Michele M, Raucoules D, Rolandone F, Briole P, Salichon J, Lemoine A, Aochi H (2011) Spatiotemporal evolution of surface creep in the Parkfield region of the San Andreas Fault (1993–2004) from synthetic aperture radar. Earth Planet Sci Lett 308(1–2):141–150

De Rose RC, Oguchi T, Morishima W, Collado M (2011) Land cover change on Mt. Pinatubo, the Philippines, monitored using ASTER VNIR. Int J Remote Sens 32(24):9279–9305

De Schutter A, Kervyn M, Canters F, Bosshard-Stadlin SA, Songo MAM, Mattsson HB (2015) Ash fall impact on vegetation: a remote sensing approach of the Oldoinyo Lengai 2007–08 eruption. J Appl Volcanol 4:15

de Sherbinin A, Chai-Onn T, Jaiteh M, Mara V, Pistolesi L, Schnarr E, Trzaska S (2015) Data integration for climate vulnerability mapping in West Africa. ISPRS Int J Geo-Inf 4(4):2561–2582

Denis G, de Boissezon H, Hosford S, Pasco X, Montfort B, Ranera F (2016) The evolution of Earth Observation satellites in Europe and its impact on the performance of emergency response services. Acta Astronaut 127:619–633

Desnos YL, Borgeaud M, Doherty M, Liebig V, Rast M (2014) The European space agency's earth observation program. IEEE Geosci Remote Sens Mag 2(2):37–46

Desnos YL, Foumelis M, Engdahl M, Mathieu PP, Palazzo F, Ramoino F, Zmuda A, IEEE (2016) Scientific exploitation of sentinel-1 within esa's seom programme element. In: 2016 IEEE international geoscience and remote sensing symposium (IGARSS), pp 3878–3881

Dey S, Sarkar S, Singh RP (2004) Anomalous changes in column water vapor after Gujarat earthquake. Adv Space Res 33:274–278. https://doi.org/10.1016/S0273-1177(03)00475-7

Douglas J (2007) Physical vulnerability modelling in natural hazard risk assessment. Nat Hazards Earth Syst Sci 7(2):283–288

Douglas J, Usländer T, Schimak G, Esteban JF, Denzer R (2008) An open distributed architecture for sensor networks for risk management. Sensors 8(3):1755–1773

Duvat VK, Magnan AK, Wise RM, Hay JE, Fazey I, Hinkel J et al (2017) Trajectories of exposure and vulnerability of small islands to climate change. Wiley Interdisc Rev Clim Change 8(6):e478

Ehrlich D, Tenerelli P (2013) Optical satellite imagery for quantifying spatio-temporal dimension of physical exposure in disaster risk assessments. Nat Hazards 68(3):1271–1289

Einarsson P, Theodórsson P, Hjartardóttir ÁR, Guðjónsson GI (2008) Radon changes associated with the earthquake sequence in June 2000 in the South Iceland Seismic Zone. In: Pérez NM, Gurrieri S, King C-Y, Taran Y (eds) Terrestrial fluids, earthquakes and volcanoes: the Hiroshi Wakita, vol III. Birkhäuser Basel, Basel, pp 63–74

Elliot et al. (this issue) Measuring earthquake hazards with Earth Observation data. Surv Geophys (in review)

Ernst GGJ, Kervyn M, Teeuw RM (2008) Advances in remote sensing of volcanoes, their activity and hazards. Int J Remote Sens 29:6687–6723

Exploitation Platform., IEEE International Geoscience and Remote Sensing Symposium (IGARSS 2019), Yokohama, Japan

Felpeto A, Marti J, Ortiz R (2007) Automatic GIS-based system for volcanic hazard assessment. J Volcanol Geoth Res 166(2):106–116

Le Cozannet G, Raucoules D, de Michele M, Benaichouche A, Gehl P, Monfort D, Negulescu C, Rohmer J, Pierdicca N, Albano M, Giovinazzi S, Foumelis M, IEEE (2018) Potential of satellite remote sensing to monitor vulnerablity of buildings to earthquakes within a semi-empirical macroseismic approach. In: IGARSS 2018—2018 IEEE international geoscience and remote sensing symposium. IEEE international symposium on geoscience and remote sensing IGARSS, pp 2956–2959

Foumelis M, Papageorgiou E, Stamatopoulos C (2016) Episodic ground deformation signals in Thessaly Plain (Greece) revealed by data mining of SAR interferometry time series. Int J Remote Sens 37(16):3696–3711

Foumelis M, Papadopoulou T, Bally P, Pacini F, Provost P, Patruno J (2019) Monitoring geohazards using on-demand and systematic services on ESA's geohazards. In: IGARSS 2019 - 2019 IEEE international geoscience and remote sensing symposium, Yokohama, Japan, pp 5457–5460

Freund F (2013) Earthquake forewarning—a multidisciplinary challenge from the ground up to space. Acta Geophys 61:775–807. https://doi.org/10.2478/s11600-013-0130-4

Fuchs S, Heiss K, Huebl J (2007) Towards an empirical vulnerability function for use in debris flow risk assessment. Nat Hazards Earth Syst Sci 7(5):495–506

Gallina V, Torresan S, Critto A, Sperotto A, Glade T, Marcomini A (2016) A review of multi-risk methodologies for natural hazards: consequences and challenges for a climate change impact assessment. J Environ Manag 168:123–132

Galve JP, Perez-Pena JV, Azanon JM, Closson D, Calo F, Reyes-Carmona C, Jabaloy A, Ruano P, Mateos RM, Notti D, Herrera G, Bejar-Pizarro M, Monserrat O, Bally P (2017) Evaluation of the SBAS InSAR service of the European Space Agency's Geohazard exploitation platform (GEP). Remote Sens 9(12):1291

Garcin M, Desprats JF, Fontaine M, Pedreros R, Attanayake N, Fernando S, Siriwardana C, De Silva U, Poisson B (2008) Integrated approach for coastal hazards and risks in Sri Lanka. Nat Hazards Earth Syst Sci 8(3):577–586

Gehl P, Quinet C, Le Cozannet G, Kouokam E, Thierry P (2013) Potential and limitations of risk scenario tools in volcanic areas through an example at Mount Cameroon. Nat Hazards Earth Syst Sci 13(10):2409–2424

Geiss C, Taubenbock H (2013) Remote sensing contributing to assess earthquake risk: from a literature review towards a roadmap. Nat Hazards 68(1):7–48

Geiss C, Taubenbock H, Tyagunov S, Tisch A, Post J, Lakes T (2014) Assessment of seismic building vulnerability from space. Earthq Spectra 30(4):1553–1583

Genovese E, Przyluski V (2013) Storm surge disaster risk management: the Xynthia case study in France. J Risk Res 16(7):825–841

Gill JC, Malamud BD (2016) Hazard interactions and interaction networks (cascades) within multi-hazard methodologies. Earth Syst Dyn 7(3):659–679

Grunthal G, Thieken AH, Schwarz J, Radtke KS, Smolka A, Merz B (2006) Comparative risk assessments for the city of Cologne—Storms, floods, earthquakes. Nat Hazards 38(1–2):21–44

Guo HD (2010) Understanding global natural disasters and the role of earth observation. Int J Digital Earth 3(3):221–230

Haasnoot M, van 't Klooster S, van Alphen J (2018) Designing a monitoring system to detect signals to adapt to uncertain climate change. Glob Environ Change Hum Policy Dimens 52:273–285

Harrison RG, Aplin KL, Rycroft MJ (2010) Atmospheric electricity coupling between earthquake regions and the ionosphere. J Atmos Solar Terr Phys 72:376–381. https://doi.org/10.1016/j.jastp.2009.12.004

Harrison RG, Aplin KL, Rycroft MJ (2014) Brief communication: earthquake–cloud coupling through the global atmospheric electric circuit. Nat Hazards Earth Syst Sci 14:773–777

Hébert H, Gailler A, Gupta H, Monnier A, Lognonné P, Occhipinti G, Rolland L, Schindelé F (this issue) Contribution of space missions to a better tsunami science: observations, models and warning. Surv Geophys (in review)

Hinkel J, Church JA, Gregory JM, Lambert E, Le Cozannet G, Lowe J, McInnes KL, Nicholls RJ, van der Pol TD, van de Wal R (2019) Meeting user needs for sea level rise information: a decision analysis perspective. Earths Future 7(3):320–337

Hoque MAA, Phinn S, Roelfsema C, Childs I (2017) Tropical cyclone disaster management using remote sensing and spatial analysis: a review. Int J Disaster Risk Reduct 22:345–354

Hussain E, Hooper A, Wright TJ, Walters RJ, Bekaert DPS (2016) Interseismic strain accumulation across the central North Anatolian Fault from iteratively unwrapped InSAR measurements. J Geophys Res Solid Earth 121(12):9000–9019

IPCC (2018) Annex I: Glossary. In: Matthews JBR (ed) Global Warming of 1.5°C. An IPCC Special Report on the impacts of global warming of 1.5°C above pre-industrial levels and related global greenhouse gas emission pathways, in the context of strengthening the global response to the threat of climate change, sustainable development, and efforts to eradicate poverty

Jäggi A, Longuevergne L, Antoine R, Lopez T, Teatini P (this issue) Hydrological hazards. Surv Geophys (in review)

Jean N, Burke M, Xie M, Davis WM, Lobell DB, Ermon S (2016) Combining satellite imagery and machine learning to predict poverty. Science 353(6301):790–794

Jenkins SF, Spence RJS, Fonseca J, Solidum RU, Wilson TM (2014) Volcanic risk assessment: quantifying physical vulnerability in the built environment. J Volcanol Geoth Res 276:105–120

Jones RN, Patwardhan A, Cohen SJ, Dessai S, Lammel A, Lempert RJ, Mirza MMQ, von Storch H (2014) Foundations for decision making. In: Field CB, Barros VR, Dokken DJ, Mach KJ, Mastrandrea MD, Bilir TE, Chatterjee M, Ebi KL, Estrada YO, Genova RC, Girma B, Kissel ES, Levy AN, MacCracken S, Mastrandrea PR, White LL (eds) Climate change 2014: impacts, adaptation, and vulnerability. Part A: global and sectoral aspects. Contribution of working group II to the fifth assessment report of the intergovernmental panel on climate change. Cambridge University Press, Cambridge, pp 195–228

Jongman B, Kreibich H, Apel H, Barredo JI, Bates PD, Feyen L, Gericke A, Neal J, Aerts J, Ward PJ (2012) Comparative flood damage model assessment: towards a European approach. Nat Hazards Earth Syst Sci 12(12):3733–3752

Joyce KE, Belliss SE, Samsonov SV, McNeill SJ, Glassey PJ (2009) A review of the status of satellite remote sensing and image processing techniques for mapping natural hazards and disasters. Prog Phys Geogr 33(2):183–207

Jozdani SE, Johnson BA, Chen DM (2019) Comparing deep neural networks, ensemble classifiers, and support vector machine algorithms for object-based urban land use/land cover classification. Remote Sens 11(14):24

Kaku K (2019) Satellite remote sensing for disaster management support: a holistic and staged approach based on case studies in Sentinel Asia. Int J Disaster Risk Reduct 33:417–432

Kaku K, Aso N, Takiguchi F (2015) Space-based response to the 2011 Great East Japan Earthquake: lessons learnt from JAXA's support using earth observation satellites. Int J Disaster Risk Reduct 12:134–153

Kappes MS, Keiler M, von Elverfeldt K, Glade T (2012) Challenges of analyzing multi-hazard risk: a review. Nat Hazards 64(2):1925–1958

Kerle N, Froger JL, Oppenheimer C, De Vries BV (2003) Remote sensing of the 1998 mudflow at Casita volcano, Nicaragua. Int J Remote Sens 24(23):4791–4816

Kervyn M, Ernst GGJ, Goossens R, Jacobs P (2008) Mapping volcano topography with remote sensing: ASTER vs. SRTM. Int J Remote Sens 29(22):6515–6538

Kienberger S, Blaschke T, Zaidi RZ (2013) A framework for spatio-temporal scales and concepts from different disciplines: the 'vulnerability cube'. Nat Hazards 68(3):1343–1369

King C-Y (1986) Gas geochemistry applied to earthquake prediction: an overview. J Geophys Res Solid Earth 91:12269–12281. https://doi.org/10.1029/JB091iB12p12269

King C-Y, Chia Y (2018) Anomalous streamflow and groundwater-level changes before the 1999 M7.6 Chi-Chi Earthquake in Taiwan: possible mechanisms. Pure Appl Geophys 175:2435–2444. https://doi.org/10.1007/s00024-017-1737-1

Klomp J (2016) Economic development and natural disasters: a satellite data analysis. Global Environ Change Hum Policy Dimens 36:67–88

Kogan FN (1997) Global drought watch from space. Bull Am Meteor Soc 78(4):621–636

Koike T, Onoda M, Cripe D, Achache J (2010) The Global Earth Observation System of Systems (GEOSS): supporting the needs of decision making in societal benefit areas. Netw World Remote Sens 38:164–169

Koks EE, Rozenberg J, Zorn C, Tariverdi M, Vousdoukas M, Fraser SA, Hall JW, Hallegatte S (2019) A global multi-hazard risk analysis of road and railway infrastructure assets. Nat Commun 10(1):1–11

Kon S, Nishihashi M, Hattori K (2011) Ionospheric anomalies possibly associated with M≥6.0 earthquakes in the Japan area during 1998–2010: case studies and statistical study. J Asian Earth Sci 41:410–420. https://doi.org/10.1016/j.jseaes.2010.10.005

Lavell A, Oppenheimer M, Diop C, Hess J, Lempert R, Li J, Muir-Wood R, Myeong S (2012) Climate change: new dimensions in disaster risk, exposure, vulnerability, and resilience. In: Field CB, Barros V, Stocker TF, Qin D, Dokken DJ, Mastrandrea MD, Mach KJ, Plattner G-K, Allen SK, Tignor M, Midgley PM (eds) Managing the risks of extreme events and disasters to advance climate change adaptation. A special report of working groups I and II of the intergovernmental panel on climate change (IPCC). Cambridge University Press, Cambridge, pp 25–64

Le Cozannet G, Bagni M, Thierry P, Aragno C, Kouokam E (2014) WebGIS as boundary tools between scientific geoinformation and disaster risk reduction action in volcanic areas. Nat Hazards Earth Syst Sci 14(6):1591

Levin N, Ali S, Crandall D (2018) Utilizing remote sensing and big data to quantify conflict intensity: the Arab Spring as a case study. Appl Geography 94:1–17

Lissak C, De Michele M, Bartsch A, Roulland T, Maquaire O, Gomez C (this issue) Remote sensing for mass movement assessment. Surv Geophys (in review)

Lorenzo-Alonso A, Utanda A, Aullo-Maestro ME, Palacios M (2019) Earth observation actionable information supporting disaster risk reduction efforts in a sustainable development framework. Remote Sens 11(1):49

Ludy J, Kondolf GM (2012) Flood risk perception in lands "protected" by 100-year levees. Nat Hazards 61:829–842. https://doi.org/10.1007/s11069-011-0072-6

Lung T, Lübker T, Ngochoch J, Schaab G (2013) Human population distribution modelling at regional level using very high resolution satellite imagery. Appl Geogr 41:36–45

Ma T (2018) Multi-level relationships between satellite-derived nighttime lighting signals and social media-derived human population dynamics. Remote Sens 10(7):1128

Mahabir R, Croitoru A, Crooks A, Agouris P, Stefanidis A (2018) News coverage, digital activism, and geographical saliency: a case study of refugee camps and volunteered geographical information. PLoS ONE 13(11):e0206825

Manna AJ (1985) 25 years of Tiros satellites. Bull Am Meteor Soc 66(4):421–423

Marrero JM, Garcia A, Llinares A, Rodriguez-Losada JA, Ortiz R (2012) A direct approach to estimating the number of potential fatalities from an eruption: application to the Central Volcanic Complex of Tenerife Island. J Volcanol Geoth Res 219:33–40

Marzocchi W, Garcia-Aristizabal A, Gasparini P, Mastellone ML, Di Ruocco A (2012) Basic principles of multi-risk assessment: a case study in Italy. Nat Hazards 62(2):551–573

Meehl GA, Tebaldi C (2004) More intense, more frequent, and longer lasting heat waves in the 21st century. Science 305(5686):994–997

Melet A, Bartsch A, Benveniste J, Conversi A, Jamet C, Le Cozannet G, Teatini P (this issue) Earth Observations for monitoring marine coastal hazards and their drivers. Surv Geophys (in review)

Mentaschi L, Vousdoukas MI, Pekel J-F, Voukouvalas E, Feyen L (2018) Global long-term observations of coastal erosion and accretion. Sci Rep 8(1):12876

Mossoux S, Kervyn M, Soule H, Canters F (2018) Mapping population distribution from high resolution remotely sensed imagery in a data poor setting. Remote Sens 10(9):1409

Mubareka S, Ehrlich D (2010) Identifying and modelling environmental indicators for assessing population vulnerability to conflict using ground and satellite data. Ecol Ind 10(2):493–503

Muck M, Taubenbock H, Post J, Wegscheider S, Strunz G, Sumaryono S, Ismail FA (2013) Assessing building vulnerability to earthquake and tsunami hazard using remotely sensed data. Nat Hazards 68(1):97–114

Myint SW, Yuan M, Cerveny RS, Giri C (2008) Categorizing natural disaster damage assessment using satellite-based geospatial techniques. Nat Hazards Earth Syst Sci 8(4):707–719

Negulescu C, Ulrich T, Baills A, Seyedi DM (2014) Fragility curves for masonry structures submitted to permanent ground displacements and earthquakes. Nat Hazards 74(3):1461–1474

Neri M, Le Cozannet G, Thierry P, Bignami C, Ruch J (2013) A method for multi-hazard mapping in poorly known volcanic areas: an example from Kanlaon (Philippines). Nat Hazards Earth Syst Sci 13(8):1929–1943

Ong C, Carrère V, Chabrillat S et al (2019) Imaging spectroscopy for the detection, assessment and monitoring of natural and anthropogenic hazards. Surv Geophys 40:431–470. https://doi.org/10.1007/s1071 2-019-09523-1

Papageorgiou E, Foumelis M, Trasatti E, Ventura G, Raucoules D, Mouratidis A (2019) Multi-sensor SAR geodetic imaging and modelling of santorini volcano post-unrest response. Remote Sens 11(3):259

Papathoma-Kohle M, Keiler M, Totschnig R, Glade T (2012) Improvement of vulnerability curves using data from extreme events: debris flow event in South Tyrol. Nat Hazards 64(3):2083–2105

Parcharidis I, Kokkalas S, Fountoulis I, Foumelis M (2009) Detection and monitoring of active faults in urban environments: time series interferometry on the cities of patras and pyrgos (Peloponnese, Greece). Remote Sens 1(4):676–696

Paris R, Wassmer P, Sartohadi J, Lavigne F, Barthomeuf B, Desgages E, Grancher D, Baumert P, Vautier F, Brunstein D, Gomez C (2009) Tsunamis as geomorphic crises: lessons from the December 26, 2004 tsunami in Lhok Nga, West Banda Aceh (Sumatra, Indonesia). Geomorphology 104(1–2):59–72

Pettinari ML, Chuvieco E, Aguado I, Salas J (this issue) Fires hazard from space. Surv Geophys (in review)

Plag HP, Amelung F, Lengert W, Marsh SH, Meertens C (2010) Supporting risk management and disaster reduction: the geohazards community of practice and the supersite initiative. Netw World Remote Sens 38:192–197

Qin K, Wu LX, Ouyang XY et al (2014) Surface latent heat flux anomalies quasi-synchronous with ionospheric disturbances before the 2007 Pu'er earthquake in China. Adv Space Res 53:266–271. https://doi.org/10.1016/j.asr.2013.11.004

Raucoules D, Maisons C, Camec C, Le Mouelic S, King C, Hosford S (2003) Monitoring of slow ground deformation by ERS radar interferometry on the Vauvert salt mine (France) - Comparison with ground-based measurement. Remote Sens Environ 88(4):468–478

Raucoules D, Le Cozannet G, Woppelmann G, de Michele M, Gravelle M, Daag A, Marcos M (2013) High nonlinear urban ground motion in Manila (Philippines) from 1993 to 2010 observed by DInSAR: implications for sea-level measurement. Remote Sens Environ 139:386–397

Rigo A (2010) Precursors and fluid flows in the case of the 1996, ML = 5.2 Saint-Paul-de-Fenouillet earthquake (Pyrenees, France): a complete pre-, co- and post-seismic scenario. Tectonophysics 480:109–118. https://doi.org/10.1016/j.tecto.2009.09.027

Rohmer J, Douglas J, Bertil D, Monfort D, Sedan O (2014) Weighing the importance of model uncertainty against parameter uncertainty in earthquake loss assessments. Soil Dyn Earthq Eng 58:1–9

Romieu E, Welle T, Schneiderbauer S, Pelling M, Vinchon C (2010) Vulnerability assessment within climate change and natural hazard contexts: revealing gaps and synergies through coastal applications. Sustain Sci 5(2):159–170

Rossetto T, Elnashai A (2003) Derivation of vulnerability functions for European-type RC structures based on observational data. Eng Struct 25(10):1241–1263

Russo S, Sterl A (2011) Global changes in indices describing moderate temperature extremes from the daily output of a climate model. J Geophys Res Atmos 116:D3

Russo S, Dosio A, Sterl A, Barbosa P, Vogt J (2013) Projection of occurrence of extreme dry-wet years and seasons in Europe with stationary and nonstationary standardized precipitation indices. J Geophys Res Atmos 118(14):7628–7639

Russo S, Dosio A, Graversen RG, Sillmann J, Carrao H, Dunbar MB, Singleton A, Montagna P, Barbola P, Vogt JV (2014) Magnitude of extreme heat waves in present climate and their projection in a warming world. J Geophys Res Atmos 119(22):12500–12512

Salichon J, Le Cozannet G, Modaressi H, Hosford S, Missotten R, McManus K, Marsh S, Paganini M, Ishida C, Plag HP, Labrecque J, Dobson C, Quick J, Giardini D, Takara K, Fukuoka H, Casagli N, Marzocchi W (2007) 2nd IGOS Geohazards Theme report, BRGM

Sedan O, Negulescu C, Terrier M, Roulle A, Winter T, Bertil D (2013) Armagedom—a tool for seismic risk assessment illustrated with applications. J Earthq Eng 17(2):253–281

Smets B, d'Oreye N, Kervyn F, Kervyn M, Albino F, Arellano SR, Bagalwa M, Balagizi C, Carn SA, Darrah TH, Fernandez J, Galle B, Gonzalez PJ, Head E, Karume K, Kavotha D, Lukaya F, Mashagiro N, Mavonga G, Norman P, Osodundu E, Pallero JLG, Prieto JF, Samsonov S, Syauswa M, Tedesco D, Tiampo K, Wauthier C, Yalire MM (2014) Detailed multidisciplinary monitoring reveals pre- and co-eruptive signals at Nyamulagira volcano (North Kivu, Democratic Republic of Congo). Bull Volcanol 76(1):787

Smolka A, Siebert A (2013) Remote sensing and earthquake risk: a (re) insurance perspective. Nat Hazards 68(1):211–212

Spence RJS, Kelman I, Baxter PJ, Zuccaro G, Petrazzuoli S (2005) Residential building and occupant vulnerability to tephra fall. Nat Hazards Earth Syst Sci 5(4):477–494

Stammer D, van de Wal RSW, Nicholls RJ, Church JA, Le Cozannet G, Lowe JA, Horton BP, White K, Behar D, Hinkel J (2019) Framework for high-end estimates of sea-level rise for stakeholder applications. Earths Future, 7:923–938

Steen M, Kuijt-Evers L, Klok J (2007) Early user involvement in research and design projects—a review of methods and practices. In: 23rd EGOS colloquium, pp 1–21

Stephens SA, Bell RG, Lawrence J (2018) Developing signals to trigger adaptation to sea-level rise. Environ Res Lett 13:104004

Surono, Jousset P, Pallister J, Boichu M, Buongiorno MF, Budisantoso A, Costa F, Andreastuti S, Prata F, Schneider D, Clarisse L, Humaida H, Sumarti S, Bignami C, Griswold J, Carn S, Oppenheimer C, Lavigne F (2012) The 2010 explosive eruption of Java's Merapi volcano-A '100-year' event. J Volcanol Geoth Res 241:121–135

Tan M, Li X, Li S, Xin L, Wang X, Li Q, Li W, Li Y, Xiang W (2018) Modeling population density based on nighttime light images and land use data in China. Appl Geogr 90:239–247

Taubenbock H, Post J, Roth A, Zosseder K, Strunz G, Dech S (2008) A conceptual vulnerability and risk framework as outline to identify capabilities of remote sensing. Nat Hazards Earth Syst Sci 8(3):409–420

te Brake B, Hanssen RF, van der Ploeg MJ, de Rooij GH (2013) Satellite-based radar interferometry to estimate large-scale soil water depletion from clay shrinkage: possibilities and limitations. Vadose Zone J. https://doi.org/10.2136/vzj2012.0098

Tellez-Arenas A, Quique R, Boulahya F, Le Cozannet G, Paris F, Le Roy S, Dupros F, Robida F (2018) Scalable interactive platform for geographic evaluation of sea-level rise impact combining high-performance computing and WebGIS Client. In: Serrao-Neumann S, Coudrain A, Coulter L (eds) Communicating climate change information for decision-making. Springer, Cham, pp 163–175

Thierry P, Stieltjes L, Kouokam E, Ngueya P, Salley PM (2008) Multi-hazard risk mapping and assessment on an active volcano: the GRINP project at Mount Cameroon. Nat Hazards 45(3):429–456

Tote C, Patricio D, Boogaard H, van der Wijngaart R, Tarnavsky E, Funk C (2015) Evaluation of satellite rainfall estimates for drought and flood monitoring in Mozambique. Remote Sens 7(2):1758–1776

Tralli DM, Blom RG, Zlotnicki V, Donnellan A, Evans DL (2005) Satellite remote sensing of earthquake, volcano, flood, landslide and coastal inundation hazards. ISPRS J Photogramm Remote Sens 59(4):185–198

Tramutoli V, Aliano C, Corrado R et al (2013) On the possible origin of thermal infrared radiation (TIR) anomalies in earthquake-prone areas observed using robust satellite techniques (RST). Chem Geol 339:157–168. https://doi.org/10.1016/j.chemgeo.2012.10.042

Tronin AA (2000) Thermal IR satellite sensor data application for earthquake research in China. Int J Remote Sens 21:3169–3177. https://doi.org/10.1080/01431160050145054

Tronin AA (2010) Satellite remote sensing in seismology. A Review. Remote Sens 2:124–150. https://doi.org/10.3390/rs2010124

United Nations (2005) Hyogo framework for action 2005–2015: building the resilience of nations and communities to disasters, world conference on disaster reduction, 18–22 January 2005, Kobe, Hyogo. https://www.unisdr.org/files/1037_hyogoframeworkforactionenglish.pdf. Accessed 1 Mar 2020

United Nations (2015) Sendai framework for disaster risk reduction 2015–2030, third United Nations world conference on disaster risk reduction, 14–18 March 2015, Sendai. https://www.preventionweb.net/files/43291_sendaiframeworkfordrren.pdf. Accessed 1 Mar 2020

Van Westen CJ (2013) Remote sensing and GIS for natural hazards assessment and disaster risk management. In: Schroder JF, Bishop MP (eds) Treatise in geomorphology. Academic Press, San Diego

Vaughan RG, Kervyn M, Realmuto V, Abrams M, Hook SJ (2008) Satellite measurements of recent volcanic activity at Oldoinyo Lengai, Tanzania. J Volcanol Geoth Res 173(3–4):196–206

Voigt S, Kemper T, Riedlinger T, Kiefl R, Scholte K, Mehl H (2007) Satellite image analysis for disaster and crisis-management support. IEEE Trans Geosci Remote Sens 45(6):1520–1528

Voigt S, Giulio-Tonolo F, Lyons J, Kucera J, Jones B, Schneiderhan T, Platzeck G, Kaku K, Hazarika MK, Czaran L, Li SJ, Pedersen W, James GK, Proy C, Muthike DM, Bequignon J, Guha-Sapir D (2016) Global trends in satellite-based emergency mapping. Science 353(6296):247–252

Vousdoukas MI, Mentaschi L, Voukouvalas E, Verlaan M, Jevrejeva S, Jackson LP, Feyen L (2018) Global probabilistic projections of extreme sea levels show intensification of coastal flood hazard. Nat Commun 9:2360

Wang XW, Xie HJ (2018) A review on applications of remote sensing and Geographic Information Systems (GIS) in water resources and flood risk management. Water 10(5):608

Wilkinson MD, Dumontier M, Aalbersberg IJ, Appleton G, Axton M, Baak A, Blomberg N, Boiten JW, da Silva Santos LB, Bourne PE, Bouwman J (2016) The FAIR Guiding Principles for scientific data management and stewardship. Sci Data 3:160018

Wu S, Qiu X, Wang L (2005) Population estimation methods in GIS and remote sensing: a review. GISci Remote Sens 42:80–96

Yu SS, Zhang ZX, Liu F (2018) Monitoring population evolution in china using time-series DMSP/OLS nightlight imagery. Remote Sens 10(2):194

Zakharenkova IE, Shagimuratov II, Tepenitzina NYu, Krankowski A (2008) Anomalous modification of the ionospheric total electron content prior to the 26 September 2005 Peru earthquake. J Atmos Solar Terr Phys 70:1919–1928. https://doi.org/10.1016/j.jastp.2008.06.003

Zhang X, Shen X, Zhao S et al (2014) The characteristics of quasistatic electric field perturbations observed by DEMETER satellite before large earthquakes. J Asian Earth Sci 79:42–52. https://doi.org/10.1016/j.jseaes.2013.08.026

Zubareva ON, Skripal'shchikova LN, Greshilova NV, Kharuk VI (2003) Zoning of landscapes exposed to technogenic emissions from the Norilsk Mining and Smelting Works. Russ J Ecol 34(6):375–380

Zuccaro G, Cacace F, Spence RJS, Baxter PJ (2008) Impact of explosive eruption scenarios at Vesuvius. J Volcanol Geoth Res 178(3):416–453

Publisher's Note Springer Nature remains neutral with regard to jurisdictional claims in published maps and institutional affiliations.

Affiliations

G. Le Cozannet[1] · M. Kervyn[2] · S. Russo[3] · C. Ifejika Speranza[4] · P. Ferrier[5] · M. Foumelis[1] · T. Lopez[6,7] · H. Modaressi[1,8]

[1] BRGM, 3 Avenue Claude Guillemin, 45060 Orléans Cedex, France

[2] Department of Geography, Vrije Universiteit Brussel, Pleinlaan 2, 1050 Brussels, Belgium

[3] Institute for Environmental Protection and Research (ISPRA), Rome, Italy

[4] Institute of Geography, University Bern, Bern, Switzerland

[5] CNES, Toulouse, France

[6] Institut de Recherche Technologique (IRT) Saint-Exupéry, Géoscience Environnement Toulouse (GET), 14 Avenue Edouard Belin, 31400 Toulouse, France

[7] International Space Science Institute (ISSI), Hallerstrasse 6, 3012 Bern, Switzerland

[8] MOD@A, 231 rue Saint-Honoré, 75001 Paris, France

Surveys in Geophysics (2020) 41:1237–1284
https://doi.org/10.1007/s10712-020-09604-6

Early Warning from Space for a Few Key Tipping Points in Physical, Biological, and Social-Ecological Systems

Didier Swingedouw[1] · Chinwe Ifejika Speranza[2] · Annett Bartsch[3,4] · Gael Durand[5] · Cedric Jamet[6] · Gregory Beaugrand[6] · Alessandra Conversi[7]

Received: 5 December 2019 / Accepted: 18 July 2020 / Published online: 3 September 2020
© Springer Nature B.V. 2020

Abstract

In this review paper, we explore latest results concerning a few key tipping elements of the Earth system in the ocean, cryosphere, and land realms, namely the Atlantic overturning circulation and the subpolar gyre system, the marine ecosystems, the permafrost, the Greenland and Antarctic ice sheets, and in terrestrial resource use systems. All these different tipping elements share common characteristics related to their nonlinear nature. They can also interact with each other leading to synergies that can lead to cascading tipping points. Even if the probability of each tipping event is low, they can happen relatively rapidly, involve multiple variables, and have large societal impacts. Therefore, adaptation measures and management in general should extend their focus beyond slow and continuous changes, into abrupt, nonlinear, possibly cascading, high impact phenomena. Remote sensing observations are found to be decisive in the understanding and determination of early warning signals of many tipping elements. Nevertheless, considerable research still remains to properly incorporate these data in the current generation of coupled Earth system models. This is a key prerequisite to correctly develop robust decadal prediction systems that may help to assess the risk of crossing thresholds potentially crucial for society. The prediction of tipping points remains difficult, notably due to stochastic resonance, i.e. the interaction between natural variability and anthropogenic forcing, asking for large ensembles of predictions to correctly assess the risks. Furthermore, evaluating the proximity to crucial thresholds using process-based understanding of each system remains a key aspect to be developed for an improved assessment of such risks. This paper finally proposes a few research avenues concerning the use of remote sensing data and the need for combining different sources of data, and having long and precise-enough time series of the key variables needed to monitor Earth system tipping elements.

Keywords Tipping point · Tipping element · Remote sensing · Earth observation · Atlantic · AMOC · SPG · Marine biology · Permafrost · Antarctic and Greenland ice sheets · Land use · Terrestrial resource use · Early warning · Bifurcation · Climate dynamics

---

✉ Didier Swingedouw
  didier.swingedouw@u-bordeaux.fr

Extended author information available on the last page of the article

1 Introduction

The analysis of different systems in biology, physics, chemistry, and environmental sciences has highlighted for more than a century that sometimes a small perturbation of some of these systems can lead to large changes. This is a key property of nonlinear systems for which the response to a perturbation is not proportional to it. The system does not necessarily return to the initial state even if the perturbation leading to the change is removed or its direction reversed. Indeed, this type of bifurcation can lead to irreversible change (where irreversible means that the recovery time scale from this state is substantially longer than the time it takes for the system to reach this perturbed state, cf. Masson-Delmotte et al. 2018).

Milkoreit et al. (2018: p. 9) define tipping point as "the point or threshold at which small quantitative changes in the system trigger a nonlinear change process that is driven by system-internal feedback mechanisms and inevitably leads to a qualitatively different state of the system, which is often irreversible. This new state can be distinguished from the original by its fundamentally altered (positive and negative) state-stabilizing feedbacks". The possible irreversible characteristic of this change of state is related to hysteresis behaviour, meaning that a return to the same value of the driving parameter does not necessarily lead to a return of the system to its former state. This involves the existence of multiple steady states for the same driving parameter values, which is at the origin of the potential irreversibility of the changes following a tipping point. This behaviour is related to positive feedback loops by which a perturbation is amplified until the system reaches another steady state, in which the system is kept by novel feedback mechanisms. The time required for the transition depends on the system considered and its inertia properties and can vary from about days to millennia. The abruptness of the transition thus depends on the system, but is classically defined by being faster than the forcing that leads to the transition.

Dynamical system theory from the mathematical field has analysed in detail the attributes of this type of systems and highlighted that the systems that are nonlinear are not necessarily complex and can be driven by relatively simple differential equations (e.g., Scheffer et al. 2009). A key aspect of these systems is constituted by the existence of thresholds in some driving parameters after which, when they are crossed, the systems can totally change in their state and nature. This threshold is also called bifurcation in the mathematical field, a word first introduced by Poincaré (1885) in a seminal paper showing such a behaviour in some mathematical objects.

The Earth system is constituted of several subcontinental-scale subsystems, crucial for climate mitigation, which are suspected of being tipping elements (Lenton et al. 2008), for example the Amazon rainforest, where crossing a tipping point would not only mean the Amazon rainforest turning locally into dry savanna, but modifying the global rain patterns, and possibly becoming a source of CO_2 as opposed to being a sink (Lovejoy and Nobre 2018).

Tipping elements are Earth subsystems at least subcontinental in scale, which can be switched into a qualitatively different state by relatively small perturbations (Lenton et al. 2008, 2019; Milkoreit et al. 2018), i.e. they contain a *tipping point*, or *critical threshold*, past which a bifurcation in the system leads to a large reorganization (Good et al. 2018).

This concept implies a radically different view on the potential effects of global change on such systems, and on associated costs and management options (Lemoine and Traeger 2014; Lontzek et al. 2015). Global warming is considered a key driver towards reaching tipping points in multiple ecological and physical systems as well as in subcontinental-scale subsystems. Furthermore, there is growing evidence that these tipping elements are not isolated, and a tipping point in a subsystem can have cascading effects on the others, carrying huge impacts on human societies (Cai et al. 2016; Lenton et al. 2019). The risk of crossing thresholds in these subsystems was therefore a crucial argument to try to keep global warming in reasonable amplitude as initiated by the target of the Paris Agreement in 2016, and highlighted at about 1.5 °C in the recent special report from the Intergovernmental Panel on Climate Change (Hoegh-Guldberg et al. 2018).

A few examples of known key tipping elements will be depicted in the present paper and are represented in Fig. 1. They are covering the different realms of ocean, land, and cryosphere. These examples were chosen because of their potential high impact on societies and include ocean circulation structures like the Atlantic Meridional Overturning Circulation (AMOC) and the Subpolar Gyre (SPG) systems, the permafrost in the boreal regions, marine ecosystems, the Antarctic and Greenland ice sheets, and terrestrial resource use systems. We use the term terrestrial resource use systems to refer to the coupled human–nature interactions in the human use of land as a spatial resource, soil, water and plant, and animal biodiversity resources.

Ice cores analysis from Greenland has suggested the existence of abrupt climate changes in the past, happening in less than a decade (Dansgaard et al. 1993; Steffensen et al. 2008). The exact mechanisms causing these rapid events are still the subject of intense research, and a large number of potential processes have already been proposed (e.g., Clement and Peterson 2008). While ocean circulation changes certainly played a role, robust palaeoclimate data are still not sufficient to provide a definite view on the exact suite of events that leads to these rapid variations. Other subsystems of the climate system have also potentially changed state relatively rapidly in the past. An iconic example is the Green Sahara, showing that the south part of the Sahara was covered by vegetation 6000 years ago, but changing climate transformed it in just about 3 centuries

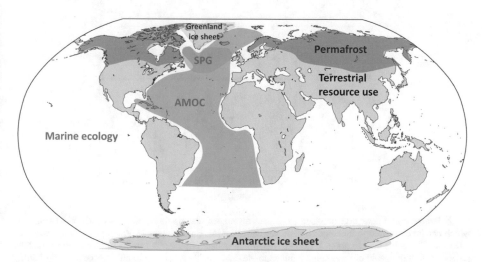

Fig. 1 Schematic of the tipping elements assessed in the paper. The systems found in the ocean realm are written in green, the ones from the cryosphere in dark blue and the one from the land in black. "Terrestrial resource use" includes land use systems and land cover

(Demenocal et al. 2000) into the desert that we know now, among the driest land on Earth (Claussen 1997; Hoelzmann et al. 1998).

While rapid transition and bifurcation behaviours have been found in a number of simplified models of tipping elements (Bathiany et al. 2016), more comprehensive models do not always exhibit such instabilities, highlighting that these models may be able to incorporate some stabilizing feedbacks. Drijfhout et al. (2015) led a systematic analysis of rapid transition of a number of variables in the comprehensive climate model database CMIP5 at a regional scale and found a number of rapid transitions (a few decades) in some key tipping elements. These include potential convection collapse in the North Atlantic and Southern Ocean, Amazon forest dieback, Arctic permafrost thawing, and Eastern Sahel greening. This study highlighted that some complex climate models can identify nonlinear behaviour in tipping elements, which may be key to forecasts in a changing climate.

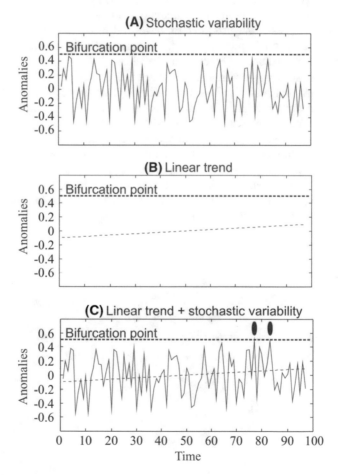

Fig. 2 Example of interaction between stochastic resonance and tipping points. **a** Stochastic variability, which can be compared to inherent climatic variability in this fictive example. Alone, this source of variability cannot reach a tipping point. **b** Long-term (low-frequency) linear trend in the same climatic parameter related to anthropogenic climate change (an example could be the warming trend). **c** Stochastic variability and trend. Red ovals indicate that the tipping (or bifurcation) point is reached. Stochasticity was modelled here by adding a Gaussian random variable of small intensity, following the method of Benzi (2010). From Beaugrand (2015)

Nevertheless, these simulations do not yet provide enough information on the time scale or the irreversibility of the forecasted events.

Stochastic resonance is an important mechanism by which anthropogenic climate change may interact with climatic variability to trigger acute species mortality at the species level or regime shift at the ecosystem level (Fig. 2). In the fictive example of Fig. 2, neither the stochastic (year-to-year) variability nor the long-term trend in climatic forcing can reach a tipping point on the given time window of 100 years (Fig. 2a, b). However, when both signals are combined, a critical threshold may sometimes be reached and trigger a major phase transition (Fig. 2c). It is extremely difficult to forecast the time at which a tipping point is crossed. The best we can, perhaps, do is to evaluate the probability to reach this point, as the bifurcation point itself is rarely known with accuracy. Once the tipping point has been crossed, the system can go in quite surprising directions and changes are often difficult to reverse. There are many ways to investigate stochastic resonance (Wellens et al. 2004). Here, we used in Fig. 2 a simple model by adding to the trend a Gaussian random variable (Benzi 2010).

In order to anticipate the occurrence of an abrupt change in a given system, detecting early warnings has become an intense field of research. Based on the findings that many systems modify their variability before shifting to another state (e.g., Scheffer et al. 2009; Beaulieu et al. 2012), this field analyses the statistical properties of key variables in various systems. It has been shown that rather different systems, at the proximity of a tipping point (e.g., lakes turning turbid, rise in an epidemic), exhibit common signals, for example increase in the variance and autocorrelation of some key variables (Bathiany et al. 2016; Dakos et al. 2015; Lenton 2011; Scheffer et al. 2009). These phenomena do not occur in all systems, and multidecadal time series are necessary to correctly assess the natural variability of the system and its changes near a tipping point. Despite these limitations, early warning is a very promising field, which can become a major element of disaster risk reduction (Melet et al. 2020) and deserves more research to become entirely applicable.

Remote sensing and Earth observations (EO) may strongly contribute to the knowledge about the different tipping elements of a system. Although the length of observations is currently limited, satellites are providing unprecedented knowledge about the dynamics and processes that govern the different tipping elements depicted in this review. These different pieces of knowledge, the utility they may have for early warning of any kind, and their caveats will be shown for the different case studies analysed in this paper and represented in Fig. 1.

In this paper, we address tipping points from an interdisciplinary perspective. We provide a review of a few key tipping points in physical and biological systems and in a societal terrestrial context, and assess the research avenues that need to be followed to improve our understanding of these elements and the way to set up early warning signals. We focus on the use of satellite observations for this purpose: how they can contribute to providing early warning signals, and how they can help to better understand the different systems analysed here.

2 Oceanic Tipping Elements

2.1 Ocean Circulation

2.1.1 AMOC Stability

The AMOC has become an iconic tipping element of the climate system. The first guess that this circulation can tip in response to changes in the atmospheric hydrological cycle was highlighted as early as 1961 by Henry Stommel in an influential paper (Stommel 1961). In this paper, Stommel showed—using a two-box model of the ocean circulation, one box for the equatorial area and the other for the northern polar region, connected by a current proportional to the density gradient between the two boxes—that the AMOC response to surface freshwater fluxes (FW) is highly nonlinear. A hysteresis diagram drives its steady-state variations with the possibility, for a range of FW forcing, of two AMOC stable states. He also found a bifurcation for a critical value of the FW forcing where the AMOC system jumps from one state to another. This result was then found in more sophisticated ocean models (e.g., Rahmstorf 1996) and is still true in three-dimensional coupled ocean–atmosphere climate model at the eddy-permitting resolution in the ocean (e.g., Mecking et al. 2016).

In parallel to these theoretical investigations, palaeoclimate data reveal that the North Atlantic climate was very unstable in the last glacial period (Dansgaard et al. 1993), as well as the Holocene (Renssen et al. 2001) and the last interglacial (Galaasen et al. 2014), i.e. for warm periods. This sub-millennial variability is difficult to reconcile with very slow changes in insolation and have been therefore associated with rapid changes in the ocean circulation and the instability of the AMOC (Rahmstorf 2002).

Worldwide climate impacts associated with AMOC variability are also deciphered from palaeodata (Collins et al. 2019). When the AMOC weakens, the associated decrease in northward heat transport strongly cools the North Atlantic regions by up to 10 °C (Jackson et al. 2015). The hydrological cycle is also perturbed with a southward migration of the intertropical convergence zone that strongly modifies the precipitation pattern throughout the world including the monsoons in southeast Asia (Monerie et al. 2019). Numerous other climatic impacts have been reported in response to a weakening of the AMOC, including a strengthening of the storm track position and intensity in the North Atlantic (Gastineau et al. 2016) or an increase in sea-level rise in the North Atlantic (Ezer 2015), among many others.

The main processes that explain the nonlinearity of the AMOC are related to the salt advection feedback. Indeed, when the AMOC weakens, it conveys less salty waters from the tropical area towards the North Atlantic, decreasing the sea surface salinity (SSS) there and further limiting oceanic convection that feeds the strength of the AMOC. This positive feedback is key for the AMOC's nonlinear behaviour and its stability (Weijer et al. 2019). It has been argued (Rahmstorf 1996) that the main driver of this stability is related to the salt transport by the overturning circulation inside the Atlantic Ocean at around 34°S. If the overturning transports freshwater, the feedback is negative and the AMOC is in a monostable state, if it transports saline water, then the feedback is positive and the AMOC is in a bi-stable state. Observations seem to indicate that present-day overturning transports salt at 34°S and thus the AMOC may be in a bi-stable state (Deshayes et al. 2013). Most of the CMIP5 models do not properly represent this transport (Weaver et al. 2012); hence, they may be too stable. Nevertheless, it should be highlighted here that the stability of the AMOC is a concept suited for steady-state analysis, which usually means hundreds of years of transient response for the ocean (Sgubin et al. 2015). Hence, if the AMOC is close to a bifurcation point, it can shift from one state to another for a small perturbation, but the time scale of changes may still be relatively long, of the order of a century.

2.1.2 On the Possibility of Monitoring of Early Warning

The observations of the salinity transport both at 34°S and elsewhere in the Atlantic remain crucial to estimate ongoing changes in the AMOC. There exist a few in situ hydrographic arrays that allow such a monitoring: at 16°N (Send et al. 2011), 26°N (McCarthy et al. 2015), in the North Atlantic subpolar gyre (Lozier et al. 2019), between Portugal and the tip of Greenland (Mercier et al. 2015), and at 34.5°S (Meinen et al. 2018), among others. They allow regular evaluation of the strength of the AMOC as well as its characteristics and evolution.

While these observations are crucial to indicate some early warnings of strong weakening of the AMOC, which may precede changes in sea surface temperature (SST) by a few years (Knight et al. 2006), the analysis of time series variations may not allow per se to an estimate of whether the system is getting close to a bifurcation. Indeed, an analysis from Boulton et al. (2014) within a climate model showed that, to detect in advance the approach of a bifurcation in the AMOC, it will be necessary to decipher its statistical characteristic, hence to have hundreds of years of knowledge of AMOC variations. Since the AMOC has been directly measured continuously only since 2004, we are far from being able to use time series analysis to have any insights on the proximity of a bifurcation. Thus, the detection of early warning for an AMOC collapse is beyond our capability and knowledge at the moment, and we cannot state if we might have already crossed the threshold nowadays. Still, measurements of the AMOC are crucial to monitor its evolution and possible ongoing changes, which may allow us to slightly anticipate any rapid changes and also possibly help to detect early warning. The use of palaeodata would be helpful and necessary to reach this goal. Monitoring of the Atlantic remains crucial for knowing lower amplitude variations that are very useful to improve the initial conditions of decadal prediction system (e.g., Keenlyside et al. 2008; Swingedouw et al. 2013), which might help to anticipate abrupt transitions.

These hydrographic arrays also serve as references to evaluate the capacity of remote sensing to estimate oceanic variations through the use of key fingerprints of the AMOC variations. For instance, Frajka-Williams (2015) used the RAPID array at 26°N to evaluate the capacity of remote sensing altimetry using AVISO (Archiving, Validating and Interpretation of Satellite Oceanographic) data to capture changes in the overturning. They found a good correlation between AMOC variations measured at 26°N and changes in sea level at 30°N–70°W. They use this location to propose a reconstruction of the AMOC further back in time since the beginning of remote sensing altimetry in 1993. Mercier et al. (2015) also used altimetry (and in situ ocean velocity measurements) to propose a continuous reconstruction of the overturning along the OVIDE array that are measured in situ every two years in spring since 2002. Finally, Landerer et al. (2015) proposed a very innovative approach to evaluate the variations in ocean density over the whole water column and its link with circulation through geostrophy. They used the GRACE (Gravity Recovery And Climate Experiment) remote sensing products to infer the density variations in the deep ocean and then succeeded in reconstructing most of the observed variations from the RAPID array. These methods allow the estimation of the density variations with a wider spatial sampling than just arrays (although with coarse grid resolution of a few hundreds of kilometres), which offers an unprecedented view of the adjustment of the deep ocean density to changes in the circulation. This new type of measurements of the deep ocean, which is currently poorly sampled by ARGOS floats, may offer some avenues to better understand the deep ocean variations that are still controversial (e.g., Lozier et al. 2019). It may indeed permit to understand the signal propagation between different arrays, and notably between the new OSNAP (Overturning in the Subpolar North Atlantic Program) array in the subpolar (Lozier et al. 2019) and the RAPID one at 26°N for instance.

Using SST as a proxy of the real AMOC, Caesar et al. (2018) suggested that the AMOC may have already weakened by around 15% over the last six decades, suggesting that this system may be already changing. Recent projections of the AMOC changes in the coming centuries including Greenland ice sheet melting show a weakening of the AMOC of 37% [15%, 65%; 90% probability] in 2100 in RCP8.5 scenario and of 18% [3%, 34%] in RCP4.5 as compared to year 2006 (Bakker et al. 2016). On the longer term (2300), there is a 44% likelihood of an AMOC collapse in RCP8.5 scenario, while it resumes in RCP4.5; RCP4.5 and RCP8.5 are two representative concentration pathways, i.e. two greenhouse gas concentration trajectories adopted by the Intergovernmental Panel for Climate Change (IPCC). This highlights the strong benefit of greenhouse gases mitigation concerning the possibility of crossing a threshold of the AMOC. Furthermore, present-day climate models are suspected to be far too stable for the AMOC (e.g., Liu et al. 2017), so that this estimate is believed to be too conservative.

2.1.3 The Subpolar Gyre as a Faster and Closer Tipping Element

A subsystem of the AMOC is constituted by the subpolar gyre (SPG) system. This circulation is found at latitudes just below the tip of Greenland. It is a cyclonic gyre that turns around two intermittent convection sites in the Labrador Sea and the Irminger Sea. It has been shown that this system may be a tipping element of the climate system, involving different processes than the large-scale AMOC. While the salt-advection feedback also plays a key role, it acts on more local spatial scale (Born et al. 2016) so that the SPG state change can occur on a faster time scale than the AMOC (about a decade compared to a century, e.g., Sgubin et al. 2017). In the "on" state, the convection in the Labrador and Irminger seas is active, leading to strong density in the middle of the gyre that intensifies the circulation and the arrival of warm and salty water from the subtropical gyre. When a critical threshold is reached for the stratification (due to increase in SST or decrease in SSS), the convection is not permitted anymore and the gradient of density between the centre and the periphery of the gyre decreases. This reduces the flow of the gyre and the import of warm and salty water from the subtropical gyre, which further decreases convection, i.e. a positive feedback. The climatic impact of such a collapse of convection mainly affects SST and air temperature in the subpolar gyre area. The transition occurs in less than 10 years and can induce a cooling of 2–3 °C of SST in the gyre vicinity, affecting the rate of warming in the neighbouring regions (Sgubin et al. 2017, 2019). It has also been proposed that such a cold blob in the SPG can modify the atmospheric circulation, leading to more heat waves in summer over Europe (Duchez et al. 2016).

Using some CMIP5 climate models that do show such an abrupt (< 10 years) shift (here, 4 models, with the abrupt collapse of the SPG occurring at different years within the 21st century), we can estimate the critical stratification in the subpolar gyre just preceding the collapse of convection that would shift the system in an "off" state. This critical stratification is presented in Fig. 3, together with the present-day stratification and the stratification

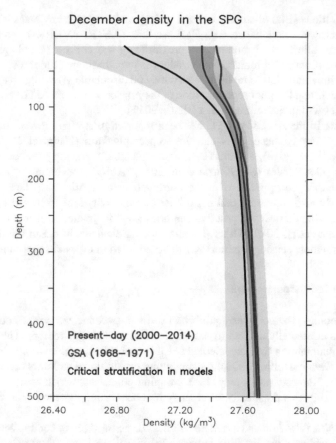

Fig. 3 Critical stratification in the subpolar gyre just preceding the collapse of convection. Density in the subpolar gyre (62–27°W, 45:70°W) in 4 models from CMIP5 (in black) showing a collapse of the convection (CESM1-CAM5, GFDL-ESM2G, GISS-E2-R, MIROC5). The density profile is shown as an average for the three months of December just before the year of the convection collapse in the different models (which occur in different years, depending on the model). The overlap stands for the one standard deviation of the 4 models. The present-day density profile (in red) is shown from EN4 data, an objective analysis of most of the observational datasets available (Good et al. 2013), averaged over the December month between 2000 and 2014, with the envelope representing the associated standard deviation over this time period. Finally, the density profile of the Great Salinity Anomaly (GSA) occurring between about 1968 to 1971 in the subpolar gyre is shown in blue, once again as an average for the month of December and associated standard deviation as an envelope. A 20-m running mean has been applied to all profile to improve readability

during the Great Salinity Anomaly in the late 1960s (Belkin et al. 1998). From this figure, we can see that the present-day stratification in the Labrador Sea seems relatively far from the collapse threshold, which will require a further decrease in surface density of about 0.5 kg m^{-3} compared to the ensemble mean of the 4 models. Nevertheless, the error bar (defined here as one standard deviation) of models and of present-day observations overlaps, and even the average of observed density at the surface is very close to the models' uncertainty. From this, we can estimate that the SPG may be relatively close to the stratification threshold exhibited by some climate models, although the Labrador Sea is still showing vigorous convection in the recent years (Yashayaev and Loder 2017). Thus,

Earth observations (EO) of surface density, which can be achieved by measuring both SST (through microwave radiometer, e.g., GMI satellite, SLSTR, AVHRRs) and SSS (SMOS satellite), may allow us to monitor the proximity of the convection-collapse threshold. However, this proposition remains valid under the hypothesis that the threshold quantified by CMIP5 climate models is realistic, which may be disputable given the coarse resolution of these models and the poor representation of the impact of oceanic eddies on re-stratification processes within the Labrador Sea (Heuze 2017).

Weakening in the AMOC or SPG has been reported to strongly impact the ocean biogeochemistry. For instance, SPG changes impact plankton (Hatun et al. 2016), fishes (Miesner and Payne 2018), seabirds (Descamps et al. 2013), and top predators such as tuna, billfish, and pilot whales (MacKenzie et al. 2014). AMOC weakening is also suspected to explain marine deoxygenation in the northwest coastal Atlantic (Claret et al. 2018). A recent study also suggested that net primary productivity has decreased by $10 \pm 7\%$ in the subarctic Atlantic over the past two centuries possibly related to recent changes in the AMOC (Osman et al. 2019). Other potential tipping elements in ocean biology and their linkage with remote sensing monitoring will be explored in the next subsection.

2.2 Marine Ecosystems

In marine ecology, the complex equilibrium of the interconnected species composing the ecosystem is continually subject to natural and anthropogenic stressors. The ecosystem's capacity of maintaining the characteristics within its original regime as conditions change is called *resilience* (Holling 1973). As continuing or additional stressors erode the ecosystem resilience, the ecosystem may reach a tipping point, past which even a smooth additional stress triggers a drastically different system behaviour, a phenomenon called regime shift.

Regime shifts (also known as *critical transitions, phase shifts* in benthic ecology, *abrupt shifts* when feedback mechanisms are not identified), are large, abrupt (less than a few years in many ecosystems), persistent changes in the structure and function of a system, corresponding to a profound restructuration of the trophic web into a novel, often alternative state (Biggs et al. 2012; Scheffer et al. 2009, 2001; Scheffer and Carpenter 2003). While in some cases (*abrupt shifts*) the system can switch back to the original state, in other cases (*discontinuous shifts*) feedback mechanisms maintain the novel state. In mathematical language, the system response to stressors forms a fold bifurcation with hysteresis and the possibility of reversal to the previous state is extremely difficult as the forcing needed to switch back to the original state must be higher than the original one. Examples of discontinuous phase shifts in the ocean are the worldwide collapses of productive and scenic kelp bed habitats to persistent sea-urchin-dominated barren grounds (Ling et al. 2015), and the worldwide shifts from coral state to the alternate state of fleshy algae (Bellwood et al. 2004).

Multiple studies in the marine environment indicate that tipping points are common (Beaugrand et al. 2019; Hunsicker et al. 2016; Möllmann et al. 2015)—in fact regime shifts have been found in all marine ecosystems where long time series are available, often in concurrent periods. In the pelagic environment, abrupt (about a year) shifts were found in the North Pacific in 1976–1977, 1988–1989 and 1998–1999, in the South Pacific (Humboldt Current) in 1968–1970 and 1984–1986, in the north-west Atlantic in 1989–1990 (see review in Beaugrand 2015). Regime shifts were observed in the late 1980s also around all European seas: the North Sea, the Baltic Sea, the north-west European shelf, the eastern

(Adriatic) Mediterranean Sea, and the Black Sea (see review in Conversi et al. 2010). In the benthic environment, widely publicized shifts examples include the worldwide, dramatic phase shifts from coral reef to algal barren (Bellwood et al. 2004, 2019; Hughes 1994; Hughes et al. 2017, 2010) and from kelp reefs to sea urchin barrens (Ling et al. 2015).

2.2.1 Drivers of Marine Ecological Shifts

Marine ecological shift drivers pertain to both the exogenous/physical and the endogenous/biological (trophic predator–prey interactions) environments (Boada et al. 2017; Conversi et al. 2015; de Young et al. 2008). Major, large-scale stressors eroding the ecosystem resilience include climate change and overfishing. Overfishing of top predators affects all species linked to it (e.g., their preys, the predators and preys of those, and so on), in a phenomenon called trophic cascading that can trigger regime shifts over entire marine basins (Beaugrand et al. 2015; Casini et al. 2009; Daskalov et al. 2007; Jackson et al. 2001; Llope et al. 2011). Global change, on the other hand, affects both the physical environment (e.g., circulation changes, marine oxygen decline, ocean acidification, etc.), and the species physiology and resilience through temperature increase, simultaneously affecting multiple trophic levels in the marine food chain (Kirby and Beaugrand 2009; Conversi et al. 2015). Global change impacts on marine communities include poleward shifts, spatial homogenization, phenology changes, bleaching phenomena, etc. (Beaugrand et al. 2009; Burrows et al. 2011; Magurran et al. 2015; Hughes et al. 2017). In addition to these, local drivers (chemical pollution, eutrophication, alien species introduction) can be important for triggering shifts from local to regional scale, such as bays, coastlines, basins. There is the possibility that multiple drivers add synergistically to bring an ecosystem beyond a tipping point. In fact, there are many examples indicating regime shifts triggered by more than one driver (Bruno et al. 2019; Mackinson et al. 2009; Möllmann and Diekmann 2012; Papworth et al. 2016); and that shifts in a Earth climate subsystem can have cascading effects on the others (Lenton et al. 2019). On the other hand, there are several studies indicating that a single driver, temperature, can be used to predict ecological shifts (Beaugrand 2015; Beaugrand et al. 2019), which makes this variable extremely valuable for predictions. In many cases, regime shifts can have large impacts on ecosystem services and on the societies utilizing them (Bellwood et al. 2019; Sguotti and Cormon 2018). For example, the phase shifts from coral to algal dominance affect also all ecosystem services provided by the coral reefs (biodiversity, tourism, fishing, and coastal zone protection). Shifts involving major fishery collapses can have disastrous economic consequences on the fishing industry (Blenckner et al. 2015).

2.2.2 Ecological Shifts Predictions

Understanding and foreseeing the proximity of tipping points is of major importance not only from an ecological point of view, but also from a socio-economic point of view (de Young et al. 2008; Hewitt and Thrush 2019). The early detection of incoming shifts is a field of study on its own, which has evolved a lot in recent years. Studies focus on temporal, spatial, or mixed spatial–temporal indicators. In fact, temporal (local) variance and autocorrelation often rise when approaching an unstable equilibrium prior to a critical transition; spatial variance may increase and spatial correlation may change before a catastrophic shift (Carpenter and Brock 2006; Lindegren et al. 2012). An analysis of false/true indications in temporal variance and autocorrelation shows, on the other hand, a majority of false indications. Thus, a multivariate approach on a suite of potential indicators is best (Burthe et al. 2016).

Scientists and resource managers often use methods and tools that assume ecosystem components respond linearly to environmental drivers and human stressors. This mental model is probably inadequate, and tipping points should be considered an ordinary, rather than an exceptional, possibility. In fact, the pervasivity of nonlinear relationships is shown by Hunsicker et al. (2016): these authors conducted a wide literature review on the degree of nonlinearity in single-driver-response relationships in pelagic marine systems, and found that nonlinearities comprised at least 52% of all driver-response relationships—which they actually consider an underestimate.

2.2.3 How Can Remote Sensing Help Understanding Processes Behind Ecological Shifts, and Provide Early Warning Signals?

Even if most animal marine communities are hidden from EO tools, the latter can provide crucial information for understanding, and possibly predicting, tipping points. The first crucial factor is the temporal/spatial scale. Because of budget and logistic constraints, monitoring programs usually sample near the coast (there are very few open-ocean time series), with low sampling frequencies (rarely daily or weekly, often monthly, sometimes seasonally or yearly)—in other words, they collect data that are not very representative of the world ocean. However, since some monitoring collections have been running for several decades, they do provide inestimable baseline information. In particular, decadal collection is necessary for the retrospective analysis of ecosystem shifts, as regimes, by definition, last many years. On the other hand, EOs provide precious high temporal frequency (hours, days), high spatial frequency (small-meso-large scale), global coverage, in some cases (e.g., SST, ocean colour radiometry) over several decades. Thus, EO data are vital for extrapolating information beyond the range of in situ monitoring and for large-scale modelling.

The second crucial factor is the multivariate information that EO provide: for example, pressure, wind intensity and direction, wave and current data, all help to define the large-scale physical environment that constrain marine animals and plants, both pelagic and benthonic. Temperature data provide the baseline metric for global change, as well as defining the life conditions of the individual species. However, we have to keep in mind that abundance of data does not translate necessarily in needed data, and that we need to continue to find novel ways to use EO data without falling prey to "reification fallacy", which in this case would be treating convenient but incomplete indicators as inclusive (Bush et al. 2017).

2.2.4 Chlorophyll

In the case of phytoplankton, EO can provide rather good information. Phytoplankton is at the basis of the food web; in fact, it produces directly or indirectly the food for all marine animals. Although it accounts for $< 1\%$ of the photosynthetic biomass on Earth, it contributes to almost half of the world's total primary production, and has a major role in regulating carbon dioxide sequestration (Falkowski 2012).

Climate change can modify the seasonal cycle of the oceanic primary productivity by changing SST and the quantity of nutrients and light available to phytoplankton growth. This could lead to shifts in the phenology of the phytoplankton, and, vice versa, changes in phytoplankton abundance can modify the climate (e.g., changes in the depth of light penetration, Mignot et al. 2013). Thus, it is important to monitor phytoplankton in order to understand the impact of natural and anthropogenic factors on its variability.

Chlorophyll-a is a common pigment of phytoplankton, and since this pigment enables phytoplankton to absorb blue to green light, it can be measured by satellite colour sensors. This is an extremely developed field, and algorithms linking colour to phytoplankton are continually developed, for example, to separate chlorophyll into different phytoplankton components (e.g., nanoeukaryotes, *Prochlorococcus* spp., diatoms, Synechococcus-like cyanobacteria, and phaeocystis-like), using specific radiance spectra measured by ocean-colour sensors (Alvain et al. 2005; Mouw et al. 2017).

Several studies have been undertaken in the past decades to study the potential changes of phytoplankton in the global ocean due to anthropogenic factors. The first study was from Polovina et al. (2008) using a 9-year time series of SeaWiFS data. They focused on the most oligotrophic areas (with values of chlorophyll-a concentration lower than 0.07 mg m^{-3}) and showed that those areas have expanded at averaged annual rates of 0.8 to 4.3% per year. Their findings were in accordance with global warming scenarios (increase in the SST leading to warmer subtropical gyres) showing an increase in the heat content and vertical stratification. Furthermore, they suggested that these areas will keep increasing with the continuation of global warming.

Other studies have focused on the variations in phytoplankton phenology (timing, amplitude, duration of the blooms), since these can impact the entire trophic chain. These changes are controlled by the physical environment (SST, wind regime, water column stratification, solar radiation), which are routinely measured by satellites (Friedland et al. 2018; Henson et al. 2013; Holt et al. 2016). For instance, Zhang et al. (2017) studied the change in time and amplitude of phytoplankton blooms in the North Pacific (NP) and North Atlantic (NA) over 2002–2015. They showed that in the NA the spring blooms occurred earlier and their magnitude decreased, while it was the contrary for the autumn blooms. In NP, delays and increased magnitude were observed for the spring and autumn blooms. These changes in the timing and magnitude of the biannual blooms in these two regions can have a huge impact on the marine food web and fisheries.

The case of phytoplankton is particular, since it is not simply affected by climate (e.g., via changes in SST, ocean circulation, nutrient availability, ocean acidification), but it has in turn the potential to affect climate. In fact, its huge blooms can change the Earth albedo and radiative processes, via dimethyl sulphide production (e.g., phaeocystis) which contributes to cloud condensation nuclei, or via direct sunlight reflection (e.g., coccolithophores) (Charlson et al. 1987; Hays et al. 2005; Kim et al. 2018). Phytoplankton variations in the CO_2 oceanic pump can affect global warming. This field is extremely interesting, especially if associated with tipping points that could destabilize the negative feedbacks that keep Earth's balance. Abrupt shifts in marine phytoplankton abundance or species composition can obviously have vast consequences both on the marine trophic chain and on Earth's climate. At the moment, our knowledge is not advanced enough to know whether we are near these kinds of tipping points; hence, monitoring phytoplankton, its drivers and its effects (e.g., albedo, dimethyl sulphide) on a global scale is extremely important (Kokhanovsky et al. 2020), and EO can be extremely useful.

While chlorophyll can be detected from space relatively well offshore and provide good estimates for phytoplankton, EO detection of zooplankton, or of any non-superficial organism remains constrained by EO limits (Behrenfeld et al. 2019). However, indirect, interlinked measures are continually investigated. For example, the size of productivity fronts estimated from the horizontal gradient of chlorophyll-a content appears to be directly linked to mesozooplankton biomass (Druon et al. 2019).

More specifically, the EO potential for detecting regime shifts or for providing early warnings is still undeveloped, but continually growing. A recent example is given in Beaugrand et al. (2019). Using METAL (MacroEcological Theory on the Arrangement of Life) predictions based on a single stressor—temperature—these authors have shown that abrupt shifts occurred in multiple pelagic ocean communities over the last decades, and have increased in magnitude and extent in the last few years. Their work suggests that as global temperature continues to increase, we should expect intensification of the phenomena we are already experiencing: ecosystem regime shifts, as well as biogeographic shifts, poleward migrations, local extinctions. In general, EO global SST information can be associated with ecological models for forecasts related to ecosystems shifts.

Public services with applications to tipping points are being developed. For example, EO SST data are incorporated into algorithms that provide near-real-time global maps of coral bleaching thermal stress risk (http://coralreefwatch.noaa.gov) (see also Melet et al. 2020). These are just a few examples. As these applications increase, and the observation time span reaches a few decades, EO data will provide more invaluable information for management, models, and forecasts.

3 Cryospheric Tipping Elements

3.1 Permafrost

3.1.1 Monitoring Permafrost

Permafrost is defined as ground, which stays frozen for at least two consecutive years. Monitoring of permafrost itself is a major challenge. The identification of its existence below ground relies on in situ measurements and boreholes. A global network exists but is still scarce (Biskaborn et al. 2015). Only a fraction of sites provides records long enough for identification of trends related to climate change. A continuous increase can be, however, reported from those which are available (Biskaborn et al. 2019). Satellite data cannot be used to directly monitor permafrost. Surface changes are used as proxy, and related observable parameters such as land surface temperature and snow are used as input for models (Trofaier et al. 2017). Only the latter can provide global information. So far, the extent of terrestrial permafrost on the northern hemisphere has been estimated from traditional sources to be 24–25% of the total land area. A new estimate, which utilized satellite records (land surface temperature and land cover), comes to 22% (Obu et al. 2019). The state of submarine permafrost is even less known as very little observations are existing. Overduin et al. (2019) estimate that about 2.5×10^6 km^2 of the Arctic shelves is underlain by permafrost of which 97% are currently thinning.

3.1.2 Status of Tipping Element Discussion and Drivers

Permafrost as a tipping element is quantified by its volume according to Lenton et al. (2008) and is controlled by soil temperature. Critical values of thresholds are rarely defined. Key impacts are release of methane (CH_4) and carbon dioxide (CO_2), which lead to further warming at the global scale and therefore potentially further permafrost thawing (i.e. a positive feedback related to carbon cycle). Soil temperatures increase, and the seasonal thaw layer on top of permafrost (active layer) increases. Lenton et al. (2008) suggested that projections of permafrost thaw are expected to be quasi-linear and do not exhibit threshold behaviour. This was attributed to the fact that the effect of permafrost thawing on the global carbon positive feedback is rather weak. No study convincingly demonstrated that permafrost is a tipping element according to Lenton et al. (2008). Schuur et al. (2015) also suggested gradual release of greenhouse gas emissions with permafrost thaw. However, an additional positive feedback has been recently described in the literature at a more local scale. This feedback is related to the local heat production, while the permafrost is thawing, which is caused by microbial decomposition of the soil and may further enhance permafrost thaw (Bathiany et al. 2016). Its effect is expected to become of relevance after a certain rate of local warming is exceeded (Luke and Cox 2011). This would suggest a rate-induced tipping point (Bathiany et al. 2016) which cannot be represented with a bifurcation diagram. A further issue is subsea permafrost and methane hydrates at the sea floor, for which little observations exist in general (e.g., Shakhova et al. 2017). It has been so far rarely discussed in the context of tipping elements. It is expected that methane from the permafrost affected sea floor will be gradually released rather than through an abrupt release event (Duarte et al. 2012; Lenton 2012, Schuur et al. 2015). Recent laboratory experiments, however, indicate that when the permafrost exceeds a critical temperature, rapid dissociation of gas hydrates can produce active methane emission, which means that minor warming of subsea permafrost may lead to hazardous dissociation in the Arctic shelf (Chuvilin et al. 2019).

3.1.3 Predictions

Loss of permafrost carbon is one of the expected changes which are irreversible on time scales that matter to contemporary societies (Steffen et al. 2018). This assumption is based on the strength of the permafrost carbon positive feedback by 2100 in the order of 0.09 (0.04–0.16) °C for a RCP4.5 scenario. Values do, however, increase to 0.29 ± 0.21 °C for RCP8.5 (Schaefer et al. 2014). Bathiany et al. (2016) argued that the permafrost carbon-associated feedback is not strong enough to lead to self-acceleration of permafrost thaw. Steffen et al. (2018), however, found large-scale permafrost loss a self-perpetuating response after crossing a tipping point. Loss of permafrost is expected to occur when 5 °C of warming is exceeded (Steffen et al. 2018). Model simulations from Yumashev et al. (2019) indicate that the permafrost carbon feedback is increasingly positive in warmer climates and nonlinear.

3.1.4 Potential Added Value of Satellite Data

A major issue is the limited knowledge on carbon stocks, which are vulnerable to thaw, especially regarding subsea permafrost (Bathiany et al. 2016). Accounts for storage of organic matter on land exist but are also lacking detail and consistency across the Arctic.

Interpolated maps from in situ records are currently the main source, but land cover and land surface properties, as identified from space, can be partially used to fill the gaps (Hugelius et al. 2013, Bartsch et al. 2016a). The latter specifically helps to identify near surface carbon but cannot account for larger depths as of high importance in case of peatlands.

A challenge is the heterogeneity of tundra landscapes. The mosaic of anoxic and oxic conditions adds to the complexity of processes relevant for reactions to permafrost thaw. Thaw under anoxic conditions leads to the establishment of methane producing microbial communities (Knoblauch et al. 2018). After their stabilization, the production of methane and CO_2 from these sites is equal. This leads eventually to a twice as high production of CO_2 at anoxic sites than for oxic sites. Soil wetness is determined largely by terrain, which is changing with permafrost thaw. Novel approaches are needed to represent such heterogeneity in climate models. Knowledge on current day conditions is required in a first step. Satellite data have been so far of limited applicability for this purpose due to lack of spatial resolution at pan-arctic scale to accurately represent wetland patterns (e.g., Bartsch et al. 2016b). The use of satellite data for terrestrial ecosystem modelling has been limited to coarse resolution inundation datasets in this context (Oh et al. 2020; Watts et al. 2014). Airborne observations of methane concentrations demonstrate the importance of heterogeneity in Arctic landscapes specifically related to water bodies (Elder et al. 2020). Upscaling of methane emissions based on satellite data has been so far only demonstrated at the local scale. A further aspect in this context is that thaw of soils gives also access to additional nutrients, which can increase aquatic macrophyte biomass and total CH_4 emission by 54% and 64%, respectively, in tundra wetlands (Lara et al. 2018).

Knowledge on soil type and snow cover is crucial for modelling of sub-ground temperatures as they determine heat conductivity. Changes in snow conditions affect ground thaw. This is amplified by vegetation. Shrubs advance the timing of snowmelt when they protrude through the snow surface, thereby exposing the active layer to thawing earlier in spring (Wilcox et al. 2019). Their height and thus potential snow trapping capabilities can be regionally estimated with remote sensing. Snow water equivalent can also be derived from satellite data, even producing time series over several decades, but the spatial resolution and data quality are too low to be applicable in heterogeneous permafrost landscapes (Trofaier et al. 2017).

Land cover derived from satellite data is commonly used to make assumptions about soil types, but this usually is only applicable locally (Bartsch et al. 2016b). Terrain height is changing over large areas due to the high ice content in the ground. The ice melts with increasing temperature, leading to irreversible lowering of the terrain. This can be monitored with methods such as InSAR (Interferometry using Synthetic Aperture Radar). Variations in seasonal behaviour of heave and subsidence can in addition reveal the sensitivity of certain soil types to increasing temperatures. Areas which are characterized by an organic layer on top show little change in warm years (Bartsch et al. 2019). The seasonally unfrozen layer thickness remains stable.

Earth observation has been shown to provide insight into related land surface change but cannot be directly used to identify changes of the tipping element permafrost. Consistent land surface observations and products (by satellites) of the vast land area underlain by permafrost are also still lacking to date. Satellite observations of methane concentration over the Arctic may contribute to observations of methane transport from sub-seabed sediments to the atmosphere (Angelopoulos et al. 2020; Yurganov et al. 2019) as well as the terrestrial environment but are to date not exploited.

3.2 Ice Sheets and Shelves System

3.2.1 Ice Sheets Mass Balance

Ice sheets form when snow accumulation over a continent remains from one winter to the other. Buried successive snow layers compact and turn into ice, which progressively flows towards the periphery of the continent. Surface melt might occur (mainly at the ice sheet margin) if surface temperatures allow it and, ultimately, the ice reaches the ocean, possibly comes afloat to form ice shelves and finally calves. (The limit between the grounded ice sheet and the floating ice shelves is called the grounding line.) Ice sheets retain a huge amount of water out of the ocean and have been the pacemaker of the large-scale sea-level changes during the Quaternary. For instance, during the Last Glacial Maximum, as a consequence of two massive ice sheets resting over North America (Laurentide) and Scandinavia (Fennoscandia), the global sea level was approximately 130 m below the observed current level (Dutton et al. 2015). Rapid loss of these ice sheets during the deglaciation occurred, and raised sea-level to present level in less than 10,000 years, with contribution to global mean sea-level rise (SLR) up to 4 m/century during the melt water pulse 1A (Dutton et al. 2015). Such rapid and massive contribution might have been due to large dynamical changes in the ice sheets. Nowadays, there remains approximately 58 m of equivalent SLR over Antarctica (Fretwell et al. 2013) and 7 m over Greenland (Morlighem et al. 2017). Sea level has been relatively stable during the last centuries, and ice sheets are believed to have been roughly in balance during that period, precipitation balancing mass loss by surface melt at the surface, or below ice shelves and iceberg discharges (Church et al. 2013). Since the early 1990s, Greenland and Antarctica are losing mass at an increasing rate (300 Gt year^{-1}, van den Broeke et al. 2017; and 71 Gt year^{-1}, The IMBIE team 2018, respectively), contributing together to currently raise the global mean sea level by approximately 1 mm year^{-1} (Shepherd et al. 2012).

Remote sensing has been crucial in determining the first sign of imbalance and constraining mass loss. For instance, the retreat of the grounding line of the Pine Island Glacier in West Antarctica by about a kilometre per year has been detected using InSAR interferometry (Rignot 1998), the collapse of the Larsen B ice shelves in the Antarctic Peninsula has been recorded in March 2002 (Scambos et al. 2004), massive retreats of the front of Greenland outlet glaciers, associated with an increase in their surface velocities, have been recorded (e.g., the ice tongue of the Jakobshavn Glacier almost completely collapsed in 2003 and ice surface velocity increased from 5.7 km year^{-1} in 1992 to 12.6 km year^{-1} in 2003, cf. Joughin et al. 2004). In general, mass balance of ice sheets has been progressively monitored through (i) surface altimetry (e.g., Pritchard et al. 2009), (ii) InSAR measurements of surface velocities to estimate ice outflow (e.g., Rignot and Kanagaratnam 2006) or (iii) mass gravimetry (e.g., Velicogna and Wahr 2006). All these three methods progressively converged (Cazenave 2006), all highlighting similar regions of ice sheet primary concerned by mass loss, before converging on the quantity of ice flushed into the ocean (Shepherd et al. 2012). Nowadays, evolution of ice sheets, Antarctica in particular, is the main uncertainty in SLR projection for the coming centuries (Church et al. 2013). In particular, in the context of a warming environment, there is a risk that both ice sheets cross tipping points (Pattyn et al. 2018), which may lead to self-sustained massive release of ice into the ocean up to a 1-m contribution from Antarctica alone in 2100 (DeConto and Pollard 2016). Ice sheet and climate models have been decisive in understanding the underlying mechanisms behind these instabilities (e.g., Charbit et al. 2008; Favier et al.

2014; Robinson et al. 2012; Schoof 2007; Weertman 1974) which are further described for Greenland and Antarctica in the next paragraphs.

3.2.2 Instability of Ice Sheets: Driving Mechanisms

Greenland ice sheet experiences intense seasonal surface melting at its margin. In a rapidly warming environment enhanced by the Arctic amplification ($\sim +2$ °C in summer since the mid-1990s, cf. Hanna et al. 2012), melting at the surface of the Greenland ice sheet has substantially increased (Trusel et al. 2018). Large imbalance of the Greenland ice sheet is observed since the early 1990s, rising up to a 300 Gt year^{-1} mass loss in average between 2011 and 2015, 61% of this loss being ascribed to the decrease in surface mass balance, the remaining being attributed to increase in ice discharge (van den Broeke et al. 2017). This extensive surface melting induces two intertwined positive feedbacks: (i) melting reduces the surface albedo and therefore enhances further melting, (ii) mass loss of the ice sheet leads to a progressive decrease in surface elevation, a subsequent increase in local temperature and an associated increase in melting. This might progressively lead to an obvious threshold for ice sheet sustainability: if surface mass balance becomes negative in average (i.e. no perennial accumulation of snow), the ice sheet would irremediably shrink and ultimately disappear. An ensemble of numerical simulations of the Greenland ice sheet evolution indicates that such a threshold could occur with an increase in global mean temperature above preindustrial between +1.9 and +5.1 °C (95% confidence interval) with a best estimate of +3.1 °C (Robinson et al. 2012).

Owing to much colder surface temperatures, melting of the Antarctic surface is far more limited, and outlet glaciers feed large floating ice shelves. Ice shelves exert a buttressing over upstream glaciers, limiting their outflow. A change in the geometry of the ice shelf, induced by an increase in sub-marine melt or calving, might reduce the buttressing and therefore enhance the outflow (Pattyn et al. 2018). The loss of the Larsen B ice shelf has been a prominent natural example with an up to eightfold increase in upfront glacier velocity as observed after its collapse in 2002 (Scambos et al. 2004). Following an enhanced sub-ice shelf melt (Pritchard et al. 2012), a large acceleration of the Amundsen Sea outlet glaciers has been recorded (Mouginot et al. 2014). This region is currently driving the Antarctic mass loss (The IMBIE team 2018), and it will most probably remain so in the coming century (Ritz et al. 2015). A key aspect of this region explains this fact: most of the outlet glaciers flowing into the Amundsen Sea embayment rest over a bedrock below sea level and present a retrograde bed slope (Fretwell et al. 2013). In the theoretical and simplified flow-line hypothesis, it has been demonstrated that at the grounding line, the flux of ice is a power function of the ice thickness (Schoof 2007). As a consequence, when resting on a retrograde bed slope, initial grounding line retreat leads to an increase in the ice thickness which translates into an enhanced outflow, leading to further retreat in turn: this is the Marine Ice Sheet instability (MISI, see Fig. 4, Schoof 2007; Weertman 1974). However, on an actual three-dimensional geometry, the mechanical impact of the ice shelf is crucial as buttressing can stabilize the grounding line retreat (Gudmundsson et al. 2012) and geometrical consideration on the bedrock is not sufficient enough to undoubtedly consider the glacier as unstable. Accurate ice sheet modelling is required to make progress on this complex issue. Modelling experiments of the Pine Island Glacier (Favier et al. 2014) and Thwaites Glacier (Joughin et al. 2014), the two main glaciers of the Amundsen Sea region, indicate that these two glaciers might have already initiate a MISI. Furthermore, if important increase in temperature of the atmosphere and surface melt occurs in the future, enhanced

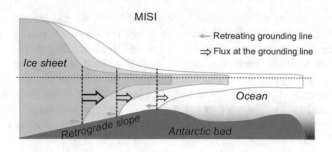

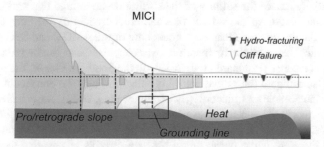

Fig. 4 Depiction of MISI (top) and MICI (bottom), adapted from Pattyn et al. (2017). Ice discharge generally increases with increasing ice thickness at the grounding line. For a bed sloping down towards the interior (retrograde bed slop) this may lead to unstable grounding-line retreat (MISI), as increased flux (for example, due to reduced buttressing) leads to thinning and eventually flotation, which moves the grounding line into deeper water where the ice is thicker. Thicker ice results in increased ice flux, which further thins (and eventually floats) the ice, which results in further retreat into deeper water (and thicker ice), and so on. MICI is the result of collapse of exposed ice cliffs (after the ice shelf collapses due to hydrofracturing) under their own weight. MISI applies for a retrograde slope bed, while MICI can also apply for prograde slopes. Both MISI and MICI are thus superimposed for retrograde slopes. The red colour beneath the ice shelf suggests that the deeper the ice the higher it is subjected to melt and grounding-line retreat (after DeConto and Pollard 2016)

fracture propagation by water (hydrofracturing) might initiate and strongly weaken the mechanical integrity of the ice shelves, therefore limiting the buttressing they exert on the upstream grounded glaciers. Ultimately, if hydrofracturing leads to ice shelf break up and formation of ice cliffs higher than approximately 100 m, they would mechanically not support their own weight and collapse, leading to the formation of even higher cliffs upstream which in turn could not sustain themselves (Fig. 4). This process named Marine Ice Cliff Instability (MICI) could strongly increase the rate of retreat, possibly leading to a 1-m contribution to SLR from Antarctica in 2100 under a RCP8.5 scenario (DeConto and Pollard 2016).

3.2.3 Intertwining Remote Sensing and Modelling: A Corner Stone to Better Apprehend Ice Sheet Tipping Points

In recent decades, remote sensing observations have highlighted the vulnerability of ice sheets, their ability to respond faster to environmental changes than previously thought and now provide a continuous monitoring of their evolution. In the meantime, owing to the improved understanding of the previously described feedbacks that could lead to self-sustained large-scale retreat or collapse, ice sheets have been suspected to have tipping points. This has been progressively confirmed by improved understanding of processes through step changes in ice sheet numerical modelling, particularly in the ability of adequately tracking the grounding line movement and initializing the ice sheet models through assimilation of EO (Pattyn et al. 2017). Because the dynamics of the outlet glaciers has changed and might enter into a so far unobserved regime, direct extrapolation of current trends in their contribution to SLR might be misleading. Furthermore, if changes in both ice sheets mass balance will considerably affect the sea-level height, they also have the potential to strongly affect the oceanic currents. Indeed, the increase in freshwater released in the ocean might substantially reduce the salinity of the surface ocean around this region, which may increase the stratification of the surface that can crucially modify the oceanic conditions. In the North Atlantic, this may lead to a weakening of the AMOC and SPG as mentioned in Sect. 2.1, which may cool the temperature over Greenland and can be seen as a negative feedback for the ice loss (Pritchard et al. 2012). In the Southern Ocean, the input of freshwater might also mitigate the global warming effect through the enhancement of sea ice production by the oceanic stratification (Swingedouw et al. 2008). Nevertheless, this stratification also increases the subsurface warming of the ocean, which might further enhance the melting of the ice shelf at depth. It is not clear up to now, which of the processes might dominate and therefore if the ocean might act as a positive or negative feedback for the Antarctic ice sheet.

A better anticipation of ice sheet contribution to future SLR and impact on the climate system requires an improved intertwining of remote sensing and modelling. Indeed, the huge increase in the amount of data available from remote sensing since 2013 now allows high spatial and temporal survey of all the outlet glaciers of both ice sheets. Assimilation of these new data into ice sheet models will progressively allow reanalysing past ice sheet mass loss, therefore improving our understanding of the underlying processes, the part of annual and decadal variability into the observed trends, and initiate attribution studies. This will greatly improve the ability of ice sheet models to reproduce past changes and therefore give more confidence in the reliability of their projections. Integrating dynamical ice sheet models into Earth system models is a mandatory effort to further investigate the feedbacks between ice sheets, ocean, and atmosphere. This will be an unavoidable prerequisite to quantify the various tipping points, the climate conditions required to cross the identified thresholds, and determine early warning variables that could be directly observed.

4 Tipping Elements in Land and Resource Use Systems

Land and resource use systems are social-ecological systems characterized by interactions of land resources (e.g., land as space, soil, water, and vegetation), the natural environment, and human activities, such as agriculture. Tipping elements in land systems differ from the other realms because of nature-human interactions and the associated multiple drivers and complex feedbacks. Hence, there are many tipping points in land systems, which differ, depending on the land use and land management goals or focus species, and cannot be generalized (Johnson 2013). Land system change is a major factor implicated in biophysical processes affecting the functioning and resilience of the Earth system (Steffen et al. 2015). For these reasons, we deviate in this section from framing tipping elements according to realms to framing them according to social-ecological interactions using a land use systems lens.

Many land use systems are local and have not been identified as tipping elements as such, but together with the increasingly globalized land systems (Popp et al. 2017), their combined impacts have global implications (Rockstrom et al. 2009). Land system change is a major process contributing to carbon emissions. According to the Intergovernmental Panel on Climate Change (IPCC), only 420 Gt of CO_2 can still be added to the atmosphere until around 2050 to stay below the 1.5 °C temperature threshold (Rogelj et al. 2018: 96). Steffen et al. (2015) estimate global average boundaries of phosphorous flows from fertilizers to erodible soils at 6.2 Tg per year, which have been surpassed by current values of about 14 Tg per year. Regarding nitrogen (N), Steffen et al. (2015) highlight the need to limit global average of introduced reactive N to 62 Tg N per year, although this limit is already surpassed by the current estimated value of about 150 Tg N per year. For land system change, the authors propose increasing globally the "area of forested land as a percentage of original forest cover" or for biomes "area of forested land as percentage of potential forest" (Steffen et al. 2015). Globally, they estimate retaining 75% of forested area as being "safe", although, currently only 62% remain forested. While these thresholds are challenging to apply at local and regional scales (Hossain and Speranza 2020), countries have begun introducing and incentivising measures to reduce the adverse impacts of land system change on biodiversity and human wellbeing (United Nations 2015; Ellison and Ifejika Speranza 2020).

To illustrate tipping elements in land and resource use systems and to fit the scope of this article, we present the case of (1) forests and (2) agricultural land use. We chose these two examples as they capture many challenges facing human use of land and natural resources and their impacts for Earth system processes.

4.1 Forests

Forest systems are social-ecological systems that provide various ecosystem services—they serve as habitat for biodiversity, influence local–regional rainfall patterns, serve as carbon sinks, and regulate hydrological flows, among others (Lenton et al. 2008; Nobre et al. 2016; Ellison and Ifejika Speranza 2020). Their status depends on the land use system, which itself can irreversibly change ecosystems such as through severe land degradation including deforestation, habitat loss and biodiversity loss or extinction. Besides climate change, land use change is a key driver, and both drivers can interact to fasten ecosystem degradation. The dieback of the Amazon forest, the greening of the Sahel and linked collapse of the West African Monsoon, and the dieback of boreal forests are among the tipping elements of the Earth's climate system identified by Lenton et al. (2008).

Boreal forests depend on a complex interplay of permafrost, tree physiology, and fire (Lenton et al. 2008). The Boreal forest dieback is associated with decreasing winter temperature, increasing summer temperature, linked to rising CO_2 emissions. These changes expose the forests to fires, insect pests, biodiversity loss (Lenton and Ciscar 2013) and their boreal transition to continental steppe grasslands (Lenton et al. 2008).

For the Amazon rainforest, land use and land cover change (LULCC) driven largely by agricultural development, causes deforestation and biodiversity loss. Nepstad et al. (2008) showed that global demand for biofuels, cash crops (soya, sugar, corn), and beef could lead to a tipping point resulting in Amazon forest degradation and forest dieback. They identified global warming, logging, drought, and forest fire to cause tree mortality and subsequently lead to a grass and herb invasion and Amazon forest dieback (Nepstad al. 2008). The Amazon rainforest dieback is also associated with large-scale transformation of forests to pasture, which have increased atmospheric temperature and decreased evapotranspiration rates and regional rainfall (Lovejoy and Nobre 2018; Nobre et al. 1991). This in turn can result in further droughts, fires, and biodiversity loss (Lenton and Ciscar 2013). Thus, deforested areas can influence regional climate (e.g., Lenton et al. 2019).

Changes in temperature and rainfall also influence forest conditions. Nobre et al. (2016) showed that the temperature in the Amazon region increased by 1 °C in 60 years (from 2016 backwards) and with deforestation of about 20%. Modelling studies have identified two tipping points for a transition of part of the Amazon forest to a savannah landscape—a 4 °C increase in global temperature (Lenton et al. 2008; Lovejoy and Nobre 2018; Nobre et al. 2016; Salazar and Nobre 2010) or deforestation of over 40% (Sampaio et al. 2007; Nobre et al. 2016). Lenton et al. (2008), drawing on Zeng et al. (1996), reported that part of Amazonian rainfall is recycled; hence, its deforestation causes about -30% precipitation reduction in some areas, the prolongation of the dry season and raised summer temperatures that make it difficult for the forest to recuperate (Kleidon and Heimann 2000). Among others, the El Niño–Southern Oscillation-driven fluctuations in SST have been associated with droughts affecting large forest areas (e.g., Amazon; Nobre et al. 2016). With global warming, the incidence of severe droughts and longer dry seasons may increase, thereby adversely affecting forests due to water loss and increase in forest fires (Abatzoglou and Williams 2016; Duffy et al. 2015). On the other hand, CO_2-enriched atmosphere can contribute to faster growth of trees due to CO_2 fertilization. Other factors and processes may also play a role in the fate of this complex system and are now better and better observed thanks to remote sensing (Sellers et al. 2018). An example of a threshold for the land realm is desertification, a persistent decrease in the productivity of dry lands. Indeed, an increasing literature now shows land systems can be subject to climatic hazards such as droughts and flooding, and climate change through global warming (Biggs et al. 2018). Aridification triggered by climate change may also drive desertification (Bachelet et al. 2016; Prince et al. 2018), but overgrazing can also cause desertification (e.g., semi-arid northern Eurasian agricultural frontier; Horion et al. 2016). The roles played by agricultural land use in regime shifts thus need closer attention.

4.2 Agricultural Land Use Systems

Being a social-ecological system, agriculture is dependent on biophysical conditions of soil, water, and various natural cycles, which interact with human agricultural practices aimed at producing food and fibre, among others. Agricultural production has affected land resources such as forests, wetlands, water, and soil causing biodiversity loss, water scarcity and pollution and interfering with biogeochemical cycles (e.g., carbon, nitrogen, phosphorous). In turn, as land systems are coupled social-ecological systems, agricultural practices, socio-economic, and political factors can directly and indirectly affect land use systems (e.g., through choice of production systems; agricultural subsidies) but also through land system outcomes (e.g., crop yields and incomes).

High crop yields, high chemical and fertilizer inputs, and low labour inputs are common features of intensive agricultural systems. Large-scale soy production such as in South America has established at the expense of forests. Yet, they are also exposed to various drivers, many being socioeconomic and political such as the high demand for soy as animal feed or favourable international prices for such agricultural products (Ramankutty and

Coomes 2016). These external drivers can keep the production system in a basin of attraction over decades even increasing the land areas under production (at the detriment of other land uses). Yet global price collapse or policies change (e.g., EU policy response to Bovine spongiforme Enzephalopathie by banning the use of animal protein to feed livestock) incentivised expansion of soy cropping, thus displacing other land uses (Ramankutty and Coomes 2016). Since the goal of such production systems is to maximize productivity, practices (e.g., fertilizer and herbicide application), to achieve this goal can adversely affect the underlying ecological and biophysical systems through interfering with nutrient recycling or causing soil degradation for instance. In the following, we elaborate the different ways tipping elements might evolve in agricultural land use systems.

4.2.1 Fragmented Agriculture-Forest Landscapes

Through human use of land for agriculture, landscapes in many world regions have become fragmented agriculture-forest mosaics. Features of a landscape system include habitat quality and size, connectivity, and heterogeneity, among others (Donaldson et al. 2016). Human population and activities continue to reduce animal habitats, transforming about 40% of terrestrial ecosystems to agricultural landscapes from which run-off pollute surface and underground water, thereby contributing to altering atmospheric and ocean chemistry (Barnosky et al. 2012). These have been linked to global-scale forcings such as global warming that can cause shifts in other ecosystems. Hence, local- and planetary-scale drivers interplay to drive critical transitions at planetary scales (Barnosky et al. 2012).

Although it is difficult to identify a tipping point a priori, changes in a landscape such as large-scale deforestation and fire in the Amazon can affect its hydrological cycle and trigger a shift from forest to non-forest ecosystems (Brando et al. 2014; Nobre et al. 2016). Landscape tipping points such as land cover and habitat fragmentation also result in adaptation failures whereby habitat size is no longer adequate to support certain species, hence leading to biodiversity loss (Fernández-Giménez et al. 2017; Khishigbayar et al. 2015).

4.2.2 Salt-Affected Agricultural Soils

Agricultural practices can significantly affect soil fertility in ways that critically reduce yields—a tipping point in an agri-food system. Associated land degradation processes relate to practices such as leaving soils uncovered and prone to erosion, changing the nutrient content of the soil, such as not adding nutrients to the soil (mining soils) or adding the wrong or too many nutrients to soils.

Salt-affected agricultural soils occur in many world regions, in particular in arid, semiarid and sub-humid conditions, under rainfed and irrigated agriculture. Salinity in agricultural land has been a major factor for the collapse of past civilizations (e.g., Mesopotamia, Shahid et al. 2018). Areas affected by salt-induced land degradation include, for example, the Aral Sea Basin in Central Asia, the Indo-Gangetic Basin in India, the Indus Basin in Pakistan, the Yellow River Basin in China, the Euphrates Basin in Syria and Iraq, the Murray-Darling Basin in Australia, and the San Joaquin Valley in the USA (Zdruli et al. 2017).

Saline soils, that is, soils with electrical conductivity equal or exceeding 4 deci Siemens per metre (dS m^{-1}) at 25 °C (Shahid et al. 2018) affect crops through crop water stress, thereby influencing crop growth (for less salt-tolerant crops) and raising crop canopy temperature (Ivushkin et al. 2018). Soil salinization, that is, "salt accumulation in the root zone", can reduce yields by up to 30% (Cherlet et al. 2018; Shahid et al. 2018). Shahid et al. (2018) estimate that globally about 2000 ha is lost to salinization daily. Saliniza-tion can be through primary (natural processes, e.g., weathering of rock parent material) and secondary sources (human-induced—e.g., inappropriate irrigation, overuse of fertiliz-ers, restricted drainage and replacing deep-rooted trees with shallow-rooted annual crops) (Cherlet et al. 2018). In extreme cases, salinization can lead to desertification (Daliako-poulos et al. 2016). Soil salinization can disrupt biological, biochemical, and hydrologi-cal cycles (Daliakopoulos et al. 2016), thereby affecting among others agricultural produc-tion and by extension human well-being. These changes in agricultural productivity thus have consequences for food availability but also feedback to the livelihoods and incomes of dependent social actors, which in turn can affect the choice of agricultural practices in the following season.

4.2.3 Groundwater-Based Agricultural Systems

To raise and maximize yields per unit land, agriculture has been intensified by add-ing external inputs such as inorganic fertilizers and extracting river and ground water for irrigation. The Central Valley in California, a semi-arid but one of the most productive agricultural regions globally, depends largely on irrigation using surface and groundwa-ter water abstraction. In some cases, water withdrawals exceed groundwater recharge. This mode of production led to groundwater drawdowns and subsidence of about 54 cm from 2008 to 2010 (Faunt et al. 2016). Realizing the possibility of ground-water drawdown, the State of California promulgated the 2014 Sustainable Groundwater Management Act to protect its aquifers and promote sustainable groundwater use (Faunt et al. 2016). Although not framed in the language of tipping elements, the drawdown of aquifers will likely neces-sitate a reduction in irrigated croplands.

4.2.4 Agricultural Fertilizer Management

Fertilizer application can increase crop yields, but it can also adversely influence nutri-ent cycles. Agriculture contributes about 60% to global anthropogenic emissions of nitrous oxide (N_2O), a major greenhouse gas (Syakila and Kroeze 2011). Changes in the global nitrogen (N) cycles are associated with the persistent increase in use of N fertilizer, slurry, and low nitrogen use efficiency (NUE). N and phosphorus fertilizer application have been linked to eutrophication of lakes in agricultural landscapes (Le Moal et al. 2019; Schindler et al. 2016). Reactive nitrogen is also implicated in soil degradation and loss of biodiversity and diminished quality of drinking water (Le Moal et al. 2019). While low food production in many world regions, in particular Africa, has been attributed to low fertilizer use and negative nitrogen (N) balances, high use of inorganic fertilizer adds N to soils and can lead to nonlinear increases of N_2O emissions (Hickman et al. 2015). Lu et al. (2019) identified thresholds in N fertilizer use rate for corn and winter wheat, over which NUE starts declin-ing, in other words, when crop yields slow down. Dari et al. (2018) propose using the soil P storage capacity (SPSC) and soil P saturation ratio (PSR) to determine the amount of P to add to a soil and in that way reduce P runoff to the wider environment. Improving NUE (in fertilizer application and livestock production), including integrated fertilizer management, is thus critical (Fageria and Baligar 2005).

4.2.5 Multiple Feedback in Land and Natural Resource Use Systems

Because natural and human systems are coupled, a change in an environmental tipping point can lead to a change in ecosystem services such as food and changes in socio-economic variables such as food prices and human well-being. In response to climate change, biomass-based climate mitigation technologies are increasingly promoted. Biomass Energy with Carbon Capture and Storage (BECCS) may affect food production such as planting trees that compete for land and water resources with food production. In places where climate change causes water scarcity, it may limit the extent to which BECCS can be deployed to maintain carbon emission budgets (Séférian et al. 2018). In turn, the water demand from BECCS is likely to cause declines in available water, thereby raising water scarcity and further constraining the deployment of BECCS-based mitigation (Séférian et al. 2018).

Changes in land use systems can disrupt ecosystem processes such as primary production, nutrient and water cycling, or soil formation, among others (Biggs et al. 2018). Through land use practices, biodiversity and ecosystem services (e.g., food crops and fibre, water regulation, soil erosion regulation) may change in ways that adversely affect human well-being (food and nutrition, livelihoods, and land use). This in turn, may feedback to intensify adverse land use practices (e.g., deforestation, leaving soil bare, overgrazing) that positively feedback in further disruption or collapse of certain ecological processes. Such tipping land system elements can be at a local scale (e.g., a degraded watershed) or at a regional scale (e.g., tropical forests such as the Amazon), but they are also known to connect across scales (e.g., many local level deforestation and land degradation contributing to global greenhouse gas emissions) (Rockstrom et al. 2009). Timescales of regime shifts thus depend on the social-ecological system and can range from weeks to centuries (Biggs et al. 2018).

4.2.6 Reversibility in Agricultural Land Use Systems

Some regime shifts are irreversible while some show hysteresis. The degradation and loss of land resources can be (ir)reversible depending on the resource and the time scale. For instance, draining peatlands for crop farming has often led to an irreversible change in the amount of soil organic matter, soil biodiversity, and caused soil subsidence (Könönen et al. 2018). Studies on irrecoverable carbon show that peatlands and mangroves are most at risk as the average time it takes to recover carbon in these ecosystems is at least 100 years (Goldstein et al. 2020). Recovery can be slow, thereby requiring long-term monitoring. McCrackin et al. (2017) in a meta-analysis of 89 studies found recovery periods in lakes and coastal marine ecosystems affected by eutrophication to range from 10 to 34 years for phosphorous- and nitrogen-cycles variables after a complete reduction in nutrients application.

4.3 Using Remote Sensing Data for Monitoring and Identifying Tipping Elements in Land and Resource Use Systems

Very few studies approach land and resource use systems directly from a tipping element lens, although one can argue that monitoring can provide early warning about tipping elements. Generally, combining data from optical sensors and radar instruments helps use the synergies between these different sources. Multi-sensor (combining optical and radar information) and multi-methods approach that combine statistical, object-oriented or machine learning methods are increasingly applied (cf. Ali et al. 2016; Chlingaryan et al. 2018). As land and resource use systems are heterogeneous, we provide illustrative examples of how remote sensing products or Earth observation data are used for monitoring and early warning for identifying tipping points.

Depending on availability of long-term time series, four statistical measures can be used as early warning of an approaching tipping point—increase in autocorrelation, skewness, variance and exceeding a threshold (Krishnamurthy and Krishnamurthy 2016). However, a long-term time series against which a tipping point can be identified relative to baseline is often missing for some variables.

Most approaches applying remote sensing data for monitoring or to identify tipping points use historical time series data; hence, remotely sensed data often serve as a basis for subsequent statistical analysis and simulation modelling of conditions/drivers that are likely to trigger regime shifts. Change detection, that is comparing remote sensing images of different dates or time series to identify changes or differences in land cover and land use, is often a first step to identifying tipping points in social-ecological systems. Change detection such as through time series data analysis is often followed by spatial statistics, numerical modelling or scenario planning to identify potential tipping points in the system. While remote sensing studies have often examined the spatial and temporal variations in a particular phenomenon, e.g., increases and decreases in vegetation greenness (Eckert et al. 2015), few of them have explicitly applied the concept of "tipping points" to identify thresholds (e.g., a shift from grasslands to forests or vice versa) in a land use system.

The use of indices to monitor the spectra of various land cover remains widespread. Often, similar indices are applied for vegetation monitoring, whether for forests, crops or for grasslands. Such indices include the normalized difference vegetation index (NDVI), enhanced vegetation index (EVI), soil-adjusted vegetation index (SAVI), and the normalized difference salinity index (NDSI). Break points in NDVI time series have been used to monitor vegetation changes (Browning et al. 2017; Burrell et al. 2017). At global scale, root-zone soil moisture observations have been derived from the Soil Moisture and Ocean Salinity (SMOS) mission (Dumedah et al. 2015).

4.3.1 Tipping Points in Vegetation (Forests, Grasslands, Croplands) Conditions

Image analysis in land use and resource systems usually involve identifying a time series of several years (often more than 10 years) and calculating NDVI as a proxy for vegetation biomass production. Decreases in NDVI or "negative slopes in NDVI residuals" are then interpreted as vegetation degradation, while "positive slopes in the NDVI residuals" indicate improved vegetation (Fernández-Giménez et al. 2017). Tipping points in vegetation conditions can be inferred from monitoring trends in baseline vegetation (Barnosky et al. 2012), identifying sharp declines in vegetation greenness indices (e.g., NDVI) and Land Surface Water Index (LSWI) (Zhou et al. 2017), from solar-induced chlorophyll fluorescence (SIF) values (Sun et al. 2017) and from evapotranspiration (ET) levels (Fisher et al. 2017).

Fernández-Giménez et al. (2017) used remote sensing data to monitor forage use and increased grazing pressure on rangelands, cropping density, shifts in plant species/vegetation and other changes in biomass conditions. Fernández-Giménez et al. (2017) conducted a trend and tipping point analyses of time series Advanced Very-High-Resolution Radiometer (AVHRR) and Moderate Resolution Imaging Spectroradiometer (MODIS) NDVI data of Mongolian rangelands and identified significant increases in the AVHRR NDVI over

1982–2012. Using residual trend analysis, no trends in the residuals of human-induced change were detected after filtering out the effects of climate from human activities except in Western Mongolia, or when considering the combined effect of temperature and precipitation. A residual trend analysis of MODIS NDVI (2000–2013) also showed no significant trends. Using the coefficient of variation (CV), and 5-year moving windows for standard deviation and autocorrelation, they identified increasing trend in variability of forage production, which they interpreted as a potential tipping point in rangeland conditions (Fernández-Giménez et al. 2017).

Often, other non-remote sensing data (e.g., in situ data) are required for a comprehensive analysis of social-ecological tipping points. For example, Fernández-Giménez et al. (2017) also integrated data on livestock population dynamics and percentage forage use over time to strengthen their identification of potentially approaching tipping points for the Mongolian rangelands.

Using MODIS NDVI and RADAR Vegetation Optical Depth (VOD) monthly data time series of evergreen tropical forests across Africa, Southeast Asia and South America, Verbesselt et al. (2016) showed that declining rates of recovery captured by temporal autocorrelation can be used to identify thresholds for collapse of tropical forests facing drought and high temperatures. Using additive regression models, they analysed the dynamics of tropical forests as a function of mean annual precipitation (MAP) using TRMM data and showed that temporal autocorrelation steeply increases as MAP decreases to levels identified as being critical (e.g., 1500 mm) for tropical forests. They suggest that "critical slowing down" identified in satellite time series data can also be applied as an indicator in other tipping elements such as boreal ecosystems, lakes and drylands.

Similarly, Xu et al. (2016) used vegetation structure (e.g., tree cover and canopy height) as a composite indicator of a tipping point. Xu et al. (2016) distinguished three vegetation states at global scale, namely tropical forest, savannah, and treeless landscapes according to mean annual precipitation tipping points of 600, 1500 and 2000 mm. Using LiDAR estimates of canopy heights and MODIS-derived tree cover data and MAP data, Xu et al. (2016) identified three canopy heights ($\sim 40, \sim 12$, and ~ 2 m) to correspond to "forest, savannah and treeless landscapes". They show that tropical forests respond to rainfall in a discontinuous manner as forests rapidly decline at 1500 mm MAP, while at 600 mm, trees no longer characterize the landscape (Xu et al. 2016). According to the authors, the identified modes align with the hypothesis of a "forest-savannah bistability" in areas with an average yearly rainfall ranging from ~ 1500 to ~ 2000 mm.

4.3.2 Land Use Driven Forest Loss and Fragmentation

Land use-driven forest loss and fragmentation increase the vulnerability of species to environmental influences and act as forerunners of a tipping point characterized by the extinction of species (Drechsler and Surun 2018; Taubert et al. 2018). Factors driving forest and land use systems towards tipping points are largely socio-economic and political although a natural hazard such as a hurricane or a flood can fasten the process (Drechsler and Surun 2018; Nepstad et al. 2008). Here remote sensing can provide insights by identifying and monitoring such threatening processes. Applying percolation theory to a global forest cover map made from high-resolution remote sensing imagery and simulation modelling of forest fragmentation, Taubert et al. (2018) identified patterns that indicate fragmentation of tropical and sub-tropical forests is near a critical point of percolation. Their analysis of forest fragmentation dynamics showed the number of forest remnants initially increased slowly but their numbers abruptly increased (in a few decades) after passing a critical point with many more forest remnants of decreased sizes (Taubert et al. 2018). Observations from remote sensing confirmed the authors' simulation of forest fragmentation.

4.3.3 Monitoring Non-vegetation Variables in Agriculture

Remote sensing data have also been used in monitoring other variables in land and resource use systems, to assess soil salinity (Ivushkin et al. 2018), aquifer volume and groundwater drought (Thomas et al. 2017; Vasco et al. 2019), and temperature warming over landscapes and oceans (FEWS NET 2018a, b). It has been used to identify hotspot areas of high N inputs and to model biogeochemical processes of N flows in agricultural systems (Liu et al. 2020). Threshold values have been applied for phosphorous monitoring to reduce eutrophication in lakes (Schindler et al. 2016).

Thomas et al. (2017) used NASA's GRACE satellite mission data to quantify groundwater storage deficits in California's central Valley, enabling them to capture groundwater drought. Using InSAR, Vasco et al. (2019) could better match the identified change in aquifer volume in the valley with point observations, an improvement compared to using only the GRACE data.

4.3.4 Monitoring Food Insecurity and Famine

Remote sensing-data analysis also informs exposure to natural hazards of different intensities (Helderop and Grubesic 2019), hazard planning and preparedness. Slow-onset natural hazards such as droughts can cause the drying up of vegetation, crop loss, food insecurity, and famines. Remote sensing data are thus a critical data source for famine early warning systems. The Famine Early Warning Systems Network combines the Climate Hazards Infrared Precipitation with Stations (CHIRPS), Rainfall Estimates Version 2.0 (RFE2), and Africa Rainfall Climatology Version 2 (ARC2) data and mean annual eMODIS NDVI data in a tool to assess agro-climatology, rainfall, and vegetation conditions as a season progresses (FEWS Network 2018a, b). It uses the coefficient of variation as compared to normal standard deviation to monitor food security (FEWS Network 2018b). Interpretation of these and other data such as the water requirements and phenology of crops, and regional seasonal forecasts provide early warning on impending food crises, hence contributing to forestalling tipping points in food systems. It then uses the Integrated Food Security Phase Classification (IPC) scale to identify food insecurity conditions (FEWS Network 2018a, b).

Choularton and Krishnamurthy (2019) found the Famine Early Warning Systems Network food security projections to be generally of high accuracy in Ethiopia. However, they identified decreased accuracy of FEWS Network food security early warning during extreme events such as the 2015/16 El Niño and in data-scarce areas in Ethiopia. From a tipping point perspective, they also found mixed forecasting accuracy in the transition phase from "food security to food crises" (Choularton and Krishnamurthy 2019). Other remote sensing-based crop monitoring systems include among others, China's global crop-monitoring system (CropWatch; Wu et al. 2014), the European Commission AGRI4CAST program (European Union 2019) or the FAO Global Information and Early Warning System (GIEWS) (FAO 2019).

A fused image multispectral SPOT imagery 10 m and panchromatic SPOT of 2.5 m have been used to derive population vulnerability to natural hazards (Zeng et al. 2012). Ahmed et al. (2017) used multi-temporal MODIS NDVI time series and Landsat-8 OLI derived data to assess the impact of a 2017 flash flood on rice production in Bangladesh. By delineating the areas affected by the flash flooding, estimates were made about the acreage of crop damage and impact on food security, which if extreme, may indicate a tipping point in inability of dependent land users to meet their food security. Such information can be further processed with other spatial data to identify the proportion of population affected, thereby providing insights on likely social tipping points.

Most of the literature focussing on use of remote sensing data for identifying tipping points in land and resource systems infer tipping points or regime shifts from monitoring time series—an approach that rather looks back instead of looking forward. Hence, most remote sensing-based analysis for tipping points in land and resource use systems either use remote sensing data as baseline or draw on them to confirm results from simulation modelling. Steps usually include identifying a potential driver, a change-point in time such as detecting a significant change in the mean, using the coefficient of variation, nonparametric statistical tests of difference between periods, trend analysis, applying principal component analysis and chronological clustering (Andersen et al. 2009; Fernández-Giménez et al. 2017). Other analyses apply spatial modelling in Geographic Information Systems, statistics or scenario analysis to identify potential tipping points in land and resource systems (cf. Ali et al. 2016; Thellmann et al. 2018).

Using remote sensing as an early warning for tipping points in land systems thus often draws on additional data such as census and livelihoods data to determine, for example, food security status and potential population at risk. The assumption is that if such adverse environmental (land cover) and resources (e.g., water availability) conditions continue, biophysical drivers (e.g., rainfall, drought, hurricanes) might lead to a socio-economic collapse of dependent livelihoods and economies. Remote sensing data thus provide critical data for tipping point analysis in land and resource use systems.

5 Synergies for Cascading Tipping Points

The different tipping elements of the climate system may interact with each other, potentially leading to a cascade of tipping points. For instance, a large weakening in the AMOC in the future may induce a southward shift of the Inter Tropical Convergence Zone (ITCZ) which may strongly affect the precipitation in the Sahelian region, potentially triggering social tipping points in these regions. The study by Defrance et al. (2017) evaluated that the decrease in precipitation associated with a ~70% weakening in the AMOC can reduce the available cultivable area for sorghum down to a million of square kilometres, thereby affecting tens of millions of inhabitants of this region. To reach such estimates, this study coupled off-line—integrating one model after the other, not a two-way coupling—a climate model with a crop model and then a demographical estimate for the near future. This example illustrates the potential interactions between tipping points explored in Sect. 2.1 and 4 of this paper. We argue here that further research will be certainly useful to improve these first rough estimates and start thinking of some adaptation scenario to such an event.

Such a catastrophic scenario drawing attention to the potential interaction between different tipping elements has also been highlighted in Cai et al. (2016). They suggested that a rapid melting of the Greenland ice sheet can trigger a collapse of the AMOC, itself modifying the ITCZ in the tropical area which may amplify the Amazon dieback in its northern part, warm the Southern Ocean, push the West Antarctic ice shelf beyond a critical threshold, and induce large sea-level rise in the coming centuries. The impact of an AMOC collapse on marine biology would be also large, changing primary production (e.g., Mariotti et al. 2012), while it may provide a negative feedback for the permafrost loss. Nevertheless, the lack of precise estimates of the different thresholds and interactions between tipping elements prevent to have a clear idea on the likelihood of such a suit of events.

Research on such cascading events remains very limited at the moment. Nevertheless, even low probability events remain key for adaptation measures especially if they carry potentially huge consequences. As suggested by Sutton (2018), the scientific community needs to further improve its risk assessment by also evaluating the exact impact of these low probability yet rapid events to start preparing appropriate adaptation and mitigation measures,

Additional examples of huge impacts on human society when crossing key thresholds include the potential implication of sea-level rise from shrinking ice sheets, even on a long time scale. The impacts on populations and urban areas could be enormous as most megacities are located along the coasts (IPCC 2019). Also, the melt of glaciers in the mountains means a critical change in livelihood activities in certain areas (Collins et al. 2019). These few examples clearly highlight that crossing some tipping points might have crucial implication on human livelihood and ecosystems survival. As Lenton et al. (2019) state, these events are "too risky to bet against" and a wait and see approach is not appropriate at all.

6 Conclusions

In this paper, we have reviewed a few pieces of knowledge concerning different tipping elements of the Earth system with a specific focus on how remote sensing data can help to make progress in their understanding and how they can provide some early warning signals of an approaching tipping point. The main findings and assessments are summarized in Table 1.

We have highlighted that two systems of ocean circulation in the Atlantic are considered to be at risk of crossing a tipping point. These are the large-scale Atlantic Meridional Overturning Circulation (AMOC) and the more local North Atlantic subpolar gyre (SPG). The risk of an AMOC collapse is considered to be low at the end of the twenty-first century whatever the emission scenario (Collins et al. 2019), while there is a chance of about one in two that of an AMOC collapse in 2300 if we follow a strong emission pathway (RCP8.5, Bakker et al. 2016). The impacts of an AMOC collapse are potentially huge so that, even though the risk is low at the end of the twenty-first century, it is a scenario to be considered and preparation to adaptation may be useful, notably in the Sahelian region. Concerning the SPG, the risk of a collapse in the coming century is larger than for the AMOC, since the proximity to a convection threshold is potentially close to present-day conditions as shown in Sect. 2.1. The time scale of the response is faster (one decade) relatively to AMOC. However, the climatic and societal impacts are far lower and may mainly moderate the warming of the North Atlantic region. A few studies have used remote sensing to improve estimates of AMOC variations in the past through the use of oceanic fingerprints, but a lot still remains to be done and some original ideas are emerging by coupling data from different types of satellites. In any case, the use of in situ measurements

Table 1 Synthesis of the knowledge concerning the different tipping elements, in terms of time scale of abrupt change, irreversibility potential of the change, regions impacted, and a rough assessment of the likelihood of bifurcation (i.e. system switching to an alternative state), depending on the emission scenario

Realm	System	Time scale of abrupt change	Irreversible	Regions impacted	Likelihood of bifurcation in the 21st century	Use of EO
Ocean circulation	AMOC	Centuries	Yes	Global, including monsoon systems and phytoplankton production	< 10% whatever the scenario, but larger for high emission scenario	In its infancy but promising
	SPG	Decade	Yes	Atlantic bordering region	~50% whatever the scenario	In its infancy but promising
Marine ecosystems	Coral reefs	Decade	Probably	Reefs worldwide	Not clarified	Advanced
	Pelagic ecosystems	Decade	Unlikely	Worldwide	Not clarified	Advanced for phytoplankton. Indirect for zooplankton but promising
	Fish communities	Decade	Possibly	Worldwide	Not clarified	None
	Species extinctions	Decades to century	Yes	Worldwide	Not clarified	None
Permafrost	Terrestrial (carbon cycle)	Gradual, but under discussion	Yes, regarding permafrost carbon	Arctic and Global through carbon cycle	Not clarified (rather gradual/ rate induced change)	In its infancy but promising
	Submarine (carbon cycle)	Gradual, but under discussion	Yes, regarding permafrost carbon	Arctic shelves	Not clarified	In its infancy but promising
Ice sheets	GrIS and AIS	Centuries to millennia	Yes	Global through SLR	high for global mean temperature increase 1.5-2°C above pre-industrial average.	Advanced
Terrestrial resource use	Various – forests (Amazon, boreal)	Varying	Varying	Forest biomes, global	Possible but varies	Advanced
	Agricultural land use systems	Varying	Varying	Various local-regional scales	Possible but varies	Progressing

The last column indicates the use of EO for monitoring the systems: *Low* means that EO of these systems are not yet developed; however, indirect measurements can be developed; *Medium* means that indirect measurements are being developed for monitoring these systems; *High* means that direct or indirect measurements are already key in the monitoring protocols

appear to be still necessary at the moment to correctly assess variations of these complex three-dimensional oceanic circulations, notably beyond 2000 metres depths, which is the present-day limit of ARGOS floats.

The ocean circulation is still poorly observed from space. The coupling between in situ observations and different remote sensing data like altimetry and gravimetry can be very useful to assess the three-dimensional changes in circulation and density properties. Also, the recent acquisition of SSS is very promising, but still need to be evaluated, notably in the high latitude where the resolution and precision might be very poor (Estella-Perez et al. 2020). Coupling this different source of information, using data assimilation techniques may be promising and may help to reduce the large uncertainty that is found for the moment in large-scale ocean circulation from oceanic reanalysis (Karspeck et al. 2017).

The inclusion of remote sensing data within climate models through assimilation is a key avenue to make progress on early warnings based on models. It has to be noted that decadal prediction systems are still in their infancy and are confronting the difficult challenges avoiding the effect of model bias on their predictions. These decadal prediction systems use initialized forecast in the ocean and explore the diversity of potential responses due to chaos and noise through relatively large ensemble of simulations. Once refined, they will be of great help evaluating the risk and proximity of crossing a threshold in ocean circulation systems.

With regard to marine communities, regime shifts have occurred at some point in every habitat where long-term data exist. These shifts have so far been identified by historical time series observations because marine animals (be that transparent zooplankton, or deep-sea fishes) are mostly invisible to satellites, and also because the Earth observations (EOs) time series did not go back in time enough. Detecting tipping points in marine animals can be potentially done via proxy, and since temperature has been found to be a primary driver (or a co-driver) in most ecological shifts, global scale, long-term SST satellite measures continue to be essential.

The trends of change of phytoplankton concentration (as indicated by EO of ocean colour) over the past twenty-plus years are still uncertain. Current trends indicate an increase in the size of most oligotrophic areas although more observations are needed to confirm these results. Combining ocean colour radiometry observations with data assimilation will help to merge different sensors and to distinguish trends. For the detection of trends in ocean colour (phytoplankton mainly) and tipping points, it is necessary to continue the development of new sensors and to plan them in advance (such as the Sentinel-3 program from the European Space Agency) for avoiding holes in the measurements (as it happened in Europe between MERIS (2002–2012) and OLCI (2016–)). A huge gap could be filled in by merging standard passive ocean colour radiometry measurements with active observations such as Lidar. Lidar can provide information on the water column (up to 50 metres) and coupled with standard ocean colour sensors and bio-Argo floats could lead to a three-dimensional view of the ocean (Hostetler et al. 2018; Jamet et al. 2019).

With regard to the topic of tipping points in marine ecosystems, the future belongs to studies linking these variables to ecological communities and to applications for management, such as the coral reef heat risk alert maps (see Melet et al. 2020) and to improved algorithms related to early warning for tipping points, for example, changes in variance and autocorrelation in multiple variables (Carpenter and Brock 2006; Lindegren et al. 2012). New types of biological global-scale models based on EO are being implemented to anticipate as early as possible regime shifts in the pelagic ocean. They are operational, and their implementation may help in the real-time detection of regional regimes shifts (Henson et al. 2010). Nevertheless, at least thirty years of continuous EO are needed to observe the impact of climate change and to observe tipping points. Particular attention must be paid to the Arctic Ocean, an area strongly impacted by global climate change, where severe biological shifts are currently expected.

Springer

In recent decades, there have been huge investments at European and international scale aiming to develop and implement effective ocean observing capacity, some including in situ observatories and EO (among which EOOS (http://www.eoos-ocean.eu), Jerico, Jerico next (http://www.jerico-ri.eu; mostly in situ data but also radar), and of course the ESA Sentinels), which are providing an enormous amount of new knowledge. What is needed now, with regard to tipping points, is to incorporate tipping point knowledge in their algorithms and products, so in addition to global patterns and trends, they can start providing early signals for tipping points and relative alerts (taking into consideration that this is a field under development).

Permafrost monitoring relies largely on in situ data which are scarce, specifically long time series. There are different views on whether permafrost is a tipping element, whether it exhibits threshold behaviour. Recently, described processes such as heat production by microbial decomposition lead to a different view: a rate-induced tipping point. The strength of the permafrost carbon feedback effect is debated. It is argued that it is rather weak, but the loss of carbon is irreversible on time scales that matter to contemporary societies. Different views also exist on self-perpetuating response after crossing a tipping point. Remote sensing data can be used as proxies for permafrost-related processes to address some of these issues but has not been extensively used yet in this context. Knowledge on soils and landscape heterogeneity is crucial. The fusion of land cover information and terrain change is expected to be the key to utilization of satellite data.

Inconsistent acquisition strategies for high-to-medium resolution satellite data are one of the main issues for land applications. This applies specifically to permafrost-related monitoring, a phenomenon which occurs over 22% of the Northern Hemisphere terrestrial environment. Also, here, merging active and passive observation methods may provide the required information, especially to address the issue of the permafrost carbon feedback.

EO of the Greenland and the Antarctic ice sheets revealed during the last two decades an increasing trend in their mass loss. This resurrects a long-standing hypothesis on possible instabilities of ice sheets resting on a bedrock below sea level, the so-called Marine Ice Sheet Instability (MISI). Recent progress in ice sheet modelling tends to confirm that the Pine Island and Thwaites Glacier flowing into the Amundsen Sea might be engaged in a self-sustained retreat which might possibly initiate a massive collapse of the West Antarctic Ice Sheet on centennial to millennial time scales. However, this remains to be confirmed by further modelling efforts to better constrain the threshold before the retreat and subsequent rate of mass loss. Undoubtedly, increasing the duration of remote sensing observations is mandatory to better constrain the models and validate the simulations.

Land use can drive tipping elements in terrestrial resource use systems such as the Amazon forest dieback, the greening of the Sahel and the dieback of boreal forests, and influence climate-land feedbacks. As land and resource use systems result from interactions between social and ecological systems, tipping points in land use systems differ depending on land management goals and type of terrestrial systems, making their monitoring difficult. EO has mainly been used for change detection analysis in these tipping elements. Identifying tipping points in land and resource use systems is still largely based on analysis of historical observations (time series) with identification of likely future tipping points mainly dependent on additional statistical analysis, modelling and increasingly machine learning approaches.

Remote sensing can aid early warning for identifying tipping points in landscape systems, land cover and land use fragmentation, as well as the linkages between fast and slow onset natural hazards with land and resource systems. However, this is mainly through a retrospective "historical approach" and in some cases with only a short lead time between an ecological condition (e.g., drought) and its consequences (food scarcity). The advent of data from new sensors such as the greenhouse gas sensors (cf. Sellers et al. 2018) can complement data derived from optical sensors to infer how climate affects the carbon cycle especially during El Niño event, and by extension, potential dynamics in land and resource use systems. However, complex statistical and numerical modelling is likely additionally needed for assessing tipping points in the complex climate-land coupled system.

As some features are not directly observable from Earth observations, the interacting links between components (e.g., SST and precipitation or vegetation-rainfall connections such as for the Amazon) indicate that a variable may serve as a proxy indicator. Hence, observing such factors (e.g., SST) can provide complementary information on the potential dynamics of land cover and vegetation condition. One way of looking forward in addition to using historical time series EO data is therefore to monitor other related variables associated with land conditions such as is done in drought monitoring. Hence, instead of monitoring land conditions using normalized difference vegetation index (NDVI), monitoring SST and ocean circulation, which predate drought occurrences on land even by a few weeks, can provide early warning of impending drought or floods. Also, Sellers et al. (2018) report that satellite data on greenhouse gases can be related to terrestrial photosynthesis through solar-induced fluorescence (SIF), thus providing information on the net ecosystem exchange (NEE) of CO_2, photosynthesis of gross primary production (GPP), and the burning of fossil and biomass.

Furthermore, vegetation–climate feedbacks as well as the influence of global warming and links to large-scale circulations (e.g., the AMOC) mean that land-based mitigation and adaptation measures may be limited in their effectiveness in the face of collapse of such large-scale circulations—this raises the question of combining spatial and temporal scales of EO. Land-based actions (e.g., reforestation) can be effective in the short term and at a regional scale to ameliorate the impacts of climate change and climate variability on ecosystems and dependent livelihoods. Yet, the prospect of the collapse of the AMOC indicates a further need for other responses that adequately consider the global ramifications and highlight the uncertain outcomes of the interactions between multiple drivers (e.g., vegetation–climate interactions, global warming, SST and changes in regional or large-scale circulations) and cascading effects on other tipping elements. Hence, more studies are needed that explore the links between the different factors or multiple stressors and how well they serve as early warning of tipping points.

Generally speaking, remote sensing observations are still covering a short time frame. This is a crucial issue to correctly identify anthropogenic forced variations from natural variability. Also, classical early warning methods that necessitate long time frames can hardly be applied to the different tipping elements analysed here, based only on remote sensing EO. Nevertheless, the knowledge gained from EO concerning the different systems analysed is clear and may help to improve the modelling of these systems. Indeed, given the limit of the statistical methods to provide early warnings, modelling is an alternative to estimate from large ensembles the risk of crossing a threshold for a given tipping element. Currently, the uncertainty concerning the value of key variables and the thresholds of tipping elements is usually very large, necessitating improvements in the modelling of these elements. The interactions between the different systems have been highlighted here as another key aspect of development. Coupling effectively numerical and process-based models of the different realms will help to better evaluate the risk of cascades of tipping points, which are very poorly evaluated at the moment. Such an approach will necessitate further multi-disciplinary approach, which we encourage, to go from the physical and biogeochemical information concerning nonlinearities to the potential societal nonlinear responses. Even though some of such scenarios are believed to be low-probability from present-day models (which usually exhibit strong limitations), their impacts are potentially so large that we cannot risk not accounting for them in the adaptation policies that are developed worldwide.

Acknowledgements We thank Teodolina Lopez and Anny Cazenave from ISSI who organized the Workshop leading to this review paper. We acknowledge very constructive and inspiring comments from one anonymous reviewer that clearly helped to improve the manuscript clarity. DS is supported by Blue-Action (European Union's Horizon 2020 research and innovation program, Grant Number: 727852) and EUCP (European Union's Horizon 2020 research and innovation programme under Grant Agreement No. 776613) projects. CIS' inputs contribute to the Programme on Ecosystem Change and Society (www.pecs-scien ce.org) and the Global Land Programme (www.glp.earth). AB benefited from Nunataryuk project (H2020 Research and Innovation Programme under Grant Agreement No. 773421) as well as ESA Climate Change Initiative project on Permafrost (4000123681/18/I-NB). GD was supported by the TiPACCs project (European Union's Horizon 2020 research and innovation programme, Grant Number: 820575).

References

Abatzoglou JT, Williams AP (2016) Impact of anthropogenic climate change on wildfire across western US forests. Proc Nat Acad Sci 113(42):11770–11775. https://doi.org/10.1073/pnas.1607171113

Ahmed M, Rahaman K, Kok A, Hassan Q (2017) Remote sensing-based quantification of the impact of flash flooding on the rice production: a case study over Northeastern Bangladesh. Sensors 17(10):2347. https://doi.org/10.3390/s17102347

Ali I, Cawkwell F, Dwyer E, Barrett B, Green S (2016) Satellite remote sensing of grasslands: from observation to management. J Plant Ecol 9:649–671. https://doi.org/10.1093/jpe/rtw005

Alvain S, Moulin C, Dandonneau Y, Bréon FM (2005) Remote sensing of phytoplankton groups in case 1 waters from global SeaWiFS imagery. Deep Sea Res Part I Oceanogr Res Pap 52(11):1989–2004. https://doi.org/10.1016/j.dsr.2005.06.015

Andersen T, Carstensen J, Hernández-García E, Duarte CM (2009) Ecological thresholds and regime shifts: approaches to identification. Trends Ecol Evol 24:49–57. https://doi.org/10.1016/j.tree.2008.07.014

Angelopoulos M, Overduin PP, Miesner F, Grigoriev MN, Vasiliev AA (2020) Recent advances in the study of Arctic submarine permafrost. Permafr Periglac Process. https://doi.org/10.1002/ppp.2061

Bachelet D, Ferschweiler K, Sheehan T, Strittholt J (2016) Climate change effects on southern California deserts. J Arid Environ 127:17–29. https://doi.org/10.1016/j.jaridenv.2015.10.003

Bakker P, Schmittner A, Lenaerts JTM, Abe-Ouchi A, Bi D, van den Broeke MR, Chan WL, Hu A, Beadling RL, Marsland SJ, Mernild SH, Saenko OA, Swingedouw D, Sullivan A, Yin J (2016) Fate of the Atlantic Meridional Overturning Circulation: strong decline under continued warming and Greenland melting. Geophys Res Lett 43:12252–12260. https://doi.org/10.1002/2016gl070457

Barnosky AD, Hadly EA, Bascompte J, Berlow EL, Brown JH, Fortelius M, Getz WM, Harte J, Hastings A, Marquet PA, Martinez ND, Mooers A, Roopnarine P, Vermeij G, Williams JW, Gillespie R, Kitzes J, Marshall C, Matzke N, Mindell DP, Revilla E, Smith AB (2012) Approaching a state shift in Earth's biosphere. Nature 486:52–58. https://doi.org/10.1038/nature11018

Bartsch A, Höfler A, Kroisleitner C, Trofaier AM (2016a) Land cover mapping in northern high latitude permafrost regions with satellite data: achievements and remaining challenges. Remote Sens 8:979. https://doi.org/10.3390/rs8120979

Bartsch A, Widhalm B, Kuhry P, Hugelius G, Palmtag J, Siewert MB (2016b) Can C-band synthetic aperture radar be used to estimate soil organic carbon storage in tundra? Biogeosciences 13:5453–5470. https://doi.org/10.5194/bg-13-5453-2016

Bartsch A, Leibman M, Strozzi T, Khomutov A, Widhalm B, Babkina E, Mullanurov D, Ermokhina K, Kroisleitner C, Bergstedt H (2019) Seasonal progression of ground displacement identified with satellite radar interferometry and the impact of unusually warm conditions on permafrost at the Yamal Peninsula in 2016. Remote Sens 11:1865. https://doi.org/10.3390/rs11161865

Bartsch A, Widhalm B, Leibman M, Ermokhina K, Kumpula T, Skarin A, Wilcox EJ, Jones BM, Frost GV, Höfler A, Pointner G (2020) Feasibility of tundra vegetation height retrieval from Sentinel-1 and Sentinel-2 data. Remote Sens Environ 237:111515. https://doi.org/10.1016/j.rse.2019.111515

Bathiany S, Dijkstra H, Crucifix M, Dakos V, Brovkin V, Williamson MS, Lenton TM, Scheffer M (2016) Beyond bifurcation: using complex models to understand and predict abrupt climate change. Dyn Stat Clim Syst. https://doi.org/10.1093/climsys/dzw004

Beaugrand G (2015) Theoretical basis for predicting climate-induced abrupt shifts in the oceans. Philos Trans R Soc Lond 370:20130264

Beaugrand G, Luczak C, Edwards M (2009) Rapid biogeographical plankton shifts in the North Atlantic Ocean. Glob Change Biol 15:1790–1803. https://doi.org/10.1111/j.1365-2486.2009.01848.x

Beaugrand G, Conversi A, Chiba S, Edwards M, Fonda-Umani S, Greene C, Mantua N, Otto SA, Reid PC, Stachura MM, Stemmann L, Sugisaki H (2015) Synchronous marine pelagic regime shifts in the Northern Hemisphere. Philos Trans R Soc B-Biol Sci 370:20130272. https://doi.org/10.1098/rstb.2013.0272

Beaugrand G, Conversi A, Atkinson A, Cloern J, Chiba S, Fonda-Umani S, Kirby RR, Greene CH, Goberville E, Otto SA (2019) Prediction of unprecedented biological shifts in the global ocean. Nat Clim Change 9:237

Beaulieu C, Chen J, Sarmiento JL (2012) Change-point analysis as a tool to detect abrupt climate variations. Philos Trans R Soc A Math Phys Eng Sci 370(1962):1228–1249. https://doi.org/10.1098/rsta.2011.0383

Behrenfeld MJ, Gaube P, Della Penna A, O'Malley RT, Burt WJ, Hu Y, Bontempi PS, Steinberg DK, Boss ES, Siegel DA, Hostetler CA, Tortell PD, Doney SC (2019) Global satellite-observed daily vertical migrations of ocean animals. Nature 576:257–261. https://doi.org/10.1038/s41586-019-1796-9

Belkin IM, Levitus S, Antonov J, Malmberg S-A (1998) "Great salinity anomalies" in the North Atlantic. Prog Oceanogr 41:1–68. https://doi.org/10.1016/S0079-6611(98)00015-9

Bellwood DR, Hughes TP, Folke C, Nyström M (2004) Confronting the coral reef crisis. Nature 429:827

Bellwood DR, Pratchett MS, Morrison TH, Gurney GG, Hughes TP, Álvarez-Romero JG, Day JC, Grantham R, Grech A, Hoey AS (2019) Coral reef conservation in the Anthropocene: confronting spatial mismatches and prioritizing functions. Biol Conserv 236:604–615. https://doi.org/10.1016/j.biocon.2019.05.056

Benzi R (2010) Stochastic resonance: from climate to biology. Nonlinear Process Geophys 17:431–441. https://doi.org/10.5194/npg-17-431-2010

Biggs R, Blenckner T, Folke C, Gordon L, Norström A, Nyström M, Peterson G (2012) Regime shifts. In: Hastings A, Gross L (eds) Encyclopedia of theoretical ecology. University of California Press, Berkeley, pp 609–617

Biggs R, Peterson GD, Rocha JC (2018) The regime shifts database : a framework for analyzing regime shifts in social-ecological systems. https://doi.org/10.5751/ES-10264-230309

Biskaborn BK, Lanckman J-P, Lantuit H, Elger K, Streletskiy DA, Cable WL, Romanovsky VE (2015) The new database of the Global Terrestrial Network for Permafrost (GTN-P) 7: 245–259. https://doi.org/10.5194/essd-7-245-2015

Biskaborn BK, Smith SL, Noetzli J, Matthes H, Vieira G, Streletskiy DA, Schoeneich P, Romanovsky VE, Lewkowicz AG, Abramov A, Allard M, Boike J, Cable WL, Christiansen HH, Delaloye R, Diekmann B, Drozdov D, Etzelmüller B, Grosse G, Guglielmin M, Ingeman-Nielsen T, Isaksen K, Ishikawa M, Johansson M, Johannsson H, Joo A, Kaverin D, Kholodov A, Konstantinov P, Kröger T, Lambiel C, Lanckman J-P, Luo D, Malkova G, Meiklejohn I, Moskalenko N, Oliva M, Phillips M, Ramos M, Sannel ABK, Sergeev D, Seybold C, Skryabin D, Vasiliev A, Wu Q, Yoshikawa K, Zheleznyak M, Lantuit H (2019) Permafrost is warming at a global scale. Nat Commun. https://doi.org/10.1038/s41467-018-08240-4

Blenckner T, Llope M, Möllmann C, Voss R, Quaas MF, Casini M, Lindegren M, Folke C, Chr. Stenseth N (2015) Climate and fishing steer ecosystem regeneration to uncertain economic futures. Proc R Soc B Biol Sci 282:20142809

Boada J, Arthur R, Alonso D, Pagès JF, Pessarrodona A, Oliva S, Ceccherelli G, Piazzi L, Romero J, Alcoverro T (2017) Immanent conditions determine imminent collapses: nutrient regimes define the resilience of macroalgal communities. Proc R Soc B Biol Sci 284:20162814

Born A, Stocker TF, Sando AB (2016) Transport of salt and freshwater in the Atlantic Subpolar Gyre. Ocean Dyn 66:1051–1064. https://doi.org/10.1007/s10236-016-0970-y

Boulton CA, Allison LC, Lenton TM (2014) Early warning signals of Atlantic Meridional overturning circulation collapse in a fully coupled climate model. Nat Commun 5:5752. https://doi.org/10.1038/ncomms6752

Brando PM, Balch JK, Nepstad DC, Morton DC, Putz FE, Coe MT, Silvério D, Macedo MN, Davidson EA, Nóbrega CC, Alencar A, Soares-Filho BS (2014) Abrupt increases in Amazonian tree mortality due to drought–fire interactions. Proc Nat Acad Sci 111(17):6347–6352. https://doi.org/10.1073/pnas.1305499111

  Springer

Bruno JF, Côté IM, Toth LT (2019) Climate change, coral loss, and the curious case of the parrotfish paradigm: why don't marine protected areas improve reef resilience? Annu Rev Mar Sci 11:307–334

Burrows MT, Schoeman DS, Buckley LB, Moore P, Poloczanska ES, Brander KM, Brown C, Bruno JF, Duarte CM, Halpern BS (2011) The pace of shifting climate in marine and terrestrial ecosystems. Science 334:652–655

Burthe SJ, Henrys PA, Mackay EB, Spears BM, Campbell R, Carvalho L, Dudley B, Gunn IDM, Johns DG, Maberly SC (2016) Do early warning indicators consistently predict nonlinear change in long-term ecological data? J Appl Ecol 53:666–676

Bush A, Sollmann R, Wilting A, Bohmann K, Cole B, Balzter H, Martius C, Zlinszky A, Calvignac-Spencer S, Cobbold CA (2017) Connecting Earth observation to high-throughput biodiversity data. Nat Ecol Evol 1:0176

Caesar L, Rahmstorf S, Robinson A, Feulner G, Saba V (2018) Observed fingerprint of a weakening Atlantic Ocean overturning circulation. Nature 556:191–196. https://doi.org/10.1038/s41586-018-0006-5

Cai YY, Lenton TM, Lontzek TS (2016) Risk of multiple interacting tipping points should encourage rapid CO_2 emission reduction. Nat Clim Change 6:520–525. https://doi.org/10.1038/nclimate2964

Carpenter SR, Brock WA (2006) Rising variance: a leading indicator of ecological transition. Ecol Lett 9:311–318

Casini M, Hjelm J, Molinero J-C, Lövgren J, Cardinale M, Bartolino V, Belgrano A, Kornilovs G (2009) Trophic cascades promote threshold-like shifts in pelagic marine ecosystems. Proc Natl Acad Sci 106:197–202

Cazenave A (2006) How fast are the ice sheets melting? Science 314(5803):1250–1252. https://doi.org/10.1126/science.1133325

Charbit S, Paillard D, Ramstein G (2008) Amount of CO_2 emissions irreversibly leading to the total melting of Greenland. Geophys Res Lett. https://doi.org/10.1029/2008GL033472

Charlson RJ, Lovelock JE, Andreae MO, Warren SG (1987) Oceanic phytoplankton, atmospheric sulphur, cloud albedo and climate. Nature 326:655–661. https://doi.org/10.1038/326655a0

Cherlet M, Hutchinson C, Reynolds J, Hill J, Sommer S, von Maltitz G (eds) (2018) World atlas of desertification. Publication Office of the European Union. https://doi.org/10.2760/9205

Chlingaryan A, Sukkarieh S, Whelan B (2018) Machine learning approaches for crop yield prediction and nitrogen status estimation in precision agriculture: a review. Comput Electron Agric 151:61–69. https://doi.org/10.1016/j.compag.2018.05.012

Choularton JC, Krishnamurthy PK (2019) How accurate is food security early warning? Evaluation of FEWS NET accuracy in Ethiopia. Food Security 11:333–344

Church JA, Clark PU, Cazenave A, Gregory JM, Jevrejeva S, Levermann A, Merrifield MA, Milne GA, Nerem RS, Nunn PD, Payne AJ, Pfeffer WT, Stammer D, Unnikrishnan AS (2013) Sea level change. In: Stocker TF, Qin D, Plattner G-K, Tignor M, Allen SK, Boschung J, Nauels A, Xia Y, Bex V, Midgley PM (eds) Climate change 2013: the physical science basis. Contribution of Working Group I to the Fifth Assessment Report of the Intergovernmental Panel on Climate Change. Cambridge University Press, Cambridge, pp 1137–1216. https://doi.org/10.1017/CBO9781107415324.026

Chuvilin E, Davletshina D, Ekimova V, Bukhanov B, Shakhova N, Semiletov I (2019) Role of warming in destabilization of intrapermafrost gas hydrates in the arctic shelf: experimental modeling. Geosciences 9:407. https://doi.org/10.3390/geosciences9100407

Claret M, Galbraith ED, Palter JB, Bianchi D, Fennel K, Gilbert D, Dunne JP (2018) Rapid coastal deoxygenation due to ocean circulation shift in the northwest Atlantic. Nat Clim Change 8:868–872. https://doi.org/10.1038/s41558-018-0263-1

Claussen M (1997) Modeling bio-geophysical feedback in the African and Indian monsoon region. Clim Dyn 13(4):247–257. https://doi.org/10.1007/s003820050164

Clement AC, Peterson LC (2008) Mechanisms of abrupt climate change of the last glacial period. Rev Geophys 46:RG4002. https://doi.org/10.1029/2006RG000204

Collins M, Sutherland M, Bouwer L, Cheong S-M, Frolicher T, Jacot Des Combes H, Mathew Koll R, Losada I, Mc Innes K, Ratter B, Rivera-Arriga E, Susanto RD, Swingedouw D, Tibig L (2019) IPCC special report on the ocean and cryosphere in a changing climate. Chapter 6: Extremes, abrupt changes and managing risks

Conversi A, Fonda-Umani S, Peluso T, Molinero JC, Santojanni A, Edwards M (2010) The Mediterranean Sea Regime shift at the end of the 1980s, and intriguing parallelisms with other European basins. PLoS ONE 5:e10633. https://doi.org/10.1371/journal.pone.0010633

Conversi A, Dakos V, Gardmark A, Ling S, Folke C, Mumby PJ, Greene C, Edwards M, Blenckner T, Casini M, Pershing A, Moellmann C (2015) A holistic view of marine regime shifts. Philos Trans R Soc B-Biol Sci 370:20130279. https://doi.org/10.1098/rstb.2013.0279

Dakos V, Carpenter SR, van Nes EH, Scheffer M (2015) Resilience indicators: prospects and limitations for early warnings of regime shifts. Philos Trans R Soc B Biol Sci 370:20130263. https://doi.org/10.1098/rstb.2013.0263

Daliakopoulos IN, Tsanis IK, Koutroulis A, Kourgialas NN, Varouchakis AE, Karatzas GP, Ritsema CJ (2016) The threat of soil salinity: a European scale review. Sci Total Environ 573:727–739. https://doi.org/10.1016/j.scitotenv.2016.08.177

Dansgaard W, Johnsen SJ, Clausen HB, Dahljensen D, Gundestrup NS, Hammer CU, Hvidberg CS, Steffensen JP, Sveinbjornsdottir AE, Jouzel J, Bond G (1993) evidence for general instability of past climate from a 250-kyr ice-core record. Nature 364:218–220. https://doi.org/10.1038/364218a0

Dari B, Nair VD, Sharpley AN, Kleinman P, Franklin D, Harris WG (2018) Consistency of the threshold phosphorus saturation ratio across a wide geographic range of acid soils. Agrosyst Geosci Environ 1:180028. https://doi.org/10.2134/age2018.08.0028

Daskalov GM, Grishin AN, Rodionov S, Mihneva V (2007) Trophic cascades triggered by overfishing reveal possible mechanisms of ecosystem regime shifts. Proc Natl Acad Sci 104:10518–10523

de Young B, Barange M, Beaugrand G, Harris R, Perry RI, Scheffer M, Werner F (2008) Regime shifts in marine ecosystems: detection, prediction and management. Trends Ecol Evol 23:402–409. https://doi.org/10.1016/j.tree.2008.03.008

DeConto RM, Pollard D (2016) Contribution of Antarctica to past and future sea-level rise. Nature 531:591–597. https://doi.org/10.1038/nature17145

Defrance D, Ramstein G, Charbit S, Vrac M, Famien AM, Sultan B, Swingedouw D, Dumas C, Gemenne F, Alvarez-Solas J, Vanderlinden JP (2017) Consequences of rapid ice sheet melting on the Sahelian population vulnerability. Proc Natl Acad Sci USA 114:6533–6538. https://doi.org/10.1073/pnas.1619358114

Demenocal P, Ortiz J, Guilderson T, Adkins J, Sarnthein M, Baker L, Yarusinsky M (2000) Abrupt onset and termination of the African humid period. Quat Sci Rev 19:347–361. https://doi.org/10.1016/s0277-3791(99)00081-5

Descamps S, Strøm H, Steen H (2013) Decline of an arctic top predator: synchrony in colony size fluctuations, risk of extinction and the subpolar gyre. Oecologia 173:1271–1282. https://doi.org/10.1007/s00442-013-2701-0

Deshayes J, Treguier AM, Barnier B, Lecointre A, Le Sommer J, Molines JM, Penduff T, Bourdalle-Badie R, Drillet Y, Garric G, Benshila R, Madec G, Biastoch A, Boning CW, Scheinert M, Coward AC, Hirschi JJM (2013) Oceanic hindcast simulations at high resolution suggest that the Atlantic MOC is bistable. Geophys Res Lett 40:3069–3073. https://doi.org/10.1002/grl.50534

Donaldson L, Wilson RJ, Maclean IMD (2016) Old concepts, new challenges: adapting landscape-scale conservation to the twenty-first century. Biodivers Conserv. https://doi.org/10.1007/s1053 1-016-1257-9

Drechsler M, Surun C (2018) Land-use and species tipping points in a coupled ecological-economic model. Ecol Complex 36:86–91. https://doi.org/10.1016/j.ecocom.2018.06.004

Drijfhout S, Bathiany S, Beaulieu C, Brovkin V, Claussen M, Huntingford C, Scheffer M, Sgubin G, Swingedouw D (2015) Catalogue of abrupt shifts in intergovernmental panel on climate change climate models. Proc Natl Acad Sci USA 112:E5777–E5786. https://doi.org/10.1073/pnas.15114 51112

Druon J-N, Hélaouët P, Beaugrand G, Fromentin J-M, Palialexis A, Hoepffner N (2019) Satellite-based indicator of zooplankton distribution for global monitoring. Sci Rep 9:4732

Duarte CM, Agustí S, Wassmann P, Arrieta JM, Alcaraz M, Coello A, Marbà N, Hendriks IE, Holding J, García-Zarandona I, Kritzberg E, Vaqué D (2012) Tipping elements in the arctic marine ecosystem. Ambio 41:44–55. https://doi.org/10.1007/s13280-011-0224-7

Duchez A, Courtois P, Harris E, Josey SA, Kanzow T, Marsh R, Smeed DA, Hirschi JJM (2016) Potential for seasonal prediction of Atlantic sea surface temperatures using the RAPID array at 26N. Clim Dyn 46:3351–3370. https://doi.org/10.1007/s00382-015-2918-1

Duffy PB, Brando P, Asner GP, Field CB (2015) Projections of future meteorological drought and wet periods in the Amazon. Proc Nat Acad Sci 112(43):13172–13177. https://doi.org/10.1073/pnas.14210 10112

Dumedah G, Walker JP, Merlin O (2015) Root-zone soil moisture estimation from assimilation of downscaled Soil Moisture and Ocean Salinity data. Adv Water Resour 84:14–22. https://doi.org/10.1016/j.advwatres.2015.07.021

Dutton A, Carlson AE, Long AJ, Milne GA, Clark PU, DeConto R, Horton BP, Rahmstorf S, Raymo ME (2015) Sea-level rise due to polar ice-sheet mass loss during past warm periods. Science 349:aaa4019. https://doi.org/10.1126/science.aaa4019

 Springer

Eckert S, Husler H, Liniger H, Hodel E (2015) Trend analysis of MODIS NDVI time series for detecting land degradation and regeneration in Mongolia. J Arid Environ 113:16–28

Elder CD, Thompson DR, Thorpe AK, Hanke P, Anthony KMW, Miller CE (2020) Airborne mapping reveals emergent power law of arctic methane emissions. Geophys Res Lett. https://doi.org/10.1029/2019GL085707

Ellison D, Ifejika Speranza C (2020) From blue to green water and back again: promoting tree, shrub and forest-based landscape resilience in the Sahel. Sci Total Environ 739:140002. https://doi.org/10.1016/j.scitotenv.2020.140002

Estella-Perez V, Mignot J, Guilyardi E, Swingedouw D, Reverdin G (2020) Advances in reconstructing the AMOC using sea surface observations of salinity. Clim Dyn. https://doi.org/10.1007/s00382-020-05304-4

European Union (2019) Monitoring agricultural ResourceS (MARS). https://ec.europa.eu/jrc/en/mars

Ezer T (2015) Detecting changes in the transport of the Gulf Stream and the Atlantic overturning circulation from coastal sea level data: the extreme decline in 2009–2010 and estimated variations for 1935–2012. Glob Planet Change 129:23–36. https://doi.org/10.1016/j.gloplacha.2015.03.002

Fageria NK, Baligar VC (2005) Enhancing nitrogen use efficiency in crop plants. In: Advances in agronomy. Academic Press, pp 97–185. https://doi.org/10.1016/S0065-2113(05)88004-6

Falkowski P (2012) Ocean science: the power of plankton. Nature 483:S17–S20. https://doi.org/10.1038/483s17a

FAO (2019) Global Information and Early Warning System (GIEWS)

Faunt CC, Sneed M, Traum J, Brandt JT (2016) Water availability and land subsidence in the Central Valley, California, USA. Hydrogeol J 24:675–684. https://doi.org/10.1007/s10040-015-1339-x

Favier L, Durand G, Cornford SL, Gudmundsson GH, Gagliardini O, Gillet-Chaulet F, Zwinger T, Payne AJ, Le Brocq AM (2014) Retreat of Pine Island Glacier controlled by marine ice-sheet instability. Nat Clim Change 4:117–121. https://doi.org/10.1038/nclimate2094

Fernández-Giménez ME, Venable NH, Angerer J, Fassnacht SR, Reid RS, Jamyansharav K (2017) Exploring linked ecological and cultural tipping points in Mongolia. Anthropocene 17:46–69. https://doi.org/10.1016/j.ancene.2017.01.003

FEWS NET (2018a) Agroclimatology: analyzing the effects of weather and climate on food security. http://fews.net/sites/default/files/White%20Paper%20-%20Agroclimatology%20-%20Jan212015.pdf

FEWS NET (2018b) Building rainfall assumptions for scenario development. https://fews.net/sites/default/files/documents/reports/Guidance_Document_Rainfall_2018.pdf

Fisher JB, Melton F, Middleton E, Hain C, Anderson M, Allen R, McCabe MF, Hook S, Baldocchi D, Townsend PA, Kilic A, Tu K, Miralles DD, Perret J, Lagouarde J-P, Waliser D, Purdy AJ, French A, Schimel D, Famiglietti JS, Stephens G, Wood EF (2017) The future of evapotranspiration: global requirements for ecosystem functioning, carbon and climate feedbacks, agricultural management, and water resources. Water Resour Res 53(4):2618–2626. https://doi.org/10.1002/2016WR020175

Frajka-Williams E (2015) Estimating the Atlantic overturning at 26°N using satellite altimetry and cable measurements: MOC FROM ALTIMETRY. Geophys Res Lett 42:3458–3464. https://doi.org/10.1002/2015GL063220

Fretwell P, Pritchard HD, Vaughan DG, Bamber JL, Barrand NE, Bell R, Bianchi C, Bingham RG, Blankenship DD, Casassa G, Catania G, Callens D, Conway H, Cook AJ, Corr HFJ, Damaske D, Damm V, Ferraccioli F, Forsberg R, Fujita S, Gim Y, Gogineni P, Griggs JA, Hindmarsh RCA, Holmlund P, Holt JW, Jacobel RW, Jenkins A, Jokat W, Jordan T, King EC, Kohler J, Krabill W, Riger-Kusk M, Langley KA, Leitchenkov G, Leuschen C, Luyendyk BP, Matsuoka K, Mouginot J, Nitsche FO, Nogi Y, Nost OA, Popov SV, Rignot E, Rippin DM, Rivera A, Roberts J, Ross N, Siegert MJ, Smith AM, Steinhage D, Studinger M, Sun B, Tinto BK, Welch BC, Wilson D, Young DA, Xiangbin C, Zirizzotti A (2013) Bedmap2: improved ice bed, surface and thickness datasets for Antarctica. Cryosphere 7:375–393. https://doi.org/10.5194/tc-7-375-2013

Friedland KD, Mouw CB, Asch RG, Ferreira ASA, Henson S, Hyde KJW, Morse RE, Thomas AC, Brady DC (2018) Phenology and time series trends of the dominant seasonal phytoplankton bloom across global scales. Glob Ecol Biogeogr 27:551–569. https://doi.org/10.1111/geb.12717

Galaasen EV, Ninnemann US, Irval N, Kleiven HF, Rosenthal Y, Kissel C, Hodell DA (2014) Rapid reductions in North Atlantic Deep Water during the peak of the last interglacial period. Science 343:1129–1132. https://doi.org/10.1126/science.1248667

Gastineau G, L'Heveder B, Codron F, Frankignoul C (2016) Mechanisms determining the winter atmospheric response to the atlantic overturning circulation. J Clim 29:3767–3785. https://doi.org/10.1175/jcli-d-15-0326.1

Goldstein A, Turner WR, Spawn SA, Anderson-Teixeira KJ, Cook-Patton S, Fargione J, Gibbs HK, Griscom B, Hewson JH, Howard JF, Ledezma JC, Page S, Koh LP, Rockström J, Sanderman J, Hole DG

(2020) Protecting irrecoverable carbon in Earth's ecosystems. Nat Clim Change 10:287–295. https://doi.org/10.1038/s41558-020-0738-8

Good SA, Martin MJ, Rayner NA (2013) EN4: quality controlled ocean temperature and salinity profiles and monthly objective analyses with uncertainty estimates. J Geophys Res Oceans 118:6704–6716. https://doi.org/10.1002/2013JC009067

Good P, Bamber J, Halladay K, Harper AB, Jackson LC, Kay G, Kruijt B, Lowe JA, Phillips OL, Ridley J, Srokosz M, Turley C, Williamson P (2018) Recent progress in understanding climate thresholds: ice sheets, the Atlantic meridional overturning circulation, tropical forests and responses to ocean acidification. Prog Phys Geogr 42:24–60. https://doi.org/10.1177/0309133317751843

Gudmundsson GH, Krug J, Durand G, Favier L, Gagliardini O (2012) The stability of grounding lines on retrograde slopes. Cryosphere 6:1497–1505. https://doi.org/10.5194/tc-6-1497-2012

Hanna E, Mernild SH, Cappelen J, Steffen K (2012) Recent warming in Greenland in a long-term instrumental (1881–2012) climatic context: I. Evaluation of surface air temperature records. Environ Res Lett 7:045404. https://doi.org/10.1088/1748-9326/7/4/045404

Hatun H, Lohmann K, Matei D, Jungclaus JH, Pacariz S, Bersch M, Gislason A, Olafsson J, Reid PC (2016) An inflated subpolar gyre blows life toward the northeastern Atlantic. Prog Oceanogr 147:49–66. https://doi.org/10.1016/j.pocean.2016.07.009

Hays G, Richardson A, Robinson C (2005) Climate change and marine plankton. Trends Ecol Evol 20:337–344. https://doi.org/10.1016/j.tree.2005.03.004

Helderop E, Grubesic TH (2019) Hurricane storm surge in Volusia County, Florida: evidence of a tipping point for infrastructure damage. Disasters 43:157–180. https://doi.org/10.1111/disa.12296

Henson SA, Sarmiento JL, Dunne JP, Bopp L, Lima I, Doney SC, John J, Beaulieu C (2010) Detection of anthropogenic climate change in satellite records of ocean chlorophyll and productivity. Biogeosciences 7:621–640. https://doi.org/10.5194/bg-7-621-2010

Henson S, Cole H, Beaulieu C, Yool A (2013) The impact of global warming on seasonality of ocean primary production. Biogeosciences 10:4357–4369

Heuze C (2017) North Atlantic deep water formation and AMOC in CMIP5 models. Ocean Sci 13:609–622. https://doi.org/10.5194/os-13-609-2017

Hewitt JE, Thrush SF (2019) Monitoring for tipping points in the marine environment. J Environ Manag 234:131–137

Hickman JE, Tully KL, Groffman PM, Diru W, Palm CA (2015) A potential tipping point in tropical agriculture: avoiding rapid increases in nitrous oxide fluxes from agricultural intensification in Kenya. J Geophys Res Biogeosci 120:938–951. https://doi.org/10.1002/2015JG002913

Hoegh-Guldberg O, Jacob D, Taylor M, Bindi M, Brown S, Camilloni I, Diedhiou A, Djalante R, Ebi KL, Engelbrecht F, Hijioka Y, Mehrotra S, Payne A, Seneviratne SI, Thomas A, Warren R, Zhou G, Halim SA, Achlatis M, Allen R, Berry P, Boyer C, Brilli L, Byers E, Cheung W, Craig M, Ellis N, Evans J, Fischer H, Fraedrich K, Fuss S, Ganase A, Gattuso J-P, Bolaños TG, Hanasaki N, Hayes K, Hirsch A, Jones C, Jung T, Kanninen M, Krinner G, Lawrence D, Ley D, Liverman D, Mahowald N, Meissner KJ, Millar R, Mintenbeck K, Mix AC, Notz D, Nurse L, Okem A, Olsson L, Oppenheimer M, Paz S, Petersen J, Petzold J, Preuschmann S, Rahman MF, Scheuffele H, Schleussner C-F, Séférian R, Sillmann J, Singh C, Slade R, Stephenson K, Stephenson T, Tebboth M, Tschakert P, Vautard R, Wehner M, Weyer NM, Whyte F, Yohe G, Zhang X, Zougmoré RB, Marengo JA, Pereira J, Sherstyukov B (2018) Impacts of 1.5 °C global warming on natural and human systems. In: Global warming of 1.5 °C. An IPCC special report on the impacts of global warming of 1.5 °C above pre-industrial levels and related global greenhouse gas emission pathways, in the context of strengthening the global response to the threat of climate change, sustainable development, and efforts to eradicate poverty. [IPCC report]

Hoelzmann P, Jolly D, Harrison SP, Laarif F, Bonnefille R, Pachur H-J (1998) Mid-Holocene land-surface conditions in northern Africa and the Arabian Peninsula: a data set for the analysis of biogeophysical feedbacks in the climate system. Global Biogeochem Cycles 12:35–51. https://doi.org/10.1029/97gb02733

Holling CS (1973) Resilience and stability of ecological systems. Annu Rev Ecol Syst 4:1–23

Holt J, Schrum C, Cannaby H, Daewel U, Allen I, Artioli Y, Bopp L, Butenschon M, Fach BA, Harle J, Pushpadas D, Salihoglu B, Wakelin S (2016) Potential impacts of climate change on the primary production of regional seas: a comparative analysis of five European seas. Prog Oceanogr 140:91–115. https://doi.org/10.1016/j.pocean.2015.11.004

Horion S, Prishchepov AV, Verbesselt J, de Beurs K, Tagesson T, Fensholt R (2016) Revealing turning points in ecosystem functioning over the Northern Eurasian agricultural frontier. Glob Change Biol 22:2801–2817. https://doi.org/10.1111/gcb.13267

 Springer

Hossain MS, Speranza CI (2020) Challenges and opportunities for operationalizing the safe and just operating space concept at regional scale. Int J Sustain Dev World Ecol 27:40–54. https://doi.org/10.1080/13504509.2019.1683645

Hostetler CA, Behrenfeld MJ, Hu Y, Hair JW, Schulien JA (2018) Spaceborne Lidar in the study of marine systems. Annu Rev Mar Sci 10:121–147. https://doi.org/10.1146/annurev-marine-121916-063335

Hugelius G, Bockheim JG, Camill P, Elberling B, Grosse G, Harden JW, Johnson K, Jorgenson T, Koven CD, Kuhry P, Michaelson G, Mishra U, Palmtag J, Ping C-L, O'Donnell J, Schirrmeister L, Schuur EaG, Sheng Y, Smith LC, Strauss J, Yu Z (2013) A new data set for estimating organic carbon storage to 3 m depth in soils of the northern circumpolar permafrost region. Earth Syst Sci Data. https://doi.org/10.5194/essd-5-393-2013

Hughes TP (1994) Catastrophes, phase shifts, and large-scale degradation of a Caribbean coral reef. Sci-AAAS-Wkly Pap Ed 265:1547–1551

Hughes TP, Graham NAJ, Jackson JBC, Mumby PJ, Steneck RS (2010) Rising to the challenge of sustaining coral reef resilience. Trends Ecol Evol 25:633–642

Hughes TP, Kerry JT, Álvarez-Noriega M, Álvarez-Romero JG, Anderson KD, Baird AH, Babcock RC, Beger M, Bellwood DR, Berkelmans R, Bridge TC, Butler IR, Byrne M, Cantin NE, Comeau S, Connolly SR, Cumming GS, Dalton SJ, Diaz-Pulido G, Eakin CM, Figueira WF, Gilmour JP, Harrison HB, Heron SF, Hoey AS, Hobbs J-PA, Hoogenboom MO, Kennedy EV, Kuo C, Lough JM, Lowe RJ, Liu G, McCulloch MT, Malcolm HA, McWilliam MJ, Pandolfi JM, Pears RJ, Pratchett MS, Schoepf V, Simpson T, Skirving WJ, Sommer B, Torda G, Wachenfeld DR, Willis BL, Wilson SK (2017) Global warming and recurrent mass bleaching of corals. Nature 543:373. https://doi.org/10.1038/nature21707

Hunsicker ME, Kappel CV, Selkoe KA, Halpern BS, Scarborough C, Mease L, Amrhein A (2016) Characterizing driver–response relationships in marine pelagic ecosystems for improved ocean management. Ecol Appl 26:651–663

IPCC (2019) Summary for policymakers. In: Pörtner H-O, Roberts DC, Masson-Delmotte V, Zhai P, Tignor M, Poloczanska E, Mintenbeck K, Alegría A, Nicolai M, Okem A, Petzold J, Rama B, Weyer NM (eds) IPCC special report on the ocean and cryosphere in a changing climate [IPCC report]. https://www.ipcc.ch/srocc/

Ivushkin K, Bartholomeus H, Bregt AK, Pulatov A, Bui EN, Wilford J (2018) Soil salinity assessment through satellite thermography for different irrigated and rainfed crops. Int J Appl Earth Obs Geoinf 68:230–237. https://doi.org/10.1016/j.jag.2018.02.004

Jackson JBC, Kirby MX, Berger WH, Bjorndal KA, Botsford LW, Bourque BJ, Bradbury RH, Cooke R, Erlandson J, Estes JA (2001) Historical overfishing and the recent collapse of coastal ecosystems. Science 293:629–637

Jackson LC, Kahana R, Graham T, Ringer MA, Woollings T, Mecking JV, Wood RA (2015) Global and European climate impacts of a slowdown of the AMOC in a high resolution GCM. Clim Dyn 45(11–12):3299–3316. https://doi.org/10.1007/s00382-015-2540-2

Jamet C, Ibrahim A, Ahmad Z, Angelini F, Babin M, Behrenfeld MJ, Boss E, Cairns B, Churnside J, Chowdhary J, Davis AB, Dionisi D, Duforêt-Gaurier L, Franz B, Frouin R, Gao M, Gray D, Hasekamp O, He X, Hostetler C, Kalashnikova OV, Knobelspiesse K, Lacour L, Loisel H, Martins V, Rehm E, Remer L, Sanhaj I, Stamnes K, Stamnes S, Victori S, Werdell J, Zhai P-W (2019) Going beyond standard ocean color observations: lidar and polarimetry. Front Mar Sci 6:251. https://doi.org/10.3389/fmars.2019.00251

Johnson CJ (2013) Identifying ecological thresholds for regulating human activity: effective conservation or wishful thinking? Biol Conserv 168:57–65

Joughin I, Abdalati W, Fahnestock M (2004) Large fluctuations in speed on Greenland's Jakobshavn Isbræ glacier. Nature 432:608–610. https://doi.org/10.1038/nature03130

Joughin I, Smith BE, Medley B (2014) Marine ice sheet collapse potentially under way for the Thwaites Glacier Basin, West Antarctica. Science 344:735–738. https://doi.org/10.1126/science.1249055

Karspeck AR, Stammer D, Kohl A, Danabasoglu G, Balmaseda M, Smith DM, Fujii Y, Zhang S, Giese B, Tsujino H, Rosati A (2017) Comparison of the Atlantic meridional overturning circulation between 1960 and 2007 in six ocean reanalysis products. Clim Dyn 49:957–982. https://doi.org/10.1007/s00382-015-2787-7

Keenlyside NS, Latif M, Jungclaus J, Kornblueh L, Roeckner E (2008) Advancing decadal-scale climate prediction in the North Atlantic sector. Nature 453:84–88. https://doi.org/10.1038/nature06921

Khishigbayar J, Fernandez-Gimenez ME, Angerer JP, Reid RS, Chantsallkham J, Baasandorj Y, Zumberelmaa D (2015) Mongolian rangelands at a tipping point? Biomass and cover are stable but composition shifts and richness declines after 20 years of grazing and increasing temperatures. J Arid Environ 115:100–112

Kim A-H, Yum SS, Lee H, Chang DY, Shim S (2018) Polar cooling effect due to increase of phytoplankton and dimethyl-sulfide emission. Atmosphere 9:384. https://doi.org/10.3390/atmos9100384

Kirby RR, Beaugrand G (2009) Trophic amplification of climate warming. Proc R Soc B-Biol Sci 276:4095–4103. https://doi.org/10.1098/rspb.2009.1320

Kleidon A, Heimann M (2000) Assessing the role of deep rooted vegetation in the climate system with model simulations: mechanism, comparison to observations and implications for Amazonian deforestation. Clim Dyn 16(2):183–199. https://doi.org/10.1007/s003820050012

Knight JR, Folland CK, Scaife AA (2006) Climate impacts of the Atlantic Multidecadal Oscillation. Geophys Res Lett. https://doi.org/10.1029/2006GL026242

Knoblauch C, Beer C, Liebner S, Grigoriev MN, Pfeiffer E-M (2018) Methane production as key to the greenhouse gas budget of thawing permafrost. Nat Clim Change 8(4):309–312. https://doi.org/10.1038/s41558-018-0095-z

Kokhanovsky A, Box JE, Vandecrux B, Mankoff KD, Lamare M, Smirnov A, Kern M (2020) The determination of snow albedo from satellite measurements using fast atmospheric correction technique. Remote Sens 12:234. https://doi.org/10.3390/rs12020234

Könönen M, Jauhiainen J, Straková P, Heinonsalo J, Laiho R, Kusin K, Limin S, Vasander H (2018) Deforested and drained tropical peatland sites show poorer peat substrate quality and lower microbial biomass and activity than unmanaged swamp forest. Soil Biol Biochem 123:229–241. https://doi.org/10.1016/j.soilbio.2018.04.028

Krishnamurthy L, Krishnamurthy V (2016) Decadal and interannual variability of the Indian Ocean SST. Clim Dyn 46:57–70. https://doi.org/10.1007/s00382-015-2568-3

Landerer FW, Wiese DN, Bentel K, Boening C, Watkins MM (2015) North Atlantic meridional overturning circulation variations from GRACE ocean bottom pressure anomalies. Geophys Res Lett 42:8114–8121. https://doi.org/10.1002/2015gl065730

Lara MJ, Nitze I, Grosse G, Martin P, Mcguire AD (2018) Reduced arctic tundra productivity linked with landform and climate change interactions. Sci Rep. https://doi.org/10.1038/s41598-018-20692-8

Le Moal M, Gascuel-Odoux C, Ménesguen A, Souchon Y, Étrillard C, Levain A, Moatar F, Pannard A, Souchu P, Lefebvre A, Pinay G (2019) Eutrophication: a new wine in an old bottle? Sci Total Environ 651:1–11. https://doi.org/10.1016/j.scitotenv.2018.09.139

Lemoine D, Traeger C (2014) Watch your step: optimal policy in a tipping climate. Am Econ J Econ Policy 6:137–166. https://doi.org/10.1257/pol.6.1.137

Lenton TM (2011) Early warning of climate tipping points. Nat Clim Change 1:201–209. https://doi.org/10.1038/nclimate1143

Lenton TM (2012) Arctic climate tipping points. Ambio 41:10–22. https://doi.org/10.1007/s13280-011-0221-x

Lenton TM, Ciscar J-C (2013) Integrating tipping points into climate impact assessments. Clim Change 117(3):585–597. https://doi.org/10.1007/s10584-012-0572-8

Lenton TM, Held H, Kriegler E, Hall JW, Lucht W, Rahmstorf S, Schellnhuber HJ (2008) Tipping elements in the Earth's climate system. Proc Natl Acad Sci USA 105:1786–1793. https://doi.org/10.1073/pnas.0705414105

Lenton TM, Rockstrom J, Gaffney O, Rahmstorf S, Richardson K, Steffen W, Schellnhuber H (2019) Climate tipping points—too risky to bet against. Nature 575:592–595. https://doi.org/10.1038/d41586-019-03595-0

Lindegren M, Dakos V, Gröger JP, Gårdmark A, Kornilovs G, Otto SA, Möllmann C (2012) Early detection of ecosystem regime shifts: a multiple method evaluation for management application. PLoS ONE 7:e38410

Ling SD, Scheibling RE, Rassweiler A, Johnson CR, Shears N, Connell SD, Salomon AK, Norderhaug KM, Perez-Matus A, Hernandez JC, Clemente S, Blamey LK, Hereu B, Ballesteros E, Sala E, Garrabou J, Cebrian E, Zabala M, Fujita D, Johnson LE (2015) Global regime shift dynamics of catastrophic sea urchin overgrazing. Philos Trans R Soc B-Biol Sci 370:20130269

Liu W, Xie SP, Liu ZY, Zhu J (2017) Overlooked possibility of a collapsed Atlantic Meridional Overturning Circulation in warming climate. Sci Adv. https://doi.org/10.1126/sciadv.1601666

Liu L, Zhang X, Xu W, Liu X, Li Y, Wei J, Gao M, Bi J, Lu X, Wang Z, Wu X (2020) Challenges for global sustainable nitrogen management in agricultural systems. J Agric Food Chem 68(11):3354–3361. https://doi.org/10.1021/acs.jafc.0c00273

Llope M, Daskalov GM, Rouyer TA, Mihneva V, Chan K, Grishin AN, Stenseth NC (2011) Overfishing of top predators eroded the resilience of the Black Sea system regardless of the climate and anthropogenic conditions. Glob Change Biol 17:1251–1265

Lontzek TS, Cai Y, Judd KL, Lenton TM (2015) Stochastic integrated assessment of climate tipping points indicates the need for strict climate policy. Nat Clim Change 5:441–444. https://doi.org/10.1038/nclimate2570

Lovejoy TE, Nobre C (2018) Amazon tipping point. Sci Adv 4:eaat2340. https://doi.org/10.1126/sciad v.aat2340

Lozier MS, Li F, Bacon S, Bahr F, Bower AS, Cunningham SA, de Jong MF, de Steur L, deYoung B, Fischer J, Gary SF, Greenan BJW, Holliday NP, Houk A, Houpert L, Inall ME, Johns WE, Johnson HL, Johnson C, Karstensen J, Koman G, Le Bras IA, Lin X, Mackay N, Marshall DP, Mercier H, Oltmanns M, Pickart RS, Ramsey AL, Rayner D, Straneo F, Thierry V, Torres DJ, Williams RG, Wilson C, Yang J, Yashayaev I, Zhao J (2019) A sea change in our view of overturning in the subpolar North Atlantic. Science 363:516–521. https://doi.org/10.1126/science.aau6592

Lu C, Zhang J, Cao P, Hatfield JL (2019) Are we getting better in using nitrogen?: variations in nitrogen use efficiency of two cereal crops across the United States. Earths Future 7:939–952. https://doi.org/10.1029/2019EF001155

Luke CM, Cox PM (2011) Soil carbon and climate change: from the Jenkinson effect to the compost-bomb instability 62:5–12. https://doi.org/10.1111/j.1365-2389.2010.01312.x

MacKenzie BR, Payne MR, Boje J, Høyer JL, Siegstad H (2014) A cascade of warming impacts brings bluefin tuna to Greenland waters. Glob Change Biol 20:2484–2491. https://doi.org/10.1111/gcb.12597

Mackinson S, Daskalov G, Heymans JJ, Neira S, Arancibia H, Zetina-Rejon M, Jiang H, Cheng HQ, Coll M, Arreguin-Sanchez F, Keeble K, Shannon L (2009) Which forcing factors fit? Using ecosystem models to investigate the relative influence of fishing and changes in primary productivity on the dynamics of marine ecosystems. Ecol Model 220:2972–2987. https://doi.org/10.1016/j.ecolmodel.2008.10.021

Magurran AE, Dornelas M, Moyes F, Gotelli NJ, McGill B (2015) Rapid biotic homogenization of marine fish assemblages. Nat Commun 6:8405

Mariotti V, Bopp L, Tagliabue A, Kageyama M, Swingedouw D (2012) Marine productivity response to Heinrich events: a model-data comparison. Clim Past 8:1581–1598. https://doi.org/10.5194/cp-8-1581-2012

Masson-Delmotte V, Zhai P, Portner H, Roberts D, Skea J, Shukla P, Pirani A, Moufouma-Okia W, Péan C, Pidcock S, Connors S, Matthews J, Chen Y, Zhou X, Gomis M, Lonnoy E, Maycock M, Tignor M, Waterfiled T (2018) Summary for policymakers. In: Global warming of 1.5°C. An IPCC special report on the impacts of global warming of 1.5°C above pre-industrial levels and related global greenhouse gas emission pathways, in the context of strengthening the global response to the threat of climate change, sustainable development, and efforts to eradicate poverty. World Meteorological Organization, Geneva, Switzerland, p 32

McCarthy GD, Smeed DA, Johns WE, Frajka-Williams E, Moat BI, Rayner D, Baringer MO, Meinen CS, Collins J, Bryden HL (2015) Measuring the Atlantic Meridional overturning circulation at 26 degrees N. Prog Oceanogr 130:91–111. https://doi.org/10.1016/j.pocean.2014.10.006

McCrackin ML, Jones HP, Jones PC, Moreno-Mateos D (2017) Recovery of lakes and coastal marine ecosystems from eutrophication: a global meta-analysis. Limnol Oceanogr 62:507–518. https://doi.org/10.1002/lno.10441

Mecking JV, Drijfhout SS, Jackson LC, Graham T (2016) Stable AMOC off state in an eddy-permitting coupled climate model. Clim Dyn 47:2455–2470. https://doi.org/10.1007/s00382-016-2975-0

Meinen CS, Speich S, Piola AR, Ansorge I, Campos E, Kersalé M, Terre T, Chidichimo MP, Lamont T, Sato OT, Perez RC, Valla D, van den Berg M, Le Hénaff M, Dong S, Garzoli SL (2018) Meridional overturning circulation transport variability at 34.5°S during 2009–2017: baroclinic and barotropic flows and the dueling influence of the boundaries. Geophys Res Lett 45:4180–4188. https://doi.org/10.1029/2018GL077408

Melet A, Teatini P, Le Cozannet G, Jamet C, Conversi A, Benveniste J, Almar R (2020) Earth observations for monitoring marine coastal hazards and their drivers. Surv Geophys. https://doi.org/10.1007/s11071 2-020-09594-5

Mercier H, Lherminier P, Sarafanov A, Gaillard F, Daniault N, Desbruyeres D, Falina A, Ferron B, Gourcuff C, Huck T, Thierry V (2015) Variability of the meridional overturning circulation at the Greenland-Portugal OVIDE section from 1993 to 2010. Prog Oceanogr 132:250–261. https://doi.org/10.1016/j.pocean.2013.11.001

Miesner AK, Payne MR (2018) Oceanographic variability shapes the spawning distribution of blue whiting (Micromesistius poutassou). Fish Oceanogr 27:623–638. https://doi.org/10.1111/fog.12382

Mignot J, Swingedouw D, Deshayes J, Marti O, Talandier C, Seferian R, Lengaigne M, Madec G (2013) On the evolution of the oceanic component of the IPSL climate models from CMIP3 to CMIP5: a mean state comparison. Ocean Model 72:167–184. https://doi.org/10.1016/j.ocemod.2013.09.001

Milkoreit M, Hodbod J, Baggio J, Benessaiah K, Calderon Contreras R, Donges JF, Mathias J-D, Rocha JC, Schoon M, Werners S (2018) Defining tipping points for social-ecological systems scholarship: an interdisciplinary literature review. Environ Res Lett. https://doi.org/10.1088/1748-9326/aaaa75

Möllmann C, Diekmann R (2012) Marine ecosystem regime shifts induced by climate and overfishing: a review for the Northern Hemisphere. Adv Ecol Res 47:303

Möllmann C, Folke C, Edwards M, Conversi A (2015) Marine regime shifts around the globe: theory, drivers and impacts. Philos Trans 370:20130260

Monerie P, Robson J, Dong B, Hodson DLR, Klingaman NP (2019) Effect of the atlantic multidecadal variability on the global monsoon. Geophys Res Lett 46:1765–1775. https://doi.org/10.1029/2018G L080903

Morlighem M, Williams CN, Rignot E, An L, Arndt JE, Bamber JL, Catania G, Chauché N, Dowdeswell JA, Dorschel B, Fenty I, Hogan K, Howat I, Hubbard A, Jakobsson M, Jordan TM, Kjeldsen KK, Millan R, Mayer L, Mouginot J, Noël BPY, O'Cofaigh C, Palmer S, Rysgaard S, Seroussi H, Siegert MJ, Slabon P, Straneo F, van den Broeke MR, Weinrebe W, Wood M, Zinglersen KB (2017) Bed-Machine v3: complete bed topography and ocean bathymetry mapping of greenland from multibeam echo sounding combined with mass conservation. Geophys Res Lett 44:11051–11061. https://doi.org/10.1002/2017GL074954

Mouginot J, Rignot E, Scheuchl B (2014) Sustained increase in ice discharge from the Amundsen Sea Embayment, West Antarctica, from 1973 to 2013. Geophys Res Lett 41:1576–1584. https://doi.org/10.1002/2013GL059069

Mouw CB, Hardman-Mountford NJ, Alvain S, Bracher A, Brewin RJW, Bricaud A, Ciotti AM, Devred E, Fujiwara A, Hirata T, Hirawake T, Kostadinov TS, Roy S, Uitz J (2017) A consumer's guide to satellite remote sensing of multiple phytoplankton groups in the global ocean. Fron Mar Sci. https://doi.org/10.3389/fmars.2017.00041

Nepstad DC, Stickler CM, Filho BS, Merry F (2008) Interactions among Amazon land use, forests and climate: prospects for a near-term forest tipping point. Philos Trans R Soc B Biol Sci 363:1737–1746. https://doi.org/10.1098/rstb.2007.0036

Nobre CA, Sellers P, Shukla J (1991) Amazonina deforestation and regional climate change. J Clim 4:957–988

Nobre CA, Sampaio G, Borma LS, Castilla-Rubio JC, Silva JS, Cardoso M (2016) Land-use and climate change risks in the Amazon and the need of a novel sustainable development paradigm. Proc Natl Acad Sci 113:10759–10768. https://doi.org/10.1073/pnas.1605516113

Obu J, Westermann S, Bartsch A, Berdnikov N, Christiansen HH, Dashtseren A, Delaloye R, Elberling B, Etzelmüller B, Kholodov A, Khomutov A, Kääb A, Leibman MO, Lewkowicz AG, Panda SK, Romanovsky V, Way RG, Westergaard-Nielsen A, Wu T, Yamkhin J, Zou D (2019) Northern Hemisphere permafrost map based on TTOP modelling for 2000–2016 at 1 km^2 scale. Earth-Sci Rev 193:299–316. https://doi.org/10.1016/j.earscirev.2019.04.023

Oh Y, Zhuang Q, Liu L, Welp LR, Lau MCY, Onstott TC, Medvigy D, Bruhwiler L, Dlugokencky EJ, Hugelius G, D'Imperio L, Elberling B (2020) Reduced net methane emissions due to microbial methane oxidation in a warmer Arctic. Nat Clim Change 10:317–321. https://doi.org/10.1038/s4155 8-020-0734-z

Osman MB, Das SB, Trusel LD, Evans MJ, Fischer H, Grieman MM, Kipfstuhl S, Mcconnell JR, Saltzman ES (2019) Industrial-era decline in subarctic Atlantic productivity. Nature. https://doi.org/10.1038/s41586-019-1181-8

Overduin PP, von Deimling TS, Miesner F, Grigoriev MN, Ruppel C, Vasiliev A, Lantuit H, Juhls B, Westermann S (2019) Submarine permafrost map in the arctic modeled using 1-D transient heat flux (SuPerMAP). J Geophys Res Oceans 124:3490–3507. https://doi.org/10.1029/2018JC014675

Papworth DJ, Marini S, Conversi A (2016) A novel, unbiased analysis approach for investigating population dynamics: a case study on Calanus finmarchicus and its decline in the north sea. PLoS ONE 11:e0158230

Pattyn F, Favier L, Sun S, Durand G (2017) Progress in numerical modeling of antarctic ice-sheet dynamics. Curr Clim Change Rep 3:174–184. https://doi.org/10.1007/s40641-017-0069-7

Pattyn F, Ritz C, Hanna E, Asay-Davis X, DeConto R, Durand G, Favier L, Fettweis X, Goelzer H, Golledge NR, Kuipers Munneke P, Lenaerts JTM, Nowicki S, Payne AJ, Robinson A, Seroussi H, Trusel LD, van den Broeke M (2018) The Greenland and Antarctic ice sheets under 1.5°C global warming. Nat Clim Change 8:1053–1061. https://doi.org/10.1038/s41558-018-0305-8

Poincaré H (1885) L'Équilibre d'une masse fluide animée d'un mouvement de rotation. Acta Math 7:259–380

Polovina JJ, Howell EA, Abecassis M (2008) Ocean's least productive waters are expanding. Geophys Res Lett. https://doi.org/10.1029/2007gl031745

Popp A, Calvin K, Fujimori S, Havlik P, Humpenöder F, Stehfest E, Bodirsky BL, Dietrich JP, Doelmann JC, Gusti M, Hasegawa T, Kyle P, Obersteiner M, Tabeau A, Takahashi K, Valin H, Waldhoff S, Weindl I, Wise M, Kriegler E, Lotze-Campen H, Fricko O, Riahi K, van Vuuren DP (2017)

Land-use futures in the shared socio-economic pathways. Glob Environ Change 42:331–345. https://doi.org/10.1016/j.gloenvcha.2016.10.002

Prince S, Von Maltitz G, Zhang F, Byrne K, Driscoll C, Eshel G, Kust G, Martínez-Garza C, Metzger JP, Midgley G, Moreno-Mateos D, Sghaier M, Thwin S, Bleeker A, Brown ME, Cheng L, Dales K, Ellicot EA, Wilson Fernandes G, Geissen V, Halme P, Harris J, Izaurralde RC, Jandl R, Jia G, Li G, Lindsay R, Molinario G, Neffati M, Palmer M, Parrotta J, Pierzynski G, Plieninger T, Podwojewski P, Dourado Ranieri B, Sankaran M, Scholes R, Tully K, Viglizzo EF, Wang F, Xiao N, Ying Q, Zhao C, Norbu C, Reynolds J (2018) Status and trends of land degradation and restoration and associated changes in biodiversity and ecosystem functions. IPBES, Bonn

Pritchard HD, Arthern RJ, Vaughan DG, Edwards LA (2009) Extensive dynamic thinning on the margins of the Greenland and Antarctic ice sheets. Nature 461:971–975. https://doi.org/10.1038/nature08471

Pritchard HD, Ligtenberg SRM, Fricker HA, Vaughan DG, Van Den Broeke MR, Padman L (2012) Antarctic ice-sheet loss driven by basal melting of ice shelves. Nature 484:502–505. https://doi.org/10.1038/nature10968

Rahmstorf S (1996) On the freshwater forcing and transport of the Atlantic thermohaline circulation. Clim Dyn 12:799–811. https://doi.org/10.1007/s003820050144

Rahmstorf S (2002) Ocean circulation and climate during the past 120,000 years. Nature 419:207–214

Ramankutty N, Coomes OT (2016) Land-use regime shifts: an analytical framework and agenda for future land-use research. Ecol Soc. https://doi.org/10.5751/ES-08370-210201

Renssen H, Goosse H, Fichefet T, Campin J-M (2001) The 8.2 kyr BP event simulated by a global atmosphere-sea-ice-ocean model. Geophys Res Lett 28:1567–1570. https://doi.org/10.1029/2000gl012602

Rignot EJ (1998) Fast recession of a west antarctic glacier. Science 281:549–551. https://doi.org/10.1126/science.281.5376.549

Rignot E, Kanagaratnam P (2006) Changes in the velocity structure of the greenland ice sheet. Science 311:986–990. https://doi.org/10.1126/science.1121381

Ritz C, Edwards TL, Durand G, Payne AJ, Peyaud V, Hindmarsh RCA (2015) Potential sea-level rise from Antarctic ice-sheet instability constrained by observations. Nature 528:115–118. https://doi.org/10.1038/nature16147

Robinson A, Calov R, Ganopolski A (2012) Multistability and critical thresholds of the Greenland ice sheet. Nat Clim Change 2:429–432. https://doi.org/10.1038/nclimate1449

Rockstrom J, Steffen W, Noone K, Persson A, Chapin FS, Lambin EF, Lenton TM, Scheffer M, Folke C, Schellnhuber HJ, Nykvist B, de Wit CA, Hughes T, van der Leeuw S, Rodhe H, Sorlin S, Snyder PK, Costanza R, Svedin U, Falkenmark M, Karlberg L, Corell RW, Fabry VJ, Hansen J, Walker B, Liverman D, Richardson K, Crutzen P, Foley JA (2009) A safe operating space for humanity. Nature 461:472–475. https://doi.org/10.1038/461472a

Rogelj J, Shindell D, Jiang K, Fifita S, Forster P, Ginzburg V, Handa C, Kheshgi H, Kobayashi S, Kriegler E, Mundaca L, Seferian R, Vilarino MV, Calvin K, Edelenbosch O, Emmerling J, Fuss S, Gasser T, Gillet N, He C, Hertwich E, Höglund Isaksson L, Huppmann D, Luderer G, Markandya A, McCollum D, Millar R, Meinshausen M, Popp A, Pereira J, Purohit P, Riahi K, Ribes A, Saunders H, Schadel C, Smith C, Smith P, Trutnevyte E, Xiu Y, Zickfeld K, Zhou W (2018) Chapter 2: Mitigation pathways compatible with 1.5 °C in the context of sustainable development. In: Global warming of 1.5 °C an IPCC special report on the impacts of global warming of 1.5 °C above pre-industrial levels and related global greenhouse gas emission pathways, in the context of strengthening the global response to the threat of climate change. Intergovernmental panel on climate change

Salazar LF, Nobre CA (2010) Climate change and thresholds of biome shifts in Amazonia. Geophys Res Lett. https://doi.org/10.1029/2010GL043538

Sampaio G, Nobre C, Costa MH, Satyamurty P, Soares-Filho BS, Cardoso M (2007) Regional climate change over eastern Amazonia caused by pasture and soybean cropland expansion. Geophys Res Lett. https://doi.org/10.1029/2007gl030612

Scambos TA, Bohlander JA, Shuman CA, Skvarca P (2004) Glacier acceleration and thinning after ice shelf collapse in the Larsen B embayment, Antarctica. Geophys Res Lett. https://doi.org/10.1029/2004GL020670

Schaefer K, Lantuit H, Romanovsky VE, Schuur EAG, Witt R (2014) The impact of the permafrost carbon feedback on global climate. Environ Res Lett 9:085003. https://doi.org/10.1088/1748-9326/9/8/085003

Scheffer M, Carpenter SR (2003) Catastrophic regime shifts in ecosystems: linking theory to observation. Trends Ecol Evol 18:648–656. https://doi.org/10.1016/j.tree.2003.09.002

Scheffer M, Carpenter S, Foley JA, Folke C, Walker B (2001) Catastrophic shifts in ecosystems. Nature 413:591–596

Scheffer M, Bascompte J, Brock WA, Brovkin V, Carpenter SR, Dakos V, Held H, Van Nes EH, Rietkerk M, Sugihara G (2009) Early-warning signals for critical transitions. Nature 461:53–59. https://doi.org/10.1038/nature08227

Schindler DW, Carpenter SR, Chapra SC, Hecky RE, Orihel DM (2016) Reducing phosphorus to Curb Lake eutrophication is a success. Environ Sci Technol 50:8923–8929. https://doi.org/10.1021/acs.est.6b02204

Schoof C (2007) Ice sheet grounding line dynamics: steady states, stability, and hysteresis. J Geophys Res Earth Surf. https://doi.org/10.1029/2006JF000664

Schuur EAG, McGuire AD, Schädel C, Grosse G, Harden JW, Hayes DJ, Hugelius G, Koven CD, Kuhry P, Lawrence DM, Natali SM, Olefeldt D, Romanovsky VE, Schaefer K, Turetsky MR, Treat CC, Vonk JE (2015) Climate change and the permafrost carbon feedback. Nature 520:171–179. https://doi.org/10.1038/nature14338

Séférian R, Rocher M, Guivarch C, Colin J (2018) Constraints on biomass energy deployment in mitigation pathways: the case of water scarcity. Environ Res Lett 13:054011. https://doi.org/10.1088/1748-9326/aabcd7

Sellers PJ, Schimel DS, Moore B, Liu J, Eldering A (2018) Observing carbon cycle–climate feedbacks from space. Proc Natl Acad Sci 115:7860–7868. https://doi.org/10.1073/pnas.1716613115

Send U, Lankhorst M, Kanzow T (2011) Observation of decadal change in the Atlantic meridional overturning circulation using 10 years of continuous transport data. Geophys Res Lett. https://doi.org/10.1029/2011GL049801

Sgubin G, Swingedouw D, Drijfhout S, Hagemann S, Robertson E (2015) Multimodel analysis on the response of the AMOC under an increase of radiative forcing and its symmetrical reversal. Clim Dyn 45:1429–1450. https://doi.org/10.1007/s00382-014-2391-2

Sgubin G, Swingedouw D, Drijfhout S, Mary Y, Bennabi A (2017) Abrupt cooling over the North Atlantic in modern climate models. Nat Commun. https://doi.org/10.1038/ncomms14375

Sgubin G, Swingedouw D, García de Cortázar-Atauri I, Ollat N, van Leeuwen C (2019) The impact of possible decadal-scale cold waves on viticulture over Europe in a context of global warming. Agronomy 9:397. https://doi.org/10.3390/agronomy9070397

Sguotti C, Cormon X (2018) Regime shifts–a global challenge for the sustainable use of our marine resources. In: YOUMARES 8–oceans across boundaries: learning from each other. Springer, Berlin, pp 155–166

Shahid SA, Zaman M, Heng L (2018) Introduction to soil salinity, sodicity and diagnostics techniques. In: Zaman M, Shahid SA, Heng L (eds) Guideline for salinity assessment, mitigation and adaptation using nuclear and related techniques. Springer, Cham, pp 1–42. https://doi.org/10.1007/978-3-319-96190-3_1

Shakhova N, Semiletov I, Gustafsson O, Sergienko V, Lobkovsky L, Dudarev O, Tumskoy V, Grigoriev M, Mazurov A, Salyuk A, Ananiev R, Koshurnikov A, Kosmach D, Charkin A, Dmitrevsky N, Karnaukh V, Gunar A, Meluzov A, Chernykh D (2017) Current rates and mechanisms of subsea permafrost degradation in the East Siberian Arctic Shelf. Nat Commun 8(1):15872. https://doi.org/10.1038/ncomms15872

Shepherd A, Ivins ER, Geruo A, Barletta VR, Bentley MJ, Bettadpur S, Briggs KH, Bromwich DH, Forsberg R, Galin N, Horwath M, Jacobs S, Joughin I, King MA, Lenaerts JTM, Li J, Ligtenberg SRM, Luckman A, Luthcke SB, McMillan M, Meister R, Milne G, Mouginot J, Muir A, Nicolas JP, Paden J, Payne AJ, Pritchard H, Rignot E, Rott H, Sorensen LS, Scambos TA, Scheuchl B, Schrama EJO, Smith B, Sundal AV, van Angelen JH, van de Berg WJ, van den Broeke MR, Vaughan DG, Velicogna I, Wahr J, Whitehouse PL, Wingham DJ, Yi D, Young D, Zwally HJ (2012) A reconciled estimate of ice-sheet mass balance. Science 338:1183–1189. https://doi.org/10.1126/science.1228102

Steffensen JP, Andersen KK, Bigler M, Clausen HB, Dahl-Jensen D, Fischer H, Goto-Azuma K, Hansson M, Johnsen SJ, Jouzel J, Masson-Delmotte V, Popp T, Rasmussen SO, Röthlisberger R, Ruth U, Stauffer B, Siggaard-Andersen M-L, Sveinbjörnsdóttir ÁE, Svensson A, White JWC (2008) High-resolution greenland ice core data show abrupt climate change happens in few years. Science 321:680–684. https://doi.org/10.1126/science.1157707

Steffen W, Richardson K, Rockström J, Cornell SE, Fetzer I, Bennett EM, Biggs R, Carpenter SR, de Vries W, de Wit CA, Folke C, Gerten D, Heinke J, Mace GM, Persson LM, Ramanathan V, Reyers B, Sörlin S (2015) Planetary boundaries: guiding human development on a changing planet. Science. https://doi.org/10.1126/science.1259855

Steffen W, Rockström J, Richardson K, Lenton TM, Folke C, Liverman D, Summerhayes CP, Barnosky AD, Cornell SE, Crucifix M, Donges JF, Fetzer I, Lade SJ, Scheffer M, Winkelmann R, Schellnhuber HJ (2018) Trajectories of the Earth System in the Anthropocene. Proc Natl Acad Sci 115:8252–8259. https://doi.org/10.1073/pnas.1810141115

Stommel H (1961) Thermohaline convection with two stable regimes of flow. Tellus 13(2):224–230. https://doi.org/10.3402/tellusa.v13i2.9491

Sun Y, Frankenberg C, Wood JD, Schimel DS, Jung M, Guanter L, Drewry DT, Verma M, Porcar-Castell A, Griffis TJ, Gu L, Magney TS, Köhler P, Evans B, Yuen K (2017) OCO-2 advances photosynthesis observation from space via solar-induced chlorophyll fluorescence. Science. https://doi.org/10.1126/science.aam5747

Sutton RT (2018) ESD Ideas: a simple proposal to improve the contribution of IPCC WGI to the assessment and communication of climate change risks. Earth Syst Dyn 9:1155–1158. https://doi.org/10.5194/esd-9-1155-2018

Swingedouw D, Fichefet T, Huybrechts P, Goosse H, Driesschaert E, Loutre MF (2008) Antarctic ice-sheet melting provides negative feedbacks on future climate warming. Geophys Res Lett. https://doi.org/10.1029/2008gl034410

Swingedouw D, Mignot J, Labetoulle S, Guilyardi E, Madec G (2013) Initialisation and predictability of the AMOC over the last 50 years in a climate model. Clim Dyn 40:2381–2399. https://doi.org/10.1007/s00382-012-1516-8

Syakila A, Kroeze C (2011) The global nitrous oxide budget revisited. Greenh Gas Meas Manag 1:17–26. https://doi.org/10.3763/ghgmm.2010.0007

Taubert F, Fischer R, Groeneveld J, Lehmann S, Müller MS, Rödig E, Wiegand T, Huth A (2018) Global patterns of tropical forest fragmentation. Nature 554:519–522. https://doi.org/10.1038/nature25508

The IMBIE team (2018) Mass balance of the Antarctic Ice Sheet from 1992 to 2017. Nature 558:219–222. https://doi.org/10.1038/s41586-018-0179-y

Thellmann K, Cotter M, Baumgartner S, Treydte A, Cadisch G, Asch F (2018) Tipping points in the supply of ecosystem services of a mountainous watershed in Southeast Asia. Sustainability 10:2418. https://doi.org/10.3390/su10072418

Thomas BF, Famiglietti JS, Landerer FW, Wiese DN, Molotch NP, Argus DF (2017) GRACE groundwater drought index: evaluation of California central valley groundwater drought. Remote Sens Environ 198:384–392. https://doi.org/10.1016/j.rse.2017.06.026

Trofaier AM, Westermann S, Bartsch A (2017) Progress in space-borne studies of permafrost for climate science: towards a multi-ECV approach. Remote Sens Environ 203:55–70. https://doi.org/10.1016/j.rse.2017.05.021

Trusel LD, Das SB, Osman MB, Evans MJ, Smith BE, Fettweis X, McConnell JR, Noël BPY, van den Broeke MR (2018) Nonlinear rise in Greenland runoff in response to post-industrial Arctic warming. Nature 564:104–108. https://doi.org/10.1038/s41586-018-0752-4

United Nations (2015) The Paris Agreement. https://unfccc…—Google Scholar [WWW Document], n.d. https://scholar.google.fr/scholar?hl=fr&as_sdt=0%2C5&q=United+Nations+%282015%29+The+Paris+Agreement.+https%3A%2F%2Funfccc.int%2Ffiles%2Fessential_background%2Fconvention%2Fapplication%2Fpdf%2Fenglish_paris_agreement.pdf.+Accessed+10.06.2020.&btnG=. Accessed 22 June 2020

van den Broeke M, Box J, Fettweis X, Hanna E, Noël B, Tedesco M, van As D, van de Berg WJ, van Kampenhout L (2017) Greenland ice sheet surface mass loss: recent developments in observation and modeling. Curr Clim Change Rep 3:345–356. https://doi.org/10.1007/s40641-017-0084-8

Vasco DW, Farr TG, Jeanne P, Doughty C, Nico P (2019) Satellite-based monitoring of groundwater depletion in California's Central Valley. Sci Rep 9(1):16053. https://doi.org/10.1038/s41598-019-52371-7

Velicogna I, Wahr J (2006) Measurements of time-variable gravity show mass loss in Antarctica. Science 311:1754–1756. https://doi.org/10.1126/science.1123785

Verbesselt J, Umlauf N, Hirota M, Holmgren M, Van Nes EH, Herold M, Zeileis A, Scheffer M (2016) Remotely sensed resilience of tropical forests. Nat Clim Change 6(11):1028–1031. https://doi.org/10.1038/nclimate3108

Watts JD, Kimball JS, Bartsch A, McDonald KC (2014) Surface water inundation in the boreal-Arctic: potential impacts on regional methane emissions. Environ Res Lett 9:075001. https://doi.org/10.1088/1748-9326/9/7/075001

Weaver AJ, Sedláček J, Eby M, Alexander K, Crespin E, Fichefet T, Philippon-Berthier G, Joos F, Kawamiya M, Matsumoto K, Steinacher M, Tachiiri K, Tokos K, Yoshimori M, Zickfeld K (2012) Stability of the Atlantic meridional overturning circulation: a model intercomparison. Geophys Res Lett. https://doi.org/10.1029/2012GL053763

Weertman J (1974) Stability of the junction of an ice sheet and an ice shelf. J Glaciol 13:3–11. https://doi.org/10.3189/S0022143000023327

Weijer W, Cheng W, Drijfhout SS, Fedorov AV, Hu A, Jackson LC, Liu W, McDonagh EL, Mecking JV, Zhang J (2019) Stability of the Atlantic meridional overturning circulation: a review and synthesis. J Geophys Res Oceans 124(8):5336–5375. https://doi.org/10.1029/2019JC015083

Wellens T, Shatokhin V, Buchleitner A (2004) Stochastic resonance. Rep Prog Phys 67:45–105. https://doi.org/10.1088/0034-4885/67/1/R02

Wilcox EJ, Keim D, de Jong T, Walker B, Sonnentag O, Sniderhan AE, Mann P, Marsh P (2019) Tundra shrub expansion may amplify permafrost thaw by advancing snowmelt timing. Arctic Sci 5(4):202–217. https://doi.org/10.1139/as-2018-0028

Wu B, Meng J, Li Q, Yan N, Du X, Zhang M (2014) Remote sensing-based global crop monitoring: experiences with China's CropWatch system. Int J Digit Earth 7:113–137. https://doi.org/10.1080/17538947.2013.821185

Xu XB, Rhines PB, Chassignet EP (2016) Temperature-salinity structure of the north Atlantic circulation and associated heat and freshwater transports. J Clim 29(21):7723–7742. https://doi.org/10.1175/jcli-d-15-0798.1

Yashayaev I, Loder JW (2017) Further intensification of deep convection in the Labrador Sea in 2016. Geophys Res Lett 44:1429–1438. https://doi.org/10.1002/2016gl071668

Yumashev D, Hope C, Schaefer K, Riemann-Campe K, Iglesias-Suarez F, Jafarov E, Burke EJ, Young PJ, Elshorbany Y, Whiteman G (2019) Climate policy implications of nonlinear decline of Arctic land permafrost and other cryosphere elements. Nat Commun. https://doi.org/10.1038/s41467-019-09863-x

Yurganov L, Muller-Karger F, Leifer I (2019) Methane increase over the Barents and Kara seas after the autumn pycnocline breakdown: satellite observations. Adv Polar Sci 30:382–390

Zdruli P, Lal R, Cherlet M, Kapur S (2017) New World Atlas of desertification and issues of carbon sequestration, organic carbon stocks, nutrient depletion and implications for food security. In: Erşahin S, Kapur S, Akça E, Namlı A, Erdoğan HE (eds), Carbon management, technologies, and trends in mediterranean ecosystems: the anthropocene: Politik—Economics—Society—Science. Springer, Cham, pp 13–25. https://doi.org/10.1007/978-3-319-45035-3_2

Zeng N, Dickinson RE, Zeng X (1996) Climatic impact of amazon deforestation: a mechanistic model study. J Clim 9:859–883. https://doi.org/10.1175/1520-0442(1996)009%3c0859:CIOADM%3e2.0.CO;2

Zeng J, Zhu ZY, Zhang JL, Ouyang TP, Qiu SF, Zou Y, Zeng T (2012) Social vulnerability assessment of natural hazards on county-scale using high spatial resolution satellite imagery: a case study in the Luogang district of Guangzhou, South China. Environ Earth Sci 65(1):173–182. https://doi.org/10.1007/s12665-011-1079-8

Zhang M, Zhang Y, Qiao F, Deng J, Wang G (2017) Shifting trends in bimodal phytoplankton blooms in the North Pacific and North Atlantic Oceans from Space with the holo-hilbert spectral analysis. IEEE J Sel Top Appl Earth Obs Remote Sens 10:57–64. https://doi.org/10.1109/JSTARS.2016.2625813

Zhou Y, Dong J, Xiao X, Xiao T, Yang Z, Zhao G, Zou Z, Qin Y (2017) Open surface water mapping algorithms: a comparison of water-related spectral indices and sensors. Water 9(4):256. https://doi.org/10.3390/w9040256

Publisher's Note Springer Nature remains neutral with regard to jurisdictional claims in published maps and institutional affiliations.

Affiliations

Didier Swingedouw[1] · Chinwe Ifejika Speranza[2] · Annett Bartsch[3,4] · Gael Durand[5] · Cedric Jamet[6] · Gregory Beaugrand[6] · Alessandra Conversi[7]

1 Environnements et Paléoenvironnements Océaniques et Continentaux (EPOC), UMR CNRS 5805, EPOC-OASU Université de Bordeaux, Allée Geoffroy Saint-Hilaire, 33615 Pessac, France

2 Institute of Geography, University of Bern, Hallerstrasse 12, 3012 Bern, Switzerland

3 b.geos GmbH, Industriestrasse 1, 2100 Korneuburg, Austria

4 Austrian Polar Research Institute, c/o Universität Wien, Zimmer B0410, Universitätsstrasse 7, 1010 Vienna, Austria

5 CNRS, IRD, Grenoble INP, IGE, Université Grenoble Alpes, 38000 Grenoble, France

6 Univ. Lille, CNRS, UMR 8187, LOG, Laboratoire d'Océanologie et de Géosciences, Université Littoral Côte d'Opale, 62930 Wimereux, France

7 CNR - ISMAR - Lerici, Forte Santa Teresa, Loc. Pozzuolo, National Research Council of Italy, 19032 Lerici, SP, Italy

Surveys in Geophysics (2020) 41:1285–1321
https://doi.org/10.1007/s10712-020-09611-7

Geoscientists in the Sky: Unmanned Aerial Vehicles Responding to Geohazards

R. Antoine[1] · T. Lopez[2,3] · M. Tanguy[4] · C. Lissak[5] · L. Gailler[6] · P. Labazuy[6] · C. Fauchard[1]

Received: 10 March 2020 / Accepted: 6 August 2020 / Published online: 19 September 2020
© Springer Nature B.V. 2020

Abstract

This article presents a review of the use of unmanned aerial vehicles (UAVs) in the context of geohazards. The pluri-disciplinary role of UAVs is outlined in numerous studies associated with mass earth movements, volcanology, flooding events and earthquakes. Scientific advances and innovations of several research teams around the world are presented from pre-events investigations to crisis management. More particularly, we emphasize the actual status of technology, methodologies and different applications that have emerged with the use of UAVs for each domain. It is shown that the deployment of UAVs in the geohazards context has experienced a tremendous increase during the last 10 years, with the development of more and more miniaturized, flexible and reliable systems. The use of such technology (UAV platform, instrumentation, methodologies) is different for each domain, depending on the spatial extent and the time scale of the observed phenomenon, but also on the practical constraints associated with the civil aviation agencies regulations (outside or within urban areas, before or during a crisis…). This paper also highlights the use of recent methodologies associated with semi-automatic/automatic segmentation or deep learning for the processing of important amounts of data provided by UAVs. Finally, although still sparse, the joint use of UAVs and satellite data is progressing and remains a challenge for future studies in the context of geohazards.

Keywords UAV · Geohazards · Mass earth movements · Volcanology · Flooding events · Earthquakes

✉ R. Antoine
  raphael.antoine@cerema.fr

[1]   Cerema, ENDSUM Team, 10 chemin de la Poudrière, 76121 Le Grand-Quevilly, France

[2]   Géosciences Environnement Toulouse (GET), Institut de Recherche Technologique (IRT) Saint-Exupéry, 14 avenue Edouard Belin, 31400 Toulouse, France

[3]   International Space Science Institute (ISSI), Hallerstrasse 6, 3012 Bern, Switzerland

[4]   GEOEND, Université Gustave Eiffel, Route de Bouaye - CS 5004, 44344 Bouguenais Cedex, France

[5]   UNICAEN, CNRS, LETG, Normandie Université, 14000 Caen, France

[6]   CNRS, IRD, OPGC, Laboratoire Magmas et Volcans, Université Clermont Auvergne, 63000 Clermont-Ferrand, France

1 Introduction

Robots flying through the sky are no longer a matter of science fiction. Unmanned aerial vehicles (UAVs), or drones, are usually considered as toys, mass surveillance tools or military systems. Besides, this technology is being increasingly implemented into many innovative projects worldwide, for sustainable development goals (Kitonsa and Kruglikov 2018), medical deliveries (Scott and Scott 2019), humanitarian purposes (Sandvik and Lohne 2014) or disaster mitigation (Erdelj et al. 2017). According to the United Nations, for the period 1998-2017, disasters associated with natural hazards killed 1.3 million people and affected 4.4 billion people (Wallemacq et al. 2018), along with economic losses of $2.9 trillion. In recent years, information provided by manned aircraft and satellites has demonstrated their efficiency during disasters over various temporal and spatial resolutions (from tens of m to km), due to constant up-to-date data availability and efficient GIS (geographical information system) solutions (Voigt et al. 2016). In this frame, UAVs provide an important opportunity to support disaster reliefs and make low-cost observations at the local scale, complementing aircraft and satellites when their deployment remains expensive (for instance, for kilometric size areas). Moreover, in the context of climate change, with extreme weather events increasing in frequency and amplitude, the use of UAVs may be of major interest to improve pre-disaster studies, crisis management and recovery. For example, the August 2017 extreme flood in Sierra Leone, in which the United Nations used drones to complement satellite data over landslides and flooded areas demonstrated how UAVs may assist emergency agencies (http://www.fao.org/sierra-leone/news/detail-events/en/c/1032705/). Indeed, for hazard studies, it is essential to investigate areas that are difficult to access. But, it is also necessary to easily change scales of analysis, from local to regional scales. UAV technology can be useful for the mitigation of natural hazards in several ways:

- A huge variety of civil drones as fixed-wing aircraft, multi-copters, rugged UAVs or waterproof systems may be provided for different scales studies and situations,
- Drones can collect multiple types of high-resolution data (from remote sensing to geophysical imagery) to improve early detection and crisis management,
- They can provide access to the web in damaged zones by extension of Wi-Fi,
- These systems can fly over damaged infrastructures, carrying low-weight supplies,
- They can collect massive amounts of data in a limited time (for instance, 15–20 min).

This article provides a review of the literature by examining the role of UAVs in geohazard responses and emphasizing the actual status of the technology. The recent developments from early detection to crisis management, using remote sensing, geochemical or geophysical data of several teams are presented in four different sections. The first section is dedicated to mass earth movements, the second to volcanology, the third to floods, and the last section focuses on earthquakes. The complementarity between data provided by UAVs and satellite imagery is discussed in the last part of this article.

2 UAVs Description

Numerous articles highlight the use of UAVs for detailed morphostructural studies (Gonçalves and Henriques 2015; Cawood et al. 2017; Chesley et al. 2017; Cook 2017), hazard researches (Gomez and Purdie 2016; Giordan et al. 2018) and landslide studies

(Mokhtar et al. 2014; Balek and Blahůt 2017; Casagli et al. 2017; Yu et al. 2017), as support of other remote sensing techniques and field monitoring. Indeed, UAVs provide low-cost aerial surveys, operational flexibility and a better spatial and temporal resolution (Agüera-Vega et al. 2017a). Moreover, the miniaturization of different sensors is underway (Fig. 1), as for the example of magnetic sensors for the measurement of the total magnetic field or of its three components. On-board radars are also in full development. Reflections from around mobile gravimetric sensors dedicated to density measurements are in progress (Limo-G system) (de Saint Jean 2008), with promising prospects for implementation on drones via microelectromechanical system (MEMS) technology (Middlemiss et al. 2016). Thanks to this development, many UAVs platforms as hexacopter (Hastaoğlu et al. 2019), quadcopter (Niethammer et al. 2012; Cook 2017; Peternel et al. 2017) and octocopter (e.g., Turner et al. 2015; Lindner et al. 2016) have been developed for science and civil applications (e.g., DJI Phantom and Mavic quadcopters are very famous). They can acquire a large number of high-spatial-resolution images, defined by their pixel size (0.03–0.08 m/pixel in Niethammer et al. (2012); 0.02 m/pixel in Rossi et al. (2018)) or ground sampling distance (GSD). GSD, generally expressed in cm/pixel, represents the distance between the centres of two consecutive pixels in the image on the ground that can be expressed with the simplified following function:

$$GSD = H * P/f$$

where H is the height of the UAV (m), P is the pixel size of the sensor (micro), and f is the focal length (mm). The bigger the value of the GSD, the lower the spatial resolution of the image and the less visible are details. These images are then used to obtain very quickly (a few minutes per flights), at heights generally lower than 100 m, different digital models as digital surface models (DSM), digital terrain models (DTM, which is a DSM with filtered vegetation), digital elevation models (DEM) with a spatial resolution of 0.02 m/pixel in Rossi et al. (2018) or orthophotographs with a spatial resolution ranging from 0.02 to 0.04 m/pixel (Niethammer et al. 2012; Rossi et al. 2018).

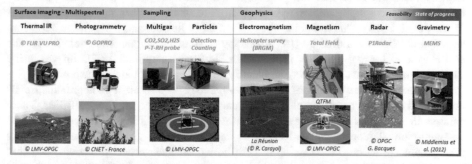

Fig. 1 Some examples of miniaturized sensors for drone applications, operational or in development in the field of imaging, sampling and geophysics

3 Methodology

One of the major interests of UAVs lies in the fact that several sensors can be simultaneously operated with RGB (red, green, blue), multi-spectral, or thermal sensors (Berni et al. 2009; Zarco-Tejada et al. 2012; Casana et al. 2014; Rossi et al. 2018; Antoine et al. 2019) and LiDAR. However, topographic surveys are possible through the simple use of a set of RGB aerial images with photogrammetric algorithms. Consequently, UAVs developed for fundamental or applied scientific research are traditionally used as a complementary tool to other ground investigations. For example, in morphometric studies, slope kinematics quantification or landslide detection, results from drone imagery interpretation must be calibrated by ground surveys. Concerning data acquisition, several protocols are possible with various parameters: number of photos, flight height, overlap between images, number of ground control points (GCPs). The choice mainly depends on the size of the study area (Chesley et al. 2017) and the expected resolution of the final 3D point cloud. Data acquisition protocols are variable in the bibliography (see Appendices 1 to 4). For 3D model construction, various Structure from Motion (SfM) software packages are now available for creating 3D models from photographs (Cook 2017). Among them, free solutions such as Visual SfM (Wu 2011, 2013) or MicMac (Rupnik et al. 2017) provide accurate solutions for data processing, but a commercial bundle such as Agisoft Metashape (2016) is the most frequently used software for image correlation and generation of 3D point clouds (Turner et al. 2015; Lindner et al. 2016; Balek and Blahůt 2017; Cook 2017). Detailed descriptions of the Metashape workflow (formerly called Photoscan) may be found in Turner et al. (2015).

Nevertheless, the geometric accuracy of 3D models and derived DEMs/DSMs also depends on the acquisition of GCPs by dGPS (Walter et al. 2009; Turner et al. 2015; Thiebes et al. 2016) or theodolite (Peternel et al. 2017). These GCPs are essential for georeferencing the images (Lucieer et al. 2014), and their homogenous distribution on both sides of the study area will influence the accuracy of the obtained point cloud (Agüera-Vega et al. 2017a). With these GCPs, photographs acquired by the drone will be georeferenced and processed in order to obtain a 3D point cloud of the area overflown, but some solutions of direct georeferencing without GCPs exist (Turner et al. 2014; Gabrlik 2015; Mian et al. 2015). Indeed, the SfM and multi-view-stereo (MVS) photogrammetric techniques for airborne or ground images processing are the most widely used photogrammetry techniques in topography studies, to generate high-resolution DEMs and orthophotographs (Turner et al. 2015). This technique allows for a fairly flexible and inexpensive 3D reconstruction of photos acquired by UAV (Westoby et al. 2012; Valkaniotis et al. 2018). The aim is to model a real object or landscape in 3D from a multitude of 2D images, using various algorithms that can detect and identify similar elements between two pictures (i.e. "scale-invariant feature transform") (Lowe 1999, 2004). As a result, a photogrammetric 3D point cloud created from drone photographs can be interpolated into DEMs/DSMs. These datasets are essential for landslide studies (Casagli et al. 2017). The SfM technique is based on multi-view stereopsis (MSV) techniques that provide 3D structure from overlapping photography (at least a 60% overlap in both horizontal and vertical directions) acquired from multiple angles and heights. SfM/MSV techniques are particularly useful and provide dense 3D point cloud and high-resolution digital terrain models with high-spatial resolution (with centimetric to sub-decimetric accuracy) with a possible high temporal resolution. Mesh models (DSMs) generated by triangulation of the dense point clouds can be created using various algorithms, such as inverse distance weighted (IDW) ones (Comert et al. 2019).

4 Mass Earth Movements

4.1 What Do We Consider as Landslides and How Can We Study Them?

Mass earth movements frequently occur in many disasters involving people and the damage or the destruction of many infrastructures. Every year, mass earth movements (whether landslides, debris flows, mudflows…) trigger in urban or rural areas in various geographical contexts (coastal, mountain). These phenomena can be extremely rapid or slow, in various geological contexts, in saturated or unsaturated materials, in steep channels or without confinement in established channels. Triggering factors are numerous, as they can be associated with external forcing (e.g., meteorological event), or internal forcing (e.g., earthquake) or combination of both (Tang et al. 2019). Nevertheless, the predominant factor is hydro-climatic forcing. In that case, instabilities can be triggered either by accumulated rainfall inducing the progressive saturation of the soil and the rise of the water tables, or by high intensity rainfall events such as thunderstorms and/or tropical storms. Therefore, in a context of global change where precipitation (rates and nature) may change, several studies have focused attention on effects of global warming on heavy rainfall frequency or tropical cyclones intensity (Edenhofer et al. 2014). Consequently, we wonder about the impact of these changes on the occurrence of hydro-gravitational phenomena (frequency, intensity, location). Several types of mass movement exist (Hungr et al. 2014) and their complexity involves a large number of approaches and methods. In order to reduce damages caused by landslides, it is therefore essential to have a precise knowledge of the hazard, especially by mapping its extension and intensity to identify vulnerable areas where elements at risk exist (Graff et al. 2019). In this context, and for operational intervention for risk management, UAVs are essential for many applications. Widely used for hazard inventory mapping and identification of unstable areas (Comert et al. 2019), as ground investigation can be time-consuming, UAVs make possible a fast intervention in post-hazard events. Thanks to high-resolution digital cameras (RGB, but also other sensors such as multi-spectral systems), we can easily obtain high detailed 3D point clouds, DEMs/DSMs and orthomosaics, orthophotographs to easily get a lot of information of the studied areas.

In landslide studies, UAVs facilitate rapid identifications of instabilities after major events with accurate data over areas of several km^2. The use of UAVs is consequently very effective during crisis situations, when quick identification of affected areas (e.g., road blockages linked to sediment flows for rescue operations) is necessary but can be tedious for very large areas (several hundred km^2). In this case, very high-resolution (VHR) satellite imagery will be favoured. Indeed, various satellite systems (e.g., Pleiades system) provide last-minute requests for images acquisition. These data are essential for precision mapping and intervention for public safety (Voigt et al. 2016; Lang et al. 2018). However, according to the geographical context (e.g., slope orientation, tropical area), the acquisition of high spatial, temporal and spectral resolution imagery just after a crisis can be useless. For instance, in tropical/subtropical areas, instabilities mainly occur during rainy (and thus cloudy) season (Saito et al. 2014), when optical imagery is difficult to implement. To overcome this problem, interferometric synthetic aperture radar (InSAR), airborne light detection and ranging (LiDAR) or airborne laser scanning (ALS) can be good alternatives for risk assessment and landslide monitoring (Delacourt et al. 2007; Jaboyedoff et al. 2012; Scaioni et al. 2014). Indeed, InSAR offers frequent and free solutions for landslide assessment (e.g., Sentinel-1) (Barra et al. 2016) with image acquisition during day and night and all weather conditions. However, this technique is not applicable everywhere because it depends on the landslide kinematics (inappropriate for slow moving landslide), and its orientation (Casagli et al. 2017). The other solution is ALS, a highly adaptable technique providing high-resolution 3D point clouds (Delacourt et al. 2007; Jaboyedoff et al. 2012), which is a great advantage for landslide detection (Mezaal et al. 2018). One of the major drawbacks of LiDAR technology is the cost of data acquisition, especially with manned aircraft. Drones are now equipped with LiDAR sensors, and also multi- and hyperspectral sensors (Morsdorf et al. 2017), but their operational use is still not straightforward (Piégay et al. 2020). In this context, UAVs equipped with optical sensors provide valuable information for hazard identification and disaster assessment with a high potential to replace expensive or time-consuming tasks. This is why the use of UAVs, among landslide monitoring techniques, is spreading widely (Giordan et al. 2020).

With a Google Scholar search for keywords "UAV, landslide," over 6000 items published since 2015 were returned. For our review, a body of 28 articles have been selected from articles found by a search by keywords in Google Scholar, and Science Direct database (i.e. "UAV, drone, photogrammetry, mass movement, landslide, remote sensing"). In the present cases, implementation of UAVs for landslide survey deals with three main scientific issues: (1) hazard mapping, (2) morphological analysis and (3) landslide physical evolution (Appendix 2).

4.1.1 Landslide Hazard Mapping and Morphological Analysis

Hazard mapping is possible at a medium or large scale, in order to carry out inventories of mass earth movements just after a specific event triggering (Valkaniotis et al. 2018; Comert et al. 2019; Tang et al. 2019), or to locate historical landslides of a region. In both cases, landslide identification can be conducted by manual analysis of orthoimages and 3D models (point clouds, 3D meshes, DEMs) from UAVs. Thus, a simple 3D visualization of a coloured point clouds can help to identify morphological features associated with landslides. Numerous studies are also based on topographic, landscape change detection performed on DEMs (Van Den Eeckhaut et al. 2012; Aditian et al. 2018; Bunn et al. 2019) and/or 3D point clouds acquired at different times. The landscape change detection is usually performed automatically using algorithms developed to compare 3D point clouds. Among them, Cloud-to-Mesh distance (C2M)/Cloud-to-Cloud distance (C2C) tools (Girardeau-Montaut et al. 2005) and M3C2 algorithm (Lague et al. 2013), implemented in the CloudCompare software (CloudCompare v.2.5.4 and newer), are the most used for 3D point cloud analysis (Valkaniotis et al. 2018). Indeed, by comparison of multi-temporal datasets (before and post-failure 3D point cloud or DEM), it is then possible to quantify the amount of removed sediments (Valkaniotis et al. 2018). Landslide inventory maps are essential for risk assessment to avoid the exposure of goods and people to hazard. Inventory maps provide quantitative and qualitative information on landslide hazard such as boundaries, main and secondary scarps, shape, size and depth. Inventories are traditionally based on manual interpretation of aerial images/orthophotographs interpretation and DEM derivatives (Görüm 2019). Several derivatives [e.g., local dominance, positive openness, negative openness, terrain ruggedness index (TRI), local relief model (Fig. 2)] can be exploited from DEM dataset, to extract information

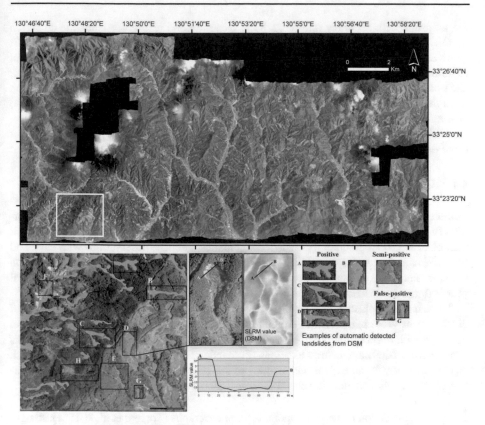

Fig. 2 UAV-based map inventory of multiple shallow landslides triggered in July 2017, Kyushu Island, Japan (photogrammetry process C. Gomez, Kobe University 2017) (Reproduced with the permission of the authors C.Lissak, V. Siccard 2020, Caen University) (Lissak et al. 2019)

on terrain morphologies from topographic anomalies. However, manual interpretation and analysis have been progressively replaced by the automation of landslide recognition with pixel-based image analysis (PBIA) of orthophotographs, or by object-based image analysis (OBIA) of orthophotographs or DEMs and DEM derivatives (Stumpf and Kerle 2011; Lahousse et al. 2011; Rau et al. 2012; Van Den Eeckhaut et al. 2012; Moosavi et al. 2014; Comert et al. 2019). OBIA is based on multi-data combination with metrics analysis with spectral (e.g., Green–Red Vegetation Index-RGVI for RGB images), geometric, contextual, and textural information of image objects defined, DEM derivatives (e.g., slopes, curvature) (Fig. 2) and seems to be a robust method for landslide recognition or microtopography assessment (Rau et al. 2012; Stumpf et al. 2013). At a local scale (1:2000–1:10,000), UAV datasets (orthophotographs, DEMs) associated with field observations (e.g., geotechnical, geophysical, (geo)morphological, geological, hydro(geo)logical surveys), provide high-resolution data to detect and characterize landslides. For example, high-resolution imagery from UAVs makes it easier to locate and provide details of cracks and fissure structures (Walter et al. 2009; Stumpf et al. 2013) by photo interpretation, image and/or 3D point clouds analysis (Akcay 2015), and gain knowledge of slope morphology associated with landslide activity. Such methods may be especially useful for large area analysis, where several hundreds of landslides must be mapped. Comert et al. (2019) show the OBIA can recognize the number and the area of the landslide with accuracy rate of ca. 83%. Despite good results for landslide detection by OBIA approach, PBIA approach is still the predominant method (Casagli et al. 2017; Bunn et al. 2019). In this context, the employment of multi-spectral images can be useful to automatically detect landslides, especially with the use of spectral vegetation indices such as NDVI values (Lin et al. 2004; Yang et al. 2013) to distinguish the vegetation from the soil (Berni et al. 2009). Whereas collecting RGB-images by UAV is simple and cost effective, integrating other remote sensing sensors on UAVs, such as thermal cameras (Guilbert et al. 2020) or micro-hyperspectral imagers, may be a good solution, but more expensive.

4.1.2 Quantification of Slope Deformation and Volumetric Changes Over Time

Remote sensing approaches are frequently applied for landslide assessment and monitoring (Petley et al. 2002; Delacourt et al. 2007; Jaboyedoff et al. 2012; Tofani et al. 2013; Casagli et al. 2017). To quantify landslide kinematics, airborne observations with UAV and SfM techniques give the possibility to investigate landslides in high temporal resolution with repeated flight campaigns from several days or months (Lindner et al. 2016) to several years (Turner et al. 2015).

To quantify displacement rates and acquire information in magnitude and direction of displacement vectors, a visual interpretation of the morphological changes with time can be useful with observation of specific objects such as, for example, the locations of blocks (Walter et al. 2009) and/or tree locations (Fernandez Galarreta et al. 2015; Peternel et al. 2017). However, for an exhaustive mapping of surface displacements and the characterization of the landslide evolution, several other techniques are possible:

1. image correlation of multi-temporal orthophotographs (Torrero et al. 2015; Hastaoğlu et al. 2019),
2. comparison of 3D point clouds (Stumpf et al. 2015) based on automatic process of Cloud-to-cloud differences (Cook 2017; Eker et al. 2018) and using an open-source algorithms such as the M3C2 algorithm,
3. by comparison of DEM (DEM of Difference, DoD) (Fernandez Galarreta et al. 2015; Lindner et al. 2016; Thiebes et al. 2016; Rossi et al. 2018).

The multi-temporal model analysis can be based on DEMs created from UAVs photogrammetric surveys (Fig. 3) (Lucieer et al. 2014; Turner et al. 2014; Lindner et al. 2016; Rothmund et al. 2017; Tanteri et al. 2017; Eker et al. 2018; Rossi et al. 2018) but also compared with models created by airborne or terrestrial LiDAR surveys (Rossi et al. 2016; Thiebes et al. 2016). The interpretation of data obtained from drones must be calibrated with ground surveys and monitoring campaigns because, according to image or 3D point cloud resolution, statistical differences are possible. Various studies combine airborne techniques for landslide kinematics and field monitoring with terrestrial laser scanner (TLS), terrestrial SfM photogrammetry combined with permanent GNSS (Thiebes et al. 2016), theodolite surveys (Peternel et al. 2017) and inclinometers (Rossi et al. 2016).

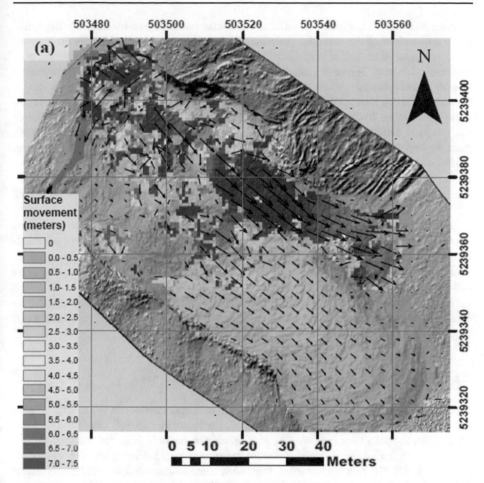

Fig. 3 Surface deformation map generated using the software COSI-corr and two different DSMs of the same area representing the surfaces at different times (from Turner et al. 2015)

5 UAVs for Volcanology

5.1 Why Use UAV Measurements to Reinforce the Monitoring of Volcanoes?

Because volcanic phenomena integrate various spatiotemporal scales, multi-disciplinary and complementary approaches are necessary. Volcanic field observations and analyses are essential, including photos, videos of eruptive phenomena, physicochemical sampling and geophysical imagery. However, in situ data acquisition remains difficult or even impossible due to complex settings (e.g., topographic barriers and danger due to eruptions). Using UAVs as instrumentation vectors will overcome these limitations by providing a more homogeneous, extensive, safer and faster prospecting. Today, this is an essential advance to improve our understanding of volcanic systems and to efficiently monitor them.

The choice of the UAV is, at the first order, a question of scale: measurements carried out at high altitude will cover regional and deep structures or atmospheric phenomena, whereas measurements close to the ground will focus on local and more shallow systems or phenomena at even higher resolution. It will also be greatly conditioned by the sensitivity and weight of the sensor(s) that should be mounted, and that depends on the volcanic issues to be addressed (e.g., measurements of surface, internal and atmospheric phenomena) (Gonzalez Toro and Tsourdos 2018). Amongst these problems, it is fundamental to map and characterize active volcanic structures on the surface, such as lava flows (De Beni et al. 2019), intrusions (Dering et al. 2019), volcano-tectonics features and deformation (Bonali et al. 2019). Repeated UAV measurements over time enable quantification of the temporal evolution of these structures (Derrien et al. 2015; Darmawan et al. 2018). Another important issue in volcanoes monitoring is to quantify heat flux associated with volcanic activity at the surface and in near-real-time through mapping the spatial distribution of thermal expressions (Walter et al. 2020). More in-depth information on fluid transfers and magmatic dynamics is now provided thanks to the development of UAV geophysical measurements such as, for example, magnetism (Catalán et al. 2014) and/or gravity (Middlemiss et al. 2016). Sampling and analysing volcanic products and gas emissions dynamics through UAV physicochemical measurements could also be envisioned with dedicated sensors (Gomez and Kennedy 2018; Rüdiger et al. 2018; Terada et al. 2018; James et al. 2020). The ultimate goal is to perform combined approaches including simultaneous and complementary measurements (e.g., Mori et al. 2016).

We now present a (non-exhaustive) list of vectors and sensors (operational or in development) commonly used or that could be used for volcanology, and some applications and developments.

5.2 UAV and Physicochemical Sensors for Studying Volcanoes

Vectors: The emergence of UAV, together with rapid and continuous technological progress, now offers an intermediate detection scale between regional satellite and local ground observations (Kreye et al. 2006). It opens the doors to high-resolution measurements, at both spatial and temporal scales, to enhance volcanic systems imaging and monitoring (Fig. 1). Different vectors exist, each with specific characteristics and performance (flight altitude, time and payload). As an example, the U.S. Department of Defense has provided a categorization into five groups (Marshall et al. 2015) and various syntheses have been proposed (Fahlstrom and Gleason 2012; Watts et al. 2012; Villa et al. 2016).

Sounding balloons have been designed to make measurements at high altitude in the stratosphere (Everaerts 2008). They are of main interest to study the volcanic products into the atmosphere (gas and particles) for meteorological, socio-economic and aeronautical issues. Fixed-wing drones such as long-range flying systems have basically limited payload but offer the possibility of flying for several hours for homogeneous investigation of large areas (Saggiani et al. 2006). Big rotor wing drones (octopters or hexacopters) have more limited flight times, but larger payload and could be used as multi-method platforms. The lightest quadcopter-type drones (see Villa et al. 2016 for details) have even more limited capacities in terms of flight time and payload, but have the advantage of being very easy and rapid to implement in the field. For given applications, a swarm of UAVs flying on similar or complementary trajectories could also be deployed (Techy et al. 2010). More recently, unmanned underwater vehicles (UUVs) have been developed to operate, among others, in submarine volcanic contexts (Caratori Tontini et al. 2019). UUV can be remotely controlled in real time (remotely operated vehicles, ROVs) or not (autonomous underwater vehicles, AUVs).

Sensors: A large range of applications may be envisioned for volcanoes imaging and monitoring (Fig. 1). For several years, UAV photogrammetry has now been fully operational for imaging volcanic structures and their evolution on the ground surface (e.g., Derrien et al. 2015; Bonali et al. 2019; De Beni et al. 2019). On-board cameras acquire photos from different points of view to derive an orthophotography (e.g., geometry, colour, texture, relief) of a selected volcanic target. One of the most interesting applications is to generate high-resolution DEMs (i.e. multi-centimetric; Harvey et al., 2016 and references therein). UAV LiDAR (light detection and ranging) surveys are also commonly used to construct DTMs at infra-metric resolution (e.g., Jones et al. 2009; Assali et al. 2014; Casini et al. 2016). The miniaturization of infrared (IR) cameras (<500 g) also offers the opportunity to carry out acquire thermal images rapidly (e.g., Wessels et al. 2013; Walter et al. 2020). Temperature information on the scale of an entire eruptive site in areas otherwise inaccessible such as active eruptive centres (Mori et al. 2016), lava flows (Spampinato et al. 2011; Blackett 2017 and references therein) and/or fracturing centres (Schneider et al. 2005) can be acquired.

UAV geophysical measurements and physical parameters characterization are also in development to monitor volcanic activity at greater depth. Magnetic sensors progressively become operational (Funaki et al. 2014); the development of gravity sensors is also in full swing for the last few years Middlemiss et al. 2016).

Regarding UAV sensors for sampling of volcanic products, in situ measurements remain challenging for physical and chemical characterizations of volcanic emissions (plumes of gas and ashes). Based on the in-depth knowledge of atmospheric teams, custom sensors (Hervo et al. 2012; Picard et al. 2019) are now being developed in order to: (1) measure the total concentration of particles and their size spectrum, (2) sample particle aggregates, (3) sample ashes with a collector for quantifying the flux of particles within the plume and their size distribution and (4) quantify major volatile components (H_2O, CO_2, H_2S and SO_2) with a MultiGas system and a [P, T, RH] (pressure, temperature, relative humidity) probe.

5.3 Some Examples of Application

At the scale of volcanic edifices, various phenomena interact at different spatiotemporal scales (e.g., stress variations, deformation, fracturing, fluids transfer, thermal variations). Those phenomena are driving a large variability of physical properties in volcanic reservoirs, shallow structures and emitted products that could be efficiently addressed thanks to UAVs measurements. Some of the most striking examples of applications are presented in the following sections.

5.3.1 Imaging Volcano-Tectonic Structures and Their Temporal Evolution

Deformations linked to volcanic activity are commonly due to magma transfers, precursors of eruptions, and require almost real-time monitoring to better follow and prevent eruptions. UAV photogrammetry has great potential in monitoring such dynamic processes (lava flows monitoring) (Favalli et al. 2018; De Beni et al. 2019). Right now, most of the drones are equipped with high-resolution cameras. The advantage of UAV photogrammetry is to repeat surveys and, therefore, to rapidly emphasize the evolution of structures with successive photogrammetric models (e.g., before and after an eruption). A recent striking

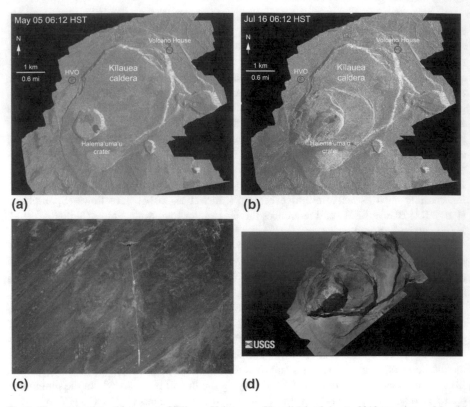

Fig. 4 Changes to the caldera area of Kilauea Volcano on Hawaii Island due to 2018 eruption: **a** May 05 and **b** July 16 (extract from GIF, JAXA). **c** Drone survey for photogrammetry and sampling. **d** 3D model from aerial photographs. Credit: U.S. Geological Survey

example of the potential of UAV photogrammetry was the monitoring of the evolution of Kilauea volcano (Hawaii), all along its main eruptive crisis in 2018 (Fig. 4a and b) (Patrick et al. 2019). Successive flights were carried out with an octocopter (Fig. 4c) by the USGS-Hawaiian Volcano Observatory scientists (https://volcanoes.usgs.gov/volcanoes/kilauea/multimedia_chronology.html), to build a 3D model and to follow the evolution of the volcano-tectonic structures at Kilauea's summit (Fig. 4d) (Patrick et al. 2019). It also enables to follow the water pond level evolution that appeared one year after the collapse event in the deepest portion of the crater.

5.3.2 Quantifying Heat Flux, Thermal Anomalies and Mass Transfers in Near Real Time

a. Infrared imagery

Mapping thermal anomalies, fluid movements, and quantifying the associated heat fluxes is commonly performed through IR thermal imaging at large scale by satellites (Bato et al. 2016), by helicopters or in situ by hand-held ground IR cameras (Harris et al. 2005; Antoine et al. 2017) and more recently on UAVs (e.g., Mori et al. 2016; Thiele et al. 2017). Following variations in the distribution and intensity of thermal anomalies thanks to the repetition of IR measurements enables to quickly and evenly image active areas: volcanic vents (e.g., Harris and Stevenson 1997), domes (e.g.,

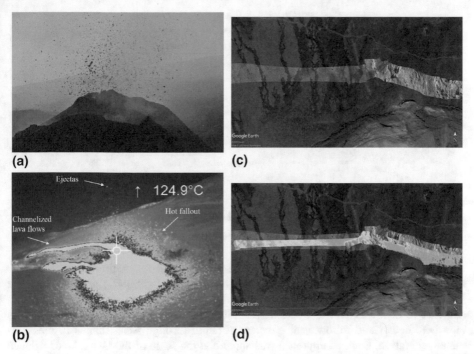

Fig. 5 **a** Eruption of Etna in July 2014 on the northeast crater (©R. Paris, LMV) and associated with **b** IR thermal imaging by drone (OPTRIS PI 450, ©Technivue and P. Labazuy, LMV-OPGC). **c** Kilometric high-resolution visible orthoimage (1.3 cm/px @ 50 m height) of the Northern Part of Dolomieu crater superimposed with Google Earth data (©R. Antoine, CEREMA/ENDSUM, SlideVOLC ANR project) and **d** IR thermal orthoimage (FLIR Vue Pro, 9.4 cm/px @ 50 m height), superimposed with the drone and Google Earth Data (©R. Antoine, CEREMA/ENDSUM, SlideVOLC ANR project)

Pallister et al. 2013), lava flows (e.g., Calvari et al. 2003; James et al. 2006) or even fumarolic zones (e.g., Harris and Baloga 2009). The thermal state of an overall eruptive site and ejectas (Turner et al. 2015) as well as the channelization of a lava flow (Patrick et al. 2017) could be accurately imaged. As an example, the IR airborne survey carried out using an octocopter during the 2014 eruption at Mount Etna has provided a high-resolution map of the active vent and associated products, otherwise invisible in surface (Fig. 5a, b) (Labazuy 2015). 3D temperature maps are now developed for IR thermal data, using the principle of photogrammetry and producing a so-called thermogravimetric model (Fig. 5c,d) (Peltier et al. 2018). This survey was done in 2018 at a kilometre scale on Piton de la Fournaise, using a DJI Phantom 4, highlighting the main thermal characteristics of the Northern part of the Dolomieu crater. Such a thermal photogrammetric approach could also be used to monitor dome eruptions (Thiele et al. 2017).

b. UAV magnetic measurements

Beyond visible and infrared measurements, UAVs are also dedicated to conduct geophysical measurements for more in-depth studies. Magnetic field measurements are particularly relevant in volcanology to map structural contacts between formations of different ages or nature. They are also a powerful tool for imaging deep thermal anomalies and intrusive systems because there is a strong influence of temperature on the

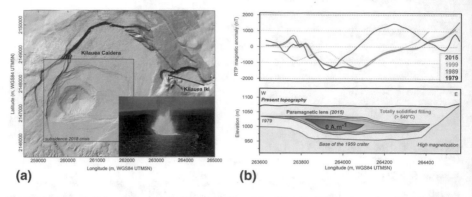

Fig. 6 Monitoring of the cooling of the Kilauea Iki lava lake (Hawaii, 1959) using reiteration of magnetic measurements along the profile located on the left. ©Photo: USGS-HVO; Composite terrain: USGS-HVO, opentopo.org

measurements of the magnetic field and associated local magnetic anomalies. Since magma and basaltic rocks above the Curie temperature (~ 580 °C) are paramagnetic (i.e. null magnetization), a loss of magnetization of volcanic formations will occur in response to heat transfers or hydrothermal alteration (Gailler et al. 2016). The advantage is to detect and follow heat transfers at depth through time, thanks to measurements reiteration. Such an approach was applied at the scale of the lava lake of Kilauea Iki (Hawaii), where the evolution of magnetic anomalies has enabled reconstructing the evolution of the still thermally active magmatic lens from 1959 to 2015 (Fig. 6) (Gailler and Kauahikaua 2017). Thanks to the miniaturization of magnetic sensors, it is now possible to conduct such magnetic surveys using UAVs in order to detect efficiently any temporal changes in magnetic signals due to volcanic activity (Catalán et al. 2014).

c. Future developments: UAV gravity measurements

UAV gravity measurements are even more challenging due to a strong effect of in-flight acceleration on such measurements. They are, however, in full development with the MEMS technology (Middlemiss et al. 2016). These measurements will complete magnetic measurements, help to better characterize magmatic systems (Magee et al. 2018) and evaluate mass transfers in depth at the scale of intrusions, conduits and reservoirs (Blaikie et al. 2014).

5.3.3 Physicochemical Measurements for Analysing Volcanic Gases

Another main natural hazard of volcanoes explosive activity concerns ash plumes and related products (gases, primary aerosols), within and in the vicinity of volcanic plumes, fumarolic zones or eruptive vents. Among them, knowledge of the composition and concentration of particles and gases from volcanic plumes is essential for understanding the processes related to their evolution in the atmosphere and associated climatic issues. Their importance was highlighted during the eruption of the Icelandic volcano Eyjafjallajökull (April–May 2010), in particular for the management of the aeronautical crisis in real time by the VAACs (Volcanic Ash Advisory Centers) (Labazuy et al. 2012; Millington et al. 2012). Such measurements are performed mainly by means of satellite or ground remote sensing in the infrared domain and may now be envisioned thanks to UAVs in situ measurements (see Gonzalez Toro and Tsourdos 2018 and references therein). Here, one of the major scientific goals is to quantify the composition and dynamics of volcanic plumes (particle concentration flux and mass, velocity field) (Pieri et al. 2002) to monitor the gas emission rate inside plumes (Rüdiger et al. 2018) and the interactions between all of these key parameters.

All these complementary techniques will considerably improve our knowledge of volcanic dynamics (deformation, fracturing, landslides, magmatic and hydrothermal transfers) to better prevent eruptive precursors, associated hazards and their management, in real time.

6 Flood Monitoring and Management

Satellite remote sensing has been extensively used over the last 30 years for detecting, monitoring or modelling flood events (Smith 1997; Klemas 2015; Domeneghetti et al. 2019). However, the temporal and spatial resolutions of data acquired by space-borne sensors are not always adapted to fast evolving events and to fine-scale analysis (Schumann and Domeneghetti 2016; Ridolfi and Manciola 2018). Satellite data may also be affected by cloud cover, often dense and persistent during flooding. Manned aircraft can overcome these constraints. Flying below clouds, they do not have revisit limitations, and produce higher accuracy and resolution data (Feng et al. 2015). Nonetheless, their important cost, the time needed to plan and validate a flight, as well as flight restrictions, have limited their implementation for flood related studies.

Conversely, UAVs have gained increased interest over the last few years (Appendix 3). Cheaper to operate and more flexible than manned aircraft, they can be deployed for rapid monitoring and mapping of flooded areas, along with routine studies (Feng et al. 2015; Popescu et al. 2017). In addition, their ability to operate at low altitude ensures the acquisition of accurate data and meets the requirements of pre- and post-crisis flood monitoring and management. This section presents examples of the use of UAVs for flood management support.

6.1 Flood Observations and Flood Extent Mapping

One of the first examples of UAV deployment in response to a flooding disaster occurred during Hurricane Katrina in 2005. Small multi-copters and fixed-wing UAVs, equipped with RGB video cameras, were deployed immediately in the aftermath of the hurricane (Murphy et al. 2016). The video feeds were used to locate people and to determine potential threats of the Pearl River. Since then, drones equipped with RGB or IR thermal sensors have been frequently used worldwide for flood observation and to help to rescue people during or right after the peak of flooding (Fernandes et al. 2018). The main interest of digital cameras on drones during the emergency lies in the supply of real-time video scenes transmitted to ground operators or crisis management services. Visual assessment of the flood extent, debris and damage is indeed often enough for experts to make the first decisions and to deploy first emergency responses (Murphy et al. 2016).

Accurate mapping of inundated areas is also particularly important to prioritize and better organize the response actions on the impacted areas. Surprisingly, to our knowledge, no operational tool, allowing for real-time or near-real time flood extent mapping using UAVs, and tested under real conditions, is available at this time. Nonetheless, results from the few studies that have specifically addressed this topic are rather promising. For instance, Popescu et al. (2017) have reached an accuracy of 99% in detecting flooded areas using an innovative segmentation approach in a set of 50 images acquired over a small rural flooded area in Romania, using a RGB digital camera mounted on a fixed-wing UAV. More recently, Gebrehiwot et al. (2019) has presented results obtained using a deep learning classification approach, based on convolutional neural networks (CNN). The approach has been tested on three sets of data (two sets of 30 images and one set of 70 images), acquired over three different urban flooded areas in the USA, using RGB cameras mounted on fixed-wing and multi-copter UAVs. It achieved about 97.5% accuracy in extracting flooded areas. Feng et al. (2015) have also used a RGB sensor to monitor serious waterlogging in a complex urban environment, in Yuyao (China). A large set of 400 images was processed using a hybrid method combining grey-level co-occurrence matrix texture features and random forest classifier. The approach showed good performance in urban flood mapping, with an overall accuracy of 87.3% (Fig. 7). Despite its high accuracy in flood extent mapping, the machine learning approach faces the issue of the time required to pre-process and analyse sets of several tens of very high-resolution images [e.g., up to thirteen hours in Gebrehiwot et al., (2019); nine hours in Feng et al. (2015)]. This is an important barrier to the operational use of such methods for crisis management. If multi-spectral visible sensors are the most frequently used over flooded areas, examples of flood detection using active microwave or infrared sensors, which have already proved useful for flood extent mapping from space or airplanes (Smith 1997; Sanyal and Lu 2004; Klemas 2015) are extremely scarce. Visible sensors being only efficient during the day, there is currently no solution allowing for flood extent mapping using UAVs at night.

Imam et al. (2019) considered an alternative approach, using data collected by UAV-based global navigation satellite systems (GNSS) passive radar sensor to detect water bodies on ground. Their approach has been tested in Northern Italy on water surfaces (rivers, lakes, ponds, etc.) from small to large sizes, using a custom-made GNSS-R sensor and a multi-copter. It proved successful at detecting the boundaries between ground and water with few tens of metres accuracy and at estimating the surface water extension with an extremely high accuracy (about 92%). However, the capacity of this approach to generate

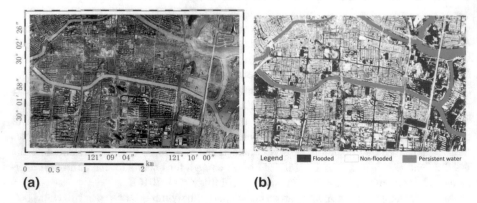

(a) (b)

Fig. 7 **a, b** Orthoimage and flood extent map (right) using a hybrid method combining grey-level co-occurrence matrix texture features and random forest classifier, applied to an orthomosaic of RGB images (left) acquired after a flood in Yuyao (China). Modified from Feng et al. (2015)

instantaneous maps, and not to provide measurements in the form of surface tracks, which have less value for flood monitoring operations, has not yet been demonstrated.

6.2 Data Acquisition for Numerical Flow Modelling

The potential of UAVs systems to provide accurate data for flood modelling at fine scale and in complex environments has also been investigated (Leitão et al. 2016; Langhammer et al. 2017; Yalcin 2018; Rinaldi et al. 2019). Their results agree on the following points: (1) topographic data derived from UAV-on-board RGB camera are comparable to DEMs generated from traditional aerial LiDAR data in term of accuracy and (2) such topographic results are appropriate for fluvial modelling in both rural and urban landscapes, which opens the way towards the use of low-cost sensors for such applications.

Research on water level estimation using UAVs systems is less documented in the literature (Bandini et al. 2017; Ridolfi and Manciola 2018). Bandini et al. (2017) have tested the capacity of three different payloads (a radar, a sonar and an in-house developed camera-based laser distance sensor) to estimate water levels on a small lake in Denmark. Water level estimations were obtained by subtracting the measured range to water surface from the vertical position retrieved by the on-board GNSS receiver. Interesting performances were obtained using the radar sensor, with measurements accuracies better than 5 cm. The approach proposed by Ridolfi and Manciola (2018) substantially differs. A RGB camera mounted on a multi-copter was used to retrieve water levels on a dam site in Italy. Water edge along the dam is extracted from a mosaic of orthorectified and calibrated images, using a supervised classification procedure accounting for edges and pixels colour, texture and contextual information. Water levels are then estimated by determining the distance between GCPs placed on the dam and the edge of water. Again, experimental results proved appropriate for flood modelling, with accuracies around 5 cm.

Lastly, it is also worth mentioning the interest given by authors to quantitative surface velocity measurements using UAVs and video cameras. The algorithms commonly used for this purpose include the large-scale particle image velocimetry (LSPIV) algorithm (e.g., Tauro et al. 2016), the Particle Tracking Velocimetry (PTV) algorithm (Eltner et al. 2020; Koutalakis et al. 2019), or the Kanade–Lucas–Tomasi (KLT) algorithm (e.g., Perks et al. 2016). The main challenge faced by these studies is the image pre-processing (co-registration and motion correction) extracted from video frames. Despite these difficulties, flow velocity estimations obtained using these algorithms are relevant and generally stay within an acceptable range of error from reference measurements, when available [around 0.15 m s^{-1} in Perks et al. (2016); around 0.25 m s^{-1} in Tauro et al. (2016) and around 0.03 m s^{-1} in Eltner et al. (2020); no reference measurements are available in Koutalakis et al. (2019)].

6.3 Monitoring of Flood Protection Structures

Fluvial and coastal structures such as levees, sea dikes and dams are frequently subjected to high loads, violent flows or debris flow impacts. On structures with pre-existing structural deficiencies, this may result in the development of major disorders and to their failure. Monitoring of such structures, before, during and after severe events enable the implementation of adequate, and often less costly repairs, before major failures occur (King et al.

2017). Several studies have recently examined the use of cameras mounted on drones to perform noninvasive and time efficient monitoring of these structures, at reasonable cost.

For instance, King et al. (2017) used a commercial multi-copter UAV to perform a survey of a 25-km length complex coastal structure at Byron Bay, Australia. A set of 455 images was processed using the SfM algorithm, helped by GCPs georeferencing. The authors reported some difficulties during the photogrammetric processing of the images, due to the presence of waves not handled by the SfM algorithm. Nevertheless, the obtained orthophotograph and DEM (RMSE of 0.068 m) were precise enough to reveal structures not observed from a classical land-based visual inspection. Brauneck et al. (2016) recorded the beginning and the evolution of a levee failure along the Elbe river (Germany), using a UAV-on-board video camera. Sets of images were extracted from five video frames and were processed into 4 DSMs, after masking their dynamic parts (flowing water areas). The main issue faced by the authors was the absence of georeferenced GCPs, which could not be positioned before the flight due to the risks involved in the area. Instead, GCPs were created by extracting the coordinates of static objects present on the images. The obtained DSMs and orthomosaics were combined to characterize the evolution of the breach and of the flow discharge within it, using a numerical 1-D hydrogeological model.

It is of note that the use of LiDAR data for DSMs generation of fluvial or sea protection structures remains less frequent. This may be explained by the important cost of LiDAR sensors compared to RGB cameras. However, optical images fail to identify disorders on a dike covered by vegetation, and accuracy of generated DSM (usually, several centimetres) are not suitable for the characterization of disorders such as dike settlement or initiation of slope sliding (Tournadre et al. 2014). For this last point, Zhou (2019) developed an innovative approach in order to generate a very high precision DSM (1 cm vertical accuracy) of such structures using a combination of oblique and nadir optical images plus embedded GNSS and only one GCP. The inclusion of oblique images helps decrease image deformations. It compensates the effects of focal lengths drifting on nadir images, which leads to more accurate camera pose estimation, and it also decreases the effects of focal lengths variations due to camera temperature changes. A camera readout time calibration method is also integrated to the approach, in order to correct the rolling shutter effect of the camera on the image measurements.

Important efforts have been made very recently for the development of operational systems for dike monitoring using UAVs. The DiDRO project (DIke monitoring by DROnes) aims at developing a solution for routine and crisis monitoring of dikes, using UAVs carrying multiple remote sensing, aquatic and geophysical equipment (Antoine et al. 2019). The routine monitoring mode includes a LiDAR, IR thermal, near-infrared and visible sensors, and provides high-resolution data of the surface. These sensors, never combined before on a UAV for this type of survey, allow the detection of many surface indicators of internal and external dike disorders. They can be carried out whatever the UAV (helicopters or multi-copters), using a structure specifically developed for these needs (Fig. 8).

Data are processed to generate 3D information of the dike (multi-spectral points clouds, DSMs, DEMSs, orthophotographs, profiles) and will be available for interpretation in a dedicated GIS web platform (Fig. 8d). The crisis monitoring mode uses a fixed-wing UAV equipped with a visible and IR thermal camera with a 360° rotation capacity. It allows day and night in-flight video transmission to the emergency services. This mode can be complemented by aquatic measurements, which consist (1) in surface velocity measurements using floating targets dropped in water by the UAV and (2) in water turbidity estimation

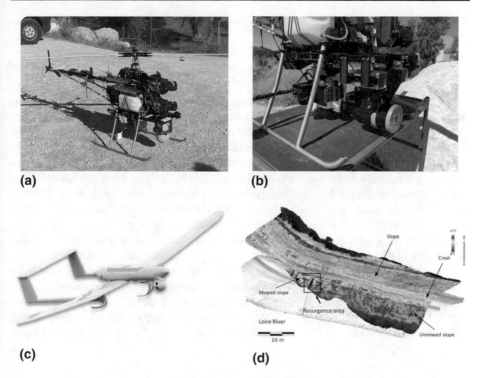

Fig. 8 **a, b** DiDRO unmanned helicopter equipped with LiDAR, HR IR thermal, near-infrared and visible cameras payload developed for routine surveys (Photo: R. Antoine ©Survey Copter/DiDRO consortium), **c** Fixed-wing drone used for crisis management (©SurveyCopter/DiDRO consortium) and **d** 3D thermal point cloud generated from IR thermal data, acquired over a dike along the Loire river (France) and super-imposed with meshed DSM. Squared area: cold thermal anomaly associated with an artificial resurgence on the river-side slope of the dike observed in summer (©R. Antoine and Jonathan Lisein/DiDRO consortium)

towed by the UAV. This system is particularly useful for the detection of erosion during a flood.

The SAFEDAM project (Weintrit et al. 2018) shares some similarities with the previously mentioned system. The preventive mode includes a multi-rotor platform, specifically developed for the purposes of the system and equipped with a LiDAR sensor and a digital camera. The data are processed to obtain a DTM and an orthomosaic, both with high accuracies, and are integrated into a dedicated GIS. The embankment condition can then be assessed through visual interpretation of the data, be helped by archived data and by simple processing tools (e.g., differential DTM, change detection, etc.). When deployed during a crisis, the UAV is equipped with a video camera providing emergency services with in-fly video streaming in the optical and in the thermal infrared spectrum. A camera also allows the acquisition of nadir optical images, transmitted to the GIS after landing of the UAV. These images are automatically processed into an orthomosaic, using only the georeferencing information provided by the on-board GNSS platform of the UAV.

7 From Tectonophysics Studies to Post-earthquakes Disaster Management

7.1 Paleoseismology and Neotectonics Pre-disaster Studies

When making an extensive paper research on Web of Science scientific database using «uav earthquakes» keywords, the use of drones in a seismic context is relatively recent and is essentially focused on post-event imagery and LiDAR collection (monitoring, damage evaluation and mapping, communication improvements, etc.). Nevertheless, some developments have been achieved for pre-disasters studies, in order to understand paleo-seismicity, map folds, faults and fractures, quantify slip rates on faults or interseismic shallow deformations. As for flood studies, photogrammetry is currently preferred to LiDAR, but relatively recent in the fields of paleoseismology and neotectonics (Johnson et al. 2014; Angster et al. 2016). Visible images are collected at various heights (< 120 m, due to regulation limitations), with diverse cameras and essentially using multi-copters. The characteristics of the flights described in this section are brought together in Appendix 4.

Bemis et al. (2014) made a review paper based on terrain and UAV observations for structural geology and paleo-seismicity. They show that centimetric 3D points clouds and orthophotographs of exposed stratigraphy or deformation features (like sigmoidal veins) may be now easily obtained using multi-copters, as they are very stable (compared to fixed-wing systems) and can fly at very low altitude along complex surfaces (vertical faces,

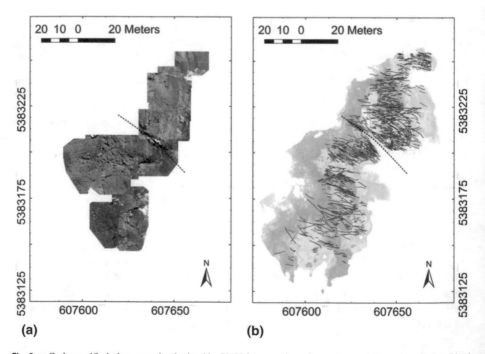

(a) **(b)**

Fig. 9 a Orthorectified photomosaic obtained by UAV for a portion of an outcrop at Piccaninny Point. Pixel resolution 1 pixel ¼ 10 mm and **b** Structural interpretation of fault and associated damage zone, showing dextral (red), sinistral (blue) and unidentified offset faults (grey) developed around a large fault (black dash), superimposed over the DEM (modified from Bemis et al. 2014)

corners, etc....). The calculated 3D models then serve for 1) sedimentary interpretation, 2) paleo-seismic offsets characterization (Gao et al. 2017) and 3) extraction of crack direction or bedding orientation to assess the geometrical characteristics of folds and faults (Fig. 9) (Vollgger and Cruden 2016; Menegoni et al. 2018). Drones can also be deployed for active tectonics studies: Giletycz et al. (2017) built a DSM of the active Hengchun fault (Southern Taiwan) using a very small DJI Mavic Pro, to analyse the deformation patterns of active fault cracks and its possible mechanisms. Deffontaines et al. (2017) use the same approach combined with field work to update the geological mapping of the active fault located at the Pingting Terraces area. In that case, the use of fixed-wing Skywalker X5-X8 UAVs was appropriated, given the large area to investigate (~ 40 km^2, see Appendix 4 for details).

If the works presented in this section produce 3D models for structural or geological mapping, kinematics studies along faults or during a seismic event have also been derived from UAVs results. Following the processing workflow of Johnson et al. (2014), Angster et al. (2016) propose a refined evaluation of the slip rate of the Pyramid Lake 50-km-long Fault Zone in Nevada (USA). The obtained DEMs permit quantification of the displaced geomorphological features (offsets) at seven chosen sites, with a resolution ranging from 5 to 9 cm (Appendix 4) and a RMS from 7 cm to 50 cm. The observed offsets (8 to 21 m) yield slip rates ranging from 0.5 to 1.9 mm yr^{-1} along the fault. If the errors calculated in this paper are significantly less than the observed offset values, it is important to keep in mind that the accuracy of such observations drastically depends on different factors affecting the photogrammetric process: quality of the photos (blurring), overlap percentage, camera calibration, acquisition angle and precision of the georeferencing strategy (target numbers and location, GPS accuracy) (Agüera-Vega et al. 2017a, b; Sanz-Ablanedo et al. 2018)

7.2 Post-earthquake Disaster Observations

7.2.1 Immediate Support During Emergency

Recent earthquakes have seen the use of single or multiple UAV systems for immediate support just after an earthquake, in complement with aircraft and satellites (Michael et al. 2014; Nedjati et al. 2016). As for flooding events, drones are usually used (1) to obtain quick local information about a situation (emergency, offset mapping during mainshocks, damage degree evaluation of facades, etc.) and (2) to complement space-borne data when the weather is too cloudy. UAVs can then be implemented with imagery and LiDAR systems for safety missions (Lee et al. 2016), geological mapping (Jiang et al. 2014) and structures and infrastructure surveying (Yamazaki and Liu 2016). For instance, the Chinese authorities used for the first time drones after the magnitude 7.9 May 2008 Sichuan earthquake to assess damages to buildings and infrastructures, or to optimize the deployment of rescue teams by observing the extent of damaged roads and/or traffic jams (https://www.wired.com/2017/01/chinas-launching-drones-fight-back-earthquakes/). Recently, the magnitude 7.5 earthquake Sulawesi earthquake and its subsequent tsunami in September 2018 caused widespread damages in the city of Palu. Under the leadership of the Indonesian emergency services, data acquired by drones were combined with satellite observations to create aerial maps evaluating damages to buildings, roads, bridges and other infrastructures. It was also possible to produce real-time HD imagery analysis and offline near-real-time 3D mapping without ground targets, using recent software platforms (DroneDeploy, Pix4D, Agisoft Metashape, etc....). Indeed, in emergency conditions, a "rough" georeferencing of the scene only taking into account the GPS metadata available in the photos (with a horizontal accuracy of 1-5 metres) may be sufficient to analyse most of the situations within a GIS software.

7.2.2 Geological Reconnaissance and Building Surveys

Various papers also investigated the utility of UAVs for geologic and geotechnical early reconnaissance after a seismic event, usually difficult to realize during the emergency phase (Gong et al. 2010; Rathje and Franke 2016; Gori et al. 2018; Saroglou et al. 2018) and damage degree evaluation of the buildings that will serve to reconstruction planning (Baiocchi et al. 2013; Sui et al. 2014; Fernandez Galarreta et al. 2015; Li et al. 2015; Yamazaki and Liu 2016; Duarte et al. 2018; Mavroulis et al. 2019). Gori et al. (2018) present field and aerial works achieved just after each of the three mainshock events that occurred in August and October 2016 in Central Italy. In this highly complex situation, geologists and remote sensing scientists worked together to locate and monitor the extent of the rupture zones. The multi-disciplinary approach using both conventional field methods and an original combination of UAV/LiDar sensing capabilities allowed (1) to find

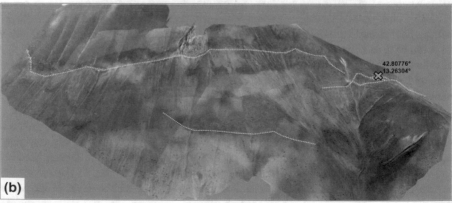

Fig. 10 **a** Surface ruptures caused by the M6.5 30 October event along the Mt. Vettore western slope, indicated by white arrows and **b** UAV-based orthomosaic model of the SW face of the Mt. Vettore Massif, showing fault traces as mapped by UAV. Yellow lines denote primary fault rupture. White line indicates rupture of the mid-slope splay (modified from Gori et al. 2018)

additional rupture zones over large areas and steeply sloping ground, (2) to observe the co-seismic vertical offsets along the fault, ranging from 0–35 cm and 70–200 cm for the 24 and 30 August, respectively (Fig. 10). Saroglou et al. (2018) present an original kinematic study of co-seismic rockfall in Greece, using a DTM, an orthophotograph and numerical modelling of the boulder movements down the slope, that successfully described the rolling and bouncing section of the trajectory. The detection of impact marks from the rock trajectory was first identified on the orthophotograph. Then, an analytical reconstruction of the rockfall is realized, with initial conditions derived from the earthquake characteristics.

Another advantage of UAVs in a post-earthquake scenario is their capacity to map the state of structures and infrastructures degraded during the seismic event, for instance, using classification methods (Xu et al. 2018) or change detection algorithms (Sui et al. 2014). Building damage detection methods traditionally focus on 2D changes detection (Chesnel et al. 2008; Brunner et al. 2010), while 3D data provide information on the height of the scene and enrich the interpretation of the experts. For instance, Fernandez Galarreta et al. (2015) propose a classification tool for the survey of buildings affected by earthquakes. They develop an object-based technique (OBIA, see Sect. 1), to evaluate the damage degree of the structures (following the European Macroseismic Scale of 1998), with photogrammetric 3D point clouds obtained by an UAV. The damaged structures are identified on the 3D point cloud and provide object-based damage indicators that are then used as auxiliary information by building analysts. Nex et al. (2019) propose a solution for autonomous building damage mapping in near real time, using a commercial UAV. The fast on-the-fly processing workflow combines photogrammetric methods with deep learning algorithms to show the location of the damages directly on an orthophotograph. In this case, the algorithms are optimized to fulfil the near-real-time conditions. Finally, when multi-resolution data are available (satellite, aerial and/or terrestrial), an interesting approach consists in merging observations to enrich classification and segmentation (Zhang 2010; Fu et al. 2017; Vetrivel et al. 2018). Duarte et al. (2018) propose deep learning (CNN) for the classification of building damages (rubble piles and debris), using fused multi-resolution imagery coming from sensors mounted on different platforms (satellite and manned/unmanned aircraft). Using thousands of images, they show that the fusing methodology is particularly powerful compared to the traditional classification of building damage, being able to capture both high-resolution degradation patterns (using UAV data) and contextual information (with satellite). The performance of the CNN drastically depends on the number of images, the quality of the fusion module, able to merge and blend the multi-resolution feature maps.

8 Discussion and Conclusions

The preceding sections share some common considerations: (1) the use of massive 3D UAV-based datasets to obtain 3D point clouds or orthophotographs, popularized by the development of user-friendly photogrammetric softwares and georeferencing methods, (2) the production of centimetric resolution photogrammetric products, also possible using low-cost UAV platforms and sensors, (3) the widespread use of small commercial low-cost multi-copters or fixed-wing drones with high safety guarantees, although self-made UAV are also used and (4) a considerable strengthening of the use of RGB data (either with photos or videos), compared to other observations (near-infrared, thermal infrared, hyperspectral, geophysical imagery or chemical sampling). Besides, some differences can be noticed:

new modes of acquisition in geophysical sampling using UAVs are tested for pre-disaster studies in volcanology, while to our knowledge, such observations are difficult to find in another natural hazard domain. This is probably due to two phenomena: (1) depending on the country, it may be easier to obtain flight permissions from the regulation authority to make experiments in an unpopulated volcanic/geothermal area and (2) given the amplitude of volcanic processes, the actual resolution and accuracy of geophysical and sampling by UAV may be sufficient to observe a phenomenon. Post-disaster methodologies have been rather developed during/after huge and frequent events (landslides, earthquakes or floods), in conjunction with authorities, to ease rescue operations or assess damages in buildings and infrastructures. This article also shows that the use of UAVs is different depending on the studied domain, as it involves phenomena with different temporal scales, spatial extents and magnitudes. For the first time, drones give access to very high-resolution spatialized observations and open new research paths. In this context, one of the most important challenges will consist in combining different observation scales to understand geological processes. Combination of UAVs and space-based data has already been done in coastal line evolutions (Nikolakopoulos et al. 2019), ground deformations (Cigna et al. 2017) and on landslide surveys (Voigt et al. 2016) and after the Mw 6.5 Lefkada earthquake (Zekkos et al. 2017). These studies obtained interesting results as (1) a good estimation of the landslide volume and enabled to precisely distinguish different generation of landslides (Zekkos et al. 2017), (2) an excellent georeferencing of coarse satellite images (Nikolakopoulos et al. 2019) and (3) information about very localized deformations (Cigna et al. 2017). However, while different applications of UAVs in preventing and reducing the risks associated with natural hazards as landslides, floods, volcanic eruptions and earthquakes exist, a few studies taking advantages of a multi-resolution data based on multi-platform sensors in post-disasters management have been made (Kakooei and Baleghi 2017; Zekkos et al. 2017; Duarte et al. 2018). The main advantage of both UAVs and space-based sensors in hazards studies is that they can provide images quickly after the disaster. Depending on the sensor (optical or radar), space-based images can be dependent on atmospheric conditions, but the high number of very high-resolution (VHR) satellites constellations increases the chance to acquire cloud-free images, while UAVs are mainly sensible to winds, depending on the type of machine (8-10 m/s for a multi-copter, 12.5 m/s for a fixed-wing aircraft). On the one hand, UAVs partial advantages rely on the multi-function portability, the high spatial and temporal resolution low-cost sensors (Casagli et al. 2017), and the possibility to consider collaboration, coordination and cooperation between UAVs when several of them are available (Pajares 2015). On the other hand, space-borne sensors are not vulnerable to the hazard itself and thus can provide pre-/co-/post-disaster images with a daily temporal resolution that cover wider areas. The drastic development of the monitoring of post-disaster areas provides precise information that may help stakeholders to react adequately. Post-disaster management may have already relied on the International Charter "Space and Major disasters," founded in 2000 by the European, French and Canadian space agencies (ESA, CNES and CSA, respectively) for major disasters (Bally et al. 2018). Once activated by national authorities, it provides and delivers at no cost a unified system of space-borne acquisition data at HR to VHR to those affected by hazards (Bally et al. 2018). One can imagine in the near-future a systematic combined workflow between space-based data that might be provided by the activation of the International Charter and UAVs based on the example provided by the use of UAVs inside the Copernicus Emergency Management Service (EMS).

Acknowledgements This paper arose from the International Workshop on "Natural and man-made hazards monitoring by the Earth Observation missions: current status and scientific gaps" held at the International Space Science Institute (ISSI), Bern, Switzerland, on April 15-18, 2019. The thermal survey of Piton de La Fournaise in 2018 was supported by the SlideVOLC French ANR project. We thank all the authors whose illustrations are presented in this article.

Appendix 1

Author	Scientific issue	Method analyse	UAV platform	UAV Equipment	Camera	Flight parameters	GSD (m)	Software
Balek & Blahut 2017	Landslide kinematics	DEM difference (DoD)	Octocopter	DJI Phantom	Nikon D100 / GoPro	20m height	0.004-0.112 m	Agisoft /Photomodeler
Comert et al. 2019	Landslide detection	OBIA	Fixed wing	Sensfly	Sensefly 20-megapixel S.O.D.A camera	212m height /1186 photos/ 70-75% overalap (ov.)	DEM 0.50m / Ortho 0.10m	Mapper
Cook 2017	Landslide kinematics	3D point Cloud comparison (PCC)	Quadrocopter	Dji Phantom 2	Canon IXUS	60 -120m height / 250 photos 27-38 GCP	Ortho 0.03m	Agisoft
Eker et al. 2018	Landslide kinematics	DoD / PCC	Octocopter	ARF MikroKopter OktoXL	Canon EOS 650 DSLR	39 - 41 m height / 94-396 photos/ 70-90% ov. / 8-9 GCP	DEM 0.10m / Ortho 0.10m	Agisoft
Fernandez et al. 2015		DoD	Octocopter	ASCTEC Falcon 8	Sony Nex-5N	90 m height	Ortho 0.03 m	Agisoft
Fernandez et al. 2016	Landslide kinematics	Visual interpretation + DoD DSM	Octocopter	Falcon 8 Asctec/Atyges FV-8	Sony Nex 5N / Canon G12	100-120m height / 72-364 photos 70-90% ov. / 8-13 GCP	DSM 0.10m / Ortho 0.5m	Agisoft
Hastaoglu et al. 2019	Landslide kinematics	Ortho comparison (pixel)	hexacopter	Dji Matrice 600 Pro	Ronin MX Gimbal and Sony A7R camera	85m height / 1867 photos 60-80% ov.		PIX4D Mapper
Lindner et al 2016	Landslide kinematics	DoD + Fissure obersations	Octocopter	ARF MikroKopter OktoXL	Canon EOS 650D DSLR	90-100 m height / 400-1700 photos / 70-90% ov. / 8-22 GCP	DSM 0.10m	Agisoft
Niethammer et al. 2012	Landslide kinematics / Fissure map	DoD (TLS/UAV)	Quadrocopter			100-250m height / 59 images 199 GCP	Ortho 0.03 - 0.08m	sfm= VSM / image matching algorithm, GOTCHA
Peternel et al. 2017	Landslide kinematics	Visual Ortho comparison (tree, boulder) + DoD	quadrocopter / hexacopter	Microdrone MD4 1000 / Survey Drone 01 hexacopter	Olympus Pen	40 m height / 155 photos/ 60 % ov.		
Peterman et al. 2015	Landslide kinematics	Point Cloud comparison	hexacopter		Olympus 2EP	40m height / 66-75 % overlap / 9 GCP		
Rau et al. 2011	Landslide detection	OBIA	Fixed wing		Canon EOS 450D	1400m height	DEM 5m / Ortho 0.20m	
Rossi et al. 2016	Landslide kinematics	PCC - TLS/UAV	Multicopter	Multicopter Saturn		68m height / 58 photos 50 60% ov. / 7 GCP	DEM 0.050m / Ortho 0.02m	Agisoft
Rossi et al. 2018	Landslide detection	DoD	Multicopter	Multicopter Saturn	Sony digital RGB camera with 8-MP	45-106 photos / 5-12 GCP	DEM 0.20m / Ortho 0.02m	
Stumpf et al. 2013	Microtopography, fissure evolution	OBIA	Quadrocopter			104 GCP	Ortho 0.03-0.10m	
Tang et al. 2019	Landslide detection / Kinematics	DoD	Fixed wing	F1000	Sony a5100	100-250m height	DSM 1m	PIX4D Mapper
Thiebes et al. 2016	Landslide kinematics	DoD (TLS/UAV)	Octocopter	SoLeon GmbH	Ricoh GR 16.2 Megapixels camera	70m height / 2033 photos 22	DSM 0.018m	PIX4D Mapper
Turner et al. 2015	Landslide kinematics	DoD	Octocopter			60-80% ov. / 40m height	Ortho 0.03-0.05m	sfm = Agisoft / Image correlation = COSI-Corr (ENVI)
Valkaniotis et al. 2018	Landslide detection	PCC - TLS/UAV	Quadrocopter	DJI Phantom 3	FC300X camera	75 photos / 8 GCP	DSM 0.08m / Ortho 0.08m	Agisoft

Bibliographic overview of case studies for mass movements

Appendix 2

	Fields of investigation	Methods	Goals	References
State of progress	Surface imaging	Photogrammetry	**Mapping and monitoring** Lava flows emplacement and volume estimation for risk mitigation	*Favalli et al. (2018) ; De Beni et al. (2019)*
			Changes in topography and volcano-tectonic fetaures of deformation	*Derrien et al. (2017) ; Darwmawan et al. (2018) ; Bonali et al. (2019) ; Patrick et al. (2019)*
			Multiscale analysis	*Dering et al. (2019)*
			Quantifying volume of volcanic ash-plume and understanding gas emissions dynamics	*Gomez and Kennedy (2017)*
			Mapping and multi-scale analysis of dyke swarms	*Dering et al. (2019)*
		Optical and radiometric infrared cameras	**Mapping** spatial distribution of thermal expressions	*Mori et al. (2016) ; Thiele et al. (2017) ; Walter et al. (2018) ; Peltier et al. (2018)*
	Combined approaches	In-situ measurements, thermal imaging, particles sampling	Origin of plume gases : high-temperature magmatic fluid of a deep origin and hydrothermal system	*Mori et al. (2016)*
	Sampling	In situ measurements of volcanogenic gases and aerosols	**Monitoring** gaz emission rate in volcanic plumes **Quantifying** composition and behaviour of such plumes from explosive volcanic eruptions	*Herva et al. (2012) ; Rüdiger et al. (2018) ; Picard et al. (2020) ; Pieri et al. (2020)*
		Water sampling	**Monitoring** volcanic unrest and water chemistry	*Terada et al. (2018)*
	Geophysical imaging	Magnetism	**Monitoring** changes of magnetic anomalies due to volcanic activity (alteration...)	*Funaki et al. (2014) ; Catalan et al. (2014)*
		Gravimetry	**Monitoring** : tracking magma and mass transfers under volcanoes (aboratory development)	*Middlemiss et al. (2016) ; Magee et al. (2018)*

Bibliographic overview on developments around sensors mounted on drone and volcanoes case studies.

Appendix 3

Case study	Application	UAV type	Sensor	Resolution	Flight height	Location
Murphy et al., 2016	Location of people and flood crest monitoring	Multi-copters and fixed wings	Video camera	?	?	Urban areas, Louisiana, USA
Popescu et al., 2017	Detection and mapping of flooded areas	Fixed-wing	RGB digital camera	?	100 m	Urban area near Bucharest, Hungary
Gebrehiwot et al., 2019	Detection and mapping of flooded areas	Fixed Wing and Multicopter	RGB digital camera	1.5 to 2.6 cm	?	Urban areas, North Carolina, USA
Feng et al., 2015	Detection and mapping of flooded areas	Fixed-wing	RGB digital camera	20 cm	350 m	Dense urban area, Yuyao, China
Imam et al., 2019	Detection and delineation of water surfaces	Multi-copter	GNSS-Reflectometry sensor	Few tens of meters	450 m	Lake, Avigliana area, Italy
Langhammer et al., 2017	DSM and DTM of a for 2D hydraulic flood modelling	Multi-copter	RGB CMOS digital camera	1.8 cm	88 m	Javoří Brook, Šumava Mountains, Czech Republic
Leitão et al., 2016	DEM for overland flow modelling in urban areas	Fixed-wing	RGB CMOS digital camera	From 2.5 cm to 10 cm	From 85 m to 310 m	Urban area, Adliswil, Switzerland
Rinaldi et al., 2019	DSM for 2D hydraulic flood simulation in urban areas	Multi-copter	RGB CMOS digital camera	?	100 m	Urban area, Tandil, Argentina
Yalcin, 2018	High-resolution DSM and land cover map for 2D modelling of flash floods	Multi-copter	RGB CMOS digital camera	4.32 cm	120 m	Dense urban area, Kirsehir city, Turkey
Bandini, 2017	Water level estimations from different sensors	Multi-copter	Radar + Sonar + Camera based laser distance sensor (CLDS)	10 cm (radar), 30 cm to 14 m (sonar), 1 m to 2 m (CLDS)	From 10 m to > 60 m	Small lake in Denmark
Ridolfi et al., 2018	Water level estimation on a dam site	Multi-copter	RGB camera	2.1 mm	15 m from the dam surface	Ridracoli lake, Italy
Tauro et al., 2016	Quantitative characterization of surface flow	Multi-copter	RGB video camera	?	?	Rio Cordon, Italy
Koutalakis et al., 2019	Estimation of water surface velocity	Multi-copter	CMOS digital camera	?	40 m	Aggitis river, Greece
Perks et al., 2016	Characterization of surface flow	Multi-copter	RGB video camera	6 cm	?	Alyth Burn, Scotland
King et al., 2017	Monitoring of coastal structures and costal areas	Multi-copter	Visible digital camera	i. 3 cm ii. 2.9 cm	i. 70 m ii. 60 m	i. Byron Bay, Australia ii. Ajman, United Arab Emirates
Brauneck et al., 2016	Monitoring of levee breaches for flow discharge estimation	Multi-copter	RGB video camera	?	?	Saale River at Breitenhagen, Germany
Weintrit et al., 2018	Routine and crisis monitoring of flood protection structures	Multi-copter	RGB digital camera	10 cm	40 m	Vistula River, Annopol, Poland
Antoine et al., 2019	Routine and crisis monitoring of flood protection structures	Helicopter, Multi-copter and fixed-wing	RGB and infrared cameras, LiDAR sensor	Depending on the sensor	70 m	Several dikes In France

Main characteristics of the studies cited on the bibliographic overview on flood monitoring and management using UAVs. Information not provided by the authors in the reviewed studies is marked by the symbol "?".

Appendix 4

Case study	Type of UAV (M, FW or H)	Sensor	Pixel size	Distance from the target	Application
Bernie et al., 2014	Oktocopter (Mikrocopter flight controller)	Canon 550D 15 Mp	10 mm	30-40 m	Mapping (faults, sigmoidal veins)
Menegoni et al, 2018	Hexacopter	Sony 6000 ILCE	4-10 mm	31-52 m	Mapping (folds, bedding, fractures)
Gao et al., 2017	Hexacopter Y6 950 CO-AXIAL	Sony NEX 5T	16 mm	60 m	Mapping (terrace risers, seismic offsets)
Saroglou et al., 2018	Quadcopter DJI Phantom 3 Pro	DJI Camera 1/2.3" CMOS, 12,3 MP	49 mm	115 m	Mapping (DSM, DTM and orthophotos)
Giletycz et al., 2017	DJI Mavic Pro		?	?	Mapping (DSM, deformation pattern, crack analysis)
Angster et al., 2016	DJI Phantom 1	Ricoh GR 8.3 MP / f=5.3 mm	30 to 80 mm	< 30 m	
Deffontaines et al., 2016	Skywalker X5 and X8	Sony DSC-QX100 and Sony ILCEQX1	77 mm	?	Mapping (DSM, orthophotos)
Gori et al., 2018	Phantom 4 Pro	DJI Camera 1" CMOS, 20 MP	?	?	Mapping just after eartquake (DSM, seismic offsets)
Fernandez Galarreta et al., 2015	Aibot X6 V1	Canon 600D	7 mm	70 m	Damage degree Evaluation(3D points clouds)

Technical overview of the flights described in Sect. 7 "From tectonophysics studies to post-earthquakes disaster management."

References

Aditian A, Kubota T, Shinohara Y (2018) Comparison of GIS-based landslide susceptibility models using frequency ratio, logistic regression, and artificial neural network in a tertiary region of Ambon, Indonesia. Geomorphology 318:101–111. https://doi.org/10.1016/j.geomorph.2018.06.006

Agisoft Metashape (2016) AgiSoft PhotoScan Professional. Version 1.2.6

Agüera-Vega F, Carvajal-Ramírez F, Martínez-Carricondo P (2017a) Assessment of photogrammetric mapping accuracy based on variation ground control points number using unmanned aerial vehicle. Measurement 98:221–227. https://doi.org/10.1016/j.measurement.2016.12.002

Agüera-Vega F, Carvajal-Ramírez F, Martínez-Carricondo P (2017b) Accuracy of digital surface models and orthophotos derived from unmanned aerial vehicle photogrammetry. J Surv Eng. https://doi.org/10.1061/(ASCE)SU.1943-5428.0000206

Akcay O (2015) Landslide fissure inference assessment by ANFIS and logistic regression using UAS-based photogrammetry. ISPRS Int J Geo-Inf 4:2131–2158. https://doi.org/10.3390/ijgi4042131

Angster S, Wesnousky S, Huang W et al (2016) Application of UAV photography to refining the slip rate on the Pyramid Lake Fault Zone, Nevada. Bull Seismol Soc Am 106:785–798. https://doi.org/10.1785/0120150144

Antoine R, Finizola A, Lopez T et al (2017) Electric potential anomaly induced by humid air convection within Piton de La Fournaise volcano, La Réunion Island. Geothermics 65:81–98. https://doi.org/10.1016/j.geothermics.2016.01.003

Antoine R, Tanguy M, Palma Lopes S, Sorin J-L (2019) DIDRO—an innovative multi-sensor UAV system for routine and crisis monitoring of dikes. AGUFM 2019:

Assali P, Grussenmeyer P, Villemin T et al (2014) Surveying and modeling of rock discontinuities by terrestrial laser scanning and photogrammetry: semi-automatic approaches for linear outcrop inspection. J Struct Geol 66:102–114. https://doi.org/10.1016/j.jsg.2014.05.014

Baiocchi V, Dominici D, Mormile M (2013) UAV application in post-seismic environment. Int Arch Photogramm Remote Sens Spat Inf Sci 1:W2

Balek J, Blahůt J (2017) A critical evaluation of the use of an inexpensive camera mounted on a recreational unmanned aerial vehicle as a tool for landslide research. Landslides 14:1217–1224. https://doi.org/10.1007/s10346-016-0782-7

Bally P, Papadopoulou T, Tinel C, Danzeglocke J, Wannop S, Kuklin A (2018) The 17th annual report: international charter space and major disasters, p 72. https://disasterscharter.org/documents/10180/188210/Annual-Report-17.pdf

Bandini F, Butts M, Jacobsen TV, Bauer-Gottwein P (2017) Water level observations from unmanned aerial vehicles for improving estimates of surface water–groundwater interaction. Hydrol Processes 31:4371–4383

Barra A, Monserrat O, Mazzanti P et al (2016) First insights on the potential of Sentinel-1 for landslides detection. Geomat Nat Hazards Risk 7:1874–1883. https://doi.org/10.1080/19475705.2016.1171258

Bato MG, Froger JL, Harris AJL, Villeneuve N (2016) Monitoring an effusive eruption at Piton de la Fournaise using radar and thermal infrared remote sensing data: insights into the October 2010 eruption and its lava flows. Geol Soc Lond Spec Publ 426:533–552. https://doi.org/10.1144/SP426.30

Bemis SP, Micklethwaite S, Turner D et al (2014) Ground-based and UAV-based photogrammetry: a multi-scale, high-resolution mapping tool for structural geology and paleoseismology. J Struct Geol 69:163–178

Berni JAJ, Zarco-Tejada PJ, Suárez L et al (2009) Remote sensing of vegetation from UAV platforms using lightweight multispectral and thermal imaging sensors. Int Arch Photogramm Remote Sens Spat Inf Sci 38:6

Blackett M (2017) An overview of infrared remote sensing of volcanic activity. J Imaging 3:13. https://doi.org/10.3390/jimaging3020013

Blaikie TN, Ailleres L, Betts PG, Cas RAF (2014) Interpreting subsurface volcanic structures using geologically constrained 3-D gravity inversions: examples of maar-diatremes, Newer Volcanics Province, southeastern Australia. J Geophys Res Solid Earth 119:3857–3878. https://doi.org/10.1002/2013JB010751

Bonali FL, Tibaldi A, Marchese F et al (2019) UAV-based surveying in volcano-tectonics: an example from the Iceland rift. J Struct Geol 121:46–64. https://doi.org/10.1016/j.jsg.2019.02.004

Brauneck J, Pohl R, Juepner R (2016) Experiences of using UAVs for monitoring levee breaches. IOP Conf Ser Earth Environ Sci 46:012046. https://doi.org/10.1088/1755-1315/46/1/012046

Brunner D, Lemoine G, Bruzzone L (2010) Earthquake damage assessment of buildings using VHR optical and SAR imagery. IEEE Trans Geosci Remote Sens 48:2403–2420. https://doi.org/10.1109/TGRS.2009.2038274

Bunn M, Leshchinsky B, Olsen M, Booth A (2019) A simplified, object-based framework for efficient landslide inventorying using LIDAR digital elevation model derivatives. Remote Sens 11:303. https://doi.org/10.3390/rs11030303

Calvari S, Neri M, Pinkerton H (2003) Effusion rate estimations during the 1999 summit eruption on Mount Etna, and growth of two distinct lava flow fields. J Volcanol Geotherm Res 119:107–123. https://doi.org/10.1016/S0377-0273(02)00308-6

Caratori Tontini F, Tivey MA, Ronde CEJ, Humphris SE (2019) Heat flow and near-seafloor magnetic anomalies highlight hydrothermal circulation at Brothers Volcano Caldera, Southern Kermadec Arc, New Zealand. Geophys Res Lett 46:8252–8260. https://doi.org/10.1029/2019GL083517

Casagli N, Frodella W, Morelli S et al (2017) Spaceborne, UAV and ground-based remote sensing techniques for landslide mapping, monitoring and early warning. Geoenviron Disasters 4:9. https://doi.org/10.1186/s40677-017-0073-1

Casana J, Kantner J, Wiewel A, Cothren J (2014) Archaeological aerial thermography: a case study at the Chaco-era Blue J community, New Mexico. J Archaeol Sci 45:207–219. https://doi.org/10.1016/j.jas.2014.02.015

Casini G, Hunt DW, Monsen E, Bounaim A (2016) Fracture characterization and modeling from virtual outcrops. AAPG Bull 100:41–61. https://doi.org/10.1306/09141514228

Catalán M, Martos YM, Galindo-Zaldívar J, Funaki M (2014) Monitoring the evolution of Deception Island volcano from magnetic anomaly data (South Shetland Islands, Antarctica). Glob Planet Change 123:199–212. https://doi.org/10.1016/j.gloplacha.2014.07.018

Cawood AJ, Bond CE, Howell JA et al (2017) LiDAR, UAV or compass-clinometer? Accuracy, coverage and the effects on structural models. J Struct Geol 98:67–82. https://doi.org/10.1016/j.jsg.2017.04.004

Chesley JT, Leier AL, White S, Torres R (2017) Using unmanned aerial vehicles and structure-from-motion photogrammetry to characterize sedimentary outcrops: an example from the Morrison Formation, Utah, USA. Sediment Geol 354:1–8. https://doi.org/10.1016/j.sedgeo.2017.03.013

Chesnel A-L, Binet R, Wald L (2008) Damage assessment on buildings using multisensor multimodal very high resolution images and ancillary data. In: IGARSS 2008—2008 IEEE international geoscience and remote sensing symposium. IEEE, Boston, MA, USA, pp 1252–1255

Cigna F, Banks VJ, Donald AW et al (2017) Mapping ground instability in areas of geotechnical infrastructure using satellite InSAR and Small UAV surveying: a case study in Northern Ireland. Geosciences 7:51. https://doi.org/10.3390/geosciences7030051

Comert R, Avdan U, Gorum T, Nefeslioglu HA (2019) Mapping of shallow landslides with object-based image analysis from unmanned aerial vehicle data. Eng Geol 260:105264. https://doi.org/10.1016/j.enggeo.2019.105264

Cook KL (2017) An evaluation of the effectiveness of low-cost UAVs and structure from motion for geomorphic change detection. Geomorphology 278:195–208. https://doi.org/10.1016/j.geomorph.2016.11.009

Darmawan H, Walter TR, Brotopuspito KS et al (2018) Morphological and structural changes at the Merapi lava dome monitored in 2012–15 using unmanned aerial vehicles (UAVs). J Volcanol Geotherm Res 349:256–267. https://doi.org/10.1016/j.jvolgeores.2017.11.006

De Beni E, Cantarero M, Messina A (2019) UAVs for volcano monitoring: a new approach applied on an active lava flow on Mt. Etna (Italy), during the 27 February–02 March 2017 eruption. J Volcanol Geotherm Res 369:250–262. https://doi.org/10.1016/j.jvolgeores.2018.12.001

de Saint Jean B (2008) Étude et développement d'un système de gravimétrie mobile. Ph.D. thesis, Observatoire de Paris

Deffontaines B, Chang K-J, Champenois J et al (2017) Active interseismic shallow deformation of the Pingting terraces (Longitudinal Valley—Eastern Taiwan) from UAV high-resolution topographic data combined with InSAR time series. Geomat Nat Hazards Risk 8:120–136. https://doi.org/10.1080/19475705.2016.1181678

Delacourt C, Allemand P, Berthier E et al (2007) Remote-sensing techniques for analysing landslide kinematics: a review. Bull Soc Géol France 178:89–100. https://doi.org/10.2113/gssgfbull.178.2.89

Dering GM, Micklethwaite S, Thiele ST et al (2019) Review of drones, photogrammetry and emerging sensor technology for the study of dykes: Best practises and future potential. J Volcanol Geotherm Res 373:148–166. https://doi.org/10.1016/j.jvolgeores.2019.01.018

Derrien A, Villeneuve N, Peltier A, Beauducel F (2015) Retrieving 65 years of volcano summit deformation from multitemporal structure from motion: the case of Piton de la Fournaise (La Réunion Island). Geophys Res Lett 42:6959–6966. https://doi.org/10.1002/2015GL064820

Domeneghetti A, Schumann GJ-P, Tarpanelli A (2019) Preface: remote sensing for flood mapping and monitoring of flood dynamics. Remote Sens 11:943. https://doi.org/10.3390/rs11080943

Duarte D, Nex F, Kerle N, Vosselman G (2018) Multi-resolution feature fusion for image classification of building damages with convolutional neural networks. Remote Sens 10:1636. https://doi.org/10.3390/rs10101636

Edenhofer O, Pichs-Madruga R, Sokona Y et al (2014) Summary for policymakers. In: Climate change 2014: mitigation of climate change. IPCC Working Group III Contribution to AR5. Cambridge University Press

Eker R, Aydın A, Hübl J (2018) Unmanned aerial vehicle (UAV)-based monitoring of a landslide: Gallenzerkogel landslide (Ybbs-Lower Austria) case study. Environ Monit Assess. https://doi.org/10.1007/s10661-017-6402-8

Eltner A, Sardemann H, Grundmann J (2020) Flow velocity and discharge measurement in rivers using terrestrial and unmanned-aerial-vehicle imagery. Hydrol Earth Syst Sci. https://doi.org/10.5194/hess-24-1429-2020

Erdelj M, Natalizio E, Chowdhury KR, Akyildiz IF (2017) Help from the sky: leveraging UAVs for disaster management. IEEE Pervasive Comput 16:24–32. https://doi.org/10.1109/MPRV.2017.11

Everaerts J (2008) The use of unmanned aerial vehicles (UAVs) for remote sensing and mapping. Int Arch Photogramm Remote Sens Sp Inf Sci 37(2008):1187–1192

Fahlstrom P, Gleason T (2012) Introduction to UAV systems. Wiley, Hoboken

Favalli M, Fornaciai A, Nannipieri L et al (2018) UAV-based remote sensing surveys of lava flow fields: a case study from Etna's 1974 channel-fed lava flows. Bull Volcanol 80:29

Feng Q, Liu J, Gong J (2015) Urban flood mapping based on unmanned aerial vehicle remote sensing and random forest classifier—a case of Yuyao, China. Water 7:1437–1455. https://doi.org/10.3390/w7041437

Fernandez Galarreta J, Kerle N, Gerke M (2015) UAV-based urban structural damage assessment using object-based image analysis and semantic reasoning. Nat Hazards Earth Syst Sci 15:1087–1101. https://doi.org/10.5194/nhess-15-1087-2015

Fernandes O, Murphy R, Adams J, Merrick D (2018) quantitative data analysis: CRASAR small unmanned aerial systems at hurricane Harvey. In: 2018 IEEE international symposium on safety, security, and rescue robotics (SSRR). IEEE, Philadelphia, PA, pp 1–6

Fu G, Liu C, Zhou R et al (2017) Classification for high resolution remote sensing imagery using a fully convolutional network. Remote Sens 9:498. https://doi.org/10.3390/rs9050498

Funaki M, Higashino S-I, Sakanaka S et al (2014) Small unmanned aerial vehicles for aeromagnetic surveys and their flights in the South Shetland Islands, Antarctica. Polar Sci 8:342–356. https://doi.org/10.1016/j.polar.2014.07.001

Gabrlik P (2015) The use of direct georeferencing in aerial photogrammetry with micro UAV. IFAC-PapersOnLine 48:380–385. https://doi.org/10.1016/j.ifacol.2015.07.064

Gailler L, Kauahikaua J (2017) Monitoring the cooling of the 1959 Kīlauea Iki lava lake using surface magnetic measurements. Bull Volcanol. https://doi.org/10.1007/s00445-017-1119-7

Gailler L-S, Lénat J-F, Blakely RJ (2016) Depth to Curie temperature or bottom of the magnetic sources in the volcanic zone of la Réunion hot spot. J Volcanol Geotherm Res 324:169–178. https://doi.org/10.1016/j.jvolgeores.2016.06.005

Gao M, Xu X, Klinger Y et al (2017) High-resolution mapping based on an Unmanned Aerial Vehicle (UAV) to capture paleoseismic offsets along the Altyn-Tagh fault, China. Sci Rep 7:1–11

Gebrehiwot A, Hashemi-Beni L, Thompson G et al (2019) Deep convolutional neural network for flood extent mapping using unmanned aerial vehicles data. Sensors 19:1486. https://doi.org/10.3390/s19071486

Giletycz SJ, Chang C-P, Lin AT-S et al (2017) Improved alignment of the Hengchun Fault (southern Taiwan) based on fieldwork, structure-from-motion, shallow drilling, and levelling data. Tectonophysics 721:435–447. https://doi.org/10.1016/j.tecto.2017.10.018

Giordan D, Hayakawa Y, Nex F et al (2018) The use of remotely piloted aircraft systems (RPASs) for natural hazards monitoring and management. Nat Hazards Earth Syst Sci 18:1079–1096. https://doi.org/10.5194/nhess-18-1079-2018

Giordan D, Adams MS, Aicardi I et al (2020) The use of unmanned aerial vehicles (UAVs) for engineering geology applications. Bull Eng Geol Environ. https://doi.org/10.1007/s10064-020-01766-2

Girardeau-Montaut D, Roux M, Marc R, Thibault G (2005) Change detection on points cloud data acquired with a ground laser scanner. Int Arch Photogramme Remote Sens Spat Inf Sci 36:W19

Gomez C, Kennedy B (2018) Capturing volcanic plumes in 3D with UAV-based photogrammetry at Yasur Volcano—Vanuatu. J Volcanol Geotherm Res 350:84–88. https://doi.org/10.1016/j.jvolgeores.2017.12.007

Gomez C, Purdie H (2016) UAV-based photogrammetry and geocomputing for hazards and disaster risk monitoring—a review. Geoenviron Disasters. https://doi.org/10.1186/s40677-016-0060-y

Gonçalves JA, Henriques R (2015) UAV photogrammetry for topographic monitoring of coastal areas. ISPRS J Photogramm Remote Sens 104:101–111. https://doi.org/10.1016/j.isprsjprs.2015.02.009

Gong J, Wang D, Li Y et al (2010) Earthquake-induced geological hazards detection under hierarchical stripping classification framework in the Beichuan area. Landslides 7:181–189. https://doi.org/10.1007/s10346-010-0201-4

Gonzalez Toro F, Tsourdos A (2018) UAV or drones for remote sensing applications. MDPI Books, Basel

Gori S, Falcucci E, Galadini F et al (2018) Surface faulting caused by the 2016 central Italy seismic sequence: field mapping and LiDAR/UAV imaging. Earthq Spectra 34:1585–1610. https://doi.org/10.1193/111417EQS236MR

Görüm T (2019) Landslide recognition and mapping in a mixed forest environment from airborne LiDAR data. Eng Geol 258:105155. https://doi.org/10.1016/j.enggeo.2019.105155

Graff K, Lissak C, Thiery Y et al (2019) Analysis and quantification of potential consequences in multirisk coastal context at different spatial scales (Normandy, France). Nat Hazards 99:637–664. https://doi.org/10.1007/s11069-019-03763-5

Guilbert V, Antoine R, Heinkele C et al (2020) Fusion of thermal and visible point clouds : application to the Vaches Noires landslide, Normandy, France. In: Accepted for the international archives of the photogrammetry, remote sensing and spatial information sciences (ISPRS Archives). Nice, France, p 6

Harris AJL, Baloga SM (2009) Lava discharge rates from satellite-measured heat flux. Geophys Res Lett. https://doi.org/10.1029/2009GL039717

Harris AJL, Stevenson DS (1997) Thermal observations of degassing open conduits and fumaroles at Stromboli and Vulcano using remotely sensed data. J Volcanol Geotherm Res 76:175–198. https://doi.org/10.1016/S0377-0273(96)00097-2

Harris A, Dehn J, Patrick M et al (2005) Lava effusion rates from hand-held thermal infrared imagery: an example from the June 2003 effusive activity at Stromboli. Bull Volcanol 68:107–117. https://doi.org/10.1007/s00445-005-0425-7

Hastaoğlu KÖ, Gül Y, Poyraz F, Kara BC (2019) Monitoring 3D areal displacements by a new methodology and software using UAV photogrammetry. Int J Appl Earth Obs Geoinf 83:101916. https://doi.org/10.1016/j.jag.2019.101916

Hervo M, Quennehen B, Kristiansen NI et al (2012) Physical and optical properties of 2010 Eyjafjallajökull volcanic eruption aerosol: ground-based, Lidar and airborne measurements in France. Atmos Chem Phys 12:1721–1736. https://doi.org/10.5194/acp-12-1721-2012

Hungr O, Leroueil S, Picarelli L (2014) The Varnes classification of landslide types, an update. Landslides 11:167–194. https://doi.org/10.1007/s10346-013-0436-y

Imam R, Pini M, Marucco G et al (2019) Data from GNSS-based passive radar to support flood monitoring operations. In: 2019 international conference on localization and GNSS (ICL-GNSS). IEEE, Nuremberg, Germany, pp 1–7

Jaboyedoff M, Oppikofer T, Abellán A et al (2012) Use of LIDAR in landslide investigations: a review. Nat Hazards 61:5–28. https://doi.org/10.1007/s11069-010-9634-2

James MR, Robson S, Pinkerton H, Ball M (2006) Oblique photogrammetry with visible and thermal images of active lava flows. Bull Volcanol 69:105–108. https://doi.org/10.1007/s00445-006-0062-9

James MR, Carr B, D'Arcy F et al (2020) Volcanological applications of unoccupied aircraft systems (UAS): developments, strategies, and future challenges. Volcanica 3:67–114

Jiang H, Su Y, Jiao Q et al (2014) Typical geologic disaster surveying in Wenchuan 8.0 earthquake zone using high resolution ground LiDAR and UAV remote sensing. In: Lidar remote sensing for environmental monitoring XIV. International Society for Optics and Photonics, p 926219

Johnson K, Nissen E, Saripalli S et al (2014) Rapid mapping of ultrafine fault zone topography with structure from motion. Geosphere 10:969–986. https://doi.org/10.1130/GES01017.1

Jones RR, Kokkalas S, McCaffrey KJW (2009) Quantitative analysis and visualization of nonplanar fault surfaces using terrestrial laser scanning (LIDAR)—The Arkitsa fault, central Greece, as a case study. Geosphere 5:465–482

Kakooei M, Baleghi Y (2017) Fusion of satellite, aircraft, and UAV data for automatic disaster damage assessment. Int J Remote Sens 38:2511–2534. https://doi.org/10.1080/01431161.2017.1294780

King S, Leon J, Mulcahy M et al (2017) Condition survey of coastal structures using UAV and photogrammetry. Australasian Coasts & Ports 2017: Working with Nature 704

Kitonsa H, Kruglikov SV (2018) Significance of drone technology for achievement of the United Nations sustainable development goals. R-Economy 4(3):115–120. https://doi.org/10.15826/recon.2018.4.3.016

Klemas V (2015) Remote sensing of floods and flood-prone areas: an overview. J Coast Res 31:1005–1013. https://doi.org/10.2112/JCOASTRES-D-14-00160.1

Koutalakis P, Tzoraki O, Zaimes G (2019) UAVs for hydrologic scopes: application of a low-cost UAV to estimate surface water velocity by using three different image-based methods. Drones 3:14. https://doi.org/10.3390/drones3010014

Kreye C, Hein GW, Zimmermann B (2006) Evaluation of airborne vector gravimetry using GNSS and SDINS observations. In: Flury J, Rummel R, Reigber C et al (eds) Observation of the earth system from space. Springer, Berlin, pp 447–461

Labazuy P (2015) Unmaned aerial vehicles (UAVs)-based remote sensing applications for studying and monitoring volcanic environments

Labazuy P, Gouhier M, Harris A et al (2012) Near real-time monitoring of the April-May 2010 Eyjafjallajökull ash cloud: an example of a web-based, satellite data-driven, reporting system. Int J Environ Pollut 48:262. https://doi.org/10.1504/IJEP.2012.049673

Lague D, Brodu N, Leroux J (2013) Accurate 3D comparison of complex topography with terrestrial laser scanner: application to the Rangitikei canyon (N-Z). ISPRS J Photogramm Remote Sens 82:10–26. https://doi.org/10.1016/j.isprsjprs.2013.04.009

Lahousse T, Chang KT, Lin YH (2011) Landslide mapping with multi-scale object-based image analysis—a case study in the Baichi watershed, Taiwan. Nat Hazards Earth Syst Sci 11:2715–2726. https://doi.org/10.5194/nhess-11-2715-2011

Lang S, Füreder P, Rogenhofer E (2018) Earth observation for humanitarian operations. In: Al-Ekabi C, Ferretti S (eds) Yearbook on space policy 2016. Springer, Cham, pp 217–229

Langhammer J, Bernsteinová J, Miřijovský J (2017) Building a high-precision 2D hydrodynamic flood model using UAV photogrammetry and sensor network monitoring. Water 9:861. https://doi.org/10.3390/w9110861

Lee S, Har D, Kum D (2016) Drone-assisted disaster management: finding victims via infrared camera and lidar sensor fusion. In: 2016 3rd Asia-Pacific world congress on computer science and engineering (APWC on CSE), pp 84–89

Leitão JP, Moy de Vitry M, Scheidegger A, Rieckermann J (2016) Assessing the quality of digital elevation models obtained from mini unmanned aerial vehicles for overland flow modelling in urban areas. Hydrol Earth Syst Sci 20:1637–1653. https://doi.org/10.5194/hess-20-1637-2016

Li S, Tang H, He S et al (2015) Unsupervised detection of earthquake-triggered roof-holes from UAV images using joint color and shape features. IEEE Geosci Remote Sens Lett 12:1823–1827. https://doi.org/10.1109/LGRS.2015.2429894

Lin C-Y, Lo H-M, Chou W-C, Lin W-T (2004) Vegetation recovery assessment at the Jou-Jou Mountain landslide area caused by the 921 Earthquake in Central Taiwan. Ecol Model 176:75–81. https://doi.org/10.1016/j.ecolmodel.2003.12.037

Lindner G, Schraml K, Mansberger R, Hübl J (2016) UAV monitoring and documentation of a large landslide. Appl Geomat 8:1–11. https://doi.org/10.1007/s12518-015-0165-0

Lissak C, Gomez C, Shimizu M et al (2019) Drifted wood distribution in Asakura (Kyushu) following the 2017 rain-triggered Debris-flows and Landslides. In: Geophysical research abstracts

Lowe DG (1999) Object recognition from local scale-invariant features. In: Proceedings of the seventh IEEE international conference on computer vision, pp 1150–1157

Lowe DG (2004) Distinctive image features from scale-invariant keypoints. Int J Comput Vis 60:91–110. https://doi.org/10.1023/B:VISI.0000029664.99615.94

Lucieer A, de Jong SM, Turner D (2014) Mapping landslide displacements using Structure from Motion (SfM) and image correlation of multi-temporal UAV photography. Progress Phys Geogr Earth Environ 38:97–116. https://doi.org/10.1177/0309133313515293

Magee C, Stevenson CTE, Ebmeier SK et al (2018) Magma plumbing systems: a geophysical perspective. J Petrol 59:1217–1251. https://doi.org/10.1093/petrology/egy064

Marshall DM, Barnhart RK, Shappee E, Most MT (2015) Introduction to unmanned aircraft systems. CRC Press, Boca Raton

Mavroulis S, Andreadakis E, Spyrou N-I et al (2019) UAV and GIS based rapid earthquake-induced building damage assessment and methodology for EMS-98 isoseismal map drawing: the June 12, 2017 Mw 6.3 Lesvos (Northeastern Aegean, Greece) earthquake. Int J Disaster Risk Reduct 37:101–169. https://doi.org/10.1016/j.ijdrr.2019.101169

Menegoni N, Meisina C, Perotti C, Crozi M (2018) Analysis by UAV digital photogrammetry of folds and related fractures in the Monte Antola Flysch Formation (Ponte Organasco, Italy). Geosciences 8:299. https://doi.org/10.3390/geosciences8080299

Mezaal M, Pradhan B, Rizeei H (2018) Improving landslide detection from airborne laser scanning data using optimized Dempster–Shafer. Remote Sens 10:1029. https://doi.org/10.3390/rs10071029

Mian O, Lutes J, Lipa G et al (2015) Direct georeferencing on small unmanned aerial platforms for improved reliability and accuracy of mapping without the need for ground control points. ISPRS Int Arch Photogramm Remote Sens Spat Inf Sci XL-1/W4:397–402. https://doi.org/10.5194/isprsarchives-XL-1-W4-397-2015

Michael N, Shen S, Mohta K et al (2014) Collaborative mapping of an earthquake damaged building via ground and aerial robots. In: Yoshida K, Tadokoro S (eds) Field and service robotics: results of the 8th international conference. Springer, Berlin, pp 33–47

Middlemiss RP, Samarelli A, Paul DJ et al (2016) Measurement of the Earth tides with a MEMS gravimeter. Nature 531:614–617. https://doi.org/10.1038/nature17397

Millington SC, Saunders RW, Francis PN, Webster HN (2012) Simulated volcanic ash imagery: a method to compare NAME ash concentration forecasts with SEVIRI imagery for the Eyjafjallajökull eruption in 2010. J Geophys Res Atmos. https://doi.org/10.1029/2011JD016770

Mokhtar MRM, Matori AN, Yusof KW et al (2014) Assessing UAV landslide mapping using unmanned aerial vehicle (UAV) for landslide mapping activity. Appl Mech Mater 567:669–674. https://doi.org/10.4028/www.scientific.net/AMM.567.669

Moosavi V, Talebi A, Shirmohammadi B (2014) Producing a landslide inventory map using pixel-based and object-oriented approaches optimized by Taguchi method. Geomorphology 204:646–656. https://doi.org/10.1016/j.geomorph.2013.09.012

Mori T, Hashimoto T, Terada A et al (2016) Volcanic plume measurements using a UAV for the 2014 Mt. Ontake eruption. Earth Planets Space 68:49. https://doi.org/10.1186/s40623-016-0418-0

Morsdorf F, Eck C, Zgraggen C et al (2017) UAV-based LiDAR acquisition for the derivation of high-resolution forest and ground information. Lead Edge 36:566–570. https://doi.org/10.1190/tle36070566.1

Murphy R, Dufek J, Sarmiento T et al (2016) Two case studies and gaps analysis of flood assessment for emergency management with small unmanned aerial systems. In: 2016 IEEE international symposium on safety, security, and rescue robotics (SSRR), pp 54–61

Nedjati A, Izbirak G, Vizvari B, Arkat J (2016) Complete coverage path planning for a multi-UAV response system in post-earthquake assessment. Robotics 5:26. https://doi.org/10.3390/robotics5040026

Nex F, Duarte D, Steenbeek A, Kerle N (2019) Towards real-time building damage mapping with low-cost UAV solutions. Remote Sens 11:287. https://doi.org/10.3390/rs11030287

Niethammer U, James MR, Rothmund S et al (2012) UAV-based remote sensing of the Super-Sauze landslide: evaluation and results. Eng Geol 128:2–11. https://doi.org/10.1016/j.enggeo.2011.03.012

Nikolakopoulos K, Kyriou A, Koukouvelas I et al (2019) Combination of aerial, satellite, and UAV photogrammetry for mapping the diachronic coastline evolution: the case of Lefkada Island. ISPRS Int J Geo-Inf 8:489. https://doi.org/10.3390/ijgi8110489

Pajares G (2015) Overview and current status of remote sensing applications based on unmanned aerial vehicles (UAVs). Photogramm Eng Remote Sens 81:281–330. https://doi.org/10.14358/PERS.81.4.281

Pallister JS, Diefenbach AK, Burton WC et al (2013) The Chaitén rhyolite lava dome: eruption sequence, lava dome volumes, rapid effusion rates and source of the rhyolite magma. Andean Geol. https://doi.org/10.5027/andgeoV40n2-a06

Patrick M, Orr T, Fisher G et al (2017) Thermal mapping of a pāhoehoe lava flow, Kīlauea Volcano. J Volcanol Geotherm Res 332:71–87. https://doi.org/10.1016/j.jvolgeores.2016.12.007

Patrick MR, Younger EF, Tollett W (2019) Lava level and crater geometry data during the 2018 lava lake draining at Kīlauea Volcano, Hawaii. U.S. Geological Survey data release. https://doi.org/10.5066/P9MJY24N

Peltier A, Fontaine FJ, Finizola A et al (2018) Volcano destabilizations: from observations to an integrated model of active deformation. In: AGU fall meeting abstracts, p 23

Perks MT, Russell AJ, Large ARG (2016) Advances in flash flood monitoring using UAVs. Hydrol Earth Syst Sci 20:4005–4015. https://www.hydrol-earth-syst-sci.net/20/4005/2016/doi:10.5194/hess-20-4005-2016

Peternel T, Kumelj Š, Oštir K, Komac M (2017) Monitoring the Potoška planina landslide (NW Slovenia) using UAV photogrammetry and tachymetric measurements. Landslides 14:395–406. https://doi.org/10.1007/s10346-016-0759-6

Petley DN, Crick WD, Hart AB (2002) The use of satellite imagery in landslide studies in high mountain areas. In: Proceedings of the 23rd Asian conference on remote sensing (ACRS 2002), Kathmandu

Picard D, Attoui M, Sellegri K (2019) B3010: a boosted TSI 3010 condensation particle counter for airborne studies. Atmos Measur Tech. https://doi.org/10.5194/amt-12-2531-2019

Piégay H, Arnaud F, Belletti B et al (2020) Remotely sensed rivers in the Anthropocene: state of the art and prospects. Earth Surf Processes Landf 45:157–188. https://doi.org/10.1002/esp.4787

Pieri D, Ma C, Simpson JJ et al (2002) Analyses of in situ airborne volcanic ash from the February 2000 eruption of Hekla Volcano, Iceland. Geophys Res Lett 29:191–194. https://doi.org/10.1029/2001GL013688

Popescu D, Ichim L, Stoican F (2017) Unmanned aerial vehicle systems for remote estimation of flooded areas based on complex image processing. Sensors 17:446. https://doi.org/10.3390/s17030446

Rathje EM, Franke K (2016) Remote sensing for geotechnical earthquake reconnaissance. Soil Dyn Earthq Eng 91:304–316. https://doi.org/10.1016/j.soildyn.2016.09.016

Rau JY, Jhan JP, Lo CF, Lin YS (2012) Landslide mapping using imagery acquired by a fixed-wing UAV. ISPRS Int Arch the Photogramm Remote Sens Spat Inf Sci XXXVIII-1/C22:195–200. https://doi.org/10.5194/isprsarchives-XXXVIII-1-C22-195-2011

Ridolfi E, Manciola P (2018) Water level measurements from drones: a pilot case study at a dam site. Water 10:297. https://doi.org/10.3390/w10030297

Rinaldi P, Larrabide I, D'Amato JP (2019) Drone based DSM reconstruction for flood simulations in small areas: a pilot study. In: Rocha Á, Adeli H, Reis LP, Costanzo S (eds) New knowledge in information systems and technologies. Springer, Cham, pp 758–764

Rossi G, Nocentini M, Lombardi L et al (2016) Integration of multicopter drone measurements and ground-based data for landslide monitoring. In: Aversa et al. (eds) Landslides and engineered slopes. Experience, theory and practice, Rome, Italy, p 6

Rossi G, Tanteri L, Tofani V et al (2018) Multitemporal UAV surveys for landslide mapping and characterization. Landslides 15:1045–1052. https://doi.org/10.1007/s10346-018-0978-0

Rothmund S, Vouillamoz N, Joswig M (2017) Mapping slow-moving alpine landslides by UAV—opportunities and limitations. Lead Edge 36:571–579. https://doi.org/10.1190/tle36070571.1

Rüdiger J, Tirpitz J-L, de Moor JM et al (2018) Implementation of electrochemical, optical and denuder-based sensors and sampling techniques on UAV for volcanic gas measurements: examples from Masaya, Turrialba and Stromboli volcanoes. Atmos Measur Tech. https://doi.org/10.5194/amt-11-2441-2018

Rupnik E, Daakir M, Pierrot Deseilligny M (2017) MicMac—a free, open-source solution for photogrammetry. Open Geospat Data Softw Stand. https://doi.org/10.1186/s40965-017-0027-2

Saggiani G, Ceruti A, Amici S et al (2006) UAV systems volcano monitoring: first test on Stromboli on October 2004. In: Geophysical research abstracts. Munich, Germany, p 1

Saito H, Korup O, Uchida T et al (2014) Rainfall conditions, typhoon frequency, and contemporary landslide erosion in Japan. Geology 42:999–1002. https://doi.org/10.1130/G35680.1

Sandvik KB, Lohne K (2014) The rise of the humanitarian drone: giving content to an emerging concept. Millennium 43:145–164. https://doi.org/10.1177/0305829814529470

Sanyal J, Lu XX (2004) Application of remote sensing in flood management with special reference to monsoon Asia: a review. Nat Hazards 33:283–301. https://doi.org/10.1023/B:NHAZ.0000037035.65105.95

Sanz-Ablanedo E, Chandler JH, Rodríguez-Pérez JR, Ordóñez C (2018) Accuracy of unmanned aerial vehicle (UAV) and SfM photogrammetry survey as a function of the number and location of ground control points used. Remote Sens 10:1606. https://doi.org/10.3390/rs10101606

Saroglou C, Asteriou P, Zekkos D et al (2018) UAV-based mapping, back analysis and trajectory modeling of a coseismic rockfall in Lefkada island, Greece. Nat Hazards Earth Syst Sci 18:321–333. https://doi.org/10.5194/nhess-18-321-2018

Scaioni M, Longoni L, Melillo V, Papini M (2014) Remote sensing for landslide investigations: an overview of recent achievements and perspectives. Remote Sens 6:9600–9652. https://doi.org/10.3390/rs6109600

Schneider DJ, Vallance JW, Logan M et al (2005) Airborne thermal infrared imaging of the 2004–2005 eruption of Mount St. Helens. In: AGU fall meeting abstracts 24

Schumann GJ-P, Domeneghetti A (2016) Exploiting the proliferation of current and future satellite observations of rivers. Hydrol Processes 30:2891–2896. https://doi.org/10.1002/hyp.10825

Scott JE, Scott CH (2019) Models for drone delivery of medications and other healthcare items. In: Unmanned aerial vehicles: breakthroughs in research and practice. IGI Global, pp 376–392

Smith LC (1997) Satellite remote sensing of river inundation area, stage, and discharge: a review. Hydrol Processes 11:1427–1439. https://doi.org/10.1002/(SICI)1099-1085(199708)11:10%3c1427:AID-HYP473%3e3.0.CO;2-S

Spampinato L, Calvari S, Oppenheimer C, Boschi E (2011) Volcano surveillance using infrared cameras. Earth-Sci Rev 106:63–91. https://doi.org/10.1016/j.earscirev.2011.01.003

Stumpf A, Kerle N (2011) Combining Random Forests and object-oriented analysis for landslide mapping from very high resolution imagery. Procedia Environ Sci 3:123–129. https://doi.org/10.1016/j.proenv.2011.02.022

Stumpf A, Malet J-P, Kerle N et al (2013) Image-based mapping of surface fissures for the investigation of landslide dynamics. Geomorphology 186:12–27. https://doi.org/10.1016/j.geomorph.2012.12.010

Stumpf A, Malet J-P, Allemand P et al (2015) Ground-based multi-view photogrammetry for the monitoring of landslide deformation and erosion. Geomorphology 231:130–145. https://doi.org/10.1016/j.geomorph.2014.10.039

Sui H, Tu J, Song Z et al (2014) A novel 3D building damage detection method using multiple overlapping UAV images. ISPRS Int Arch Photogramm Remote Sens Spat Inf Sci XL-7:173–179. https://doi.org/10.5194/isprsarchives-XL-7-173-2014

Tang C, Tanyas H, van Westen CJ et al (2019) Analysing post-earthquake mass movement volume dynamics with multi-source DEMs. Eng Geol 248:89–101. https://doi.org/10.1016/j.enggeo.2018.11.010

Tanteri L, Rossi G, Tofani V et al (2017) Multitemporal UAV survey for mass movement detection and monitoring. In: Mikos M, Tiwari B, Yin Y, Sassa K (eds) Advancing culture of living with landslides. Springer, Cham, pp 153–161

Tauro F, Porfiri M, Grimaldi S (2016) Surface flow measurements from drones. J Hydrol 540:240–245. https://doi.org/10.1016/j.jhydrol.2016.06.012

Techy L, Schmale DG, Woolsey CA (2010) Coordinated aerobiological sampling of a plant pathogen in the lower atmosphere using two autonomous unmanned aerial vehicles. J Field Robot 27:335–343. https://doi.org/10.1002/rob.20335

Terada A, Morita Y, Hashimoto T et al (2018) Water sampling using a drone at Yugama crater lake, Kusatsu-Shirane volcano, Japan. Earth Planets Space. https://doi.org/10.1186/s40623-018-0835-3

Thiebes B, Tomelleri E, Mejia Aguilar A et al (2016) Assessment of the 2006 to 2015 Corvara landslide evolution using a UAV-derived DSM and orthophoto. CRC Press, Boca Raton, pp 1897–1902

Thiele ST, Varley N, James MR (2017) Thermal photogrammetric imaging: A new technique for monitoring dome eruptions. J Volcanol Geotherm Res 337:140–145. https://doi.org/10.1016/j.jvolgeores.2017.03.022

Tofani V, Segoni S, Agostini A et al (2013) Technical note: use of remote sensing for landslide studies in Europe. Natural Hazards Earth Syst Sci 13:299–309. https://doi.org/10.5194/nhess-13-299-2013

Torrero L, Seoli L, Molino A et al (2015) The use of micro-UAV to monitor active landslide scenarios. In: Lollino G, Manconi A, Guzzetti F et al (eds) Engineering geology for society and territory -, vol 5. Springer, Cham, pp 701–704

Tournadre V, Pierrot-Deseilligny M, Faure PH (2014) UAV photogrammetry to monitor dykes-calibration and comparison to terrestrial lidar. ISPRS Int Arch Photogramm Remote Sens Spat Inf Sci XL-3/W1:143–148. https://doi.org/10.5194/isprsarchives-XL-3-W1-143-2014

Turner D, Lucieer A, Wallace L (2014) Direct georeferencing of ultrahigh-resolution UAV imagery. IEEE Trans Geosci Remote Sens 52:2738–2745. https://doi.org/10.1109/TGRS.2013.2265295

Turner D, Lucieer A, de Jong S (2015) Time series analysis of landslide dynamics using an unmanned aerial vehicle (UAV). Remote Sens 7:1736–1757. https://doi.org/10.3390/rs70201736

Valkaniotis S, Papathanassiou G, Ganas A (2018) Mapping an earthquake-induced landslide based on UAV imagery; case study of the 2015 Okeanos landslide, Lefkada, Greece. Eng Geol 245:141–152. https://doi.org/10.1016/j.enggeo.2018.08.010

Van Den Eeckhaut M, Kerle N, Poesen J, Hervás J (2012) Object-oriented identification of forested landslides with derivatives of single pulse LiDAR data. Geomorphology 173–174:30–42. https://doi.org/10.1016/j.geomorph.2012.05.024

Vetrivel A, Gerke M, Kerle N et al (2018) Disaster damage detection through synergistic use of deep learning and 3D point cloud features derived from very high resolution oblique aerial images, and multiple-kernel-learning. ISPRS J Photogramm Remote Sens 140:45–59. https://doi.org/10.1016/j.isprsjprs.2017.03.001

Villa T, Gonzalez F, Miljievic B et al (2016) An overview of small unmanned aerial vehicles for air quality measurements: present applications and future prospectives. Sensors 16:1072. https://doi.org/10.3390/s16071072

Voigt S, Giulio-Tonolo F, Lyons J et al (2016) Global trends in satellite-based emergency mapping. Science 353:247–252. https://doi.org/10.1126/science.aad8728

Vollgger SA, Cruden AR (2016) Mapping folds and fractures in basement and cover rocks using UAV photogrammetry, Cape Liptrap and Cape Paterson, Victoria, Australia. J Struct Geol 85:168–187. https://doi.org/10.1016/j.jsg.2016.02.012

Wallemacq P, Below R, McLean D (2018) UNISDR and CRED report: economic losses, poverty & disasters (1998–2017), Centre for Research on the Epidemiology of Disasters – CRED. https://urldefense.proofpoint.com/v2/url?u=https-3A__www.cred.be_unisdr-2Dand-2Dcred-2Dreport-2Deconomic-2Dlosses-2Dpoverty-2Ddisasters-2D1998-2D2017&d=DwIDaQ&c=vh6FgFnduejNhPPD0fl_yRaSfZy8CWbWnIf4XJhSqx8&r=r2aSgYn6PHMQXXmeBiKsnvfFG9T9U5fmdQ67xEVmgo0&m=ktrJ0pKA6VD69oYCBwKuReprks4SwDgxCLbHGC58tXE&s=nIlLeOALPXPx1QXrUvFduk1NN9dxWEFl_0H4nHiDH4&e=

Walter M, Niethammer U, Rothmund S, Joswig M (2009) Joint analysis of the Super-Sauze (French Alps) mudslide by nanoseismic monitoring and UAV-based remote sensing. Near Surf Geosci 27:8

Walter TR, Jousset P, Allahbakhshi M et al (2020) Underwater and drone based photogrammetry reveals structural control at Geysir geothermal field in Iceland. J Volcanol Geotherm Res 391:106282. https://doi.org/10.1016/j.jvolgeores.2018.01.010

Watts AC, Ambrosia VG, Hinkley EA (2012) Unmanned aircraft systems in remote sensing and scientific research: classification and considerations of use. Remote Sens 4:1671–1692. https://doi.org/10.3390/rs4061671

Weintrit B, Bakuła K, Jędryka M et al (2018) Emergency rescue management supported bu UAV remote sensing data. ISPRS Int Arch Photogramm Remote Sens Spat Inf Sci XLII-3/W4:563–567. https://doi.org/10.5194/isprs-archives-XLII-3-W4-563-2018

Wessels RI., Vaughan RG, Patrick MR, Coombs ML (2013) High-resolution satellite and airborne thermal infrared imaging of precursory unrest and 2009 eruption at Redoubt Volcano, Alaska. J Volcanol Geotherm Res 259:248–269. https://doi.org/10.1016/j.jvolgeores.2012.04.014

Westoby MJ, Brasington J, Glasser NF et al (2012) 'Structure-from-Motion' photogrammetry: a low-cost, effective tool for geoscience applications. Geomorphology 179:300–314. https://doi.org/10.1016/j.geomorph.2012.08.021

Wu C (2011) VisualSFM: a visual structure from motion system. http://www.cs.washington.edu/homes/ccwu/vsfm

Wu C (2013) Towards linear-time incremental structure from motion. In: 2013 International conference on 3D vision. IEEE, Seattle, WA, USA, pp 127–134

Xu Z, Wu L, Zhang Z (2018) Use of active learning for earthquake damage mapping from UAV photogrammetric point clouds. Int J Remote Sens 39:5568–5595. https://doi.org/10.1080/01431161.2018.1466083

Yalcin E (2018) Generation of high-resolution digital surface models for urban flood modelling using UAV imagery. In: WIT transactions on ecology and the environment, WIT Press. Lightning Source, UK, Great Britain, pp 357–366

Yamazaki F, Liu W (2016) Remote sensing technologies for post-earthquake damage assessment: a case study on the 2016 Kumamoto earthquake. Philippines, Cebu City

Yang W, Wang M, Shi P (2013) Using MODIS NDVI time series to identify geographic patterns of landslides in vegetated regions. IEEE Geosci Remote Sens Lett 10:707–710. https://doi.org/10.1109/LGRS.2012.2219576

Yu M, Huang Y, Zhou J, Mao L (2017) Modeling of landslide topography based on micro-unmanned aerial vehicle photography and structure-from-motion. Environ Earth Sci. https://doi.org/10.1007/s12665-017-6860-x

Zarco-Tejada PJ, González-Dugo V, Berni JAJ (2012) Fluorescence, temperature and narrow-band indices acquired from a UAV platform for water stress detection using a micro-hyperspectral imager and a thermal camera. Remote Sens Environ 117:322–337. https://doi.org/10.1016/j.rse.2011.10.007

Zekkos D, Clark M, Cowell K, Medwedeff W, Manousakis J, Saroglou H, Tsiambaos G (2017) Satellite and UAV-enabled mapping of landslides caused by the November 17th 2015 Mw 6.5 Lefkada earthquake. In Proc. 19th Int. Conference on soil mechanics and geotechnical engineering, pp. 17–22

Zhang J (2010) Multi-source remote sensing data fusion: status and trends. Int J Image Data Fusion 1:5–24. https://doi.org/10.1080/19479830903561035

Zhou Y (2019) 100% automatic metrology with UAV photogrammetry and embedded GPS and its application in dike monitoring. PhD thesis, Université Paris-Est

Publisher's Note Springer Nature remains neutral with regard to jurisdictional claims in published maps and institutional affiliations.

Surveys in Geophysics (2020) 41:1323–1354
https://doi.org/10.1007/s10712-020-09606-4

Earth Observation for the Assessment of Earthquake Hazard, Risk and Disaster Management

J. R. Elliott[1] ⓘ

Received: 6 December 2019 / Accepted: 23 July 2020 / Published online: 20 August 2020
© The Author(s) 2020

Abstract

Earthquakes pose a significant hazard, and due to the growth of vulnerable, exposed populations, global levels of seismic risk are increasing. In the past three decades, a dramatic improvement in the volume, quality and consistency of satellite observations of solid earth processes has occurred. I review the current Earth Observing (EO) systems commonly used for measuring earthquake and crustal deformation that can help constrain the potential sources of seismic hazard. I examine the various current contributions and future potential for EO data to feed into aspects of the earthquake disaster management cycle. I discuss the implications that systematic assimilation of Earth Observation data has for the future assessment of seismic hazard and secondary hazards, and the contributions it will make to earthquake disaster risk reduction. I focus on the recent applications of Global Navigation Satellite System (GNSS) and increasingly the use of Interferometric Synthetic Aperture Radar (InSAR) for the derivation of crustal deformation and these data's contribution to estimates of hazard. I finish by examining the outlook for EO in geohazards in both science and decision-making, as well as offering some recommendations for an enhanced acquisition strategy for SAR data.

Keywords Earth Observation · Earthquakes · Seismic hazard · Disaster risk reduction · InSAR · Crustal strain

Abbreviations

ASI	Agenzia Spaziale Italiana
ASTER	Advanced Spaceborne Thermal Emission and Reflection Radiometer
ALOS-PALSAR	Advanced Land Observing Satellite Phased Array type L-band Synthetic Aperture Radar
AW3D	Advance World 3-Dimensional
BIROS	Bi-spectral InfraRed Optical System
CNES	Centre National d'Études Spatiales
COMET	Centre for the Observation and Modelling of Earthquakes, Volcanoes and Tectonics
CONAE	Comisión Nacional de Actividades Espaciales

✉ J. R. Elliott
  j.elliott@leeds.ac.uk

[1] COMET, School of Earth & Environment, University of Leeds, Leeds LS2 9JT, UK

COSMO	Constellation of small Satellites for the Mediterranean basin Observation
CoV	Coefficient of variation
CSA	Canadian Space Agency
DAC	Development Assistance Committee
DEM	Digital Elevation Model
DLR	Deutsches Zentrum für Luft- und Raumfahrt
DRM	Disaster risk management
EO	Earth Observation
EROS	Earth Resources Observation Satellite
ESA	European Space Agency
GDEM	Global Digital Elevation Model
GDP	Gross Domestic Product
GEAR	Global Earthquake Activity Rate
GEO	Group on Earth Observations
GEM	Global Earthquake Model
GNSS	Global Navigation Satellite System
GSNL	Geohazard Supersites and Natural Laboratories
GSRM	Global Strain Rate Model
GTEP	Geohazards Thematic Exploitation Platform
InSAR	Interferometric SAR
INTA	Instituto Nacional de Técnica Aeroespacial
ISRO	Indian Space Research Organization
JAXA	Japan Aerospace Exploration Agency
KARI	Korea Aerospace Research Institute
LEO	Low Earth Orbit
LSTM	Land Surface Temperature Monitoring
MODIS	Moderate Resolution Imaging Spectroradiometer
MS	Multispectral
NASA	National Aeronautics and Space Administration
NEIC	National Earthquake Information Centre
NiSAR	NASA-ISRO Synthetic Aperture Radar
PRISM	Panchromatic Remote-sensing Instrument for Stereo Mapping
RCM	RADASRAT Constellation Mission
RGB	Red, green, blue
SAR	Synthetic Aperture Radar
SPOT	Satellite Pour l'Observation de la Terre
SRTM	Shuttle Radar Topography Mission
UCERF	Uniform California Earthquake Rupture Forecast
USGS	United States Geological Survey

1 Introduction

Earthquakes are a consequence of the sudden release of the accumulation of strain within the Earth's crust due to the stresses from the tectonic driving forces of plate motion and mountain building. Earthquakes produce two major sets of measurable physical phenomena. Firstly, seismic waves propagate from the earthquake source, radiating outwards, and can be measured globally by seismometers. It is these waves that are the source of the ground shaking (accelerations) that comprise the bulk of the hazard to buildings in major earthquakes. Secondly, a permanent displacement of the Earth's surface occurs due to the change in the accumulated elastic energy that results from the sudden slip across the fault surface (which in itself can be a direct hazard if buildings straddle the fault rupture). This static displacement leads in the long-term (after many earthquake cycles) to the permanent deformation of the crust, accommodating the translation of plates and crustal blocks, as well as resulting in the growth of geological structures and mountains. It is the second of these earthquake ground phenomena (permanent static displacements) that we are largely concerned with in this review because these are typically visible to Earth Observing systems that acquire images of the land surface. We seek to contribute towards better recognising the hazard associated with seismic shaking by expanding the analysis of both earthquake ground displacements and the longer-term relative interseismic displacement of the crust leading up to earthquakes. Identifying the location and rates of crustal deformation is important for constraining the potential fault sources of the seismic shaking that constitutes the hazard. In addition to the primary hazard of shaking, secondary hazards include triggered tsunamis, landsliding, liquefaction and volcanic unrest, as well as cascading events such as fires and flooding that can result following the initial earthquake (Tilloy et al. 2019), but these are not a focus of this review.

Whilst globally the death rate from natural disasters such as flooding and drought has declined dramatically in the past century, the fatality rates from earthquakes (Fig. 1) have remained persistent. It is a target of the Sendai Framework for Disaster Risk Reduction to reduce the global disaster mortality rate (along with those affected and economic losses) for this coming decade compared to the past decade. However, as populations continue to grow and cluster, an increasing number of people are exposed to the hazards from earthquakes, having gathered to live in urban centres and thus created megacities over recent decades (Fig. 2). These population centres are often clustered along fault lines (Fig. 3) resulting from historical trade routes, water supplies, fertile basins and security that exist due to the mountains created by the same tectonic processes that cause the underlying hazard (Jackson 2001, 2006). Rapid population growth in economically developing countries, many with limited economic resources or facing other pressing near-term priorities, has led to housing structures of often limited seismic resilience, thus contributing to a heightened vulnerability of those exposed. Furthermore, entrenched and deep-seated corruption tends to by-pass any attempts made to improve this situation with seismic design codes in some countries where earthquake hazard is recognised but ignored. This duplicity results in much worse outcomes in the face of an earthquake disaster than expected, given the prevailing income level (Ambraseys and Bilham 2011).

Broadly, we understand how and where earthquakes happen. However, many recent events continue to surprise us regarding the nature of their exact location, size and sometimes style of faulting. Therefore, our knowledge is incomplete and there is an advantage to be had in forensically studying each major seismic event to understand more about the overall earthquake process. This is best done globally, because typically the recurrence time interval for a major earthquake in any given area is long (hundreds of years). We can make an ergodic assumption and substituting space for time we can use global datasets to better understand the size, exact locations and rupture processes, as well as how to interpret the fault geomorphology at the surface as a clue to the underlying hazard (Fig. 3). Furthermore, we can not only use the past location of earthquakes but also examine where the strain is accumulating that eventually leads up to an earthquake. This has been done at a global scale using the Global Strain Rate Map (Kreemer et al. 2014). This makes use of

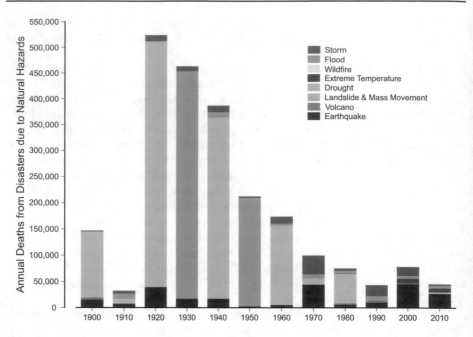

Fig. 1 Annualised death rates from disasters resulting from natural hazards, grouped by decade from the beginning of the 20th Century (e.g., 1900 is the average annual rate of deaths for the complete years 1900–1909, coloured by hazard type). A decline is observed in the number of deaths from climatological and hydrological disasters attributed to drought and flood. However deaths due to earthquake disasters have persisted. Source: International Disaster Database, EM-DAT, CRED, UCLouvain, Brussels, Belgium. http://www.emdat.be

the relatively coarse distribution of Global Navigation Satellite System (GNSS) stations to estimate the velocity fields on continental and subduction zone areas from which estimates of rates of strain can be made. It is this strain accumulation that is typically released in earthquakes. This measure of strain build-up acts as a first order proxy for the potential level of earthquake hazard that a region is exposed to. By comparing the location of capital cities for the world's economically poorer nations with this strain map, we can see that over 50 major population centres lie in regions that are accumulating a significant crustal strain, a phenomenon typically synonymous with seismic hazard (Fig. 2) as this strain is usually released violently in earthquakes. This intersection provides a targeted list of exposed populations for focusing efforts on better locating active fault traces (Fig. 3), constraining the seismic hazard, increasing the preparedness and examining the earthquake risk for major cities (Hussain et al. 2020).

For the stresses to build up on a fault (whose eventual sudden failure will lead to an earthquake), the fault surface itself is locked in the period between earthquakes, in that no displacement occurs immediately across it. However, an increasing number of observations point to the propensity for faults (in particular subduction zone interfaces) to accommodate transient slow slip within these locked zones (Jolivet and Frank 2020). This slip releases the accumulated elastic energy slowly enough that seismic shaking is not generated, but the motion can be detectable geodetically. Measuring such signals is important for determining the long-term slip budget of a fault assessing their mechanical and rheological behaviour and is particularly important for major subduction zones (Wallace et al. 2016).

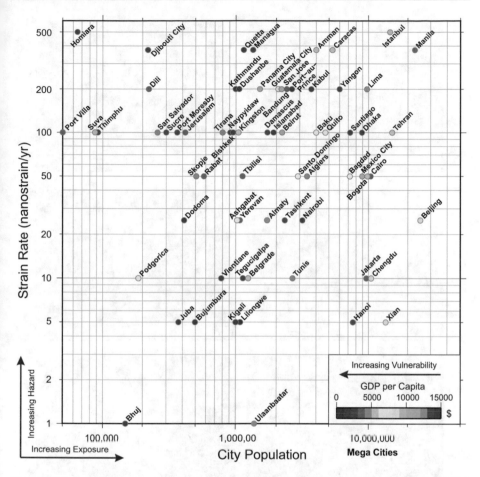

Fig. 2 Population exposure of capitals and major cities of the Development Assisted Countries (DAC) as a function of the strain rate determined from the Global Strain Rate Map (Kreemer et al. 2014), used as a proxy of the seismic hazard. Each city is coloured by the Gross Domestic Product (GDP) per capita for the country as a proxy for the first order control on the physical vulnerability

Finally, being able to forewarn of an impending earthquake through the use of earthquake precursors (Cicerone et al. 2009) has been the preoccupation of parts of the earthquake community for decades. These precursors could take the form of foreshocks in a region that indicate a build-up to the mainshock, gas emissions, electromagnetic field disturbances, ground deformation or changes to groundwater water levels. However, a reliable system for demonstrating the use of such potential precursors continues to be elusive for short-term prediction. For longer-term assessment of the location of earthquake hazard, measurements of where strain in the crust is accumulating and at what rate using GNSS (Bird and Kreemer 2015) and increasingly InSAR data offer greater potential, but does not provide short-term predictions that enable the timely evacuation of populations. Instead, efforts to focus on preparedness in the risk management cycle based upon knowledge of where hazardous faults are accumulating strain offer a tractable approach to increase resilience.

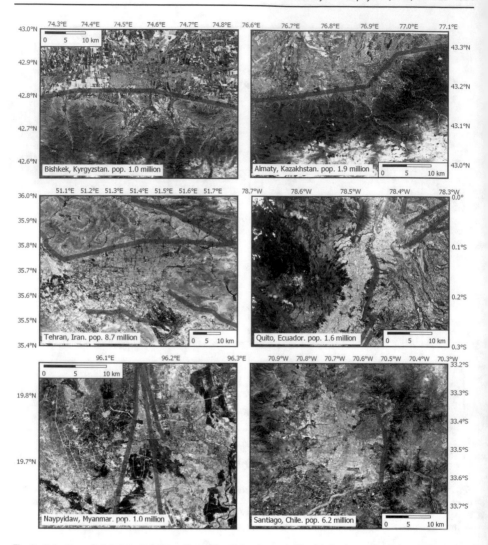

Fig. 3 Selection of global capital cities (from Fig. 2) in economically developing nations situated in regions of active tectonic strain accumulation, with major faults marked by red lines (from the Global Earthquake Model Active Fault database, Styron et al. 2010). Satellite imagery is from Sentinel 2 data and is true colour RGB composites (10 m visible bands 4, 3, 2) from cloud free images acquired in 2019

Due to the global coverage, increasing revisit frequency and wide area imaging of 100's of km width at a time, Earth Observation (EO) offers a number of contributions to the improved understanding of both the earthquake cycle in terms of a physical process and also to aid in the better management of disaster risk at various stages. In this article, I start with a review of the fleet of Earth Observing systems that currently and historically have regularly imaged the Earth's land surfaces. I then highlight the diverse contributions to disaster risk reduction that EO presently, and could potentially, provide at different points of response, recovery, mitigation and preparedness. In particular, I focus on the emerging capacity of EO data to contribute towards strain mapping and seismic hazard assessment. The recent launches of constellations of earth imaging systems have opened up new opportunities for investigating earthquake hazard. The current growth of satellite systems shows no sign of abating, and future missions are continually being planned. However, there are still improvements that could be made, in particular in terms of the imagery acquisition strategies. I therefore finish by providing some recommendations for the further optimisation of satellite observations for earthquake hazard in future mission planning and development.

2 Earth Observing Systems for Solid Earth Processes

Over the past few decades an ever-expanding fleet of Earth Observation systems have been launched in into low-earth near polar orbit, enhancing our capability to monitor solid earth hazards. This continued expansion of satellite capacity (Fig. 4) has occurred both for active radar (synthetic aperture radar—SAR) and passive optical imaging systems (multi-spectral—MS). The European Space Agency's (ESA) Copernicus programme and regular Sentinel-1/2 acquisitions have been central to making such analysis of data more systematic. The free and open data policy has greatly expanded the user base (aided by the development and provision of open software processing toolboxes). For instance, there existed a two-decade period from 1992–2014 when exploitation of SAR imagery was largely the preserve of academic specialists and small niche technical companies. The deployment of the Copernicus space segment ("the Sentinels") from late 2014 will most likely be viewed as the defining watershed moment when the wide-scale exploitation of long-time series of big EO data took off, building upon USGS/NASA's open data policy for imagery such as that from the Landsat program from 2008 (Zhu et al. 2019).

Furthermore, the type of operators of satellites is undergoing a shift. In the past, the expensive development and long-term commitment required by early satellite platforms has meant that they were the preserve of the national space agencies (and military industrial complexes). However, there has been an increasing shift (starting in the late 1990s with IKONOS at 1 m metre resolution) towards commercial operators offering the highest resolutions (increasingly sub-metre) of optical imagery both from large platforms (e.g., Satellite Pour l'Observation de la Terre SPOT6/7 1.5 m and WorldView~0.3 m) and also lower-cost microsatellites (e.g., Planetscope—3 m). For SAR systems, this shift has lagged behind optical platforms by about two decades and has only occurred with the recent deployment from 2018 onwards of ICEEYE's micro-SAR satellites.

Earth Observation satellites are typically launched into a near-polar orbit in which they pass almost directly over the poles, as this enables them to achieve near-global coverage after a certain number of orbits (175 cycles of 98.6 min orbital period in the case of Sentinel-1A/B, taking a total of 12 days each to complete full coverage). Typically, this is in a low-earth orbit (LEO) at around 700 km for many SAR systems, and the width of tracks that can be imaged is usually a couple of hundred kilometres wide (depending upon the SAR imaging mode used). This improvement to the revisit period (down to 12 days with Sentinel-1 from 35 days with Envisat, for example) is due to the increase in the width of the track imaged in the range direction (from 70 km with Envisat to 250 km with Sentinel-1, when considering their main interferometric imaging modes). This reduced revisit time both increases the volume of data (through the number of independent observations made within a given time period) and also the level of interferometric coherence (as the interval between passes is shorter, the land surface has changed less), which subsequently leads to improved measurements of time series of surface deformation when using interferometry. However, due to power limitations for radar satellites, it is possible only to actively image for about a quarter of an orbit, so not all the Earth's surface is imaged with SAR on every pass.

Earth imaging satellites are typically in a Sun Synchronous orbit so that they image the ground at the same time of day (for radar satellites this normally occurs in a dawn/dusk orbit). They consequentially pass the equator twice on each revolution, once in an ascending direction (approximately south-to-north with a retrograde orbital inclination angle of about 98 degrees to the equator, passing at around dusk) and once in the opposite descending direction (nearly north-to-south around dawn). The local time for the ascending node crossing is 18:00 for Sentinel-1. Whilst the radar instrument predominately points down towards the Earth, it is inclined off to the side and for Sentinel-1 is configured as a right-looking imaging system, so looks eastwards in an ascending direction and westwards in descending. Having images on both the ascending and descending passes increases the frequency of observing a given point and also helps to constrain the direction of ground displacement in terms of vertical versus horizontal deformation (although near polar-orbiting SAR satellites suffer from being far less sensitive to north–south motion, i.e. in the near azimuthal direction).

SAR imagery is typically acquired in one of three main wavelength bands (L 23 cm, C 6 cm, X 3 cm) with the higher-resolution imagery (1–5 m) in the past being at X-band. However, in terms of making interferometric measurements (InSAR) that are particularly useful for ground deformation, an issue of coherence arises across the time periods needed (week to months) to measure relative displacements from strain over wide areas. Phase coherence is a crucial parameter controlling the spatial extent and distribution of possible interferometric measurements with multiple SAR images. The time difference between acquisitions is one important constraint on the level of coherence achievable. In the past when revisit periods were longer (such as 35 days for Envisat) and also imagery was not acquired on every pass (with many months to year-long intervals between images), the low coherence limited the areas that were suitable for InSAR observations, depending upon land cover type and wavelength used (and this therefore created systematic variations in the level of coverage with latitude with interferometry). This means the technique succumbs to problems arising from vegetation causing decorrelation, making the data unusable, and this is particularly a problem at shorter wavelengths (Ebmeier et al. 2013) in non-urban environments. Currently, L-band measurements (such as from the Advanced Land Observing Satellite (ALOS-2) Phased Array type L-band Synthetic Aperture Radar PALSAR-2) suffer much less coherence loss at vegetated low latitudes, but the acquisition plan is not systematic enough globally to make continental-wide measurements of slow deformation easily (where regular, consistent and multiple acquisitions are required to measure the small displacement signals above the noise from differential atmospheric delay and other sources). It is anticipated that there will be an improvement in this capacity with the launch of further L-band missions such as NASA-ISRO Synthetic Aperture Radar (NISAR, Joint USA-India mission). Whilst most satellite SAR systems have focussed at X, C and L-bands, the UK's NovaSAR-1, launched in late 2018, operates in S band (~ 9 cm) to test the applications and capabilities for SAR imaging at this wavelength.

There have also been additions to the Global Navigation Satellite System (GNSS) such as with the nascent Galileo system. This will offer enhanced capability and improved accuracy of site velocities in the coming years, which will be of use both for high rate measurements of earthquake ruptures and for long-term interseismic rates of continental deformation with continuously occupied sites (and more rapid occupations of survey sites). A still burgeoning field is that of video satellites which are capable of capturing tens of seconds of high-definition imagery from steerable platforms that pivot during orbit to capture steady images of cities. Constellations, such as that based on the Carbonite-2 demonstrator (Vivid-i), are currently planned. Planet's SkySats have been able to capture high rate full motion video for 2-min durations for a number of years. These types of platforms and capabilities will play an increasingly important monitoring role in disaster response but will also be of use to scientists in a similar way to optical imagery. Use of multiple frames of images can be used to reconstruct topography through the structure-from-motion approach (Johnson et al. 2014). This could potentially play an important part in land surface change detection before and after a disaster.

Understanding the shape of the Earth and how this evolves through time underpins much of the Earth Sciences, and an important dataset for many aspects of this is topography. Globally available open datasets of digital elevation models (DEMs) are still at relatively coarse spatial resolutions compared to available optical imagery (30 m versus 10 m). Remotely sensed topography is largely derived either from radar or from optical stereo photogrammetry (and typically in the past at a national level this was often captured from aerial photography). An important milestone in 2000 was the Shuttle Radar Topography Mission (SRTM) (Farr and Kobirck 2000) that offered 90 m (and now 30 m) resolution topography for low and mid-latitudes. Open globally optically derived topography from the Advanced Spaceborne Thermal Emission and Reflection Radiometer (ASTER) Global DEM (GDEM) and from Panchromatic Remote-sensing Instrument for Stereo Mapping (PRISM) Advanced World-3D (AW3D) is also available at 30 m, with higher-resolution PRISM datasets available commercially. However, optically derived topographic datasets can exhibit greater high-frequency noise when compared to equivalent resolution radar-derived datasets. Bespoke DEMs at very high resolutions (a few metres) can be generated from in-track stereo imagery from satellite systems such as SPOT, Pleiades and Worldview. Topographic datasets often act as base-layers and background information for analysis, but they can also tell us about the physical processes of faulting, especially when looking at changes in topography (Diederichs et al. 2019).

3 International Initiatives and Organisations for the Enhanced Exploitation of EO in Disasters

The International Charter for space and major disasters provides satellite data for response and monitoring activities following major natural (e.g., flooding, earthquakes and volcanoes) or man-made disasters (such as oil spills). The International charter was initiated in 1999 and became operational on the 1 November 2000, with over 600 activations since then (http://disastercharter.org/).

The charter is activated by authorised users (representatives of a national civil protection organisation) which mobilises space agencies to obtain and quickly relay Earth Observation data from over 60 satellites on the unfolding disaster (there have been 67 earthquake activations to date in the past two decades). This can be particularly important for providing commercial data at high resolution and in a timely fashion that would otherwise be very costly to task for rapid acquisition. The main focus is on the initial response phase of a disaster with a limited timespan of activation, rather than any later phase of the cycle that may be useful for other aspects of disaster risk management (DRM).

An important organisation that seeks to coordinate the exploitation of satellite data for the benefit of society is the Committee of Earth Observing Satellites (CEOS), originally established in 1984 (https://ceos.org). This international forum brings together most space agencies with a significant EO programme in order to optimise the exchange and use of remote sensing data for the sustainable prosperity of humanity (Percivall et al. 2013). One particularly relevant part of CEOS is the working group on disasters which has coordinated pilot and demonstrator phases of multiple hazards projects (e.g.,, Pritchard et al. 2018), and one of which is focused on seismic hazards. Their aim is to broaden the user base of EO and engage with non-expert users, promoting the adoption of satellite data and products in the decision-making process for disaster risk reduction.

Another international partnership is the Group of Earth Observations (GEO) that has a similarly aligned goal of harnessing EO to better inform decisions for the benefit of humankind (www.earthobservations.org). The global network, initiated in 2003, joins governments, research organisations and businesses with data providers to target sustainable development and sound environmental management. A particularly relevant initiative was the organisation of "supersites" (Geohazard Supersites and Natural Laboratories–GSNL) that seeks to increase the openness of satellite observations–and in situ data over particular target sites, such as the San Andreas Fault Natural Laboratory and the Sea of Marmara (Istanbul) where major earthquakes are anticipated to be probable in the near future (Parsons et al. 2004).

4 Earth Observation and the Disaster Risk Management Cycle

A priority for action identified within the Sendai Framework for Disaster Risk Reduction is to understand disaster risk. Earth Observation can play an important role in achieving this in terms of characterising the sources of seismic hazard as well as the exposure of persons and assets, and the physical environment of the potential disaster. Therefore, EO can contribute to both understanding the hazard element of disaster and some of the contributors to the risk calculation, namely where populations and buildings are located relative to the sources of such hazards. Furthermore, EO can drive the enhancement of disaster preparedness for earthquakes so that a more effective response and recovery can occur by highlighting the regions exposed to earthquake hazards, thus addressing a further priority of the Sendai Framework to increase resilience. Earth Observation currently feeds into all aspects of the disaster risk reduction cycle and, with upcoming advances, has the potential to contribute even further, particularly for identifying hazard in continental straining areas (Fig. 5). I describe each stage of the cycle in turn, highlighting some examples of the exploitation and utility of such EO datasets.

4.1 During the Earthquake

A large earthquake can take many tens of seconds to minutes to rupture from the epicentre initiation point on the fault (hypocentre) to its eventual end, which may be many hundreds of kilometres away from the start. The propagation speed of surface seismic waves is of the order a few kilometres per second, so relaying early warnings over speed-of-light networks can precede the arrival of the onset of the greatest shaking at far-field distances enabling mitigation to take place (Allen and Melgar 2019). Seismometers play a crucial role in providing this early shaking information. However, an increasing number of high-rate (20 Hz) Global Navigation Satellite System (GNSS) networks exist that can measure the strong ground motion and not saturate in amplitude from the passing seismic waves, and could prove important in real-time monitoring of a developing earthquake (Fig. 5). It has been recently recognised that there is the potential that once a large earthquake initiates, it is possible to determine what the eventual size of an earthquake will become. This can be done using only the first 20 s of shaking since nucleation, rather than having to wait for the whole rupture to finish propagating to the eventual end of the fault (Melgar and Hayes 2019). If this recognition that earthquake sizes have a weak determinism holds true, it offers the chance to greatly improve early warning systems as the final magnitude (of big ruptures) can be determined before the earthquake finishes. This means it may be possible to answer the long-standing question of when a small earthquake goes on to grow into a big one. For example, the South Alpine Fault on the southern Island of New Zealand is one of the largest and fastest strain accumulating strike-slip faults in the world (about 30 mm/yr). This fault is anticipated to experience large ruptures, and the regular repeat of previous major earthquakes found in the past palaeoseismological record to be an average of 330 years (Berryman et al. 2012) means we can reasonably expect a major earthquake in the coming decades. This assessment can be done with a higher degree of certainty (Biasi et al. 2015) than exists for most faults around the world where the individual fault characteristics are often poorly known. If the rupture were to initiate at the south-western end of the Alpine fault (just along strike from earlier large earthquakes that have increased the stress in this area at the Puysegur Trench in 2003, 2006 and 2009, Hamling & Hreinsdóttir 2016), then the rupture is likely to propagate in a north-eastwards direction, towards the capital Wellington over 600 km away. If it takes 20 s to establish that this earthquake will ultimately be a magnitude 8, then there will be about 80 s to provide a warning for the arrival of the first seismic waves at Wellington, and over twice this time before the strongest surface waves reach the capital. However, this relies on the installation and real-time monitoring of a number of high-rate GNSS along or near the Alpine fault, instruments which are currently relatively sparse.

In contrast, the latency of satellite-based Earth Observations means that a seismic event of minutes is long over before imagery from space is acquired. Polar orbiting satellites have relatively long repeat intervals, typically of many days (the repeat interval for a single satellite Sentinel-2 is 10 days and for Sentinel-1 is 12 days, halved when considering the pair constellation). Whilst the polar satellite orbits have a period of about 90 min, there is a trade-off between imaging area and resolution (as well as a limited line-of-sight footprint area from 700 km orbital altitude). This represents an improvement relative to past repeat intervals in the 1990–2000s of around a month (e.g., Envisat Advanced Synthetic Aperture Radar (ASAR) was a minimum of 35 days between potential acquisitions for interferometry and typically much longer because an image was not programmed to be acquired on each pass). This reduction in latency has been achieved largely through two approaches: (1) the implementation of wider swath widths of many hundreds of kilometres (whilst often sacrificing potential technological improvements in ground spatial resolution) and (2) by having pairs of satellites, as is the case with Sentinel-1 and Sentinel-2 that have (near) identical twins A and B (and combined repeat times of 6 and 5 days, respectively). Further improvements have and are being made with slightly larger constellations (e.g., COnstellation of small Satellites for the Mediterranean basin Observation (COSMO-Skymed) and the RADARSAT Constellation Mission (RCM)). Under the ESA Copernicus programme, there are further copies of the Sentinels to be launched as replacements in the future as

Fig. 4 Timeline showing the typical major Low Earth Orbit EO satellites used for earthquakes comprising ▶ (a) optical and (b) RADAR systems (updated from Elliott et al. (2016a) to end of 2019). Optical satellites are ordered by their resolution and RADAR ones grouped by their microwave wavelength, and both sets are colour coded by their operators/agencies (an acronym list of space agency names is provided at the beginning of the paper). Anticipated launch dates for upcoming missions are shown approximately but are often subject to delay. Arrows continuing beyond present day are only indicative, and for many existing systems this is beyond their design lifespan specification (nominally 5–7 years though often exceeded). There are many other past and present satellite systems, but the ones shown here have near global or consistent systematic coverage, or they have a greater level of availability and suitability for earthquake displacement and crustal strain deformation studies. There has been a very recent increase in the number of constellations of optical microsatellites (such as Planetscopes indicated here) and also of systems from China such as Superview-1, as well as ones in commercial SAR (ICEEYE) and video (Vivid-i) that will become increasingly important to deformation studies as exploitation of these newer datasets develops

A and/or B fails or depletes its hydrazine propellant. However, pre-emptive launching of these would have the advantage of increasing constellation sizes and thus reducing latency further. Shorter observation intervals also come from repeated coverage by overlapping tracks at high latitudes and furthermore by acquisitions often occurring in both the ascending and descending directions for SAR, a feature permissible from the day and night capability of radar.

The re-visit latency is also greatly reduced from agile imaging platforms (such as Pleiades and Worldview for high resolution optical) that do not always look at a fixed ground track (as in the case of Sentinel-2 or Landsat-8 that image at nadir) but instead have steerable sensor systems or agile platforms permitting coverage over off-track areas at higher incidence angles. The waiting time is also being reduced by large constellations of cubesats such as Planet's hundred-plus individual platforms ("doves"), a constellation capable of imaging each point on the Earth at least every day (subject to cloud cover). In the future, continuously staring video satellites and SAR satellite systems from geo-stationary orbits may offer the chance to monitor an event in real time (Fig. 5).

4.2 Earthquake Response

The utility of EO systems for aiding the immediate response to earthquakes can be relatively late into this part of the DRM cycle due to the latency of systems mentioned above, limitations with cloud/night-time for optical systems and subsequent analysis time although this is continually improving. Whilst global seismological networks are able to estimate earthquake magnitudes and locations well and quickly (within minutes to hours), there is still room for EO to enhance the initial estimates of the sources of the earthquake hazards on the time scale of a couple of days. Actions by agencies such as the International Charter for space and major disasters over the past two decades have consistently improved the speed with which EO data and information products are made available to assess the earthquake impact on infrastructure. Measurements of ground deformation with either optical offsets or InSAR (Fig. 5) are particularly useful for: (1) identifying more precisely the geographic location of the earthquake, which is especially important for closest approach to cities and extent of aftershocks (e.g., 2010 Haiti earthquake, Calais et al. (2010)); (2) investigating the rupture of a large number of complex fault segments (e.g., 2008 Sichuan earthquake, de Michele et al. (2009), or the 2016 Kaikoura earthquake, Hamling and Hreinsdottir (2016)) not easy to ascertain with global seismology; (3) determining whether surface rupturing has occurred and constraining the depth of fault slip (important for estimates the degree of ground shaking, e.g., 2015 Gorkha earthquake, Avouac et al. (2015)); and (4)

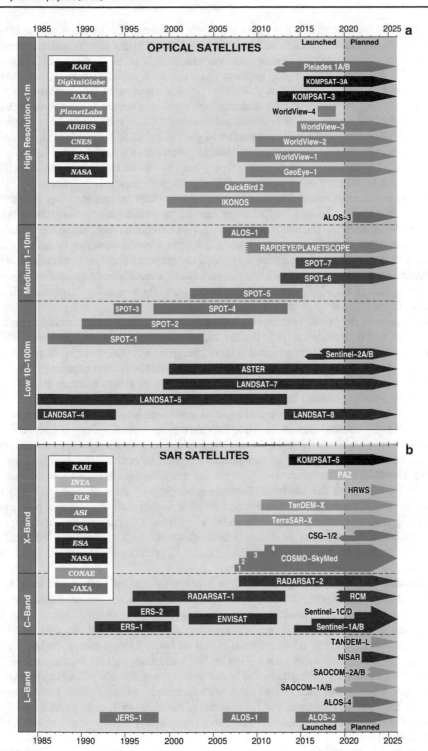

157

resolving the seismological focal plane ambiguity for earthquakes (which can be important in particular for strike slip earthquakes where the ambiguity is orthogonal in the surface strike direction) and the rupture direction which could be uni-lateral or bi-lateral and makes a larger difference for long ruptures (e.g., the 2018 earthquake in Palu, Sulawesi, Socquet et al. (2019)). Barnhart et al. (2019a) provide a review of the implementation of imaging geodesy for the earthquake response for a number of recent events where improvements in the initial estimates of fatalities and economic losses were made. They provide a workflow and framework for the integration of such EO data into their operational earthquake response efforts at the National Earthquake Information Centre (NEIC).

Whilst very high-resolution optical imagery may be of use in identifying the location of collapsed buildings (and therefore potentially trapped persons), buildings that "pancake" are not as obviously critically damaged when viewed in satellite imagery taken at nadir. Highly oblique images may be of greater utility (although may suffer from building shadowing in dense urban environments). A more recent advance using the behaviour of the coherence in SAR images (Yun et al. 2015) is a useful indicator of building collapse as it provides a good measure of ground disturbance. DEM differencing of urban environments generated before and after an earthquake from high-resolution stereo optical imagery (Zhou et al. 2015) also offers the future potential to quickly determine building height reductions, indicative of collapse.

In addition to the primary earthquake hazard of shaking, there are secondary hazards of landsliding and liquefaction (Fig. 5) that are triggered during the event (Tilloy et al. 2019). Optical imagery is used to map landslide scars (Kargel et al. 2016) and also displacements (Socquet et al. 2019), whilst SAR imagery offers the advantage of not requiring cloud free/ daytime conditions (Burrows et al. 2019). Increased use of videos (such as from SkySat and Vivid-i) and night-time imagery (such as Earth Remote Observation System EROS-B or Jilin-1) for disaster response and rapid scientific analysis of an earthquake can be expected in this coming decade. In 2016, the German Aerospace agency (DLR) launched their Bi-Spectral Infrared Optical System (BIROS) to detect high-temperature events, often forest fires as part of a Firebird constellation comprising an earlier instrument (Halle et al. 2018). This is an enhancement of ~200 m ground sampling over previous lower resolution systems such as from the Moderate Resolution Imaging Spectroradiometer (MODIS) at 1 km, offering the potential for identifying smaller fires that could be useful in post-earthquake triggered fires within cities. Other missions carrying a thermal infrared imaging payload at sub-100 m resolution are currently under development, e.g., Trishna (CNES-ISRO) and Land Surface Temperature Monitoring (LSTM, Copernicus-ESA) that will add to this capability.

Tsunamis triggered by large earthquakes beneath the oceans (subduction zone faulting) are a major contributor to fatalities in such events, and tsunami warnings can provide some level of response. Ocean pressure sensors can provide the observations of a tsunami wave passing across an ocean, where EO data have had a limited input into such processes to date. However, satellite altimeters such as that on Jason-1 that measure the sea surface topography can capture these waves (Gower 2005) and may provide future potential for constraining submarine earthquakes and tsunamis (Sladen and Herbert 2008) more rapidly. With polar orbiting satellites, this would rely on serendipity of the satellite footprint passing over the transitory tsunami, but with increased constellations, or geostationary orbits, or use of higher-resolution optical tracking, more operational direct tsunami wave tracking may be possible.

A major benefit of EO data in the immediate aftermath of a disaster is that its acquisition is not intrusive for the local population and government. The rapid deployment of research and data gathering personnel into an affected area is not always welcome or helpful, even if done with good intentions of trying to provide greater understanding of the event to improve outcomes in future disasters (Gaillard and Peek 2019). Using this advantage of EO, in cooperation and engagement with local partners, will maximise the effectiveness of its uptake. An increasingly important aim of organisations such as CEOS is to ensure that local capacity or international connections exist prior to disasters so that EO data are used effectively.

4.3 Recovery

In this period, further geohazards can perturb and interrupt the recovery process. Aftershocks typically decay in magnitude and frequency with time, but also seismic sequences can be protracted. Furthermore, the largest earthquake is not necessarily the first in a destructive sequence (Walters et al. 2018). Earth Observation data for large shallow aftershock events can provide further information on locations as for the main event. Operational Earthquake Forecasting can examine the time-dependent probabilities of the seismic hazard (Jordan et al. 2014) for communication to stakeholders. A time-dependent hazard model for seismicity following a major event can be estimated (Parsons 2005). One aspect of this is the transfer of stress from the earthquake fault onto surrounding fault structures which can be calculated from the distribution of slip on the fault found from inverting the surface InSAR and optical displacement data. In the recent Mw 6.4 and Mw 7.1 Ridgecrest earthquakes in California, using satellite InSAR and optical imagery, Barnhart et al. (2019b) were able to model the slip and stress evolution of an earthquake sequence (that occurred on faults not previously recognised as major active ones). They also observed the triggering of surface slip (creep) on the major Garlock strike-slip fault to the south of the earthquake zone. Understanding how faults such as these interact will be necessary to provide probabilistic calculations of time-dependent hazard. Additionally, major earthquakes often highlight the existence of hitherto unknown active faults or ones that were previously considered less important compared to known major faults (Hamling et al. 2017).

Postseismic processes such as afterslip, poroelasticity and viscous relaxation change the nature of stress in the earthquake area (Freed 2005). InSAR observations can provide useful constraints on the degree of postseismic behaviour and the magnitude of signals through time, which could prove useful in updating hazard assessments. Important insights regarding these postseismic processes, the role of fluids, lithology and fault friction can be gained from measuring surface displacements immediately after the earthquake. Following the 2014 Napa (California) earthquake, rapid deployment of on the ground measurements (DeLong et al. 2015) and GNSS (Floyd et al. 2016) enabled the capture of the very early rapid postseismic processes of slip and were further augmented by satellite radar in the weeks after the earthquake. The stresses involved in the earthquake can promote and impede postseismic slip and also cause triggered slip on other faults. Being able to track the motion on faults following a major event will enable a recalculation of the stress budget and potential for stressing other surrounding faults to incorporate this additional time varying postseismic phase (as well as highlight other previously unknown active fault traces from triggered slip). Reduced latency in Earth Observing systems is helping to contribute to this for remote areas where field teams are unlikely to get there quickly. Knowing the ratio of coseismic slip and postseismic slip better for entire fault lengths of major surface rupturing earthquakes will enable us to improve the interpretation of the magnitudes of past events (and therefore the likely amount of shaking), as this may be overestimated if a significant amount of surface offset is due to aseismic afterslip following major earthquakes, or inaccurate if the slip measured at the surface is not representative of the average slip at depth (Xu et al. 2016), or if there is significant off-fault deformation (Milliner et al. 2015). Much of this feeds into the longer-term aspects of DRM rather than immediate recovery.

Earthquakes often trigger landslides, in particular in regions of high relief with steep slopes, and identifying these is important in the response phase of an earthquake. However, the earthquake, subsequent shaking and triggered landslides also result in changes to the behaviour of the landscape in the recovery period, and landslides persist in being more frequent in the subsequent years post-earthquake (Marc et al. 2015). Measuring slopes that have experienced a change in their behaviour (rates or extents) using InSAR would be important to identify an acceleration in their sliding or initiation of new slope failures that would inform the recovery attempts for resettlement location plans, as well as for monitoring landslide dam stabilities for risk of catastrophic flooding.

When augmented with on the ground surveys, satellite imagery can contribute to the recovery process in mapping impacts to support various aspects of the reconstruction strategy. This can be the location of buildings that have been destroyed (and require subsequent debris removal), the location of populations in temporary, informal or emergency housing that may need to be rehoused and relocated (Ghafory-Ashtiany and Hosseini 2008), as well as some indicators of building condition (structure and roofs). This can be done through a range of visual, change detection and classification techniques in a manual to automated process (Contreras et al. 2016). Identification of new areas to build on around the city (green areas) would ideally take into account the change in hazard on surrounding areas from the earthquake, as faults along-strike or up-dip of rupture areas maybe become more prone to future rupture (Elliott et al. 2016b).

4.4 Mitigation

Strategies for reducing the risk from seismic hazards have long been known (Hu et al. 1996). The engineering requirements for buildings to survive ground accelerations from earthquakes are established and already codified into building regulations for many countries. However, in many areas these are not implemented due to increased costs relative to the status quo and/or an under appreciation of their necessity (such as in the case of hidden faults and their unrecognised proximity to cities), as well as due to corruption (and past legacies). This makes the identification of priority areas for seismic risk critical, especially where there are competing needs for the use of limited economic resources.

Earth Observation data support an array of geospatial mapping, which can provide a useful method of visualising complex spatial information, making such datasets and analyses more accessible to potential end users (Shkabatur and Kumagai 2014). When trying to communicate seismic hazard and risk, EO imagery can foster community engagement by placing the community's location in a geospatial context relative to the hazard, home, work and community buildings such as churches, as well as a family's relatives, emergency services and shelters. This could lead to the empowerment of communities to act on known solutions to earthquake hazard and support the enforcement of building codes to reduce the risk.

Ideally, the aim would be to move off the disaster risk management cycle so that a hazard no longer presents a risk and there would be no need for preparedness as the chance of disaster would have been negated (assuming the hazard was well enough characterised to know where it could occur and to what severity). Given that earthquakes are an ever-present hazard across many parts of the deforming world and will not decrease through time, risk can only be reduced by decreasing vulnerability or moving those exposed relative to the location of hazard. Risk prevention, in terms of moving the exposed population and assets relative to the active faulting and expected shaking, can be done at two spatial scales. At a local level, one approach is to establish setback distances from the surface traces of faults (Zhou et al. 2010) where peak ground accelerations are expected to be largest (as well as issues of differential offsets if buildings straddle the fault rupture itself). EO data can help by constraining the size of a suite of similar earthquake ruptures, their maximum surface offsets, and the width of the fault rupture zone (whether the zone is very localised or distributed over hundreds of metres, Milliner et al. 2015). Both visual inspection and subpixel image correlation of higher-resolution imagery is important for being able to detect the degree of on and off fault deformation due to an earthquake rupture (Zinke et al. 2014). The availability of open datasets of high-resolution imagery lags behind that of commercial operators (10 m open versus 0.3 m closed), but the higher resolution is important for pixel correlation as the measurable offsets are a fraction of the pixel size. Higher-resolution imagery would enable the detection of smaller movements and better characterisation of the width of active fault zones.

At a broader spatial scale, shifting the exposure relative to the hazard can be done to a much greater degree, such as occurred when the capital of Kazakhstan was relocated away from Almaty, in part because of seismic hazard considerations (Arslan 2014). This decision was taken because Almaty is exposed to a large seismic hazard and suffered a series of devastating earthquakes in the late nineteenth and early twentieth centuries (Grützner et al. 2017). The capital is instead based in Nur-Sultan (formerly Astana) which is both away from past earthquakes and is not in a zone rapidly straining, and is therefore considered to be of lower hazard. Such a relocation of the exposed population requires a good knowledge of where active faults are that could rupture in the future, and the potential sizes or magnitudes of future earthquakes, including what the largest credible size would be within the timeframe of human interest. Otherwise, relocating a population away from an area that has just experienced a devastating earthquake could place them in a zone that still has a significant seismic hazard or in fact a heightened hazard (such as occurs during along-strike faulting from stress changes, Stein et al. 1997). Using GNSS and InSAR data over deforming belts will enable the identification of regions within a country that are building up strain relatively slowly. For example, in some countries such as New Zealand, whilst there are regions of relatively much higher and much lower strain rate, even the slowly deforming regions are still capable of producing damaging earthquakes (Elliott et al. 2012).

At a city scale, it is important to identify the location of the faults and their segmentation. This is necessary for determining the location of faults relative to the exposure. This also allows the estimation of hazard and losses for potential earthquake scenarios, for example, by comparing rupture of a fault right beneath a city versus that expected for distant fault ruptures (Hussain et al. 2020). Such analysis is best done in conjunction with field studies and dating of fault activity from paleoseismological trenching (e.g., Vargas et al. 2014) to help ascertain the recurrence intervals and the time since the most recent event as well as potential sizes of events. Once the most exposed elements in a city are identified, mitigation by seismic reinforcement from retrofitting buildings (Bhattacharya et al. 2014) could be applied to reduce the vulnerability of certain districts. Optical imagery that captures the expansion of cities could help identify the growth of suburbs and districts onto fault scarps that leave the population more exposed. Such urban growth also hides the potential geomorphological signals of past activity from assessment today. The use of DEMs at a city scale can also help identifying relative uplift and folds beneath a city that mark out the location of active faulting (Talebian et al. 2016).

Over wider areas, previous work using EO data such as InSAR and GNSS has sought to measure the rate at which major faults are accumulating strain (Tong et al. 2013). By measuring the surface velocities across major fault zones, it is possible to determine which faults are active and how fast they accrue a slip deficit. The fastest faults are generally considered to be more hazardous, and these faster rates are more easily measured with EO techniques. Such information can then be used alongside rates of seismicity and longer-term geologic estimates of fault slip rates to make earthquake rupture forecasts (Field et al. 2014) so that regions to target for mitigation can be identified. An important component of measuring fault strain accumulation is whether there are significant creeping segments (Harris 2017) that are slipping continuously (Jin and Funning 2017) and by how much does this lower the hazard. High resolution EO imagery over wide areas provides the possibility of picking out active faults by identifying such creeping segments. Identifying the size, depth extent and persistence through time of creeping areas on major faults (Jolivet et al. 2013) can provide constraints on whether these slowly slipping portions can act as barriers to major ruptures.

4.5 Preparedness

In order to know where to prepare for earthquakes, we need to know where active faults are located and where strain is accumulating within the crust. It is necessary, but not sufficient, to also know where past earthquakes have occurred. This is because past earthquakes represent areas where strain has accumulated sufficiently to lead to rupture. A region that has previously had earthquakes is likely to be able to host them in the future. It also gives us the time since the last earthquake to determine the slip deficit that may have accumulated in that interval. But due to long recurrence times relative to the length of past seismicity records, a region that is currently devoid of past major earthquakes may still be capable of producing one if it is accumulating strain. If earthquakes were to follow a regular pattern in which they had a constant (periodic) time interval between successive ruptures (Fig. 6), then the task of forecasting them would be trivial. Some faults have enough past observations or inferences of earthquakes that it is possible to assess the degree to which they host regular earthquakes (Berryman et al. 2012), and the degree to which they deviate from periodic. A fault with a low coefficient of variation (CoV) of time intervals between successive ruptures has a pattern of earthquakes that is said to be quasi-periodic, and from which estimates of the probability of a future rupture can be updated based upon the time since the last earthquake, i.e. it is time dependent (with uncertainty ranges based in part upon the degree of variation thus far observed). This is because the mean recurrence interval is more meaningful as the relative standard deviation becomes lower. In contrast, a time-independent process follows a Poisson distribution. The time to the next earthquake does not depend on how long ago the last earthquake occurred and the coefficient of variation is 1 (Fig. 6), with the probability of rupture constant through time. This acts as a good reference case against which to compare other patterns of faulting events. Although we may attribute this to the actual behaviour of a fault, such attributions are more likely due to the fact that we do not understand the underlying physical processes that drive, govern and control the rupture characteristics of the fault. In contrast to this behaviour, if a fault zone ruptures episodically, the seismicity is said to be clustered through time, with many events close together (relative to a mean recurrence) separated by a long hiatus with no earthquakes (Marco et al. 1996). The picture becomes much more complex when considering

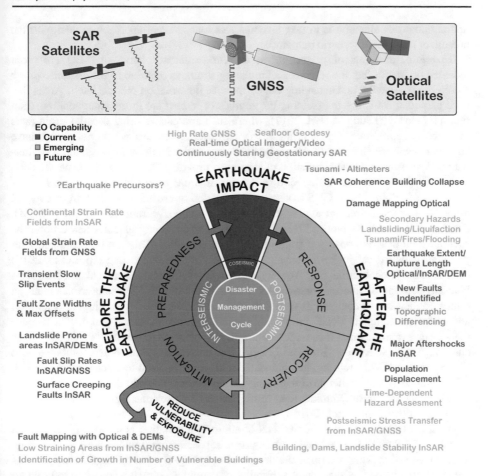

Fig. 5 Earth Observation data from SAR, optical and GNSS satellites currently contributes information to various aspects of the disaster management cycle. Emerging technologies and understanding of the physical processes of earthquakes and deformation that leads to seismic hazards mean that there are emerging and future potential avenues for EO to contribute further to the constituent elements of the cycle

interacting faults, triggered seismicity and event clustering (Scholz 2010). For most faults in the world, the number of previous events and their timings are very poorly known, and the recurrence times may be very long compared to the 100 years used as an example here. Considering Fig. 6, a more realistic time horizon for the knowledge of even major faults may only be the past 1000 years of historical records from which patterns of behaviour cannot be determined given typical recurrence intervals. Interpreting palaeoseismic records of past events and trying to calculate slip rates from previous fault offsets leads to large epistemic uncertainties when the observation window is a small multiple of only a few mean recurrence intervals (Styron 2019). Further to this, if one considers the typical life-time of an individual (and more so that of a government), the relatively long recurrence interval and variability of earthquakes may go some way to explaining the difficulties in communicating hazard and stoking an impetus for action. We usually do not know the past characteristics of a fault, nor do we know whether it is really time independent or the form of the time dependence, leading to a deep level of uncertainty (Stein and Stein 2013). One option under this situation is to take a robust risk management approach to deal with the breadth of reasonable scenario outcomes.

Therefore, in terms of preparedness, current techniques harnessing EO are more focussed on identifying the regions accumulating strain, as opposed to the response phase where EO is aimed at understanding the earthquake deformation process itself. Areas that pose particular difficulty in assessing the sources of hazard are those accumulating strain relatively slowly (Landgraf et al. 2017), where such approaches using decadal strain rates may not be appropriate (Calais et al. 2016). However, given that we do not always have enough knowledge about past earthquake behaviour and seismicity for a region, this alternative of measuring the accumulation of strain can provide a forecast of the expected seismicity (Bird and Kreemer 2015), and in the past this has been based upon GNSS measurements of strain. Using GNSS observations to derive crustal velocities from over 22 thousand locations, Kreemer et al. (2014) produced a Global Strain Rate Model (GSRM). From this, Bird and Kreemer (2015) forecast shallow seismicity globally based upon the accumulation of strain observed from these GNSS velocities. This can be combined with past seismological catalogues to enhance the capability for forecasting (Bird et al. 2015) such as in the Global Earthquake Activity Rate Model (GEAR). One current aim is to combine EO datasets of interferometric SAR time series with GNSS results to calculate an improved measure of continental velocities and strain rates (Fig. 7) at a higher resolution and with more complete global coverage of straining parts of the world that may lack dense GNSS observations. From these rates of strain, it is possible to estimate earthquake rates and forecasts of seismicity (Bird and Liu 2007) to calculate hazard. When combined with the exposure and vulnerability, this information can then be used to calculate seismic risk (Silva et al. 2014). A challenge of using surface velocities and strain derived from geodetic data is being able to differentiate long-term (secular) interseismic signals from those that are transient (Bird and Carafa 2016).

An important contribution of EO data of strain accumulation is in determining which portions of a fault are locked and building up a slip deficit, and which are stably sliding. This fraction of locking is termed the degree of coupling. By determining which portions are currently locked (i.e. the spatial variability of coupling), it may be possible to build up a range of forecasts of earthquake patterns on a fault (Kaneko et al. 2010). InSAR measurements of creeping fault segments highlight which portions of a partially coupled fault are stably sliding, and which are more fully locked, and when bursts of transient slip have occurred (Rousset et al. 2016). By determining the extent of these areas, it is possible to develop a time-dependent model of creep and establish the rate of slip deficit that builds up over time, and the potential size of earthquakes possible on a given fault section (Shirzaei and Bürgmann 2013). Determining the spatio-temporal relationship between locked regions of faults and transient slow slip portions is important to understand the potential for these stabling sliding regions to both release accumulated elastic energy and also drive subsequent failure on other portions of the fault (Jolivet and Frank 2020).

Rollins and Avouac (2019) calculate long-term time independent earthquake probabilities for the Los Angeles area based upon GNSS derived interseismic strain accumulation and the instrumental seismic catalogue. By examining the frequency–magnitude distribution of smaller earthquakes that have already occurred, they are able to estimate the maximum magnitude of earthquakes and the likelihood of certain sized events occurring in the future. Estimates of the rate of strain accumulation from EO datasets are needed to determine the rate of build-up of the moment deficit for such analysis. For fast faults that are plate bounding and relatively simple, with few surrounding faults to interact with, a time-dependent model may be appropriate. This time dependence since the last earthquake

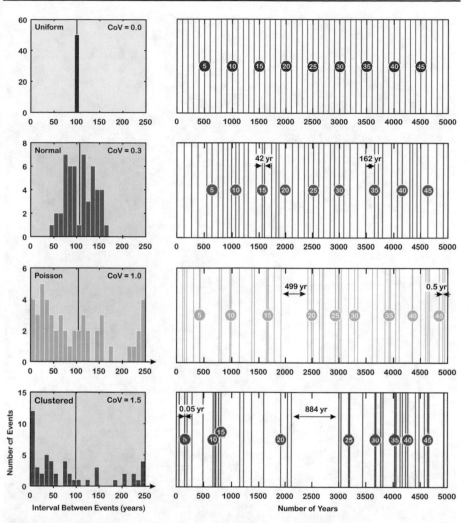

Fig. 6 Examples of realisations of earthquake occurrence patterns through time (right column) due to varying distributions of recurrence intervals (left histograms) but all with a mean interval of about 100 years (vertical black line). A uniform earthquake interval leads to a regular pattern of earthquakes (dark grey) and therefore zero coefficient of variation (CoV)—that is the standard deviation divided by the mean interval. A time-dependent model with a small standard deviation (30 years) on the recurrence interval is shown in blue (the maximum and minimum intervals in the distribution are marked by arrows). A model where the time since the last earthquake does not affect the chance of an earthquake (green) is time independent (a Poisson probability) and has a CoV = 1. A distribution that has a short interval between most events, but a long tail of intervals has a clustered pattern and subsequently a large CoV (red). Whilst 50 events in 5000 years are shown here as an example (average recurrence interval 100 years), for most faults the recurrence intervals tend to be much longer. Given that reliable records of past earthquakes generally cover a much shorter period than 5000 years, the task of determining even an average recurrence interval, let alone an accurate assessment of any variation, becomes nearly intractable

comes from the concept of the accumulation of strain within the crust. Using InSAR measurements across most of the length of the North Anatolian Fault, Hussain et al. (2018) showed that it is reasonable to use short-term geodetic measurements to assess long-term

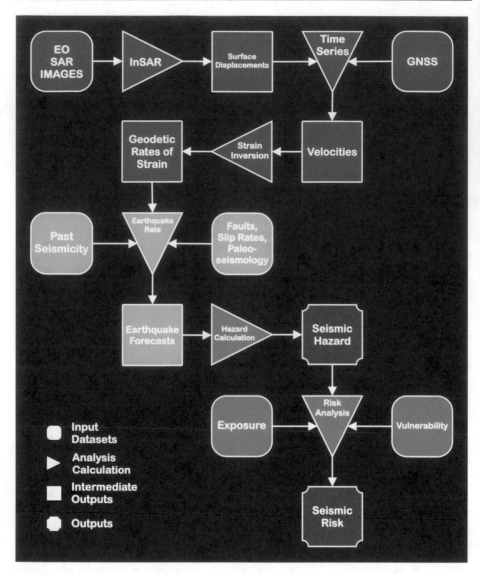

Fig. 7 Workflow for calculations of seismic hazard and strain based upon EO input datasets of SAR images and GNSS. InSAR measurements are used to calculate surface displacements and velocities through a time series approach. Rates of geodetic strain can be derived from these velocities to calculate earthquake rates (along with other seismological and geological data and constraints) and forecasts that can be used in subsequent seismic hazard and risk assessments

seismic hazard, as the measured velocities from EO are constant through time apart from during the earliest parts of the postseismic phase following major earthquakes. However, the use of a time-dependent model has been argued to have failed on the Parkfield area of the San Andreas Fault as it did not perform as expected in forecasting the repeating events on this fault (Murray and Segall 2002). The accumulation of strain through time leads to an interseismic moment deficit rate which is considered to be a guide to the earthquake potential of a fault. Being able to place bounds on this deficit is important for accurate assessment of the hazard (Maurer et al. 2017). By comparing the interseismic slip rates derived from EO with known past earthquakes and postseismic deformation, Michel et al. (2018) find that the series of past earthquakes on the Parkfield segment of the San Andreas Fault is not sufficient to explain the moment deficit. They conclude that larger earthquakes with a longer return period than the window of historical records are possible on this fault to explain the deficit. For the state of California, where there is information on active fault locations, past major ruptures, fault slip rates, seismicity and geodetic rates of strain, it has been possible to develop sophisticated time-independent (Field et al. 2014) and time-dependent forecast models (Field et al. 2015) termed the third Uniform California Earthquake Rupture Forecast (UCERF3).

An even harder problem is to characterise the accumulation of strain and slip release for offshore portions of subduction zones which host the largest earthquakes and also generate tsunamis. The expanding field of sea floor geodesy techniques opens up a huge area of the Earth's surface to future measurement of earthquake cycle processes (Bürgmann and Chadwell 2014) that can be used in seismic and tsunami hazard assessment. One of the methods for seafloor geodesy links GNSS systems with seafloor transponders to detect movement of the seafloor relative to an onshore reference station. The large costs of instrumentation and seafloor expeditions have limited the deployment of such technologies so far. Understanding such fault zone processes as slow slip at subduction zones is important as it both releases some of this strain aseismically but may also drive other areas closer to failure and is itself promoted by nearby earthquakes (Wallace et al. 2017). Such slow slip behaviour can be captured onshore with dense GNSS and InSAR.

The global earthquake model (GEM) foundation is an initiative to create a world resilient to earthquakes by enhanced preparedness, improving risk management from their collection of risk resources. Their global mosaic of seismic hazard models (Pagani et al. 2018) includes constraints using geodetic strain rates where available, such as for UCERF3 in California. This hazard mosaic will continue to be updated as hazard maps are enhanced regionally around the globe, and EO constraints on strain rates will help improve characterisations of the input hazard. In particular, GEM have compiled an Active Faults database (Fig. 3) from a compilation of regional databases (e.g., Styron et al. 2010) which is continually being added to, and can be guided by constraints from EO data using optical imagery, DEMs and geodetic measurements of faults. By combining the Global Hazard map with exposure and vulnerability models (Fig. 7), a Global Seismic Risk map (Silva et al. 2018) is produced of the geographic distribution of average annual losses of buildings and human lives due to damage from ground shaking. By sharing these in an open and transparent manner, GEM's aim is to advocate for reliable earthquake risk information to support disaster risk reduction planning.

Right at the end of the disaster management cycle is the Holy Grail of preparedness that are earthquake precursors (Cicerone et al. 2009)—the ability to forecast an impending seismic rupture and prevent a potential disaster. This has been an ever-present endeavour of a section of the community to find signals that might foretell of an upcoming earthquake. However, despite decades of research (Wyss 1991), a reliable technique accepted by the wider community has not arisen that can provide an earthquake prediction in terms of giving a temporally and spatially constrained warning of an imminent earthquake rupture that could be safely used to evacuate, or take measures to ensure the safety of, an exposed population.

A current debate surrounding processes prior to earthquake nucleation leaves open the question of whether major events may be triggered by a cascade of foreshocks or whether propagating slow slip triggers earthquake nucleation (Gomberg 2018). Measuring fore-shocks is largely the preserve of sensitive seismological equipment, and given that slow slip at depth may be small, its measurement by EO imaging systems may remain unfeasi-ble. Detecting small changes in creep rates of faults has typically been done with surface creepmeters in the past for specific locations (Bilham et al. 2016), but InSAR observa-tions can be used to measure wider area creep rates and determine longer-term background rates to measure deviations from this (Rousset et al. 2016). Perhaps identification of subtle accelerations of creep (or creep events) may be possible using EO observations of strain, but both the time resolution and sensitivity would need to be greatly enhanced beyond cur-rent capabilities.

5 Outlook and Recommendations

The uptake and exploitation of Earth Observation is expanding both in terms of under-standing the science of earthquakes and in operational disaster risk management. Measur-ing the underlying driver of earthquake hazard, in terms of the accumulation of strain in the crust, has been an important development. The use of EO has also increased in more direct application to earthquake events—both in terms of disaster response and prepar-edness. This has involved the increased uptake of SAR and its derived products, which had previously lagged behind the more intuitively interpretable optical imagery datasets. Advancements have been made by going beyond using an individual image in isolation to instead looking at change in a time series approach, be it ground deformation with InSAR, image matching and optical offsets with optical data or landscape classification for land-sliding. The power of the observational data is best harnessed when coupled with model-ling of physical processes to be able to better interpret the underlying mechanisms generat-ing earthquake hazard and the changes that occur through time.

The future lies in being able to continuously image the Earth. Much of the current use of EO data for this application comes from near polar orbiting satellites. This means con-tinuous operational monitoring or high-density time series datasets cannot be achieved nor formed due to the short time over an area and a long revisit interval. This leads to the compromise of having high spatial resolution but limited footprint datasets of coarse time resolution. In the medium term, constellations of Low Earth Orbit steerable video imag-ing satellites could provide a much-enhanced disaster response tool and improve the sci-ence of earthquakes. An unfolding disaster could be continuously monitored remotely to facilitate the response to collapsed buildings and determine the impact of subsequent after-shocks, as well as earthquake triggered landslides through analysis of the video imagery, and from constructing digital topography from the video feed to capture landscape eleva-tion changes. Such potential for harnessing space-based video feeds for the International Disaster Charter activations has been identified (Stefanov and Evans 2015) with the High Definition Earth Viewing System (HDEVS) onboard the International Space Station (ISS). Ultimately, continuously staring satellites from geostationary orbits (both optical/video and SAR based) would in the long term provide this uninterrupted monitoring mode that would greatly enhance Earth Observation in disaster risk management. Such systems have been suggested to capture the propagation of seismic waves using a geostationary optical seis-mometer (Michel et al. 2012) and for geosynchronous radar to image geohazards with SAR interferometry near-continuously (Wadge et al. 2014).

Since 2014, we have seen an important shift in the availability of SAR and optical imagery from space. Reduced latency has opened up the utility of EO data to act increasingly earlier in disaster response. This huge repository of information offers a great potential for the analysis of the underlying science of earthquakes. This will hopefully further contribute to the improved assessment of hazard and has the potential to help mitigate some of the risk, or at least allow for better preparation for it. A challenge for the community is to be able to usefully extract information from such large volumes of data to meet the aspirations of the Sendai framework for the substantial reduction of disaster risk and losses in lives, livelihoods and health.

The open availability of Sentinel-1 data through ESA's Copernicus programme has greatly enhanced the capacity for incorporating EO into disaster risk management for earthquakes as well as to enhance our understanding of the sources of seismic hazard through improved science. The availability of processing toolboxes has widened the access to exploiting such datasets to a much broader user base of varying skill levels. However, in order to further advance Copernicus' aims to provide accurate, timely and accessible information of the environment globally with Earth Observation, there are a number of recommendations to further this huge recent increase in the uptake of EO data (in terms of its application to earthquakes and tectonics for both the assessment of hazard and risk from geohazards). Whilst there have been dramatic enhancements in the latest EO systems when compared to earlier ones, there are some remaining limitations. Improvements could be made both to current systems in terms of acquisition strategy and data provision, as well as in the planning for future missions. In particular, the following recommendations pertaining to the Sentinel 1/2 satellites from ESA within the Copernicus programme may help some of the goals described here. Whilst the use of Sentinel data for earthquake science is only one of many mission objectives, outlined below are suggestions that would enhance the interferometric capability for Sentinel-1 SAR in particular, a tool useful for determining the rates of seismogenic strain accumulation on the continents.

The Sentinel-1 constellation offers two satellites in orbit and a long-term mission with spare copies currently in storage available for launch to maintain mission continuity upon failure/deorbiting of the initial pair of satellites. However, I recommend launching the two spare Sentinel-1 C/D satellites to grow the constellation, thus reducing latency for imaging an affected area following major earthquakes. If the satellites were operated in quadrature, revisit times would reduce from the current best case 6 days with Sentinel-1 A and B to 3 days with all four, and considering both ascending and descending passes, this would be down to 1.5 days. This assumes that data can be acquired all the time, which is not possible given the limited duty cycle of about 25% of an orbit due to the power constraints from the SAR and periods when in eclipse (Attema et al. 2010). Launching Sentinel-1 C & D could therefore alleviate some of the limitations of the current duty cycle by permitting more regular acquisitions globally whilst improving the acquisition repeat interval. Were acquisitions to be done on every pass (as is done over Europe) this would help both in the use of SAR for the response phase following an earthquake and also for better constraining the postseismic deformation processes that follow immediately after earthquakes. There are a number of satellites in orbit from other agencies that contribute towards disaster response (Pritchard et al. 2018), reducing times for the first pass, but Sentinel data with a large background acquisition strategy provide some of the most consistent data for interferometry and offsets. Additionally, for long-term tectonic studies measuring strain accumulation over years, there are two advantages to increasing the number of satellites: (1) a shorter interval between acquisitions improves the interferometric coherence, so InSAR data are retrievable over more vegetated areas, increasing the latitudinal coverage for such assessments with C-band SAR; (2) a greater number of observations improves the time series analysis, both for constraining the evolution of rapidly changing rates (such as postseismic deformation) and for determining smaller long-term rates above the noise (i.e. regions of slower continental deformation that can still host large earthquakes). The improvement in the uncertainty in the velocity from a time series analysis of InSAR data scales as the square root of the number of observations made (Morishita et al. 2020), so doubling the number of observations using all four satellites will reduce the uncertainty by about 30%.

For both optical cross-correlation techniques and SAR interferometry, a pre-event image is required in addition to a post-earthquake image to do the matching comparison to measure the surface displacement that has occurred in an earthquake. The longer the gap between the prior image and the newly acquired one, the greater the degradation in the quality of the displacement map, due to changes in the nature of the Earth's surface. Therefore, I recommend making sure all areas of the globe likely to be straining and exposed to earthquakes should have regular enough images acquired that no gap between acquisitions exceeds a number of months. Tracks of Sentinel-1 SAR data were not acquired at all for a whole year over tectonically important areas such as Iran and New Zealand, which occurred in 2017–2018. Such large gaps can break the time series approach of detecting deformation, and warning systems should be created to highlight the emergence of such data gaps. The uncertainty on the estimated constant velocity in a time series of InSAR data scales linearly with the length of the connected time series (Morishita et al. 2020). Any gaps in the time series result in a shorter connected network of data. Therefore, missing data acquisitions for large periods of time has a major deleterious impact on the level of uncertainty, and breaking the connected time series in half will about double the uncertainty in estimated velocities.

Regular acquisitions with consistent imaging modes are an important design consideration for optical and SAR acquisitions (such as with Landsat and Sentinel-1) for improving the potential for long-term and systematic exploitation of the archives. Such an example is Sentinel-1's Interferometric Wide Swath (IWS) model that allows for the analysis of large stacks of time series of data as the imagery is collected predominately in this single mode over land in tectonically deforming areas. This has occurred in a way that is not possible for satellite systems that have the flexibility of acquiring in a whole host of imaging modes (stripmap, wideswath, spotlight, etc.) as they do not create a consistent archive for long-term measurements for interferometry. There were also some issues arising from the shifting of acquisition frames by ESA (the start and end time of an acquisition) making processing of the data initially more difficult. This resulted in a greater amount of bookkeeping, slowing the initial uptake of the imagery, and now requires more data management. It is easier if the acquisitions are systematic and consistent such as with the ERS frame approach. This could both be usefully implemented in the current acquisition strategy and in future mission designs.

The accessibility of large volumes of the data has also been challenging. As more data are collected by the Sentinels, the existing archive is moved off the live system onto tape archives. This is necessary due to the expense of keeping such huge volumes live. Current policies such as through ESA SciHub (https://scihub.copernicus.eu/) are to limit the rate of recalls from the long-term archive (LTA). I recommend that access to LTA should be made as efficient as possible to expedite such processes; otherwise, exploitation of the archive in future may be limited. The issue of single users struggling to process large volumes of data is, however, being mitigated by cloud computing through such services as the Geohazards Thematic Exploitation Platform (G-TEP) which provides on-demand and systematic processing for both optical and SAR data. Additionally, initiatives (Meyer et al. 2017) such as the Alaska SAR Facility (https://asf.alaska.edu/) and the Centre for the Observation and Modelling of Earthquakes and Volcanoes (COMET) portal (https://comet.nerc.ac.uk/comet -lics-portal/) aim to improve the accessibility of derived data products through pre-processing of large data volumes using high-performance computing (Lazecký et al. 2020).

A current wide-ranging challenge across many applications is the availability of open access digital elevation models of the Earth's topography, which has not kept pace with the availability of SAR or optical imagery in terms of resolution (nor rate of renewal). Freely available topographic datasets are at best 30 m ground sampling in comparison with optical imagery at 10 m with Sentinel-2 and at 2–14 m with Sentinel-1 SAR. Also, the update for topographic datasets is much less frequent at a decadal interval (when considered globally for land cover areas) compared to that for optical/SAR imagery (which refreshes at least seasonally if not monthly). The lack of high-resolution topographic datasets limits some analysis of landscape change from earthquakes and secondary hazards. Such analysis is geographically limited to regions where higher-resolution datasets are available from national organisations or where the topographic datasets are made from commercial datasets at high cost and constrained openness (be it photogrammetrically from stereo optical imagery or from SAR datasets such as TanDEM-X). I therefore recommend to target a future satellite mission to provide regular and open updates of the Earth's surface in terms of changing topographic models that can match the level of analysis done with imagery datasets. Whilst a regular update of a DEM is relatively labour intensive and requires a lot of computational processing, in particular in areas of high mountains areas (Shean et al, 2016) likely associated with active tectonics, software such as the Ames Stereo Pipeline makes large area processing possible (Beyer et al., 2018).

Whilst there are a number of suggestions listed above, this should not detract from the huge volumes of high-quality data that are continuously been acquired by various space agencies, nor from the openness and availability of much of these data. This has enabled dramatic gains to be made in our observations and understanding of solid Earth deformation such as from earthquakes and faulting, which will improve our assessment of the characteristics of the processes generating seismic hazard.

Acknowledgements This work was also supported by the Centre for the Observation and Modelling of Earthquakes, Volcanoes and Tectonics (COMET) in the UK. John Elliott acknowledges support from the Royal Society through a University Research Fellowship (UF150282). This work was supported by NERC through the Looking into the Continents from Space (LiCS) large Grant (NE/K010867/1). Figure 2 contains modified Copernicus Sentinel data [2019]. I thank Susanna Ebmeier for her comments on the manuscript.

Author Contributions JRE wrote the paper and produced the figures.

References

Allen RM, Melgar D (2019) Earthquake early warning: advances, scientific challenges, and societal needs. Ann Rev Earth Planet Sci 47:361–388. https://doi.org/10.1146/annurev-earth-053018-060457

Ambraseys N, Bilham R (2011) Corruption kills. Nature 469(7329):153. https://doi.org/10.1038/469153a

Arslan M (2014) The significance of shifting capital of Kazakstan from Almaty to Astana: an evalution on the basis of geopolitical and demographic developments. Proced-Soc Behav Sci 120:98–109. https://doi.org/10.1016/j.sbspro.2014.02.086

Attema E, Snoeij P, Levrini G, Davidson M, Rommen B, Floury N, L'Abbate M (2010) Sentinel-1 mission capabilities. In Fringe 2009 Workshop-Advances in the Science and Applications of SAR Interferometry Frascati Italy

Avouac JP, Meng L, Wei S, Wang T, Ampuero JP (2015) Lower edge of locked main Himalayan thrust unzipped by the 2015 Gorkha earthquake. Nat Geosci 8(9):708. https://doi.org/10.1038/ngeo2518

Barnhart WD, Hayes GP, Wald DJ (2019a) Global earthquake response with imaging geodesy: recent examples from the USGS NEIC. Remote Sens 11(11):1357. https://doi.org/10.3390/rs11111357

Barnhart WD, Hayes GP, Gold RD (2019b) The July 2019 Ridgecrest California Earthquake Sequence Kinematics of Slip and Stressing in Cross-Fault Ruptures. Geophys Res Lett 46(21):11859–11867. https://doi.org/10.1029/2019GL084741

Berryman KR, Cochran UA, Clark KJ, Biasi GP, Langridge RM, Villamor P (2012) Major earthquakes occur regularly on an isolated plate boundary fault. Science 336(6089):1690–1693. https://doi.org/10.1126/science.1218959

Beyer RA, Alexandrov O, McMichael S (2018) The Ames Stereo Pipeline: NASA's open source software for deriving and processing terrain data. Earth Space Sci 5(9):537–548. https://doi.org/10.1029/2018EA000409

Bhattacharya S, Nayak S, Dutta SC (2014) A critical review of retrofitting methods for unreinforced masonry structures. Int J Disaster Risk Reduct 7:51–67. https://doi.org/10.1016/j.ijdrr.2013.12.004

Biasi GP, Langridge RM, Berryman KR, Clark KJ, Cochran UA (2015) Maximum-likelihood recurrence parameters and conditional probability of a ground-rupturing earthquake on the Southern Alpine Fault, South Island, New Zealand. Bull Seismol Soc Am 105(1):94–106. https://doi.org/10.1785/0120130259

Bilham R, Ozener H, Mencin D, Dogru A, Ergintav S, Cakir Z, Aytun A, Aktug B, Yilmaz O, Johnson W, Mattioli G (2016) Surface creep on the North Anatolian fault at Ismetpasa, Turkey, 1944–2016. J Geophys Res Solid Earth 121(10):7409–7431. https://doi.org/10.1002/2016JB013394

Bird P, Carafa MM (2016) Improving deformation models by discounting transient signals in geodetic data: 1. Concept and synthetic examples. J Geophys Res Solid Earth 121(7):5538–5556. https://doi.org/10.1002/2016JB013056

Bird P, Kreemer C (2015) Revised tectonic forecast of global shallow seismicity based on version 2.1 of the Global Strain Rate Map. Bull Seismol Soc Am 105(1):152–166. https://doi.org/10.1785/0120140129

Bird P, Liu Z (2007) Seismic hazard inferred from tectonics: California. Seismol Res Lett 78(1):37–48. https://doi.org/10.1785/gssrl.78.1.37

Bird P, Jackson DD, Kagan YY, Kreemer C, Stein RS (2015) GEAR1: A global earthquake activity rate model constructed from geodetic strain rates and smoothed seismicity. Bull Seismol Soc Am 105(5):2538–2554. https://doi.org/10.1785/0120150058

Bürgmann R, Chadwell D (2014) Seafloor geodesy. Annu Rev Earth Planet Sci 42:509–534. https://doi.org/10.1146/annurev-earth-060313-054953

Burrows K, Walters RJ, Milledge D, Spaans K, Densmore AL (2019) A New Method for Large-Scale Landslide Classification from Satellite Radar. Remote Sens 11(3):237. https://doi.org/10.3390/rs11030237

Calais E, Freed A, Mattioli G, Amelung F, Jónsson S, Jansma P, Hong SH, Dixon T, Prépetit C, Momplaisir R (2010) Transpressional rupture of an unmapped fault during the 2010 Haiti earthquake. Nat Geosci 3(11):794. https://doi.org/10.1038/ngeo992

Calais E, Camelbeeck T, Stein S, Liu M, Craig TJ (2016) A new paradigm for large earthquakes in stable continental plate interiors. Geophys Res Lett 43(20):10–621. https://doi.org/10.1002/2016GL070815

Cicerone RD, Ebel JE, Britton J (2009) A systematic compilation of earthquake precursors. Tectonophysics 476(3–4):371–396. https://doi.org/10.1016/j.tecto.2009.06.008

Contreras D, Blaschke T, Tiede D, Jilge M (2016) Monitoring recovery after earthquakes through the integration of remote sensing, GIS, and ground observations: the case of L'Aquila (Italy). Cartogr Geogr Inf Sci 43(2):115–133. https://doi.org/10.1080/15230406.2015.1029520

DeLong SB, Lienkaemper JJ, Pickering AJ, Avdievitch NN (2015) Rates and patterns of surface deformation from laser scanning following the South Napa earthquake. Calif Geosph 11(6):2015–2030. https://doi.org/10.1130/GES01189.1

de Michele M, Raucoules D, Lasserre C, Pathier E, Klinger Y, Van Der Woerd J, de Sigoyer J, Xu X (2009) The Mw 7.9, 12 May 2008 Sichuan Earthquake Rupture Measured by sub-pixel correlation of ALOS PALSAR amplitude images. Earth Planets Space. https://doi.org/10.5047/eps.2009.05.002

Diederichs A, Nissen EK, Lajoie LJ, Langridge RM, Malireddi SR, Clark KJ, Hamling IJ, Tagliasacchi A (2019) Unusual kinematics of the Papatea fault (2016 Kaikōura earthquake) suggest anelastic rupture. Sci Adv 5(10):eaax5703. https://doi.org/10.1126/sciadv.aax5703

Ebmeier SK, Biggs J, Mather TA, Amelung F (2013) Applicability of InSAR to tropical volcanoes: insights from Central America. Geological Society, London, Special Publications 380(1):15–37. https://doi.org/10.1144/SP380.2

Elliott JR, Nissen EK, England PC, Jackson JA, Lamb S, Li Z, Oehlers M, Parsons B, Parsons B (2012) Slip in the 2010–2011 Canterbury earthquakes, New Zealand. J Geophys Res Solid Earth 117(B3):1–6. https://doi.org/10.1029/2011JB008868

Elliott JR, Walters RJ, Wright TJ (2016a) The role of space-based observation in understanding and responding to active tectonics and earthquakes. Nat commun 7:13844. https://doi.org/10.1038/ncomms13844

Elliott JR, Jolivet R, González PJ, Avouac JP, Hollingsworth J, Searle MP, Stevens VL (2016b) Himalayan megathrust geometry and relation to topography revealed by the Gorkha earthquake. Nat Geosci 9(2):174–180. https://doi.org/10.1038/ngeo2623

Farr TG, Kobrick M (2000) Shuttle Radar Topography Mission produces a wealth of data. Eos Trans Am Geophys Union 81(48):583–585. https://doi.org/10.1029/EO081i048p00583

Field EH, Arrowsmith RJ, Biasi GP, Bird P, Dawson TE, Felzer KR, Oehlers M, Parsons B, Michael AJ (2014) Uniform California earthquake rupture forecast, version 3 (UCERF3)—The time-independent model. Bull Seismol Soc Am 104(3):1122–1180. https://doi.org/10.1785/0120130164

Field EH, Biasi GP, Bird P, Dawson TE, Felzer KR, Jackson DD, Johnson KM, Jordan TH, Madden C, Michael AJ, Milner KR (2015) Long-term time-dependent probabilities for the third Uniform California Earthquake Rupture Forecast (UCERF3). Bull Seismol Soc Am 105(2A):511–543. https://doi.org/10.1785/0120140093

Floyd MA, Walters RJ, Elliott JR, Funning GJ, Svarc JL, Murray JR, Hooper AJ, Larsen Y, Marinkovic P, Bürgmann R, Johanson IA (2016) Spatial variations in fault friction related to lithology from rupture and afterslip of the 2014 South Napa, California, earthquake. Geophys Res Lett 43(13):6808–6816. https://doi.org/10.1002/2016GL069428

Freed AM (2005) Earthquake triggering by static, dynamic, and postseismic stress transfer. Annu Rev Earth Planet Sci 33:335–367. https://doi.org/10.1146/annurev.earth.33.092203.122505

Gaillard JC, Peek L (2019) Disaster-zone research needs a code of conduct. Nature 575:440–442. https://doi.org/10.1038/d41586-019-03534-z

Ghafory-Ashtiany M, Hosseini M (2008) Post-Bam earthquake: recovery and reconstruction. Nat Hazards 44(2):229–241. https://doi.org/10.1007/s11069-007-9108-3

Gomberg J (2018) Unsettled earthquake nucleation. Nat Geosci 11(7):463. https://doi.org/10.1038/s41561-018-0149-x

Gower J (2005) Jason 1 detects the 26 December 2004 tsunami. Eos, Transactions American Geophysical Union 86(4):37–38. https://doi.org/10.1029/2005EO040002

Grützner C, Walker RT, Abdrakhmatov KE, Mukambaev A, Elliott AJ, Elliott JR (2017) Active tectonics around Almaty and along the Zailisky Alatau rangefront. Tectonics 36(10):2192–2226. https://doi.org/10.1002/2017TC004657

Halle W, Asam S, Borg E, Fischer C, Frauenberger O, Lorenz E, Richter R (2018) Firebird——Small Satellites for Wild Fire Assessment. In IGARSS 2018-2018 IEEE International Geoscience and Remote Sensing Symposium (pp. 8609-8612). IEEE. https://doi.org/10.1109/IGARSS.2018.8519246

Hamling IJ, Hreinsdóttir S (2016) Reactivated afterslip induced by a large regional earthquake, Fiordland, New Zealand. Geophys Res Lett 43(6):2526–2533. https://doi.org/10.1002/2016GL067866

Hamling IJ, Hreinsdóttir S, Clark K, Elliott J, Liang C, Fielding E, Litchfield N, Villamor P, Wallace L, Wright TJ, D'Anastasio E (2017) Complex multifault rupture during the 2016 Mw 78 Kaikōura earthquake, New Zealand. Science 356(6334):eaam7194. https://doi.org/10.1126/science.aam7194

Harris RA (2017) Large earthquakes and creeping faults. Rev Geophys 55(1):169–198. https://doi.org/10.1002/2016RG000539

Hu YX, Liu SC, Dong W (1996) Earthquake engineering. CRC Press

Hussain E, Wright TJ, Walters RJ, Bekaert DP, Lloyd R, Hooper A (2018) Constant strain accumulation rate between major earthquakes on the North Anatolian Fault. Nat commun 9(1):1392. https://doi.org/10.1038/s41467-018-03739-2

Hussain E, Elliott JR, Silva V, Vilar-Vega M, Kane D (2020) Contrasting seismic risk for Santiago, Chile, from near-field and distant earthquake sources. Nat Hazards Earth Syst Sci 20:1533–1555. https://doi.org/10.5194/nhess-20-1533-2020

Jackson J (2001) Living with earthquakes: know your faults. J Earthquake Eng 5(spec01):5–123

Jackson J (2006) Fatal attraction: living with earthquakes, the growth of villages into megacities, and earthquake vulnerability in the modern world. Philos Trans Royal Soc A: Math Phys Eng Sci 364(1845):1911–1925. https://doi.org/10.1098/rsta.2006.1805

Jin L, Funning GJ (2017) Testing the inference of creep on the northern Rodgers Creek fault, California, using ascending and descending persistent scatterer InSAR data. J Geophys Res Solid Earth 122(3):2373–2389. https://doi.org/10.1002/2016JB013535

Johnson K, Nissen E, Saripalli S, Arrowsmith JR, McGarey P, Scharer K, Williams P, Blisniuk K (2014) Rapid mapping of ultrafine fault zone topography with structure from motion. Geosphere 10(5):969–986. https://doi.org/10.1130/GES01017.1

Jolivet R, Lasserre C, Doin MP, Peltzer G, Avouac JP, Sun J, Dailu R (2013) Spatio-temporal evolution of aseismic slip along the Haiyuan fault, China: Implications for fault frictional properties. Earth Planet Sci Lett 377:23–33. https://doi.org/10.1016/j.epsl.2013.07.020

Jolivet R, Frank WB (2020) The transient and intermittent nature of slow slip. AGU Adv 1(1):e2019AV000126. https://doi.org/10.1029/2019AV000126

Jordan TH, Marzocchi W, Michael AJ, Gerstenberger MC (2014) Operational earthquake forecasting can enhance earthquake preparedness. Seismol Res Lett 85(5):955–959. https://doi.org/10.1785/02201 40143. Accessed 30 July 2020

Kaneko Y, Avouac JP, Lapusta N (2010) Towards inferring earthquake patterns from geodetic observations of interseismic coupling. Nat Geosci 3(5):363. https://doi.org/10.1038/ngeo843

Kargel JS, Leonard GJ, Shugar DH, Haritashya UK, Bevington A, Fielding EJ, Fujita K, Geertsema M, Miles ES, Steiner J, Anderson E (2016) Geomorphic and geologic controls of geohazards induced by Nepal's 2015 Gorkha earthquake. Science 351(6269):aac8353. https://doi.org/10.1126/science.aac83 53

Kreemer C, Blewitt G, Klein EC (2014) A geodetic plate motion and Global Strain Rate Model. Geochem Geophys Geosyst 15(10):3849–3889. https://doi.org/10.1002/2014GC005407

Landgraf A, Kübler S, Hintersberger E, Stein S (2017) Active tectonics, earthquakes and palaeoseismicity in slowly deforming continents. Geol Soc Lond Spec Publ 432(1):1–12. https://doi.org/10.1144/ SP432.13

Lazecký M, Spaans K, González PJ, Maghsoudi Y, Morishita Y, Albino F, Ellion J, Greenall N, Hatton E, Hooper A, Juncu D, McDougall A, Walters RJ, Watson CS, Weiss JR, Wright TJ (2020) LiCSAR: An automatic InSAR tool for measuring and monitoring tectonic and volcanic activity. Remote Sensing 12(15):2430. https://doi.org/10.3390/rs12152430

Marc O, Hovius N, Meunier P, Uchida T, Hayashi S (2015) Transient changes of landslide rates after earthquakes. Geology 43(10):883–886. https://doi.org/10.1130/G36961.1

Marco S, Stein M, Agnon A, Ron H (1996) Long-term earthquake clustering: A 50,000-year paleoseismic record in the Dead Sea Graben. J Geophys Res Solid Earth 101(B3):6179–6191. https://doi. org/10.1029/95JB01587

Maurer J, Segall P, Bradley AM (2017) Bounding the moment deficit rate on crustal faults using geodetic data: Methods. J Geophys Res Solid Earth 122(8):6811–6835. https://doi.org/10.1002/2017JB014300

Melgar D, Hayes GP (2019) Characterizing large earthquakes before rupture is complete. Sci Adv 5(5):eaav2032. https://doi.org/10.1126/sciadv.aav2032

Michel R, Ampuero JP, Avouac JP, Lapusta N, Leprince S, Redding DC, Somala SN (2012) A geostationary optical seismometer, proof of concept. IEEE Trans Geosci Remote Sens 51(1):695–703. https://doi. org/10.1109/TGRS.2012.2201487

Michel S, Avouac JP, Jolivet R, Wang L (2018) Seismic and aseismic moment budget and implication for the seismic potential of the Parkfield segment of the San Andreas fault. Bull Seismol Soc Am 108(1):19–38. https://doi.org/10.1785/0120160290

Milliner CW, Dolan JF, Hollingsworth J, Leprince S, Ayoub F, Sammis CG (2015) Quantifying near-field and off-fault deformation patterns of the 1992 Mw 7.3 Landers earthquake. Geochem Geophys Geosyst 16(5):1577–1598. https://doi.org/10.1002/2014GC005693

Morishita Y, Lazecky M, Wright TJ, Weiss JR, Elliott JR, Hooper A (2020) LiCSBAS: An Open-Source InSAR Time Series Analysis Package Integrated with the LiCSAR Automated Sentinel-1 InSAR Processor. Remote Sens 12(3):424. https://doi.org/10.3390/rs12030424

Murray J, Segall P (2002) Testing time-predictable earthquake recurrence by direct measurement of strain accumulation and release. Nature 419(6904):287. https://doi.org/10.1038/nature00984

Pagani M, Garcia-Pelaez J, Gee R, Johnson K, Poggi K, Styron R, Weatherill G, Simionato M, Viganò D, Danciu L, Monelli D (2018) Global Earthquake Model (GEM) Seismic Hazard Map (version 2018.1 - December 2018). https://doi.org/10.13117/GEM-GLOBAL-SEISMIC-HAZARD-MAP-2018.1

Parsons T (2004) Recalculated probability of M$\geq$ 7 earthquakes beneath the Sea of Marmara, Turkey. J Geophys Res Solid Earth 109(B5):1–9. https://doi.org/10.1029/2003JB002667

Parsons T (2005) Significance of stress transfer in time-dependent earthquake probability calculations. J Geophys Res Solid Earth 110(B5):10. https://doi.org/10.1029/2004JB003190

Percivall GS, Alameh NS, Caumont H, Moe KL, Evans JD (2013) Improving disaster management using earth observations—GEOSS and CEOS activities. IEEE J Sel Top Appl Earth Obs Remote Sens 6(3):1368–1375. https://doi.org/10.1109/JSTARS.2013.2253447

Pritchard ME, Biggs J, Wauthier C, Sansosti E, Arnold DW, Delgado F, Ebmeier SK, Henderson ST, Stephens K, Cooper C, Wnuk K (2018) Towards coordinated regional multi-satellite InSAR volcano observations: results from the Latin America pilot project. J Appl Volcanol 7(1):5. https://doi.org/10.1186/s13617-018-0074-0

Rollins C, Avouac JP (2019) A Geodesy-and Seismicity-Based Local Earthquake Likelihood Model for Central Los Angeles. Geophys Res Lett 46(6):3153–3162. https://doi.org/10.1029/2018GL080868

Rousset B, Jolivet R, Simons M, Lasserre C, Riel B, Milillo P, Çakir Z, Renard F (2016) An aseismic slip transient on the North Anatolian Fault. Geophys Res Lett 43(7):3254–3262. https://doi.org/10.1002/2016GL068250

Scholz CH (2010) Large earthquake triggering, clustering, and the synchronization of faults. Bull Seismol Soc Am 100(3):901–909. https://doi.org/10.1785/0120090309

Shean DE, Alexandrov O, Moratto ZM, Smith BE, Joughin IR, Porter C, Morin P (2016) An automated, open-source pipeline for mass production of digital elevation models (DEMs) from very-high-resolution commercial stereo satellite imagery. ISPRS J Photogramm Remote Sens 116:101–117. https://doi.org/10.1016/j.isprsjprs.2016.03.012

Shirzaei M, Bürgmann R (2013) Time-dependent model of creep on the Hayward fault from joint inversion of 18 years of InSAR and surface creep data. J Geophys Res Solid Earth 118(4):1733–1746. https://doi.org/10.1002/jgrb.50149

Silva V, Crowley H, Pagani M, Monelli D, Pinho R (2014) Development of the OpenQuake engine, the Global Earthquake Model's open-source software for seismic risk assessment. Nat Hazards 72(3):1409–1427. https://doi.org/10.1007/s11069-013-0618-x

Silva V, Amo-Oduro D, Calderon A, Dabbeek J, Despotaki V, Martins L, Rao A, Simionato M, Viganò D, Yepes-Estrada C, Acevedo A, Crowley H, Horspool N, Jaiswal K, Journeay M, Pittore M (2018) Global Earthquake Model (GEM) Seismic Hazard Map (version 2018.1). https://doi.org/10.13117/GEM-GLOBAL-SEISMIC-RISK-MAP-2018.1. Accessed 30 July 2020

Sladen A, Hébert H (2008) On the use of satellite altimetry to infer the earthquake rupture characteristics: application to the 2004 Sumatra event. Geophys J Int 172(2):707–714. https://doi.org/10.1111/j.1365-246X.2007.03669.x

Socquet A, Hollingsworth J, Pathier E, Bouchon M (2019) Evidence of supershear during the 2018 magnitude 7.5 Palu earthquake from space geodesy. Nat Geosci 12(3):192. https://doi.org/10.1038/s41561-018-0296-0

Stefanov WL, Evans CA (2015) Data collection for disaster response from the international space station. Int Arch Photogramm Remote Sens Spatial Inf Sci XL-7/W3:851–855. https://doi.org/10.5194/isprsarchives-XL-7-W3-851-2015

Stein RS, Barka AA, Dieterich JH (1997) Progressive failure on the North Anatolian fault since 1939 by earthquake stress triggering. Geophys J Int 128(3):594–604. https://doi.org/10.1111/j.1365-246X.1997.tb05321.x

Stein S, Stein JL (2013) Shallow versus deep uncertainties in natural hazard assessments. Eos Trans Am Geophys Union 94(14):133–134. https://doi.org/10.1002/2013EO140001

Styron R (2019) The impact of earthquake cycle variability on neotectonic and paleoseismic slip rate estimates. Solid Earth 10(1):15–25. https://doi.org/10.5194/se-10-15-2019

Styron R, Taylor M, Okoronkwo K (2010) Database of active structures from the Indo-Asian collision. Eos Trans Am Geophy Union 91(20):181–182. https://doi.org/10.1029/2010EO200001

Talebian M, Copley AC, Fattahi M, Ghorashi M, Jackson JA, Nazari H, Sloan RA, Walker RT (2016) Active faulting within a megacity: the geometry and slip rate of the Pardisan thrust in central Tehran, Iran. Geophys J Int 207(3):1688–1699. https://doi.org/10.1093/gji/ggw347

Tilloy A, Malamud BD, Winter H, Joly-Laugel A (2019) A review of quantification methodologies for multi-hazard interrelationships. Earth-Sci Rev 196:102881. https://doi.org/10.1016/j.earscirev.2019.102881

Tong X, Sandwell DT, Smith-Konter B (2013) High-resolution interseismic velocity data along the San Andreas fault from GPS and InSAR. Journal of Geophysical Research: Solid Earth 118(1):369–389. https://doi.org/10.1029/2012JB009442

Vargas G, Klinger Y, Rockwell TK, Forman SL, Rebolledo S, Baize S, Lacassin R, Armijo R (2014) Probing large intraplate earthquakes at the west flank of the Andes. Geology 42(12):1083–1086. https://doi.org/10.1130/G35741.1

Wadge G, Guarnie AM, Hobbs SE, Schul D (2014) Potential atmospheric and terrestrial applications of a geosynchronous radar. In: 2014 IEEE geoscience and remote sensing symposium (pp 946–949). IEEE

Wallace LM, Webb SC, Ito Y, Mochizuki K, Hino R, Henrys S, Schwartz SY, Sheehan AF (2016) Slow slip near the trench at the Hikurangi subduction zone. N Z Sci 352(6286):701–704. https://doi.org/10.1126/science.aaf2349

Wallace LM, Kaneko Y, Hreinsdóttir S, Hamling I, Peng Z, Bartlow N, D'Anastasio E, Fry B (2017) Large-scale dynamic triggering of shallow slow slip enhanced by overlying sedimentary wedge. Nat Geosci 10(10):765. https://doi.org/10.1038/ngeo3021

Walters RJ, Gregory LC, Wedmore LNJ, Craig TJ, McCaffrey K, Wilkinson M, Chen J, Li Z, Elliott JR, Goodall H, Iezzi F (2018) Dual control of fault intersections on stop-start rupture in the 2016 Central Italy seismic sequence. Earth Planet Sci Lett 500:1–14. https://doi.org/10.1016/j.epsl.2018.07.043

Wyss M (1991) Evaluation of proposed earthquake precursors. Eos Trans Am Geophys Union. 72(38):411–411. https://doi.org/10.1029/90EO10300

Xu X, Tong X, Sandwell DT, Milliner CW, Dolan JF, Hollingsworth J, Leprince S, Ayoub F (2016) Refining the shallow slip deficit. Geophys J Int 204(3):1867–1886. https://doi.org/10.1093/gji/ggv563

Yun SH, Hudnut K, Owen S, Webb F, Simons M, Sacco P, Gurrola E, Manipon G, Liang C, Fielding E, Milillo P (2015) Rapid Damage Mapping for the 2015 M w 7.8 Gorkha Earthquake Using Synthetic Aperture Radar Data from COSMO–SkyMed and ALOS-2 Satellites. Seismol Res Lett 86(6):1549–1556. https://doi.org/10.1785/0220150152

Zhou Y, Parsons B, Elliott JR, Barisin I, Walker RT (2015) Assessing the ability of Pleiades stereo imagery to determine height changes in earthquakes: A case study for the El Mayor-Cucapah epicentral area. J Geophys Res Solid Earth 120(12):8793–8808. https://doi.org/10.1002/2015JB012358

Zhou Q, Xu X, Yu G, Chen X, He H, Yin G (2010) Width distribution of the surface ruptures associated with the Wenchuan earthquake: implication for the setback zone of the seismogenic faults in post-quake reconstruction. Bull Seismol Soc Am 100(5B):2660–2668. https://doi.org/10.1785/0120009 0293

Zhu Z, Wulder MA, Roy DP, Woodcock CE, Hansen MC, Radeloff VC, Healey SP, Schaaf C, Hostert P, Strobl P, Pekel JF (2019) Benefits of the free and open Landsat data policy. Remote Sens Environ 224:382–385. https://doi.org/10.1016/j.rse.2019.02.016

Zinke R, Hollingsworth J, Dolan JF (2014) Surface slip and off-fault deformation patterns in the 2013 MW 7.7 Balochistan, Pakistan earthquake: Implications for controls on the distribution of near-surface coseismic slip. Geochem Geophys Geosyst 15(12):5034–5050. https://doi.org/10.1002/2014GC0055 38

Publisher's Note Springer Nature remains neutral with regard to jurisdictional claims in published maps and institutional affiliations.

Surveys in Geophysics (2020) 41:1355–1389
https://doi.org/10.1007/s10712-020-09608-2

Earth Observation for Crustal Tectonics and Earthquake Hazards

J. R. Elliott[1] · M. de Michele[2] · H. K. Gupta[3]

Received: 6 December 2019 / Accepted: 6 August 2020 / Published online: 28 August 2020
© The Author(s) 2020

Abstract

In this paper, we illustrate some of the current methods for the exploitation of data from Earth Observing satellites to measure and understand earthquakes and shallow crustal tectonics. The aim of applying such methods to Earth Observation data is to improve our knowledge of the active fault sources that generate earthquake shaking hazards. We provide examples of the use of Earth Observation, including the measurement and modelling of earthquake deformation processes and the earthquake cycle using both radar and optical imagery. We also highlight the importance of combining these orbiting satellite datasets with airborne, in situ and ground-based geophysical measurements to fully characterise the spatial and timescale of temporal scales of the triggering of earthquakes from an example of surface water loading. Finally, we conclude with an outlook on the anticipated shift from the more established method of observing earthquakes to the systematic measurement of the longer-term accumulation of crustal strain.

Keywords Earth observation · Earthquakes · InSAR · Geophysical modelling · Seismic hazard · Deformation

Abbreviations

AGGM	Airborne gravity gradient and magnetic
ASTER	Advanced spaceborne thermal emission and reflection radiometer
ALOS-PALSAR	Advanced land-observing satellite phased-array-type L-band synthetic aperture radar
AW3D	Advance world 3-dimensional
BEM	Bare earth model
BRGM	Bureau de Recherches Géologiques et Minières
CNES	Centre National d'Etudes Spatiales
COMET	Centre for the Observation and Modelling of Earthquakes, Volcanoes and Tectonics

✉ J. R. Elliott
  j.elliott@leeds.ac.uk

1 COMET, School of Earth and Environment, University of Leeds, Leeds LS2 9JT, UK

2 BRGM, 3 Avenue Claude Guillemin, 45060 Orléans Cedex 2, France

3 Council of Scientific and Industrial Research (CSIR), National Geophysical Research Institute, Hyderabad 500007, India

DEM	Digital elevation model
EO	Earth observation
ESA	European space agency
GBIS	Geodetic Bayesian Inversion Software
GCMT	Global centroid moment tensor
GTEP	Geohazards thematic exploitation platform
GNSS	Global navigation satellite system
ICDP	International continental drilling programme
ICA	Independent component analysis
InSAR	Interferometric SAR
ISC	International seismological centre
ISRO	Indian Space Research Organization
LEO	Low Earth Orbit
LiDAR	Light detection and ranging
LoS	Line of sight
M_w	Moment magnitude
NASA	National Aeronautics and Space Administration
PALSAR-2	Phased-array-type L-band synthetic aperture radar-2
PRISM	Panchromatic remote sensing instrument for stereo mapping
RGB	Red, Green, Blue
RTS	Reservoir-triggered seismicity
SAR	Synthetic aperture radar
SLC	Single look complex
SPOT	Satellite Pour l'Observation de la Terre
SRTM	Shuttle radar topography mission
TIR	Thermal infrared
USGS	United States Geological Survey
VLF	Very low frequency

1 Introduction

In the previous paper (Elliott 2020), we reviewed the different Earth Observing systems for measuring solid Earth processes and discussed at what stages they could contribute to improving the assessment of earthquake hazard, risk and disaster management. We consider earthquake hazard to constitute damaging seismic events that lead to loss of life, and the economic and social damages that can occur to exposed vulnerable populations. We aim to mitigate these losses by improving our understanding of the physical processes that generate the earthquake hazard, as well as the characteristics of the sources of earthquakes in terms of faulting. The examples presented here do not aim to measure the direct earthquake hazard of strong ground motion (such as peak ground acceleration) that is the primary concern for engineering solutions designed to mitigate earthquake losses. Instead, we aim to improve hazard assessment by measuring and understanding the deformation of the Earth's crust to illuminates the location, potency and rate of major earthquake-generating sources—information and knowledge that can be subsequently used to better estimate future strong ground motion when combined with other sources of information and models. Here we provide examples of the use of Earth Observation (EO) and associated geophysical datasets (airborne and ground based), to measure the deformation associated with failure of the crust on faults, a consequence of the earthquake cycle.

The earthquake cycle encapsulates our current understanding of the physical solid Earth process of earthquake generation in terms of deformation (Savage and Prescott 1978) measurable with geodetic techniques using satellite systems. The cycle comprises an increase in strain across faults occurring in the interseismic period, often taking hundreds or thousands of years to build up, depending on the tectonic environment and faulting. This may result in only millimetres of relative displacement across distances of hundreds of kilometres on either side of a major fault over the course of a year (e.g., Bell et al. 2011). However, this very small gradient of displacement is detectable from Earth Observation satellites using Interferometric Synthetic Aperture Radar (InSAR) and the Global Navigation Satellite System (GNSS) (Wright 2002). The accumulation of strain continues until the resulting stress on the fault overcomes the friction resisting it, leading to the initiation of earthquake rupture (termed the coseismic period, lasting seconds to minutes). This coseismic period involves the relative displacement of large volumes of crustal material across the fault in the short time of an earthquake rupture. For a large earthquake, these displacements may be many metres in the near-field close to the fault, and they decay away to centimetres and millimetres at tens to hundreds of kilometres away in the far-field. The fault trace itself may also be many tens or hundreds of kilometres long, so the total area over which the permanent ground deformation is measurable with sensitive satellite observations for even a moderate earthquake maybe as much as 100–100,000 sq. km. This coseismic phase is immediately followed by the postseismic period, in which deformation continues to occur around and below the fault zone and may last (and be detectable) for months to decades (Bürgmann and Dresen 2008). This period involves both seismic processes (in the form of a decaying prevalence and magnitude of aftershocks), and aseismic processes (Wright et al. 2013) comprising afterslip and deeper viscoelastic relaxation (e.g., Hearn et al. 2002), and potentially poroelastic rebound (e.g., Jónsson et al. 2003). Postseismic deformation is typically some fraction of the coseismic deformation but is at a higher rate than the long-term interseismic rates normally observed. For major earthquakes, this postseismic deformation may be visible from Earth Observing systems, particularly after time series analysis of satellite data corrected for atmospheric noise (e.g., Li et al. 2009). The cycle for the given fault then returns to the background rate of strain accumulation during the next interseismic period. Each of these stages of deformation (when onshore) is measurable by Earth Observing systems. InSAR/GNSS data are applicable to observing and modelling all three parts of the cycle, whilst optical imagery is predominately used when large coseismic offsets have occurred.

EO derived measurements of coseismic ground deformation (coupled with other a priori constraints from seismological, optical, topographic and field mapping) can be used as inputs for geophysical inverse models (e.g., Elliott et al. 2016). These models aim to approximate the Earth, typically as an elastic medium in a half-space, from which the parameters of the earthquake source (Okada 1985) such as fault location and orientation (e.g., Bagnardi and Hooper 2018) and the distribution of slip are determined (e.g., Simons et al. 2002; Feng et al. 2010; Amey et al. 2018). More complicated models can include other rheologies such as viscoelastic processes (e.g., Pollitz et al. 2000), in particular when looking at postseismic deformation (e.g., Deng et al. 1998). From these models, further calculations of stress transfer through the crust can be made (e.g., Barnhart et al. 2019b), as well as interpretations of active tectonic process such as the growth of geological structures and the deformation of the continents.

The study of earthquake sources and their relationship to the active fault structures seen at the surface has made a significant contribution to our understanding of plate tectonics, as well as determining where strain is currently building up in the crust. Geodetic systems are particularly suited to determining how much of a fault surface is locked and accumulating strain, as opposed to releasing it aseismically (a ratio termed the degree of fault coupling). Establishing this is a critical constraint in assessing the seismic potential for a fault system (Avouac 2015). The relatively new field of seafloor geodesy is opening up our ability to use GNSS to observe deformation underwater (Bürgmann and Chadwell 2014), which is particularly important for the Earth's subduction zones that cause the largest earthquakes and associated tsunamis (e.g., Lay et al. 2005; Watanabe et al. 2014). Seafloor multibeam bathymetric surveys also offer the chance to capture the displacement associated with major earthquakes, although areas with sufficient pre-existing high-resolution coverage, such as the Japan trench (Fujiwara et al. 2011), are limited due to the expense of acquisition.

In terms of both earthquake hazard and risk, there is a contrast between coastal areas exposed to subduction zone events (those at plate boundaries) versus those in continental interiors (within deforming "plates"). The location of fault interfaces for subduction zones are relatively well determined (Hayes et al. 2018). The shallow, up-dip part of these faults is located offshore where there is no population exposure to generate a risk. By the time the earthquake rupture is beneath land on the down-dip portion of the fault, the slip interface is typically located relatively deep, affecting the type of shaking experienced in earthquakes. Such faults tend to rupture relatively frequently as the fault interfaces accumulate strain more rapidly, and they are often relatively well studied as a fault system and seismic cycle, such as along the Chilean subduction zone (e.g., Chlieh et al. 2004), although the potential size and chances of major events may not be recognised (Kagan and Jackson 2013). Conversely, the locations of faults onshore are not always well known, and their relatively infrequent rupture means that less information can be drawn from past events and their recurrence. The immediate proximity of onshore (shallow) faults to population centres results in earthquakes in continental interiors causing more fatalities despite being smaller (England and Jackson 2011). There is also a contrast in what EO data can offer in these two domains. In the case of subduction zones, EO normally only provides measurements onshore and therefore of the far-field effects of deformation associated with these types of convergent arcas. Conversely, for continental earthquakes (in particular for large or shallow processes), EO provides information on the detail of near-field deformation right at, or above, the fault. In terms of preparedness there are therefore some differences in the EO approaches to assessing the hazard applied to these two contrasting domains. For subduction zones, the potential for major hazard is more widely recognised and the location of the fault that is going to host the big earthquakes is more of a known quantity. The priorities are to determine the degree of interface coupling and the mechanical behaviour of the zone, as well as seismic gaps left behind from previous events to characterise the hazard (e.g., Métois et al. 2012), as well as the distribution of earthquake magnitudes and maximum credible size of rupture. The identification of these characteristics is made more tractable by much (but by no means all) of the accumulating strain being released on the single dominant offshore fault structure. Conversely for continental interiors, the fault that will eventually break in an earthquake will not necessarily have been identified before the event, as many active faults are not identified until they rupture due to their hidden or subtle expression in the landscape geomorphology. The continental lithosphere is much older than oceanic crust and hence much more complicated in terms of inherited structures, fabrics and weaknesses. Furthermore, the way in which the strain in the crust is

Table 1 Imaging characteristics of the main satellite systems illustrated within this paper for measuring earthquake deformation (or their newer active equivalent)

Satellite	Sentinel-1 A/B	ALOS-2 (PALSAR-2)	Sentinel-2 A/B	Landsat-8	SPOT-6/7
Radar/optical	Radar	Radar	Optical	Optical	Optical
Pixel size (m)	2.3×14	3–10, 100	10, 20, 60	15, 30, 100	1.5, 6
Repeat cycle (days)	12 (6)	14	10 (5)	16	26 (13)
Swath width (km)	250	50–70, 350	290	180	60–120
Wavelength	5.6 cm	22.9 cm	440–2200 nm	430–1250 nm	450–890 nm
Day/night	Day/night	Day/night	Day	Day	Day
Launched (dd/mm/yy)	03/04/2014 25/04/2016	24/05/2014	23/06/2015 07/03/2017	11/02/2013	09/09/2012 30/06/2014

For constellations consisting of a pair of (near) identical satellites, the repeat cycle is halved and shown in parenthesis. For active radar imaging systems, a range of imaging modes is possible leading to a variety of pixel spacings and swath widths (a subset of which are given here). For Sentinel-1 this is Interferometric Wide Swath (IWS) mode (Torres et al. 2012) and for ALOS-2 this is Stripmap and ScanSAR mode (Arikawa et al. 2014). For passive optical systems, multispectral images are acquired and have increasing pixel sizes with increasing wavelength (Drusch et al. 2012; Roy et al. 2014)

eventually partitioned onto the array of distributed faults is not fully understood. However, where there are interactions between both subduction fault interfaces and upper crustal faults, the earthquake and rupture pattern can result in extremely complicated displacement patterns, as was the case in the 2016 Kaikoura earthquake in New Zealand (e.g., Hamling et al. 2017; Furlong and Herman 2017).

In this article, we present some individual examples of the use of such Earth Observation data (some current imaging systems are provided in Table 1) for understanding earthquakes, using satellite radar for a recent small earthquake in western Turkey (2019) and optical systems for the large Palu earthquake (Sulawesi) in 2018 and Sichuan earthquake (China) in 2008. These examples illustrate EO measurements for an earthquake at the limit of detectability with InSAR, as compared to a pair of major earthquakes with very large geodetic signals and a major human impact. None of this work can be done in isolation without an understanding of the local context: in situ geophysical and field measurements can be critical in interpreting or constraining satellite observations. To this end we present an example of seismic risk from triggered earthquakes at a reservoir and dam in Koyna (India) that additionally makes use of non-Earth Observation data. We finish with conclusions on the recent and potential for use of EO in better constraining the sources of earthquake hazards, in particular the use of EO for measurements of long-term strain accumulation.

2 InSAR Observations and Modelling of an Earthquake

Measurements of tectonic deformation have increased dramatically over the past two decades from the growth in, and quality of, satellite-geodetic measurements, which have in turn improved our understanding of active faulting. Prior to the emergence of EO data, the predominant constraint on earthquakes was from seismological measurements and field observations. By combining multiple sets of observations (e.g., Fielding et al. 2013), it is possible to better constrain shallow continental earthquake locations, fault segmentation and ruptures. These are all critical parameters for the assessment of seismic hazard because they capture the characteristics of potential seismic sources and the relative positions of fault structures likely to generate strong motion.

A large number of small and shallow earthquakes have occurred beneath the continents in the time window of Sentinel-1 radar satellite imaging (October 2014–present). With systematic and regular coverage we expect to capture with EO data most of the continental earthquakes with magnitude greater than moment magnitude (M_w) 5.5 that are shallow (< 20 km), something that was less achievable with previous Synthetic Aperture Radar (SAR) systems (Lohman and Simons 2005) due to the lack of data. However, it is not always clear whether these earthquakes are readily visible in interferograms produced over the epicentral areas or not (Funning and Garcia 2018), as the ground deformation can be masked by atmospheric noise for these small events, or their magnitude and depth are beyond the ability of the imaging systems to detect them. Phase decorrelation associated with vegetation also means that the longer-wavelength L-band data is more suitable at equatorial latitudes to capture earthquake deformation signals (Morishita 2019) than the C-band of Sentinel-1, but the volume of data (in terms of both temporal and spatial coverage) is not yet as great at this longer SAR wavelength. However, by using atmospheric corrections and time series approaches (Hooper et al. 2012), modelling of an increasing number of earthquakes should be possible using EO techniques (Tian et al. 2018). Other signal processing approaches such as Independent Component Analysis (ICA) can extract masked signals from InSAR data (Ebmeier 2016) for improved detection of even smaller seismic and aseismic events (Maubant et al. 2020). Machine learning also offers the prospect to improve detection of hidden or slowly deforming events within large data sets of automatically processed InSAR products (Anantrasirichai et al. 2018).

From such an analysis, the locations of active faults can be better identified based upon precise locations of seismic activity, especially in regions of low seismological instrumentation, where solutions can be biased in terms of earthquake location and depth (e.g., Elliott et al. 2010). Links can then be established between geodetically derived fault locations and the expression of the active fault in the geomorphology of the surface. Here, we illustrate this workflow by providing our analysis of a specific example from a recent small earthquake in Turkey and demonstrate the use of other satellite-derived data such as digital elevation models from stereo optical imagery to supplement the interpretation made with Sentinel-1 InSAR data.

On the 20 March 2019, a moment magnitude (M_w) 5.7 earthquake struck the south-west corner of Turkey, 10 km east of the town of Acipayam (population ~ 11,000) and just west of Lake Salda (Fig. 1). This normal faulting event occurred in a region of distributed deformation, with rates of extension across the whole region of 20 mm/yr according to GNSS measurements (Aktug et al. 2009). Major earthquakes have also occurred in the area, the largest of which was the 1914 magnitude 7.0 event 80 km to the east near the city of Burdur, which killed 4000 people (Ambraseys 1988). Other major normal faulting earthquakes in the past half century (Fig. 1) were the 1971 magnitude 6.2 earthquake sequence also near Burdur (Taymaz and Price 1992), the 1995 M_w 6.2 Dinar earthquake (Wright et al. 1999) and the 2017 M_w 6.6 Kos-Bodrum earthquake (Karasözen et al. 2018). These earthquakes are the release of extensional strain that accumulated over centuries. Major normal faulting scarps are visible in the geomorphology across the whole region, and large mountains strike approximately perpendicular to the extensional direction.

Sentinel-1 imagery over Turkey are acquired by the European Space Agency (ESA) every 6 days (achieved using both identical copies of Sentinel 1A and 1B currently in orbit). This occurs along ground tracks (numbered by orbit) that are normally about 250 km wide (Table 1) imaged from a low Earth orbit at about 700 km altitude (at the boundary of the Earth's outermost pair of atmospheric layers of the thermosphere and exosphere). As the satellites are launched into a near-polar orbit, they image the Earth in two directions on every orbit—in the ascending direction when travelling south-to-north and in the descending direction when completing the orbit north to south. We use data from both the ascending and descending tracks to better constrain the surface deformation. This is because multiple look directions are differently sensitive to the ground motion in the vertical and horizontal (both look directions image vertical uplift as a shortening in range, but eastward motion is measured with opposite signs). This can be particularly useful in constraining some earthquake fault parameters, which can easily trade-off against each other in fault inversions if only a single component of displacement is measured (Funning et al. 2005).

The short interval between data acquisitions of 6 days (due to imaging from both Sentinel-1A and 1B) from immediately before (17 March 2019) and after the earthquake (23 March 2019) means the interferometric results are coherent even in regions of dense vegetation or intensive cultivation. Additionally, the sooner after the earthquake the second image is acquired, the less postseismic deformation contaminates the coseismic signal, which can often be a disadvantage of the latency of the EO polar satellites for InSAR. The regular repeat also helps to separate out any potential additional deformation from

Fig. 1 a Map of SW Turkey with past earthquakes overlaid on hillshaded topography. Focal mechanisms ▶ of shallow (<40 km) earthquakes of magnitude 5+from 1976 to 2020 from the Global Centroid Moment Tensor (GCMT) catalogue are denoted by the blue and white circles and demonstrate the extension of the upper crust by normal faulting. Major faults from the Global Earthquake Model (GEM) Active Fault Database are denoted by red lines, based largely in this region upon Woessner et al. (2015). Seismicity prior to 1976 is from the USGS, based upon the International Seismological Centre ISC-GEM Global Instrumental Earthquake Catalogue (Storchak et al. 2013). Seismic catalogue locations of the 2019 M_w 5.7 mainshock and M_w 5.1 aftershock near Acipayam are denoted for GCMT and USGS. Dashed box indicates the extent of the area show in **b–g**. Earthquake deformation observations: **b–c** example of InSAR data from Sentinel-1 interferograms (displacement of the Earth's surface relative to the satellite in the line of sight, with negative motion (blue) indicating motion away from the satellite (predominately subsidence in this case) and **d, e** modelling of a small earthquake—20 March 2019 M_w 5.7 earthquake at Acipayam, Turkey. InSAR data are available in **b** ascending (track 58) and **c** descending (track 138) directions (Az indicate direction of satellite, los is the line of sight look direction). These data are then modelled as elastic dislocations to determine the best fitting fault parameters, and the residual difference between these two datasets **f, g** indicates the noise in the interferograms and any mis-modelling of the data from the simplified assumptions of modelling a fault as a single rectangular dislocation of constant slip. The area of slip on the two possible fault planes is indicated by the rectangles and the up-dip projections of these faults to the surface are indicated by the lines (with the ticks indicating the down-dip direction). There remains a focal ambiguity where it is not possible to determine if the fault dips to the east (black outline) or west (grey outline) as both these solutions yield almost the same pattern of deformation at the surface (the east dipping solution is shown here in the model). This ambiguity is because the fault slip is buried at depth and there is no surface rupture to resolve this. The red stars indicate the epicentral location reported by the USGS (37.408° N, 29.531° E) and centroid location by GCMT (37.37° N, 29.38° E). The black dashed rectangle indicates the spatial extent of Fig. 3

aftershocks, such as the M_w 5.1 that occurred in the same area on the 31 March 2019. The earthquake deformation is visible in the processed InSAR data with 3–4 cm of motion away from the satellite (Fig. 1), which indicates elastic subsidence typical of normal faulting (as the sign of the signal is the same in both look directions from ascending and descending passes, this indicates predominantly vertical motion). The smoothness of the signal demonstrates that the faulting did not reach to the surface and that the slip remains buried at depth. If there had been surface faulting or fissuring, then discontinuities in the signal would be visible in the InSAR data. Such signals provide a useful method for identifying off-fault deformation (Xu et al. 2016) or other active fault strands that may have been induced (Fialko et al. 2002) and triggered in the main earthquake (Xu et al. 2020), as well as any triggered landsliding (e.g., Kargel et al. 2016; Huang et al. 2017; Delorme et al. 2020). The regular pattern or 'blotchy' signal in the data (particularly for ascending track 58, Fig. 1) is indicative of atmospheric noise due to the difference in tropospheric path delays between the two SAR image dates and most likely due to variations in water vapour (Jolivet et al. 2014a). This is one of the main limiting factors for detecting small tectonic signals as it can often obscure the geophysical ground displacement. Further corrections to such data could be made using weather model data, such as from the European Centre for Medium-Range Weather Forecasts (ECMWF) which removes part of the atmospheric noise (e.g., Jolivet et al. 2011; Yu et al. 2018). An improved signal-to-noise ratio improves subsequent elastic dislocation modelling that is often implemented on such data and in turn this will reduce uncertainties in the derived fault parameters.

The next step is to downsample the data in order to exclude redundant data points and speed up the subsequent modelling analysis. The very high resolution of the InSAR data (typical pixel spacing less than 100 m) is not required to characterise most features of deformation especially in this case of a small buried event such as this. In this case, 750 datapoints for each interferogram are sufficient to capture the magnitude, shape and displacement gradient within the data. Modelling the surface deformation data points is

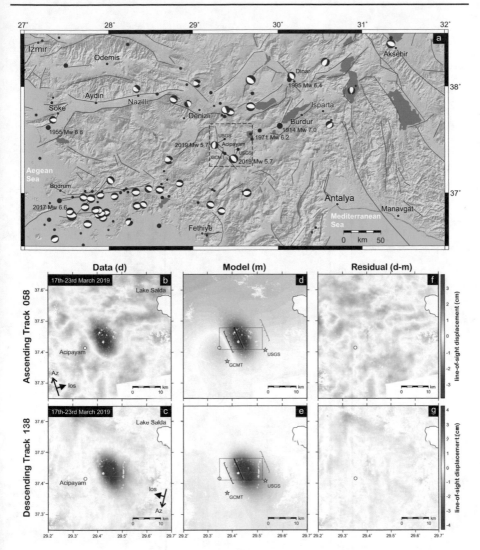

Fig.

usually achieved with a form of elastic dislocation (e.g., Okada 1985) code to solve for the fault parameters of location, depth and orientation, as well as magnitude of slip. A large range of modelling software and approaches have been developed to estimate sub-surface fault parameters based upon inversions of surface displacement data (e.g., Jónsson et al. 2002; Funning et al. 2007; Barnhart and Lohman 2010; Minson et al. 2013). Here we use the Geodetic Bayesian Inversion Software (GBIS) to perform inversions of the InSAR data using a Markov-chain Monte Carlo algorithm that finds the posterior probability distribution of the fault location (x, y, depth), size (length, width), orientation (strike, dip) and slip (Bagnardi and Hooper 2018). Therefore, not only do we find a model that fits the data well, but we also get a sense of the range of uncertainty and trade-offs between the various fault parameters based upon how noisy the input datasets are assessed to be (Table 2).

The fault parameter modelling results (Figs. 1, 2) indicate that the fault strikes NNW-SSE and is predominately normal dip-slip. This corroborates the seismological solutions

Table 2 Fault parameters from the geodetic solution from both the east dipping nodal plane solution (NP1) and for the westward dipping solution (NP2)

Solution	Strike (°)	Dip (°)	Rake (°)	Longitude (°E)	Latitude (°N)	Depth (km)	Moment × 10^{17} Nm	M_w	Slip (m)	Length (km)	Width (km)
InSAR (NP1)	336	58	−70	29.426	37.444	6.1	3.12	5.6	2.9	8.1	0.41
C.I.	330–340	52–62	−94 to −59	±0.5 km	±1 km	5.2–6.3	2.74–3.65	5.56–5.64	0.5–3	6.8–9.0	0.4–2.7
InSAR (NP2)	156	33	−79	29.433	37.437	6.0	3.15	5.6	1.9	8.1	0.63
C.I.	149–161	27–37	−99 to −71	±0.4 km	±0.5 km	5.6–6.4	2.73–3.66	5.56–5.64	0.7–2.9	6.9–9.1	0.42–1.8
USGS W (NP1)	326	50	−87	29.531	37.408	17.5	4.57	5.71			
USGS W (NP2)	140	40	−94								
USGS BW (NP1)	320	50	−88			6.0	2.48	5.53			
USGS BW (NP2)	137	40	−93								
GCMT (NP1)	321	42	−87	29.38	37.37	12	4.04	5.67			
GCMT (NP2)	137	48	−93								

The confidence interval (C.I) from the Bayesian analysis (Fig. 2) is shown for the percentile range 2.5–97.5%. Also shown are seismological catalogue earthquake solutions from the USGS (W-phase moment tensor solution and BodyWave (BW) moment tensor solution) and the Centroid Moment Tensor from GCMT. Note depth for the InSAR solution is the midpoint of the upper part of the slipping fault plane, for the seismological solutions it is the centroid depth. The moment has been calculated for the InSAR solution assuming a rigidity of 3.2×10^{10} Pa

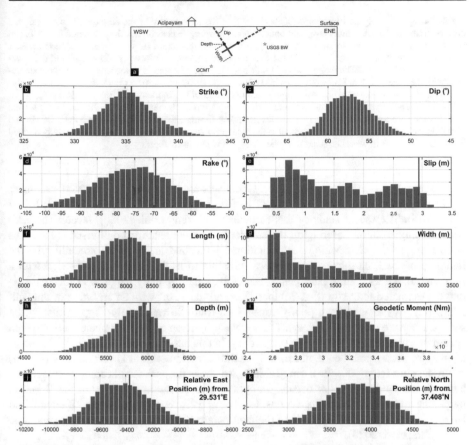

Fig. 2 a Schematic diagram of the relative location of the two possible fault plane solutions for the Acipayam earthquake. In one solution, the fault plane dips steeply to the ENE, in the other it dips more shallowly to the WSW (solid red lines). The up-dip projections of the fault planes to the surface are denoted by dashed red lines. As the modelled slip increases, the width of these fault lines collapse to approximately the same point in the modelling due to trade-offs between these two parameters for buried sources. The relative locations of the seismological solutions are indicated by red stars. **b–k** Modelling results for fault parameters (location, size, orientation and slip) and parameter uncertainties (posterior probability distributions) through Bayesian analysis using GBIS (Bagnardi and Hooper 2018) for the eastward dipping fault plane solution based upon the data in Fig. 1. The best fit fault model for each parameter is shown by the red line, with the histogram of the distributions of solutions shown in blue for 800,000 model iterations (Table 2 has the best fit model and range of solutions). The fault plane location (upper midpoint of slip area) is given relative to a reference point 29.531° E, 37.408° N which is the USGS epicentre

in terms of fault orientation (Table 2), but the location of the earthquake from the InSAR modelling (denoted by the rectangles in Fig. 1) is more towards the town of Acipayam than in the USGS seismological solution (USGS 2019) by almost 8 km (indicated by the red star in Fig. 1). As the fault slip remains buried, a focal plane ambiguity remains (as it does for the seismological solution), in which it is not possible to discriminate whether the fault dips to the ENE or WSW (Fig. 2a). If the rupture were to dip eastwards, the model indicates the fault would project up to the surface near to the town of Acipayam (2 km to the west of it). If the fault instead dips westwards, the up-dip projection of the fault plane is about halfway between Acipayan and Lake Salda. Further time series analysis may be

Fig. 3 Examples of Digital Elevation Models (DEMs) of the topography of the earthquake epicentral area ▶ derived from radar (left column **a–c**) and optical stereo imagery (right column **d–f**) depicted as hillshaded relief illuminated from the southeast. The rows go from high resolution (top) to lower-resolution datasets (bottom). **g** Depicts a Sentinel-2 RGB image over the same region. The town of Acipayam is in the lower left corner of each panel. The white arrows in **a** denote the edge of a step in topography, most likely associated with the edge of a fan. The white lines in **d** indicate the surface trace (ticks indicating direction of dip) and subsurface fault plane of the east-dipping solution found in the modelling of the InSAR data (Fig. 1). **h** Profile of topography perpendicular to the surface projection of the fault trace (X–X' and Y–Y') for the east-dipping solution shown in **d**. The profiles are taken through the TanDEM-X and SPOT DEMs and highlight the differences in resolution but also the greater noise present in the photogrammetrically derived SPOT DEM (note 3 m was subtracted from the SPOT DEM height to align it)

able to resolve this ambiguity, but usually local seismic networks and relocated seismicity are needed when there is no evidence of surface rupture (de Michele et al. 2013; Elliott et al. 2015). The recent study of this earthquake by Yang et al. (2020), favours the eastward dipping solution based upon the supplementary evidence from aftershock locations and reported field observations. An additional issue that arises when modelling buried fault slip in small earthquakes in that it is possible to fit the data as equally well with a near line source as a finite fault plane. Therefore, what typically happens in such cases is that the estimated fault slip (Fig. 2e) increases and trades off with a narrowing of the fault width (Fig. 2g) leaving both poorly constrained (but still maintaining approximately the same earthquake moment—Fig. 2i). The surface projection of the fault planes at depth then appear nearly as lines in map view (Fig. 1d, e). In this case the optimal slip is very high (3 metres) for such a small earthquake, and the estimated fault width is very narrow (400 m) given the fault length is over 8 km (Table 2), so stress drops would become unreasonably large. When such a trade-off occurs, an option is to constrain the prior in the Bayesian modelling approach within what is considered physically reasonable bounds (Bagnardi and Hooper 2018), or alternatively one of the fault parameters such as the fault width could be fixed (Yang et al. 2020).

An interplay exists between the active tectonics associated with faulting and the subsequent change in topography, erosion and deposition that is revealed in the analysis of the tectonic geomorphology of a landscape, which can be useful in earthquake-prone areas to identify the location of active fault structures (Burbank and Anderson 2009). Combining geodetic observations of earthquake cycle deformation with geomorphic observations of faulting can be important for assessing seismic hazard, in particular in regions with large, infrequent earthquakes (Hodge et al. 2015). To augment the analysis performed here with the Sentinel-1 InSAR data and the subsequent modelling, we also examine the topography in the area to try to link any surface geomorphic features with the causative fault plane. The availability and resolution of topographic data are highly variable globally, but it often acts as an important underlying dataset for geomorphic analysis of active faulting (e.g., Arrowsmith and Zielke 2009) and for landscape evolution in mountainous terrain (e.g., Boulton and Stokes 2018). Open global datasets from the Shuttle Radar Topography Mission (SRTM) (Farr and Kobrick 2000) provide 30 m (1 arc second) topographic data (Fig. 3b) for most of the mid and low latitude areas of the world (60 degrees north to 56 degrees south, Farr et al. 2007) and these are typically used in the InSAR processing chain to correct for topography (originally prior to 2015 only 90 m (3 arc second) data was available outside of the USA—Fig. 3c). Higher-resolution datasets exist commercially or may be available from national agencies but are not openly available. Here we compare the hierarchy of both SAR derived and optically stereo-derived Digital Elevation Models (DEMs) for this region (Fig. 3).

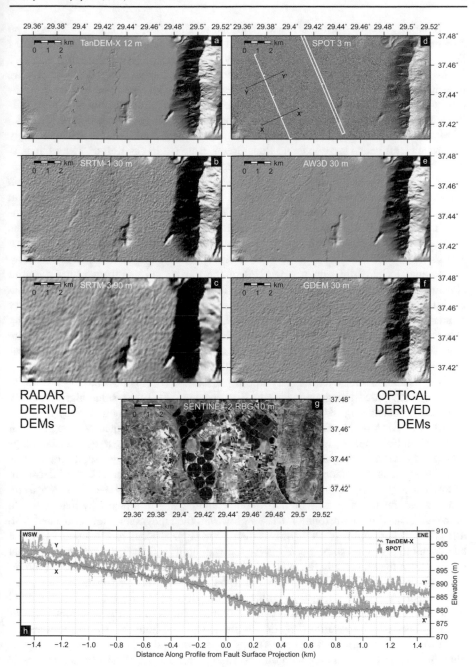

Low-resolution topographic datasets (often in conjunction with interpreting optical satellite imagery from Landsat) have been useful in the past for mapping out major fault structures (Taylor and Yin 2009) and can contribute towards building databases across the globe of active fault traces (Styron et al. 2010). However, for more subtle geomorphic traces of activity, high-resolution datasets are required (with Light Detection and Ranging (LiDAR) being the best (Prentice et al. 2009), but least available). The Advanced Spaceborne Thermal Emission and Reflection Radiometer (ASTER) GDEM (Fig. 3f) offers 30 m resolution near-globally as well as the freely available Advance World 3-Dimensional (AW3D) (Fig. 3e) based upon the Panchromatic Remote sensing Instrument for Stereo Mapping (PRISM). These openly available relatively low-resolution topography datasets enable analysis of major tectonic features, but it is more difficult to discriminate subtle features such as those within the basin in this case (Fig. 3).

Many modern optical imaging satellite instruments are multispectral with relatively narrow spectral bands across visible Red, Green, Blue (RGB) as well as wider bands at Near-Infrared (NIR) and beyond. However, they commonly also have a wider band that crosses most of the visible spectrum in a single channel termed the panchromatic, which is usually double the resolution (half the pixel size/spacing) of the visible multi-spectral bands. This band is typically used to derive topography because of its higher resolution, and usually is done when acquired as an in-track stereo mode when two images are taken in quick succession (as opposed to cross track which is separated in space and in time). The baseline separation of images (often a couple of hundreds of kilometres apart) yields different perspectives of the Earth's surface from which a digital elevation model can be derived from using the process of photogrammetry (Noh and Howat 2015). Satellite systems that are able to acquire in-track panchromatic imagery at medium and high resolutions, and that have been applied to active tectonic faulting observations include the WorldView series (Barnhart et al. 2019a), the pair of Satellite Pour l'Observation de la Terre (SPOT6/7) systems (Zhou et al. 2016) and the two Pleiades satellites (Zhou et al. 2015; Hodge et al. 2019).

Using panchromatic stereo imagery (1.5 m) from SPOT6, it is possible to derive relatively high-resolution topography at about 3 m spacing, albeit with some high-frequency noise and artefacts (Fig. 3d, h), especially over flat-lying areas. Commercially available TanDEM-X WorldDEM data (Krieger et al. 2007) provides good quality topography at 12 m resolution (Fig. 3a, h). In both these latter two datasets, the streets and buildings within the town of Acipayam start to become visible, as well as more subtle fluvial features. The potential location of the projection of the east-dipping fault that is inferred to be immediately up-dip of the modelled InSAR data is 2 km east of the town of Acipayam and is approximately aligned with north–south running step in topography that is visible in the hill-shaded higher-resolution DEMs. However, the topographic step associated with this potential fault uplift is subtle at less than 10 m (Fig. 3h) and such features can often be difficult to discriminate between relative footwall uplift and terrace edges due to fluvial incision from the drainage in this valley or from alluvial fans out washing from the mountains. Improved quality DEMs are required to be able to better interrogate the landscape geomorphology and its interaction with tectonics. However, such ambiguities highlight the necessity to supplement EO data with field observations to improve the robustness of remotely inferred interpretations. Whilst for many major onshore shallow earthquakes EO data captures the deformation and topography of an area well, it is still important to combine this remotely derived data with other seismological and geophysical datasets as well as field observations to fully constrain and understand the earthquake deformation process and potential for future hazard (Hamling 2019).

The requirements for high-resolution DEMs also highlight the discrepancy of the closed nature of such datasets versus the more open policies for recently acquired optical and SAR imagery, which are much more freely available and regularly updated at 10 m. In contrast, topography datasets are more often static (infrequently updated) and not openly available at such resolutions of 10 metres. Most aspects of active landscape evolution involve changes in the Earth's topography, be it migrating fluvial systems and knickpoints, retreating glacial streams, uplifting landscapes associated with earthquake faulting or landsliding resulting from seismic shaking. Additionally, the changing built environment associated with urbanisation also alters the local topography and is an important measure of the exposure and potential vulnerability of a population to a hazard. The lack of high-resolution topography that keeps pace with this rate of change presents a challenge to understanding the processes driving the evolving shape of our Earth and the parts of our society exposed to hazards.

The initial seismological epicentre from the USGS (USGS 2019) placed the earthquake further to the east than found here, amongst the high topography southwest of Lake Salda (Fig. 1). Without further information and investigation, such events might be attributed to one of the obvious major faults in the area that has been previously mapped (Emre et al. 2013). However, seismological locations of earthquakes (determined teleseismically) can be incorrect by many kilometres to tens of kilometres, although this can be greatly reduced with local networks and earthquake relocation techniques (e.g., Elliott et al. 2015). The geodetic data and modelling in this particular case indicate the faulting is further to the west but not along the major known fault immediately south and west of Acipayam (Emre et al. 2013, 2018). Comparison of the modelled fault location with the subtle surface geomorphology expressed in the higher-resolution topographic data, points to a previously unidentified active fault within the basin as a likely candidate for this earthquake, much nearer the town of Aipayam than suggested by the original USGS epicentre location. The main topography in the east runs north–south (Fig. 3), whereas the fault plane solution indicates a fault striking more NNW-SSE (Fig. 1). Whilst there appears to be a subtle raised portion of topography on the western edge of the basin with a similar strike (Fig. 3a–f), and a line of greener vegetation that may be associated with a spring line running along the proposed fault (Fig. 3g), the nonlinear shape of this feature along strike (Fig. 3a) indicates more likely that the step in topography is associated with the edge of an alluvial fan emanating from the mountains west of Acipayam, rather than a fault scarp, despite the surface projection of the modelled fault running approximately along this line of topography (Fig. 3d). The profiles through the highest resolution elevation data (Fig. 3h) do not show a consistent step in topography at the location of the surface projection of the fault trace, indicating that the fault plane has not accumulated significant enough cumulative offset that has propagated to shallow depths to present a clear scarp in the geomorphology. Robustly identifying the location of such active faults near to towns and cities is an important part of improving our knowledge of the seismic hazard. The proximity of the fault to the exposed buildings affects the estimated magnitude of ground accelerations and the subsequent calculated losses (Hussain et al. 2020). Knowing the location of particular scarps is also important at a local scale to avoid building infrastructure that straddles a fault. In this particular case, the surface projection of the fault plane is 2 km further into the basin (Fig. 3d) than the existing mapped fault that is behind the town of Acipayam, and is an example of the spatial migration of activity that might be associated with the breakup of a hanging wall block (e.g., Biggs et al. 2010). Furthermore, examining the faulting related to such minor earthquakes (as done here) acts as a motivator for targeting both future research and also raising societal awareness of the active faulting in the area that may be capable of hosting much larger earthquakes from ruptures on the previously identified major faults.

3 Sub-pixel Cross-Correlation of SAR-Amplitude Imagery and Optical Imagery

Sub-pixel correlation of both radar amplitude and optical images is now a commonly used technique for the measurement of surface deformation. It has been proven complementary to InSAR in several geophysical studies (e.g., Klinger et al. 2006; de Michele and Briole 2007; de Michele et al. 2010). Subpixel offsets of optical data, along with Synthetic Aperture Radar (SAR) and InSAR, are also used to constrain models of neo-rifting episodes (e.g., Barisin et al. 2009; Grandin et al. 2010).

In the Radar domain, the cross-correlation method was proven reliable with SAR amplitude data by Michel et al. (1999). A SAR system sends radar pulses to the ground and measures both the amplitude and the phase of the backscattered signal. The phase is used to perform the synthetic aperture. The phase difference is used to construct differential interferograms (DInSAR, InSAR) as used in the previous section, while the radar amplitude data can be used to map subpixels offsets following the correlation methodology firstly described in Michel et al. (1999). These methods, often called "offset tracking" complement InSAR particularly when the ground displacement is larger than half an interferometric fringe per pixel. This was the case for the major 2008 M_w 7.9 Sichuan earthquake, China, due to large slip of a fault that ruptured all the way to the surface (Fig. 4). At this rate of ground deformation, interferometric fringes become indistinguishable and

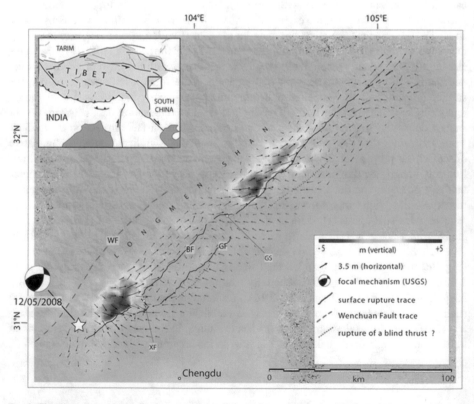

Fig. 4 The three dimensional displacement field of the Sichuan earthquake (2008 M_w 7.8, China) retrieved by cross-correlation of SAR data. Modified from de Michele et al. 2010

the InSAR signal becomes incoherent, thus unusable. This is one of the main reasons why SAR offset tracking is today a widely used technique for retrieving coseismic surface displacements of large earthquakes (e.g., Peltzer et al. 2001; Fialko et al. 2005; Pathier et al. 2006; de Michele et al. 2009, 2010; Yan et al. 2013; Hamling et al. 2017). Additionally, the use of a SAR system presents multiple other advantages. Firstly, SAR pulses penetrate through clouds. Secondly, they are independent of solar illumination, since the SAR antenna emits his own source of illumination. Thirdly, correlograms obtained from SAR-amplitude images contain different sets of information with respect to optical correlograms; since a SAR system acquires data along the Line of Sight direction of the satellite (LoS) whilst travelling in the azimuthal direction of the satellite, the SAR amplitude correlogram contains contributions from both horizontal (in the azimuth direction) and vertical (in LoS direction) ground motion. Significant East–West horizontal motion is recorded in the LoS component as well. This information, both from ascending and descending orbits, can be combined to calculate the 3D displacement field of an earthquake, such as that obtained for the M_w 7.9 Sichuan earthquake (de Michele et al. 2010), shown in Fig. 4.

The use of cross-correlation to measure displacement fields of the Earth's surface was first conceptualised by Crippen and Blom (1991, 1992) and Crippen (1992) and applied to optical spaceborne imagery. They called the method "imageodesy" and applied it to measure the displacement field of the Landers earthquake (1992 M_w 7.2, California) and the displacement field of a landslide with CNES (Centre National d'Etude Spatiales—French Space Agency) SPOT satellite imagery. The method assumes that image distortions due to mass movements can be measured with high precision as "errors" in the resampling field between orthorectified pre- and post-event images. The application of the methodology in earthquake studies was further developed in Michel (1997) and published in Van Puymbroeck et al. (2000), using SPOT data. The authors successfully implemented the methodology and applied it to measure the displacement field of the Landers Earthquake. In the method, images acquired before and after a deforming event are first resampled to a common geometry, typically using a Digital Elevation Model (DEM). Offsets are commonly calculated by differentiating the phases of the Fast Fourier Transforms on a moving window basis. Then, subpixel offset is achieved by interpolating the correlation peak within the moving window. The residual offset between the two images is expected to be due to surface deformation that occurred within the images' acquisition period. The theoretical precision is 1/10 of the pixel size, but this value largely depends on the image noise.

The method has been widely used, alone or as a complement to other geodetic techniques, to improve our knowledge of how the Earth's crust deforms. It is typically called either offset tracking, image-correlation or offset method but it relies on the same principles. There are many studies using the method and its subsequent modifications. As an example, Michel and Avouac (2002) applied the offsets method to measure the displacement field of the Izmit Earthquake (1999, M_w 7.5, Turkey) using SPOT satellite. Dominguez et al. (2003) used SPOT to measure the horizontal displacement field of the Chi–Chi earthquake (1999, M_w 7.6, Taiwan). Coupling their results with an elastic dislocation model (Okada 1985), they understood that the deeper portion of the fault was not activated during the Chi–Chi earthquake.

Klinger et al. (2006) used cross-correlated SPOT data before and after the Kokoxili earthquake (2002, M_w 7.8, Tibet) to extract features suggesting a rupture model with fault segments separated by strong persistent geometric barriers. Michel and Avouac (2006) used sub-metric resolution aerial photographs to precisely map the coseismic displacement field of the Landers earthquake (1992, M_w 7.3, California) and studied in detail the Kickapoo step over. The use of aerial photographs was crucial to measure volcano deformation at Piton de La Fournaise (La Reunion Island) by de Michele and Briole (2007). The method is further used in Leprince et al. (2007), Ayoub et al. (2009) and also in Milliner et al. (2015). Avouac et al. (2006) analyzed the 2006 M_w 7.6 Kashmir earthquake by modelling both seismic waveforms and using sub-pixel offset of the Advanced Spaceborne Thermal Emission and Reflection Radiometer (ASTER) images to measure ground deformation. SPOT is used to map the horizontal displacement field of the Denali earthquake (2002, M_w 7.9, Alaska) by Taylor et al. (2008). If the images are not acquired from exactly the same point of view, the methodology requires that the pre- and post-event images be perfectly resampled to the same geometry by means of a Digital Elevation Model (DEM). This requires a robust sensor focal plane model. If in the case that the latter information is undisclosed (which is often the situation for commercial satellites), then de Michele et al. (2008) suggested a method to extract the displacement field of an earthquake from non-orthorectified images coming from different image sources, by means of Principal Component Analyses. This concept of avoiding the use of a DEM to extract the displacement field of an earthquake is further expressed into the Perpendicular to Epipolar Offset method (Hollingsworth et al. 2012; Ayoub 2014).

In the past two decades, the subpixel correlation technique has become indispensable to map the surface ruptures associated with major earthquakes. The earthquake rupture geometry at the surface can be used as a hint for understanding rupture style and velocity at depth. Jointly with InSAR, GPS and seismological data, Konca et al. (2010) reveal a supershear behaviour of the 1999 Duzce earthquake (M_w 7.1, Turkey). On the other hand, sometimes the surface rupture geometry is particularly simple with respect to the complex source geometry, as highlighted by Wei et al. (2011) for the 2010 El Mayor Cucapah earthquake (M_w 7.2, Mexico). Potentially, if we could acquire images at very high frequency from space, from a geostationary telescope, subpixel offsets could be used as a seismometer, provided that the images' geometry is exactly the same, as highlighted in Michel et al. (2012). The advent of a new generation of satellites, with improved repetition frequency, allows more and more studies of earthquake ruptures from space. For instance, Landsat 8 has been used by Jolivet et al. (2014b) and Avouac et al. (2014) to show that a geological thrust fault can respond to neo-tectonic stress by slipping with a strike slip mechanism during the Balochistan earthquake (2013, M_w 7.7). In this earthquake, Vallage et al. (2015, 2016) used cross-correlation of SPOT5 images to precisely map the fault rupture at the surface, inferring non-elastic properties of the shallow fault section and structural control on its geometry. In the study of Xu et al. (2016), subpixel offsets of optical data have been used as a hint to challenge a well-established model (the "shallow slip deficit") pointing out the impact of data resolution on fault process analyses. During the 2016 M_w 7.8 Kaikōura earthquake, New Zealand, Hollingsworth et al. (2017) and Hamling et al. (2017) used Landsat 8, Sentinel 1 and ALOS2 respectively to map the intricate surface ruptures and the displacement field of this complex earthquake. Klinger et al. (2018) derived coseismic horizontal displacements in the Papatea–Jordan–Kekerengu triple junction area using high resolution optical satellite image correlation, with Pleiades and SPOT6. They found evidence for significant off-fault deformation. With this earthquake, Zinke et al. (2019) pushed the methodology further by combining cross-correlation and ray tracing to stereo World View images to retrieve the detailed 3D displacement field, without having to differentiate pre- and post-earthquake DEMs, thus increasing the precision of the measurements. Optical image correlation and SAR offset tracking are today well-established and robust techniques. The advent of the EU Copernicus Sentinel program (particularly

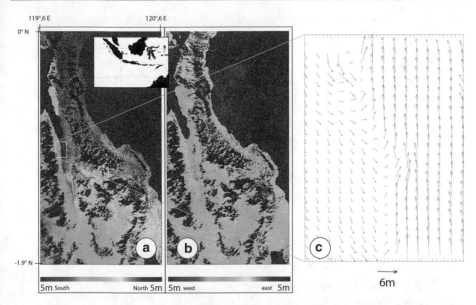

Fig. 5 The displacement field of the Palu earthquake (2018, M_w 7.3, Sulawesi) from Sentinel 2. Ground displacement is presented here as sub-pixel offsets i.e. a fraction of the Sentinel-2 pixel size (plus or minus 5 m in the North–South direction and in the East–West direction (**a** and **b** respectively). **c** A zoom on the Palu city area; offsets are shown as displacement vectors in an area where the North–South displacement is very sharp, indicating the surface trace of the fault rupture. Contains modified Copernicus data. Modified after Bacques et al. (2020)

Sentinel 1 and 2) brings an unprecedented amount of data, available with high repetition frequency (2 satellites per mission, between 5 and 12 day image acquisitions) and available at no cost. This improves the chances of imaging an earthquake with the offset-tracking methods. At the time of writing, the methodology is routinely used alone (as shown in Fig. 5), where the Palu earthquake (2018, M_w 7.5, Soulawesi) ruptured the surface generating a pluri-metric displacement field (e.g., Scott et al. 2019; Bacques et al. 2020), and in combination—or as a complement—with InSAR (e.g., Marchandon et al. 2018; Scott et al. 2019).

4 Artificial Water Reservoir-Triggered Earthquakes

Earth Observation data typically allow for wide area coverage and measurements of major deformation events, as illustrated in the previous examples. However, in many cases, the geophysical phenomena of interest are small in magnitude, or the background rates from which we try to detect a deviation are themselves small. Thus in situ and local measurements made using sensitive ground-based instruments are often required to fully characterised the evolution of the geophysical process. Additionally, the kind of information sought may require inferences of subsurface structures, which are more suited to be determined from airborne data using differing geophysical tools than perhaps that available from high altitude orbiting satellites. For this case study, we highlight a range of remotely derived observations as applied to understanding the potential for triggering of earthquakes from artificial reservoirs. Artificial water reservoirs are created all over the world for flood

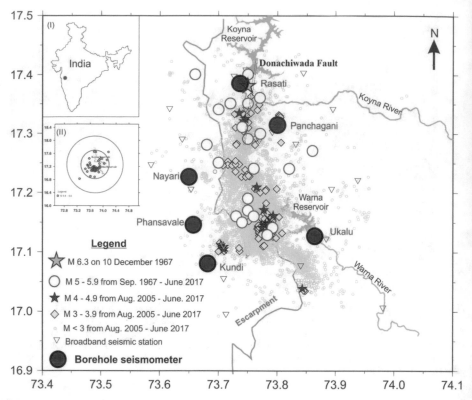

Fig. 6 Location of the Koyna Warna region in the vicinity of west coast of India; epicentres of the 10 December 1967 *M* 6.3, earthquakes of *M* 5.0 to 5.9 and smaller events for the period August 2005 to December 2017; locations of surface and bore-well seismic stations; the Western Ghat Escarpment is shown by the green line. (Inset) Location of Koyna in western India. (inset II) Epicentres of *M* ≥ 3.7 earthquakes during 1967–2015 (USGS) within 50 km (inner) and 100 km (outer circle) of the Koyna Dam. There are almost no seismic events detected outside the Koyna region. Modified from Gupta (2017). The Donachiwada Fault that hosted the 10 December 1967 earthquake and several *M* ~ 5 and smaller earthquakes is shown by parallel black lines

control, irrigation and power generation. Under certain geological conditions, the filling of these reservoirs can trigger earthquakes. To date, at least four sites globally have experienced triggered earthquakes exceeding magnitude 6 (Hsingfengkiang in China, Kariba on the Zambia–Zimbabwe border, Kremasta in Greece and Koyna in India). Koyna, located near the west coast of India, is a classic example of such reservoir-triggered seismicity (RTS), whereby triggered earthquakes started soon after the impoundment of the Koyna Dam in 1962. The creation of another (Warna dam), just 20 km south of Koyna in 1985 gave further rise to RTS. So far in the last 57 years, 22 earthquakes of magnitude *M* ≥ 5, some 200 earthquakes of *M* ≥ 4 and several thousand smaller earthquakes have occurred in the region (Fig. 6). Detailed studies of RTS events carried out in 1970s has led to the identification of certain characteristics of RTS sequences that delineate them from normal earthquake sequences. The association between water level changes and RTS in Koyna–Warna region is well established. However, the part played by reservoirs in the triggering of earthquakes is not well understood due to the lack of near field studies. The earthquakes are shallow (mostly 2 to 9 km depth), confined to a region of some 30 km by

20 km, and moreover, there is no other source of earthquakes within 100 km of Koyna Dam. To better characterise these processes, the suitability of the Koyna region for setting up a deep borehole laboratory was examined during the International Continental Drilling Program (ICDP) workshop held at Hyderabad and Koyna in 2011. However, prior to committing to an expensive project, a range of datasets were acquired to assess the tectonics of the region to enable the suitable design and deployment of a deep borehole. These included airborne gravity, gravity gradient and magnetic surveys and LiDAR coverage of the Koyna region, amongst several others (Gupta 2017, 2018). A 3 km deep pilot borehole has so far been drilled, with observations currently being made to design the final ~ 7 km deep borehole laboratory.

Airborne gravity-gradient and magnetic (AGGM) surveys provide measurements from which density structures and magnetic anomalies can be delineated. Using such surveys, we unravelled the sub-surface structure in the Koyna–Warna region to address

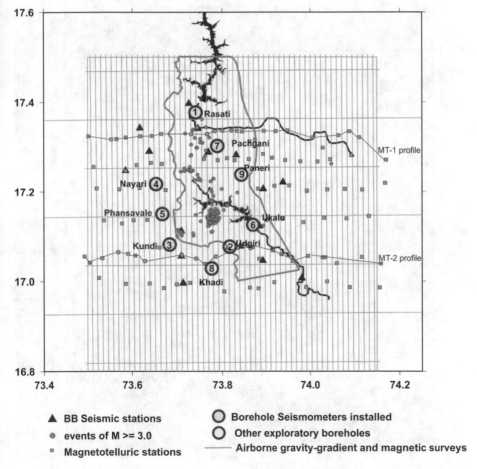

Fig. 7 Location of geophysical investigations in the Koyna–Warna region. These included deployment of 6 borehole seismometers; airborne gravity-gradient—magnetic surveys (green lines), LiDAR coverage (orange polygon), the western limit of LiDAR surveys coinciding with the Western Ghat Escarpment; magneto-telluric profiling. Epicentres of earthquakes of $M \geq 3$ for the period August 2005 through December 2015 are also depicted (after Gupta et al. 2016)

the potential controls on the pattern and distribution of observed RTS in the area. The region covered is shown in Fig. 6 and is at a draped surface about 120 m above ground level at an interval of ~ 1 km. The magnetic anomalies are predominately caused by the basaltic layer, which varies in thickness from 400 to 1600 m (Gupta et al. 2016; Mishra et al. 2017). The magnetic data were filtered with a cut-off wavelength of 10 km to focus on the deeper sources of these anomalies. The resulting map broadly depicts the subsurface structure (Fig. 7). It is characterised by a prominent NW–SE trend in the southern region and a NNE-SSW trend in the centre, flanked by a prominent negative anomaly. This anomaly is centred close the Udgiri borehole (Figs. 6, 7). An east–west trend of the anomaly is observed close to the southern end of the Koyna Dam. It has been also found

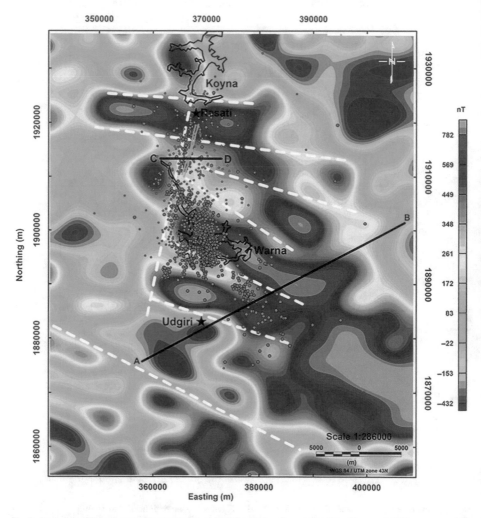

Fig. 8 Low-pass filtered magnetic anomaly map of the region (location of the Koyna and Warna reservoirs are indicated by black outlines). $M \geq 2$ earthquakes for the period 2005–2015 are shown by blue dots. AB is a section across the most prominent magnetic anomalies. CD is section across the Donachiwada causative fault (green lines)

that this east–west trending anomaly disappears when data are filtered with a higher cut-off wavelength leading to the inference of the source of these anomalies being shallower compared to the north–south and northwest–southeast trending anomalies. The northwest-southeast and north–south trending features are prominently identified in the central area of RTS in the filtered anomaly plots. This is consistent with the earthquake cluster in the vicinity of the Western Ghat Escarpment. The subsurface structure below the Donachiwada fault (considered to be the main causative fault of RTS at Koyna) has been inferred using the AGGM data as depicted in Fig. 8. It corresponds to a low in the observed GDD (Vertical Gravity Gradient) and magnetic field, as the top most layer is basalt with high magnetic susceptibility and density increases with depth. The delineated vertical block structure is consistent with the fault inferred from earthquake data.

Topographic data acquired at very high resolutions using LiDAR (Light Detection and Ranging) has become a fundamental remote sensing tool for the Earth sciences (Krishnan et al. 2011). LiDAR data are particularly useful for fault ruptures and for identification of geomorphic markers associated with active faulting (Zielke and Arrowsmith 2012), especially in the presence of forest canopy and vegetation cover that can obscure the subtle geomorphology from observation by optical satellite imagery. There were surface traces of the faulting caused by the Koyna earthquake of 10 December 1967 (Gupta et al. 1999). However, neither the M 5.8 earthquake of 13 September 1967, nor the later $M \sim 5$ earthquakes in the Koyna–Warna region, provided any surface evidence of faulting at depth. One question that remained unanswered was whether the faulting was confined to the basement only or propagated through the basalt cover as well. A combination of rugged topography and dense vegetation makes fieldwork looking for possible traces of faulting difficult. To overcome this difficulty LiDAR surveys were carried out in the area covering the region of RTS in the Koyna–Warna area. The area covered 1064 sq. km and is shown in Fig. 9 where LiDAR and orthophotograph data were acquired in April 2014 (Arora et al. 2017). A total of 21 ground control points were established before acquisition of multiple return and waveform LiDAR and orthophotograph data. The NNE-SWW trending lineament, starting just south of the Koyna Reservoir and running through the Warna Reservoir is the surface expression of the Donachiwada Fault responsible for most of RTS in the region, including the 10 December 1967 M 6.3 Koyna main earthquake.

The 1967 earthquake occurred more than two decades before the first observations of ground deformation associated with major seismic events were possible with SAR satellites. However, the magnitude 5 events in 2009 occurred in the era of ALOS SAR data and were just large enough to have surface deformation associated with them. Arora et al. (2018) have focused on the lineaments within the Deccan Traps in the area of RTS in the Koyna Warna region and have shown their connection with the subsurface basement using LiDAR and Synthetic Aperture Radar (SAR) interferometry. Interferometric measurements conducted for the Koyna–Warna region indicated displacement associated with two $M \sim 5$ earthquakes that occurred on 14 November 2009 and 12 December 2009. Both these earthquakes had normal faulting dominated movement. The size of these events is beyond the detectability limit of InSAR data using single interferograms (as used in the earlier section) and instead a time series approach is required to use larger data volumes before and after the earthquakes to reduce the noise due to atmosphere in the data. As depicted in Fig. 10, a LoS displacement of up to ~ 12 mm between March 2009 and September 2010 was observed. Arora et al. (2018) note that incremental LoS displacement calculated for the total time period of SAR coverage from 12 January 2007 to 10 March 2011 shows that all the displacement occurred within the period March 2009 to September 2010 (i.e. from just before to just after the occurrence of two $M \sim 5$ RTS events). Direct evidence of

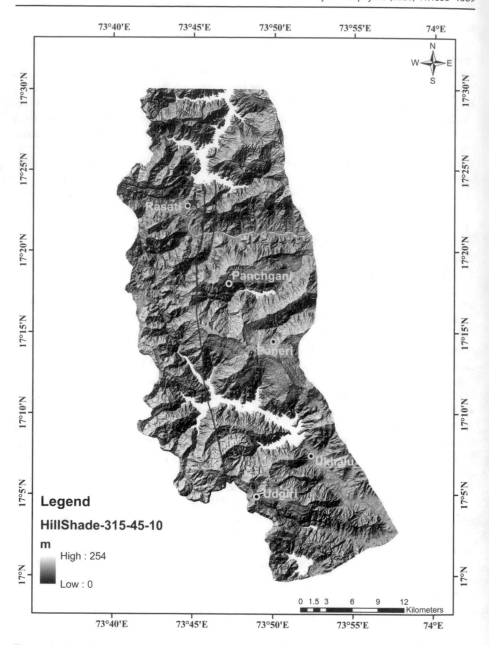

Fig. 9 LiDAR hillshade DEM of the Koyna region. A major regional north–south fracture across the seismic region and three NNE-SSW trending fracture zones (red lines) are seen from south of the Koyna Reservoir to north of the ridge to the left of the Warna Reservoir. The western most NNE-SSW trend coincides with the trend of the Donachiwada fault (after Gupta et al. 2016)

faulting was also provided by the slicken sides in the deep bore holes drilled in the region for various scientific investigations and setting up of the borehole seismic network. Arora et al. (2018) conclude that their work supports an inheritance model for basement faulting

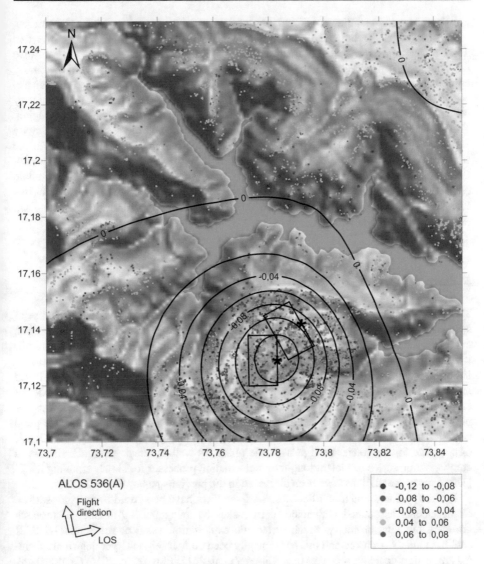

Fig. 10 Comparison of total line of sight (LoS) displacements (coloured points, dm) estimated by the Advanced Land Observation Satellite Phased-Array-type L-band Synthetic Aperture Radar (ALOS-PAL-SAR) images in the Warna area from ascending track 536 (frame 330) for the period from March 2009 to September 2010 with synthetic LoS displacements (isolines, in dm) calculated using fault-plane models of the two $M \geq 5$ earthquakes. Positive LoS displacements are directed towards the satellite. Direction of flight and LoS are shown by arrows below the figure. Rectangles are projections to the surface of the fault planes. The asterisks show the epicentres of the two earthquakes (after Arora et al. 2018)

with repeated earthquakes causing upward propagation of faulting, which in turn causes fractures permitting water from the reservoirs to percolate into faults and consequently triggering earthquakes.

Near-field investigations are of utmost importance to comprehend the initiation of an earthquake in a fault zone and observe what proceeds and follows the nucleation process.

For these near field studies, it is crucial to locate a suitable place for setting up the necessary near field observation laboratory. At Koyna, India, due to a thick basalt cover and dense vegetation, it was very difficult to trace the surface manifestation of the causative fault of the 1967 M 6.3 earthquake and the continued RTS in the Koyna–Warna region. By combining a range of datasets we have shown that a range of remotely sensed data from airborne gravity-gradient and magnetic surveys, through to LiDAR and SAR Interferometry have provided evidence for the nature the subsurface structure in the Koyna–Warna region. This has helped in tracing the causative Donachiwada fault, and the association of gravity and magnetic anomalies with hypocenters, and that the basement faulting extends to the surface. These inputs have been helpful in locating the site for placing the 3 km deep Pilot Borehole, which was completed in June 2017 (Gupta 2018). Drilling of this Pilot Borehole confirmed that the inferred location of the Donachiwada fault is correct. The work at this Pilot Borehole is in progress and will provide the necessary inputs for designing the ~ 7 km deep main Borehole laboratory.

5 Conclusions

The recent increase in the number of Earth Observation satellites has expanded the use of such data in characterising earthquake deformation. The large data volumes and an open data policy of the Sentinel missions have been an important development for accessibility, and this change has democratised the application and exploitation of such datasets. It has widened the uptake from users for both scientific and commercial use, and it has increased the range of scientific questions that can be addressed, as well as the number of environmental, commercial and industrial applications. Here we have selected a few examples where the data coverage, quality and availability have been optimal to make for good case studies. However, sometimes there is a suboptimal response to an earthquake in terms of coverage for the longer-term postseismic phase, as well as a lack of regular data for a complete time series to look at long-term deformation processes for slowly straining areas. (These recommendations have been discussed in the previous paper, Elliott 2020.)

For the past two and a half decades, SAR satellites have been used to measure surface displacements associated with major earthquakes by an increasing number of research groups using interferometry. Much systematic exploitation using newer Sentinel-1 SAR data has already occurred and this has typically been on focused on larger deformation signals from significant earthquakes (e.g., Lindsey et al. 2015; Floyd et al. 2016; Grandin et al. 2016; Xu et al. 2020), although this is expanding to cover ever-increasing smaller events (Funning and Garcia 2018). As the data coverage and ease of access has improved, the number of potential events to study has greatly expanded. Here we have shown an example of measuring deformation associated with a small earthquake in Turkey and its relationship with the surrounding geomorphology. Whilst the earthquake itself did not pose a particular major shaking hazard to the region, it illuminates a potential source fault in an area that previously contained only a few identified active fault traces. Although ambiguities in interpreting remotely derived datasets still remain, such observations provide an impetus to carefully study an area in terms of the active tectonics. Combining InSAR observations with remotely derived high-resolution topography will continue to improve our assessment for the potential of earthquake shaking near cities. We also described examples of making surface observations using pixel offsets from both optical EO systems and from SAR using the technique of sub-pixel cross-correlation. This method is complementary to InSAR and is particularly suited to observing large offsets associated with major surface rupturing continental earthquakes. Using ever-increasing high-resolution imagery from optical satellites, it is possible to provide huge amounts of detail of the fault rupture at the surface. Important questions of the amount of on and off fault deformation can start to be addressed, as well as to determine the differing behaviour of the crust in terms of elastic and non-elastic behaviour (Diederichs et al. 2019). Finally, we illustrated the importance of combining space-based EO with other airborne remotely derived datasets using an example of triggered seismicity from Konya, India. Using a single source of remotely sensed data is unlikely to give the complete overview of a geophysical process; as is the case here in understanding the impact of reservoir loading on the evolution of seismicity. The importance of a complete synthesis is highlighted for determining the controls and interaction of the subsurface geology and surface loads with the locations and rates of subsequent seismicity.

Improvements in our understanding of earthquakes and faulting will enable better assessments of the locations of future earthquake hazards, as well as help constrain estimates of the sizes and characteristics of shaking sources. Understanding the relative location, depth and extent of earthquake-generating sources near cities is particularly relevant when determining the risk, as the ground accelerations quickly drop off with distance from the fault. Whilst identifying new active fault traces is an important outcome, it is also necessary to identify which portions of a fault did not rupture in a given earthquake because those portions may now be prone to future failure. Here we have focused on illustrating the use of EO with examples of the deformation associated with earthquakes (i.e. the process at the end of the earthquake cycle). However, where newer satellite systems such as Sentinel-1 offer a game-changing approach is in understanding the other parts of the earthquake cycle (Salvi et al. 2012) described in the introduction—in particular that of the long-term strain accumulation building up to earthquakes. Whilst studies of strain rates using Sentinel-1 have focused on deformation over a few 100's km scale (Shirzaei et al. 2017; Morishita et al. 2020), and have begun to encompass whole countries such as Iceland (Drouin and Sigmundsson 2019), large (1000's km +) systematic attempts to do so are now only beginning to be achieved (Weiss et al. 2020) as the archive of data becomes suitably long to improve the signal to noise ratio. To make accurate measurements of small deformation signals over long wavelengths requires a continuity of data that is important for establishing a long time series. Also, widespread, consistent coverage is required, with almost all the tectonic zones now being covered regularly by Sentinel-1 offering the chance to capture whole continent-scale deformation. This ability to measure solid Earth processes will be greatly enhanced with the upcoming joint NASA-ISRO SAR mission (NISAR) in a couple of years which will image land surface changes globally.

Acknowledgements This work contains modified Copernicus Sentinel data 2019. Original SAR data is available to download from https://scihub.copernicus.eu/. Geodetic Bayesian Inversion Software (GBIS) used in the modelling of InSAR data is available here: https://comet.nerc.ac.uk/gbis/. John Elliott acknowledges support from the Royal Society through a University Research Fellowship (UF150282). This work was supported by NERC through the Looking into the Continents from Space (LiCS) large Grant (NE/K010867/1). This work was also supported by the Centre for the Observation and Modelling of Earthquakes, Volcanoes and Tectonics (COMET) in the United Kingdom. Harsh Gupta acknowledges support from the National Geophysical Research Institute, Hyderabad, India and thanks the Geological Society of India (GSI) for use of Fig. 6 from J-GSI, and the Seismological Society of America (SSA) for use of Fig. 10 from the Bulletin SSA. We thank Susanna Ebmeier for proof-reading the final draft. We also thank two anonymous reviewers for helpful comments in improving the manuscript.

Author Contributions All authors contributed to the writing of the paper and production of the figures.

ReferencesATTACHED PDF OF PROOFS INCLUDES ALMOST ALL DOIs TO PLEASE BE INSERTED INTO REFERENCES

Aktug B, Nocquet JM, Cingöz A, Parsons B, Erkan Y, England P et al (2009) Deformation of western Turkey from a combination of permanent and campaign GPS data: limits to block-like behavior. J Geophys Res Solid Earth 114(B10). https://doi.org/10.1029/2008JB006000

Ambraseys NN (1988) Engineering seismology: part I. Earthq Eng Struct Dynam 17(1):1–50

Amey RMJ, Hooper A, Walters RJ (2018) A Bayesian method for incorporating self-similarity into earthquake slip inversions. J Geophys Res Solid Earth 123(7):6052–6071. https://doi.org/10.1029/2017JB015316

Anantrasirichai N, Biggs J, Albino F, Hill P, Bull D (2018) Application of machine learning to classification of volcanic deformation in routinely generated InSAR data. J Geophys Res Solid Earth 123(8):6592–6606. https://doi.org/10.1029/2018JB015911

Arikawa Y, Saruwatari H, Hatooka Y, Suzuki S (2014) ALOS-2 launch and early orbit operation result. In: 2014 IEEE geoscience and remote sensing symposium, pp 3406–3409. IEEE. https://doi.org/10.1109/IGARSS.2014.6947212

Arora K, Chadha RK, Srinu Y, Selles A, Srinagesh D, Smirnov V, Ponomarev A, Mikhailov VO (2017) Lineament fabric from airborne LiDAR and its influence on triggered earthquakes in the Koyna–Warna region, western India. J Geol Soc India 90:670–677. https://doi.org/10.1007/s12594-017-0774-9

Arora K, Srinu Y, Gopinadh D, Chadha RK, Raza H, Mikhailov VO, Ponomarev A, Kiseleva E, Smirnov V (2018) Lineaments in deccan basalts: the basement connection in the Koyna–Warna RTS region. Bull Seismol Soc Am 108(5B):2919–2932. https://doi.org/10.1785/0120180011

Arrowsmith JR, Zielke O (2009) Tectonic geomorphology of the San Andreas fault zone from high resolution topography: an example from the Cholame segment. Geomorphology 113(1–2):70–81

Avouac JP (2015) From geodetic imaging of seismic and aseismic fault slip to dynamic modeling of the seismic cycle. Annu Rev Earth Planet Sci 43:233–271. https://doi.org/10.1016/j.geomorph.2009.01.002

Avouac JP, Ayoub F, Leprince S, Konca O, Helmberger DV (2006) The 2005, Mw 7.6 Kashmir earthquake: sub-pixel correlation of ASTER images and seismic waveforms analysis. Earth Planet Sci Lett 249(3–4):514–528. https://doi.org/10.1016/j.epsl.2006.06.025

Avouac J-P, Ayoub F, Wei S, Ampuero J-P, Meng L et al (2014) The 2013, Mw 7.7 Balochistan earthquake, energetic strike-slip reactivation of a thrust fault. Earth Planet Sci Lett 391:128–134. https://doi.org/10.1016/j.epsl.2014.01.036

Ayoub F (2014) Monitoring morphologic surface changes from aerial and satellite imagery, on Earth and Mars. PhD thesis in applied geology. Universite Toulouse III Paul Sabatier, 2014. English

Ayoub F, Leprince F, Avouac JP (2009) Co-registration and correlation of aerial photographs for ground deformation measurements. ISPRS J Photogram Remote Sens 64(6):551–560. https://doi.org/10.1016/j.isprsjprs.2009.03.005

Bacques G, de Michele M, Foumelis M et al (2020) Sentinel optical and SAR data highlights multi segment faulting during the 2018 Palu-Sulawesi earthquake (Mw 7.5). Sci Rep 10:9103. https://doi.org/10.1038/s41598-020-66032-7

Bagnardi M, Hooper A (2018) Inversion of surface deformation data for rapid estimates of source parameters and uncertainties: a Bayesian approach. Geochem Geophys Geosyst 19(7):2194–2211. https://doi.org/10.1029/2018GC007585

Barisin I, Leprince S, Parsons B, Wright T (2009) Surface displacements in the September 2005 Afar rifting event from satellite image matching: asymmetric uplift and faulting. Geophys Res Lett 36(7). https://doi.org/10.1029/2008GL036431

Barnhart WD, Lohman RB (2010) Automated fault model discretization for inversions for coseismic slip distributions. J Geophys Res Solid Earth 115(B10). https://doi.org/10.1029/2010JB007545

Barnhart WD, Gold RD, Shea HN, Peterson KE, Briggs RW, Harbor DJ (2019a) Vertical coseismic offsets derived from high-resolution stereogrammetric DSM differencing: the 2013 Baluchistan, Pakistan earthquake. J Geophys Res Solid Earth 124(6):6039–6055. https://doi.org/10.1029/2018JB017107

Barnhart WD, Hayes GP, Gold RD (2019) The July 2019 Ridgecrest, California Earthquake Sequence: kinematics of slip and stressing in cross-fault ruptures. Geophys Res Lett 46(21):11859–11867. https://doi.org/10.1029/2019GL084741

Bell MA, Elliott JR, Parsons BE (2011) Interseismic strain accumulation across the Manyi fault (Tibet) prior to the 1997 Mw 7.6 earthquake. Geophys Res Lett 38(24). https://doi.org/10.1029/2011GL049762

Biggs J, Nissen E, Craig T, Jackson J, Robinson DP (2010) Breaking up the hanging wall of a rift-border fault: the 2009 Karonga earthquakes, Malawi. Geophys Res Lett 37(11). https://doi.org/10.1029/2010GL043179

Boulton SJ, Stokes M (2018) Which DEM is best for analyzing fluvial landscape development in mountainous terrains? Geomorphology 310:168–187. https://doi.org/10.1016/j.geomorph.2018.03.002

Burbank DW, Anderson RS (2009) Tectonic geomorphology. Wiley, London

Bürgmann R, Chadwell D (2014) Seafloor geodesy. Annu Rev Earth Planet Sci 42:509–534. https://doi.org/10.1146/annurev-earth-060313-054953

Bürgmann R, Dresen G (2008) Rheology of the lower crust and upper mantle: evidence from rock mechanics, geodesy, and field observations. Annu Rev Earth Planet Sci 36:531–567. https://doi.org/10.1146/annurev.earth.36.031207.124326

Chlieh M, De Chabalier JB, Ruegg JC, Armijo R, Dmowska R, Campos J, Feigl KL (2004) Crustal deformation and fault slip during the seismic cycle in the North Chile subduction zone, from GPS and InSAR observations. Geophys J Int 158(2):695–711. https://doi.org/10.1111/j.1365-246X.2004.02326.x

Crippen RE (1992) Measurement of subresolution terrain displacements using SPOT panchromatic imagery. Episodes 15(1):56–61. https://doi.org/10.18814/epiiugs/1992/v15i1/009

Crippen RE, Blom RG (1991) Concept for the subresolution measurement of earthquake strain fields using SPOT panchromatic imagery. In: Earth and atmospheric remote sensing, vol 1492, pp 370–377. International Society for Optics and Photonics. https://doi.org/10.1117/12.45870

Crippen RE, Blom RG (1992) The first visual observations of fault movements from space—the 1992 Landers earthquake: EOS. Trans Am Geophys Union 73:364

de Michele M, Briole P (2007) Deformation between 1989 and 1997 at Piton de la Fournaise volcano retrieved from correlation of panchromatic airborne images. Geophys J Int 169(1):357–364. https://doi.org/10.1111/j.1365-246X.2006.03307.x

de Michele M, Raucoules D, Aochi H, Baghdadi N, Carnec C (2008) Measuring coseismic deformation on the northern segment of the Bam-Baravat escarpment associated with the 2003 Bam (Iran) earthquake, by correlation of very-high-resolution satellite imagery. Geophys J Int 173:459–464. https://doi.org/10.1111/j.1365-246X.2008.03743.x

de Michele M, Raucoules D, Lasserre C, Pathier E, Klinger Y, Van Der Woerd J, de Sigoyer J, Xu X (2009) The Mw 7.9, 12 May 2008 Sichuan Earthquake Rupture Measured by sub-pixel correlation of ALOS PALSAR amplitude images. Earth Planets Space. https://doi.org/10.5047/eps.2009.05.002

de Michele M, de Raucoules D, Sigoyer J, Pubellier M, Chamot-Rooke N (2010) Three-dimensional surface displacement of the 2008 May 12 Sichuan earthquake (China) derived from Synthetic Aperture Radar: evidence for rupture on a blind thrust. Geophys J Int 183:1097–1103. https://doi.org/10.1111/j.1365-246X.2010.04807.x

de Michele M, Briole P, Raucoules D, Lemoine A, Rigo A (2013) Revisiting the shallow Mw 5.1 Lorca earthquake (southeastern Spain) using C-band InSAR and elastic dislocation modelling. Rem Sens Lett 4(9):863–872. https://doi.org/10.1080/2150704X.2013.808777

Delorme A, Grandin R, Klinger Y, Pierrot-Deseilligny M, Feuillet N, Jacques E et al (2020) Complex deformation at shallow depth during the 30 October 2016 Mw6. 5 Norcia earthquake: interference between tectonic and gravity processes?. Tectonics 39(2):e2019TC005596. https://doi.org/10.1029/2019TC005596

Deng J, Gurnis M, Kanamori H, Hauksson E (1998) Viscoelastic flow in the lower crust after the 1992 Landers, California, earthquake. Science 282(5394):1689–1692. https://doi.org/10.1126/science.282.5394.1689

Diederichs A, Nissen EK, Lajoie LJ, Langridge RM, Malireddi SR, Clark KJ et al (2019) Unusual kinematics of the Papatea fault (2016 Kaikōura earthquake) suggest anelastic rupture. Sci Adv 5(10):eaax5703: https://doi.org/10.1126/sciadv.aax5703

Dominguez S, Avouac J-P, Michel R (2003) Horizontal coseismic deformation of the 1999 Chi-Chi earthquake measured from SPOT satellite images: implications for the seismic cycle along the western foothills of central Taiwan. J Geophys Res 108:2083. https://doi.org/10.1029/2001JB000951

205

Drouin V, Sigmundsson F (2019) Countrywide observations of plate spreading and glacial isostatic adjustment in iceland inferred by Sentinel-1radar interferometry, 2015–2018. Geophys Res Lett 46(14):8046–8055. https://doi.org/10.1029/2019GL082629

Drusch M, Del Bello U, Carlier S, Colin O, Fernandez V, Gascon F et al (2012) Sentinel-2: ESA's optical high-resolution mission for GMES operational services. Remote Sens Environ 120:25–36. https://doi.org/10.1016/j.rse.2011.11.026

Ebmeier SK (2016) Application of independent component analysis to multitemporal InSAR data with volcanic case studies. J Geophys Res Solid Earth 121(12):8970–8986. https://doi.org/10.1002/2016JB013765

Elliott JR (2020) Earth observation for the assessment of earthquake hazard, risk and disaster management. In: Cazenave A et al (eds) Geohazards and risks from earth observations. Surveys in geophysics

Elliott JR, Walters RJ, England PC, Jackson JA, Li Z, Parsons B (2010) Extension on the Tibetan plateau: recent normal faulting measured by InSAR and body wave seismology. Geophys J Int 183(2):503–535. https://doi.org/10.1111/j.1365-246X.2010.04754.x

Elliott JR, Bergman EA, Copley AC, Ghods AR, Nissen EK, Oveisi B et al (2015) The 2013 Mw 6.2 Khaki-Shonbe (Iran) Earthquake: insights into seismic and aseismic shortening of the Zagros sedimentary cover. Earth Space Sci 2(11):435–471. https://doi.org/10.1002/2015EA000098

Elliott JR, Walters RJ, Wright TJ (2016) The role of space-based observation in understanding and responding to active tectonics and earthquakes. Nat Commun 7:13844. https://doi.org/10.1038/ncomms13844

Emre Ö, Duman TY, Özalp S, Elmacı H, Olgun Ş, Şaroğlu F (2013) Active fault map of turkey with and explanatory text. General Directorate of Mineral Research and Exploration, Special Publication Series-30. Ankara-Turkey

Emre Ö, Duman TY, Özalp S, Şaroğlu F, Olgun Ş, Elmacı H, Can T (2018) Active fault database of Turkey. Bull Earthq Eng 16(8):3229–3275. https://doi.org/10.1007/s10518-016-0041-2

England P, Jackson J (2011) Uncharted seismic risk. Nat Geosci 4(6):348. https://doi.org/10.1038/ngeo1168

Farr TG, Kobrick M (2000) Shuttle Radar Topography Mission produces a wealth of data. EOS Trans Am Geophys Union 81(48):583–585. https://doi.org/10.1029/EO081i048p00583

Farr TG, Rosen PA, Caro E, Crippen R, Duren R, Hensley S et al (2007) The shuttle radar topography mission. Rev Geophys 45(2). https://doi.org/10.1029/2005RG000183

Feng G, Hetland EA, Ding X, Li Z, Zhang L (2010) Coseismic fault slip of the 2008 Mw 7.9 Wenchuan earthquake estimated from InSAR and GPS measurements. Geophys Res Lett 37(1). https://doi.org/10.1029/2009GL041213

Fialko Y, Sandwell D, Agnew D, Simons M, Shearer P, Minster B (2002) Deformation on nearby faults induced by the 1999 Hector Mine earthquake. Science 297(5588):1858–1862. https://doi.org/10.1126/science.1074671

Fialko Y, Sandwell D, Simons M, Rosen P (2005) Three-dimensional deformation caused by the Bam, Iran, earthquake and the origin of shallow slip deficit. Nature 435:295–299. https://doi.org/10.1038/nature03425

Fielding EJ, Lundgren PR, Taymaz T, Yolsal-Çevikbilen S, Owen SE (2013) Fault slip source models for the 2011 M7.1 Van earthquake in Turkey from SAR interferometry, pixel offset tracking, GPS and seismic waveform analysis. Seismol Res Lett 84(4):579–593. https://doi.org/10.1785/0220120164

Floyd MA, Walters RJ, Elliott JR, Funning GJ, Svarc JL, Murray JR et al (2016) Spatial variations in fault friction related to lithology from rupture and afterslip of the 2014 South Napa, California, earthquake. Geophys Res Lett 43(13):6808–6816. https://doi.org/10.1002/2016GL069428

Fujiwara T, Kodaira S, No T, Kaiho Y, Takahashi N, Kaneda Y (2011) The 2011 Tohoku-Oki earthquake: displacement reaching the trench axis. Science 334(6060):1240. https://doi.org/10.1126/science.1211554

Funning GJ, Garcia A (2018) A systematic study of earthquake detectability using Sentinel-1 Interferometric Wide-Swath data. Geophys J Int 216(1):332–349. https://doi.org/10.1093/gji/ggy426

Funning GJ, Parsons B, Wright TJ, Jackson JA, Fielding EJ (2005) Surface displacements and source parameters of the 2003 Bam (Iran) earthquake from Envisat advanced synthetic aperture radar imagery. J Geophys Res Solid Earth 110(B9). https://doi.org/10.1029/2004JB003338

Funning GJ, Parsons B, Wright TJ (2007) Fault slip in the 1997 Manyi, Tibet earthquake from linear elastic modelling of InSAR displacements. Geophys J Int 169(3):988–1008. https://doi.org/10.1111/j.1365-246X.2006.03318.x

Furlong KP, Herman M (2017) Reconciling the deformational dichotomy of the 2016 Mw 7.8 Kaikoura New Zealand earthquake. Geophys Res Lett 44(13):6788–6791. https://doi.org/10.1002/2017GL074365

Grandin R, Socquet A, Jacques E, Mazzoni N, de Chabalier J-B, King GCP (2010) Sequence of rifting inAfar, Manda-Hararo rift, Ethiopia, 2005–2009: time-space evolution and interactions between dikes from interferometricsynthetic aperture radar and static stress change modeling. J Geophys Res 115:B10413. https://doi.org/10.1029/2009JB000815

Grandin R, Klein E, Métois M, Vigny C (2016) Three-dimensional displacement field of the 2015 Mw8.3 Illapel earthquake (Chile) from across-and along-track Sentinel-1 TOPS interferometry. Geophys Res Lett 43(6):2552–2561. https://doi.org/10.1002/2016GL067954

Gupta HK (2017) Koyna, India, an ideal site for Near Field Earthquake Observations. J Geol Soc India 90(6):645–652

Gupta HK (2018) Review: reservoir Triggered Seismicity (RTS) at Koyna, India over the Past 50 years. Bull Seismol Soc Am 108(5B):2907–2918. https://doi.org/10.1785/0120180019

Gupta HK, Rao RUM, Srinivasan R, Rao GV, Reddy GK, Dwivedi KK, Banerjee DC, Mohanty R, Satyasaradhi YR (1999) Anatomy of surface rupture zones of two stable continental region earthquakes, 1967 Koyna and 1993, Latur, India. Geophys Res Lett 26(13):1985–1988. https://doi.org/10.1029/1999GL900399

Gupta HK, Kusumita Arora N, Rao Purnachandra, Sukanta Roy VM, Tiwari Prasanta K, Patro HVS Satyanarayana, Shashidhar D, Mahato CR, Srinivas KNSSS, Srihari M, Satyavani N, Srinu Y, Gopinadh D, Raza Haris, Jana Monikuntala, Akkiraju Vyasulu V, Goswami Deepjyoti, Digant Vyas CP, Dubey DCh, Raju V, Borah Ujjal, Kashi Raju K, Reddy Chinna, Narendra Babu BK, Bansal Shailesh Nayak (2016) Investigations of continued reservoir triggered seismicity at Koyna, India. Geol Soc Lond Special Publ 445:151–188. https://doi.org/10.1144/SP445.11

Hamling IJ (2019) A review of the 2016 Kaikōura earthquake: insights from the first 3 years. J R Soc N Z 1–19. https://doi.org/10.1080/03036758.2019.1701048

Hamling IJ, Hreinsdóttir S, Clark K, Elliott J, Liang C, Fielding E et al (2017) Complex multifault rupture during the 2016 Mw 7.8 Kaikōura earthquake, New Zealand. Science 356(6334):eaam7194. https://doi.org/10.1126/science.aam7194

Hayes GP, Moore GL, Portner DE, Hearne M, Flamme H, Furtney M, Smoczyk GM (2018) Slab2, a comprehensive subduction zone geometry model. Science 362(6410):58–61. https://doi.org/10.1126/science.aat4723

Hearn EH, Bürgmann R, Reilinger RE (2002) Dynamics of Izmit earthquake postseismic deformation and loading of the Duzce earthquake hypocenter. Bull Seismol Soc Am 92(1):172–193. https://doi.org/10.1785/0120000832

Hodge M, Biggs J, Goda K, Aspinall W (2015) Assessing infrequent large earthquakes using geomorphology and geodesy: the Malawi Rift. Nat Hazards 76(3):1781–1806. https://doi.org/10.1007/s11069-014-1572-y

Hodge M, Biggs J, Fagereng Å, Elliott A, Mdala H, Mphepo F (2019) A semi-automated algorithm to quantify scarp morphology (SPARTA): application to normal faults in southern Malawi. Solid Earth 10(1):27–57. https://doi.org/10.5194/se-10-27-2019

Hollingsworth J, Leprince S, Ayoub F, Avouac JP (2012) Deformation during the 1975–1984 Krafla rifting crisis, NE Iceland, measured from historical optical imagery. J Geophys Res Solid Earth 117(B11). https://doi.org/10.1029/2012JB009140

Hollingsworth J, Ye L, Avouac J-P (2017) Dynamically triggered slip on a splay fault in the Mw 7.8, 2016 Kaikoura (New Zealand) earthquake. Geophys Res Lett 44:3517–3525. https://doi.org/10.1002/2016GL072228

Hooper A, Bekaert D, Spaans K, Arıkan M (2012) Recent advances in SAR interferometry time series analysis for measuring crustal deformation. Tectonophysics 514:1–13. https://doi.org/10.1016/j.tecto.2011.10.013

Huang MH, Fielding EJ, Liang C, Milillo P, Bekaert D, Dreger D, Salzer J (2017) Coseismic deformation and triggered landslides of the 2016 Mw 6.2 Amatrice earthquake in Italy. Geophys Res Lett 44(3):1266–1274. https://doi.org/10.1002/2016GL071687

Hussain E, Elliott JR, Silva V, Vilar-Vega M, Kane D (2020) Contrasting seismic risk for Santiago, Chile, from near-field and distant earthquake sources. Nat Hazards Earth Syst Sci 20:1533–1555. https://doi.org/10.5194/nhess-20-1533-2020

Jolivet R, Grandin R, Lasserre C, Doin MP, Peltzer G (2011) Systematic InSAR tropospheric phase delay corrections from global meteorological reanalysis data. Geophys Res Lett 38(17). https://doi.org/10.1029/2011GL048757

Jolivet R, Agram PS, Lin NY, Simons M, Doin MP, Peltzer G, Li Z (2014a) Improving InSAR geodesy using global atmospheric models. J Geophys Res Solid Earth 119(3):2324–2341. https://doi.org/10.1002/2013JB010588

Jolivet R, Duputel Z, Riel B, Simons M, Rivera L, Minson SE, Zhang H et al (2014b) The 2013 Mw 7.7 Balochistan earthquake: seismic potential of an accretionary wedge. Bull Seismol Soc Am 104(2):1020–1030. https://doi.org/10.1785/0120130313

Jónsson S, Zebker H, Segall P, Amelung F (2002) Fault slip distribution of the 1999 Mw 7.1 Hector Mine, California, earthquake, estimated from satellite radar and GPS measurements. Bull Seismol Soc Am 92(4):1377–1389. https://doi.org/10.1785/0120000922

Jónsson S, Segall P, Pedersen R, Björnsson G (2003) Post-earthquake ground movements correlated to pore-pressure transients. Nature 424(6945):179–183. https://doi.org/10.1038/nature01776

Kagan YY, Jackson DD (2013) Tohoku earthquake: a surprise? Bull Seismol Soc Am 103(2B):1181–1194. https://doi.org/10.1785/0120120110

Karasözen E, Nissen E, Büyükakpınar P, Cambaz MD, Kahraman M, Kalkan Ertan E et al (2018) The 2017 July 20 Mw 6.6 Bodrum–Kos earthquake illuminates active faulting in the Gulf of Gökova, SW Turkey. Geophys J Int 214(1):185–199. https://doi.org/10.1093/gji/ggy114

Kargel JS, Leonard GJ, Shugar DH, Haritashya UK, Bevington A, Fielding EJ et al (2016). Geomorphic and geologic controls of geohazards induced by Nepal's 2015 Gorkha earthquake. Sci 351(6269):aac8353

Klinger Y, Michel R, King GCP (2006) Evidence for an earthquake barrier model from Mw 7.8 Kokoxili (Tibet) earthquake slip-distribution. Earth Planet Sci Lett 242(3–4):354–364. https://doi.org/10.1016/j.epsl.2005.12.003

Klinger Y, Okubo K, Vallage A, Champenois J, Delorme A, Rougier E, Lei Z, Knight EE, Munjiza A, Satriano C, Baize S, Langridge R, Bhat HS (2018) Earthquake damage patterns resolve complex rupture processes. Geophys Res Lett 45:10279–10287. https://doi.org/10.1029/2018GL078842

Konca AO, Leprince S, Avouac J-P, Helmberger DV (2010) Rupture process of the 1999 Mw 7.1 Duzce earthquake from joint analysisof SPOT, GPS, InSAR, strong-motion, and teleseismic data: a supershear rupture with variable rupture velocity. Bull Seismol Soc Am 100(1):267–288. https://doi.org/10.1785/0120090072

Krieger G, Moreira A, Fiedler H, Hajnsek I, Werner M, Younis M, Zink M (2007) TanDEM-X: a satellite formation for high-resolution SAR interferometry. IEEE Trans Geosci Remote Sens 45(11):3317–3341. https://doi.org/10.1109/TGRS.2007.900693

Krishnan S, Crosby C, Nandigam V, Phan M, Cowart C, Baru C, Arrowsmith R (2011) OpenTopography: a services oriented architecture for community access to LIDAR topography. In: Proceedings of the 2nd international conference on computing for geospatial research and applications, pp 1–8. https://doi.org/10.1145/1999320.1999327

Lay T, Kanamori H, Ammon CJ, Nettles M, Ward SN, Aster RC et al (2005) The great Sumatra-Andaman earthquake of 26 December 2004. Science 308(5725):1127–1133. https://doi.org/10.1126/science.1112250

Leprince S, Barbot S, Ayoub F, Avouac JP (2007) Automatic and precise orthorectification, coregistration, and subpixel correlation of satellite images, application to ground deformation measurements. IEEE Transact Geosci Remote Sens 45(6):1529–1558. https://doi.org/10.1109/TGRS.2006.888937

Li Z, Fielding EJ, Cross P (2009) Integration of InSAR time-series analysis and water-vapor correction for mapping postseismic motion after the 2003 Bam (Iran) earthquake. IEEE Trans Geosci Remote Sens 47(9):3220–3230. https://doi.org/10.1109/TGRS.2009.2019125

Lindsey EO, Natsuaki R, Xu X, Shimada M, Hashimoto M, Melgar D, Sandwell DT (2015) Line-of-sight displacement from ALOS-2 interferometry: Mw 7.8 Gorkha Earthquake and Mw 7.3 aftershock. Geophys Res Lett 42(16):6655–6661. https://doi.org/10.1002/2015GL065385

Lohman RB, Simons M (2005) Locations of selected small earthquakes in the Zagros mountains. Geochem Geophys Geosyst 6(3). https://doi.org/10.1029/2004GC000849

Marchandon M, Vergnolle M, Sudhaus H, Cavalie O (2018) Fault geometry and slip distribution at depth of the 1997 Mw 7.2 Zirkuh earthquake: Contribution of near-field displacement data. J Geophys Res Solid Earth 123:1904–1924. https://doi.org/10.1002/2017JB014703

Maubant L, Pathier E, Daout S, Radiguet M, Doin MP, Kazachkina E ct al (2020) Independent component analysis and parametric approach for source separation in InSAR time series at regional scale: application to the 2017–2018 Slow Slip Event in Guerrero (Mexico). J Geophys Res Solid Earth 125(3):e2019JB018187. https://doi.org/10.1029/2019JB018187

Métois M, Socquet A, Vigny C (2012) Interseismic coupling, segmentation and mechanical behavior of the central Chile subduction zone. J Geophys Res Solid Earth 117(B3). https://doi.org/10.1029/2011JB008736

Michel R (1997) Les mesures de mouvements par imagerie sar et leur exploitation en glaciologie et en sismotectonique. Thèse de doctorat en Terre, océan, espace, sous la direction de Jean Taboury. Soutenue en 1997 à l'Université Paris 11

Michel R, Avouac J-P (2002) Deformation due to the 17 August 1999 Izmit, Turkey, earthquake measured from SPOT images. J Geophys Res. https://doi.org/10.1029/2000jb000102

Michel R, Avouac JP (2006) Coseismic surface deformation from air photos: the Kickapoo step over in the 1992 Landers rupture. J Geophys Res Solid Earth 111(B3). https://doi.org/10.1029/2005JB003776

  Springer

Michel R, Avouac J-P, Taboury J (1999) Measuring ground dis-placements from SAR amplitude images: application to the Landers earthquake. Geophys Res Lett 26:875–878. https://doi.org/10.1029/1999G L900138

Michel R, Ampuero JP, Avouac JP, Lapusta N, Leprince S, Redding DC et al (2012) A geostationary optical seismometer, proof of concept. IEEE Trans Geosci Remote Sens 51(1):695–703. https://doi.org/10.1109/TGRS.2012.2201487

Milliner CWD, Dolan JF, Hollingsworth J, Leprince S, Ayoub F et al (2015) Quantifying near-field and off-fault deformation patterns of the 1992 Mw 7.3 Landers earthquake. Geochem Geophys Geosyst 16(5):1577–1598. https://doi.org/10.1002/2014GC005693

Minson SE, Simons M, Beck JL (2013) Bayesian inversion for finite fault earthquake source models I—Theory and algorithm. Geophys J Int 194(3):1701–1726. https://doi.org/10.1093/gji/ggt180

Mishra S, Bartakke V, Athavale G, Akkiraju VV, Goswami D, Roy S (2017) Granite-gneiss Basement below Deccan Traps in the Koyna Region, Western India: outcome from Scientific Drilling. J Geol Soc India 90(6):776–787. https://doi.org/10.1007/s12594-017-0790-9

Morishita Y (2019) A systematic study of synthetic aperture radar interferograms produced from ALOS-2 data for large global earthquakes from 2014 to 2016. IEEE J Sel Topics Appl Earth Obs Remote Sens 12(7):2397–2408. https://doi.org/10.1109/JSTARS.2019.2921664

Morishita Y, Lazecky M, Wright TJ, Weiss JR, Elliott JR, Hooper A (2020) LiCSBAS: an open-source InSAR time series analysis package integrated with the LiCSAR automated sentinel-1 InSAR processor. Remote Sens 12(3):424. https://doi.org/10.3390/rs12030424

Noh MJ, Howat IM (2015) Automated stereo-photogrammetric DEM generation at high latitudes: surface Extraction with TIN-based Search-space Minimization (SETSM) validation and demonstration over glaciated regions. GISci Remote Sens 52(2):198–217. https://doi.org/10.1080/15481603.2015.10086 21

Okada Y (1985) Surface deformation due to shear and tensile faults in a half-space. Bull Seismol Soc Am 75:1135–1154

Pathier E, Fielding EJ, Wright TJ, Walker R, Parsons BE, Hensley S (2006) Displacement field and slip distribution of the 2005 Kashmir earthquake form SAR imagery. Geophys Res Lett. https://doi.org/10.1029/2006GL027193

Peltzer, G., Crampé, F., & Rosen, P. (2001). The Mw 7.1, Hector Mine, California earthquake: surface rupture, surface displacement field, and faultslip solution from ERS SAR data. Comptes Rendus de l'Académie des Sci-Ser IIA-Earth Planet Sci 333(9), 545-555.https://doi.org/10.1016/S1251 -8050(01)01658-5

Pollitz FF, Peltzer G, Bürgmann R (2000) Mobility of continental mantle: evidence from postseismic geodetic observations following the 1992 Landers earthquake. J Geophys Res Solid Earth 105(B4):8035–8054. https://doi.org/10.1029/1999JB900380

Prentice CS, Crosby CJ, Whitehill CS, Arrowsmith JR, Furlong KP, Phillips DA (2009) Illuminating Northern California's active faults. EOS Trans Am Geophys Union 90(7):55. https://doi.org/10.1029/2009E O070002

Roy DP, Wulder MA, Loveland TR, Woodcock CE, Allen RG, Anderson MC, Scambos TA (2014) Landsat-8: Science and product vision for terrestrial global change research. Remote sensing of Environment 145:154–172. https://www.sciencedirect.com/science/article/pii/S003442571400042X

Salvi S, Stramondo S, Funning GJ, Ferretti A, Sarti F, Mouratidis A (2012) The Sentinel-1 mission for the improvement of the scientific understanding and the operational monitoring of the seismic cycle. Remote Sens Environ 120:164–174. https://doi.org/10.1016/j.rse.2011.09.029

Savage JC, Prescott WH (1978) Asthenosphere readjustment and the earthquake cycle. J Geophys Res Solid Earth 83(B7):3369–3376. https://doi.org/10.1029/JB083iB07p03369

Scott C, Champenois J, Klinger Y, Nissen E, Maruyama T, Chiba T, Arrowsmith R (2019) The 2016 M7 Kumamoto, Japan, earthquake slip field derived from a joint inversion of differential Lidar topography, optical correlation, and InSAR surface displacements. Geophys Res Lett, 46. https://doi.org/10.1029/2019GL082202

Shirzaei M, Bürgmann R, Fielding EJ (2017) Applicability of Sentinel-1 terrain observation by progressive scans multitemporal interferometry for monitoring slow ground motions in the San Francisco Bay Area. Geophys Res Lett 44(6):2733–2742. https://doi.org/10.1002/2017GL072663

Simons M, Fialko Y, Rivera L (2002) Coseismic deformation from the 1999 Mw 7.1 Hector Mine, California, earthquake as inferred from InSAR and GPS observations. Bull Seismol Soc Am 92(4):1390–1402. https://doi.org/10.1785/0120000933

Storchak DA, Di Giacomo D, Bondár I, Engdahl ER, Harris J, Lee WH et al (2013) Public release of the ISC–GEM global instrumental earthquake catalogue (1900–2009). Seismol Res Lett 84(5):810–815. https://doi.org/10.1785/0220130034

Styron R, Taylor M, Okoronkwo K (2010) Database of active structures from the Indo-Asian collision. EOS Trans Am Geophys Union 91(20):181–182. https://doi.org/10.1029/2010EO200001

Taylor M, Yin A (2009) Active structures of the Himalayan-Tibetan orogen and their relationships to earthquake distribution, contemporary strain field, and Cenozoic volcanism. Geosphere 5(3):199–214. https://doi.org/10.1130/GES00217.1

Taylor MH, Leprince S, Avouac JP, Sieh K (2008) Detecting co-seismic displacements in glaciated regions: an example from the great November 2002 Denali earthquake using SPOT horizontal offsets. Earth Planet Sci Lett 270(3–4):209–220. https://doi.org/10.1016/j.epsl.2008.03.028

Taymaz T, Price S (1992) The 1971 May 12 Burdur earthquake sequence, SW Turkey: a synthesis of seismological and geological observations. Geophys J Int 108(2):589–603. https://doi.org/10.1111/j.1365-246X.1992.tb04638.x

Tian Y, Liu-Zeng J, Luo Y, Li Y, Zhang J (2018) Deformation Related to an M < 5 Earthquake Sequence on Xiangyang Lake-Burog Co Fault in Central Xizang, China, Observed by Sentinel-1 Data. Bull Seismol Soc Am 108(6):3248–3259. https://doi.org/10.1785/0120180066

Torres R, Snoeij P, Geudtner D, Bibby D, Davidson M, Attema E et al (2012) GMES Sentinel-1 mission. Remote Sens Environ 120:9–24. https://doi.org/10.1016/j.rse.2011.05.028

USGS (2019) Earthquake Hazards Program, Latest Earthquakes Catalogue Event Page. [Online]. Accessed 28 May 2020. https://earthquake.usgs.gov/earthquakes/eventpage/us1000jj23/executive

Vallage A, Klinger Y, Grandin R, Bhat HS, Pierrot-Deseilligny M (2015) Inelastic surface deformation during the 2013 Mw 7.7 Balochistan, Pakistan, earthquake. Geology 43:1079–1082. https://doi.org/10.1130/G37290.1

Vallage A, Klinger Y, Lacassin R, Delorme A, Pierrot-Deseilligny M (2016) Geological structures control on earthquake ruptures: the Mw 7.7, 2013 Balochistan earthquake. Geophys Res Lett, Pakistan. https://doi.org/10.1002/2016gl070418

Van Puymbroeck N, Michel R, Binet R, Avouac JP, Taboury J (2000) Measuring earthquakes from optical satellite images. Appl Opt 39:3486–3494. https://doi.org/10.1364/AO.39.003486

Watanabe SI, Sato M, Fujita M, Ishikawa T, Yokota Y, Ujihara N, Asada A (2014) Evidence of viscoelastic deformation following the 2011 Tohoku-Oki earthquake revealed from seafloor geodetic observation. Geophys Res Lett 41(16):5789–5796. https://doi.org/10.1002/2014GL061134

Wei S, Fielding E, Leprince S, Sladen A, Avouac JP, Helmberger D et al (2011) Superficial simplicity of the 2010 El Mayor-Cucapah earthquake of Baja California in Mexico. Nat Geosci 4(9):615. https://doi.org/10.1038/ngeo1213

Woessner J, Laurentiu D, Giardini D, Crowley H, Cotton F, Grünthal G et al (2015) The 2013 European seismic hazard model: key components and results. Bull Earthq Eng 13(12):3553–3596. https://doi.org/10.1007/s10518-015-9795-1

Wright TJ (2002) Remote monitoring of the earthquake cycle using satellite radar interferometry. Philos Trans R Soc Lond Ser A Math Phys Eng Sci 360(1801):2873–2888. https://doi.org/10.1098/rsta.2002.1094

Wright TJ, Parsons BE, Jackson JA, Haynes M, Fielding EJ, England PC, Clarke PJ (1999) Source parameters of the 1 October 1995 Dinar (Turkey) earthquake from SAR interferometry and seismic bodywave modelling. Earth Planet Sci Lett 172(1–2):23–37. https://doi.org/10.1016/S0012-821X(99)00186-7

Wright TJ, Elliott JR, Wang H, Ryder I (2013) Earthquake cycle deformation and the Moho: implications for the rheology of continental lithosphere. Tectonophysics 609:504–523. https://doi.org/10.1016/j.tecto.2013.07.029

Xu X, Tong X, Sandwell DT, Milliner CWD, Dolan JF, Hollingsworth J et al (2016) Refining the shallow slip deficit. Geophys J Int 204(3):1867–1886. https://doi.org/10.1093/gji/ggv563

Xu X, Sandwell DT, Smith-Konter B (2020) Coseismic displacements and surface fractures from sentinel-1 InSAR: 2019 Ridgecrest Earthquakes. Seismol Res Lett 91:1979–1985. https://doi.org/10.1785/0220190275

Weiss JR, Walters RJ, Morishita, Y, Wright T, Lazecky, M, Wang H et al (2020) High-resolution surface velocities and strain for Anatolia from Sentinel-1 InSAR and GNSS data, Geophys Res Lett, e2020GL087376. https://doi.org/10.1029/2020GL087376

Yan Y, Pinel V, Trouvé E, Pathier E, Perrin J, Bascou P, Jouanne F (2013) Coseismic displacement field and slip distribution of the 2005 Kashmir earthquake from SAR amplitude image correlation and differential interferometry. Geophys J Int. https://doi.org/10.1093/gji/ggs102

Yang J, Xu C, Wang S, Wang X (2020) Sentinel-1 observation of 2019 Mw 5.7 Acipayam earthquake: a blind normal-faulting event in the Acipayam basin, southwestern Turkey. J Geodyn 135:101707. https://doi.org/10.1016/j.jog.2020.101707

Yu C, Li Z, Penna NT, Crippa P (2018) Generic atmospheric correction model for Interferometric Synthetic Aperture Radar observations. J Geophys Res Solid Earth 123(10):9202–9222. https://doi.org/10.1029/2017JB015305

Zhou Y, Parsons B, Elliott JR, Barisin I, Walker RT (2015) Assessing the ability of Pleiades stereo imagery to determine height changes in earthquakes: a case study for the El Mayor-Cucapah epicentral area. J Geophys Res Solid Earth 120(12):8793–8808. https://doi.org/10.1002/2015JB012358

Zhou Y, Walker RT, Hollingsworth J, Talebian M, Song X, Parsons B (2016) Coseismic and postseismic displacements from the 1978 Mw 7.3 Tabas-e-Golshan earthquake in eastern Iran. Earth Planet Sci Lett 452:185–196. https://doi.org/10.1016/j.epsl.2016.07.038

Zielke O, Arrowsmith JR (2012) LaDiCaoz and LiDARimager—MATLAB GUIs for LiDAR data handling and lateral displacement measurement. Geosphere 8(1):206–221. https://doi.org/10.1130/GES00686.1

Zinke R, Hollingsworth J, Dolan JF, Van Dissen R (2019) Three-dimensional surface deformation in the 2016 MW 7.8 Kaikōura, New Zealand, earthquake from optical image correlation: implications for strain localization and long-term evolution of the Pacific-Australian plate boundary. Geochem Geophys Geosyst 20:1609–1628. https://doi.org/10.1029/2018GC007951

Publisher's Note Springer Nature remains neutral with regard to jurisdictional claims in published maps and institutional affiliations.

Surveys in Geophysics (2020) 41:1391–1435
https://doi.org/10.1007/s10712-020-09609-1

Remote Sensing for Assessing Landslides and Associated Hazards

Candide Lissak[1] · Annett Bartsch[2,3] · Marcello De Michele[4] · Christopher Gomez[5,6] · Olivier Maquaire[1] · Daniel Raucoules[4] · Thomas Roulland[1]

Received: 6 December 2019 / Accepted: 6 August 2020 / Published online: 3 September 2020
© Springer Nature B.V. 2020

Abstract

Multi-platform remote sensing using space-, airborne and ground-based sensors has become essential tools for landslide assessment and disaster-risk prevention. Over the last 30 years, the multiplicity of Earth Observation satellites mission ensures uninterrupted optical and radar imagery archives. With the popularization of Unmanned Aerial Vehicles, free optical and radar imagery with high revisiting time, ground and aerial possibilities to perform high-resolution 3D point clouds and derived digital elevation models, it can make it difficult to choose the appropriate method for risk assessment. The aim of this paper is to review the mainstream remote-sensing methods commonly employed for landslide assessment, as well as processing. The purpose is to understand how remote-sensing techniques can be useful for landslide hazard detection and monitoring taking into consideration several constraints such as field location or costs of surveys. First we focus on the suitability of terrestrial, aerial and spaceborne systems that have been widely used for landslide assessment to underline their benefits and drawbacks for data acquisition, processing and interpretation. Several examples of application are presented such as Interferometry Synthetic Aperture Radar (InSAR), lasergrammetry, Terrestrial Optical Photogrammetry. Some of these techniques are unsuitable for slow moving landslides, others limited to large areas and others to local investigations. It can be complicated to select the most appropriate system. Today, the key for understanding landslides is the complementarity of methods and the automation of the data processing. All the mentioned approaches can be coupled (from field monitoring to satellite images analysis) to improve risk management, and the real challenge is to improve automatic solution for landslide recognition and monitoring for the implementation of near real-time emergency systems.

Keywords Landslide · Remote sensing · Machine learning · Monitoring · Inventory

---

✉ Candide Lissak
candide.lissak@unicaen.fr

Extended author information available on the last page of the article

1 Introduction

Despite several decades of research, landslide detection is still a challenging task due to the wide variety of sizes, shapes and morphologies that those events can take and due to the variability of the area they trigger. Consequently, a broad array of methods has been tested in remote-sensing with lately combinations of those at different-scales using multi-platform methodologies. For decades, Remote-Sensing techniques are thus widely employed for landslides studies (e.g., Petley et al. 2002; Delacourt et al. 2007; Jaboyedoff et al. 2012, 2019; Tofani et al. 2013; Casagli et al. 2017; Huang and Zhao 2018). The appeal of multi-platform remote-sensing—from space, airborne- to ground-based sensors—originates in the possibility to cater for the difficulties of various contexts (coastal landslides, mountain debris flows, rockfalls and mudflows in periglacial environments…). Furthermore, the possibility of complementary techniques for data acquisition in various environments is a major asset for landslide studies with real-time and near real-time data acquisition for virtually any place in the world, especially with the growing demand for detailed and accurate landslide maps and inventories around the globe (Ghorbanzadeh et al. 2019). Among all available methods and data, however, it can be difficult to select the most appropriate approach for a given study. And the choice of the most appropriate approach can depend on the accessibility of study areas, its geographical context, orientation (especially for the satellites), the vegetation cover, the landslide velocity, its size, and other morphological and geographic parameters.

From those technological and methodological developments, the question of how to choose the most effective method then arises. The objective of the present paper is therefore to define different ways of studying landslides with two main objectives: hazard inventory/mapping and the quantification of surface deformations using a combination of several remote-sensing methods. Inherited from the era of landslide inventories construction, the definition of the geometry of landslides and their change over time still dominates one of the foci of remote sensing applied for landslides (Jaboyedoff et al. 2012; Wasowski and Bovenga 2014; Telling et al. 2017; Huang and Zhao 2018). As a contribution to this tradition, this paper is focused on the main spaceborne, aerial and terrestrial remote-sensing methods currently used for mass movement surveys in various geographical contexts and scales, for landslides that have already occurred. Within this field of research, the authors investigated the multiple platforms that come with multiple sensors—optical imagery (3 bands), multispectral imagery (4–12 bands) or LiDAR and Radar imagery, each providing different yet complementary data.

2 The Range of Remote-Sensing Platforms for Landslide Detection

2.1 Suitable Systems for Numerous Scientific Purposes

The main asset of remote-sensing techniques is the variety of applications, and so in the various contexts of landslide observations. Remote sensing is applicable from steep-slopes landslides to sub-horizontal deformation; from extremely rapid to slow movements; in saturated and unsaturated materials; in confined steep-channel to open-slope landslides (Hungr et al. 2014). Another source of variability originates from the environment landslides and the related hazards occur. Among the case studies presented in this contribution, we can find landslides in periglacial environments, which are characterized by moderate slope terrain in morainal material where landslides are associated to the presence of ground ice, permafrost conditioned by freeze/thaw cycling (Lewkowicz 2007; Jorgenson and Grosse 2016; Lewkowicz and Way 2019). In such context, unconsolidated sediments currently frozen can be easily mobilized under exceptionally warm conditions, when the depth of the seasonal thaw layer exceeds normal conditions in such years (Bartsch et al. 2019). The analysis of such landslides in permafrost notably is of particular importance in the Arctic, as landslides are proxies to understand the carbon cycle, as it has been demonstrated that carbon-rich landslides can even contribute to ocean acidification (Zolkos et al. 2019). Other examples have been given in subtropical areas where rainstorm disasters, cyclone and earthquake, or the combination of triggering factors (Chigira et al. 2004, 2010; Ingles et al. 2006; Yin et al. 2009; Xu 2015; Shafique et al. 2016; Marc et al. 2017; Ko and Lo 2018) can induce the concomitant reduction of effective shear-strength and trigger hundreds or thousands of simultaneous landslide (Chigira et al. 2004; Huang and Li 2009; Yin et al. 2009). This kind of event disturbs the sediment budget with transfer of thousands of cubic meters of sediment into highly urbanized watersheds. Finally, coastal landslides have a particular significance in the light of climate change, rapid source-to-sink concepts and related hazards. It is therefore essential to quantify the regressive dynamics of coastal cliffs and slopes induced by the sea erosion as a predominant parameter of slope stability (Letortu et al. 2015a, b) notably in association with groundwater flows (Lissak et al. 2014).

One of the main drivers of landslide observation is hazard and disaster-risk assessment and management, for which it is essential to define the spatial and temporal evolution of landslides, especially when instabilities occur close to settlements and infrastructures (e.g., roads, bridges…) and disrupting ecosystems. Also, a single remote-sensing approach can be effective to answer several scientific questions (e.g., the use of laser scanner in Lissak et al. 2014 or Letortu et al. 2019 to investigate landslide morphology, hazard mapping, deformations measurement or multitemporal satellite images for landslide detection and deformation monitoring).

Remote-sensing techniques for landslide assessment can be classified according to the scale of observation, ranging from spaceborne platforms—single or swarms of satellites—, airborne platforms—airplanes and drones—, to ground-based and close-range platforms—terrestrial laser-scanners hand-held cameras. The plurality of systems for Earth observation (ground, aerial and satellite-based) is now so widely developed that it is possible to assess landslide virtually anywhere in the world, at any frequency, to quantify the seasonality of the kinematics (Delacourt et al. 2007) as well as the long-term patterns of surface motion with image correlation techniques. Researchers and practitioners have thus immediate access to imagery after a disaster for instance (Proy et al. 2013).

2.2 Satellite Systems

Over the last 30 years, the multiplicity of Earth Observation (EO) satellites mission ensures uninterrupted optical imagery archives (e.g., Landsat 1–8 ~ 1972, SPOT 1–7 ~ 1986, Ikonos 1999/2015, RapidEye ~ 2008, Sentinel ~ 2014), and radar images acquisition (e.g., ERS 1991/2001, JERS 1992/1998, Envisat 2002/2012, TerraSAR-X ~ 2008, Sentinel-1 ~ 2014). Earth observation satellites (Table 1) are largely used for crisis management (Voigt et al. 2016; Lang et al. 2018) and especially for landslide investigation and related issues (e.g., hazard identification, spatial extension delineation, volume estimation, displacement measurement in Table 1). The interest for satellite imagery can be explained by the availability of many open-source data with high-resolution images and regular information updates.

Table 1 Non-exhaustive list of various study cases of landslide assessment and satellite systems employed

Satellite	Bands	Radar	Spectral (P/MS) spatial (m) resolution	Launch date	Revisiting time (days)	Displacement field	Land use or cover/band ratio for landslide susceptibility or detection	Landslide inventory
ALOS	1 P/4 MS/SAR- L	X	P 2,5/MS 10/100	2006–2011	46	de Michele et al. (2010)		Xu et al. (2014), Casagli et al. (2016)
ALOS-2	1 P/4 MS/SAR- L	X	P 2,5/MS 10/100	~2014	14	Fan et al. (2017)		
ENVISAT		X			35	Casagli et al. (2016), Singleton et al. (2014), de Michele et al. (2010)		
ERS 1/2		X			35	Michel et al. (1999), Casagli et al. (2016)		Roessner et al. (2001)
GeoEye-1	1 P/4 MS		P 0,46/MS 1,84	~2008	<3			Stumpf and Kerle (2011a, b), Casagli et al. (2016), Moosavi et al. (2014)
JERS		X	18	1992–1998	44	Delacourt et al. (2009)		

216

Springer

Table 1 (continued)

Satellite	Bands	Radar	Spectral (P/MS) spatial (m) resolution	Launch date	Revisiting time (days)	Displacement field	Land use or cover/band ratio for landslide susceptibility or detection	Landslide inventory
Ikonos	1 P/4 MS		P 0,82/MS 3,28	1999–2015	3			Petley et al. (2002), Nichol and Wong (2005a, b), Stumpf and Kerle (2011a, b), Xu et al. (2014), Casagli et al. (2016)
Landsat 5	P 1/MS 7			~1984	16			Yang and Chen (2010), Lewkowicz and Way (2019)
Landsat 7	1 P/7 MS		P 15/MS 30	~1999	16		Hong et al. (2016), Mohammady et al. (2019), Fang et al. (2020),	Petley et al. (2002), Lewkowicz and Way (2019)
Landsat 8	8 P/11 MS		P 15/MS 30	~2013	16	Lacroix et al. (2019)	Kalantar et al. (2018), Arabameri et al. (2019), Bui et al. (2019, 2020), Achour and Pourghasemi 2019), Hu et al. (2020),	Huang and Zhao (2018), Huang et al. (2020), Lewkowicz and Way (2019)

Table 1 (continued)

Satellite	Bands	Radar	Spectral (P/MS) spatial (m) resolution	Launch date	Revisiting time (days)	Displacement field	Land use or cover/band ratio for landslide susceptibility or detection	Landslide inventory
Pléiades 1A/1B	1 P/4 MS		P 0,5/MS 2	~2011/~2012	1	Lacroix et al. (2015), Stumpf et al. (2017)		Fan et al. (2018), Zhou et al. (2018)
QuickBird-2	1 P/4 MS		P 0,65/MS 2,62	2001–2015	1–3.5	Jiang et al. (2016)		Zhang et al. (2014), Ciampalini et al. (2015)
RADARSAT1/2/RMC		X		1995/2007/2016/2017	12	Delacourt et al. (2009), Casagli et al. (2016), Paquette et al. (2020)		
RapidEye	1 P/4 MS		5–6.5 m	~2008	5.5		Ghorbanzadeh et al. (2019)	Behling et al. (2014), Kurtz et al. (2014)
Sentinel 1A/1B	SAR- C	X		~2014/~2016	12–6	Dai et al. (2016), Fan et al. (2017), Carlà et al. (2019)		
Sentinel 2A/2B	13 MS		MS 10–60	~2015/~2016	10–5	Lacroix et al. (2018)		Svennevig (2019)

Table 1 (continued)

Satellite	Bands	Radar	Spectral (P/MS) spatial (m) resolution	Launch date	Revisiting time (days)	Displacement field	Land use or cover/band ratio for landslide susceptibility or detection	Landslide inventory
SPOT 1 to SPOT 3	1 P/3MS		P 10/MS 20	1986–2003 / 1990–2009 / 1983–1996	1–4 OFF nadir/26 repeat cycle	Crippen (1992), Van Puymbroeck et al. (2000), Yamaguchi et al. (2003), Delacourt et al. (2004), Nichol and Wong (2005a, b)		
SPOT 4			P 10/MS 20	1998–2013	5	Klinger et al. (2006)		
SPOT 5			P 5/MS 10	~2002	2–3	Le Bivic et al. (2017)	Yang et al. (2013)	Sato and Harp (2009), Yang et al. (2013), Xu et al. (2014), Casagli et al. (2016), Fan et al. (2018)
SPOT 6	1 P/4 MS		P 1,5/MS 6	~2012	1			Fan et al. (2018)
Terra (EOS AM-1)	14 MS		MS VNIR 15 SWIR 30 TIR 90	~1999	16		Yang et al. (2013)	Yang and Chen (2010), Yang et al. (2013), Xu et al. (2014)

Table 1 (continued)

Satellite	Bands	Radar	Spectral (P/MS) spatial (m) resolution	Launch date	Revisiting time (days)	Displacement field	Land use or cover/band ratio for landslide susceptibility or detection	Landslide inventory
TerraSAR-X		X		~2007	2,5–11	Li et al. (2011), Raucoules et al. (2013), Singleton et al. (2014), Casagli et al. (2016), Zwieback et al. (2018)	Stettner et al. (2018)	
WorldView-2	1 P/8 MS		P 0,46/MS 2,4	~2009	1	Jiang et al. (2016)		Zhang et al. (2014), Fan et al. (2018)

Recent programs such as Copernicus Sentinel missions or USGS/NASA Landsat Program provide free optical and radar imagery with high revisit time (Table 1). Among all existing satellite systems, many are those which propose daily revisit capability. But in a context of risk management and emergency response, it is also essential that some systems (such as Pleiades systems) allow last-minute request for images acquisition in order to precision mapping and intervention.

Spatial and spectral resolutions of satellite systems are various (e.g., Sentinel 2 provides 13 spectral bands including 3 bands for atmospheric correction, and a spatial resolution from 10 to 60 m, Table 1). Consequently, the selection of the most suitable system for landslide assessment will highly depend on the scientific purpose and the required scale analysis. For example, for inventory and landslide detection it is more suitable to use Very High-Resolution (VHR, 0–5 m) and High-resolution (5–20 m) optical data (Table 1) with the possibility of Pan-sharpening (Nichol and Wong 2005b) or to improve historical and recent satellite images with the use of super-resolution algorithms (Lanaras et al. 2018) and VHR orthoimages. But, for an exhaustive inventory, it is sometimes necessary to combine sources of images because of the incomplete spatial coverage (Shafique et al. 2016), multi-temporal images (Fan et al. 2018) and historical inventories (Catani et al. 2005; Ardizzone et al. 2007; Arabameri et al. 2019; Pánek et al. 2019). The interest of satellite images time series is to detect examples of past landslides which may be remodelled by anthropogenic action or progressively hidden by vegetation. Numerous archives are available, and several computing online platforms such as Google Earth Engine (https://earthengine.googl e.com/), Google Earth, EOS platform for Earth Observation imagery (https://eos.com/platform/) or Sentinel Hub (https://www.sentinel-hub.com) are free to use for the visualization and running simple radiometric analyses of various remote-sensing data (Sato and Harp 2009; Yang and Chen 2010; Pham et al. 2019; Fang et al. 2020; Hu et al. 2020). Radar images are also viewable on these platforms and widely used for landslide monitoring. Indeed, to detect and quantify surface deformation, Synthetic Aperture Radar (SAR) with an interferometric approach (i.e., spaceborne InSAR and Ground-Based InSAR, GB-InSAR), or a non-interferometric approach with image matching (GBSAR), are more useful than an analysis of optical images with 3 bands, or multispectral images from 4 to 12 bands. But the feasibility of this depends highly on the orientation and size of the landslide. In complement, offsets by correlation of both panchromatic and radar-amplitude images are also commonly used technique for the measurement of surface deformation (Crippen 1992; Michel et al. 1999; Van Puymbroeck et al. 2000) complementary to InSAR (Klinger et al. 2006; de Michele and Briole 2007; de Michele et al. 2010).

2.3 Airborne Systems

Airborne systems can complement spaceborne techniques. They can be based on a range of platforms and sensors (Red–Green–Blue, multispectral sensors, radar, thermal…). The most affordable and flexible ones are balloons, blimps balloon and small Unmanned Aerial Vehicles (UAVs). UAVs are widely used for landslide studies (Rau et al. 2011, Niethammer et al. 2012). If we combine the two key words ('UAV' and 'landslide') in Google Scholar, more than 5000 items are identified since 2016. Less expensive than manned aircraft (ultralight trikes, helicopters, planes), these techniques provide high-resolution measurements of landslides. Two types of data from airborne surveys are frequently used for landslide assessment:

(1)   Aerial photographs and orthoimages in the visible domain (RGB). These are essential for landslide detection especially for a historical reconstruction of the slope deformation using aerial images time series. The detection of landslides is based on the potential visibility of specific morphological features where the vegetation cover is sparse (e.g., major and minor scarps, hummocks…),

(2)   3D models with point clouds, 3D meshes, Digital Elevation Models (DEM)/Digital Terrain Model (DTM) and Digital Surface Model (DSM). DEMs (the generic term DEM will be retained for the paper) are exploited quasi-systematically for morphological and topographical analysis. DEMs are essential for morphological analyses, landslides detection (van Westen et al. 2008) and deformation quantification (Casson et al. 2005). Especially if we consider that landslide distribution varies considerably with the slope aspect and value (Chen et al. 2014).

Airborne-based LiDAR (Light Detection And Ranging), also mentioned Airborne Laser Scanning (ALS), provides very high-resolution data with several million georeferenced 3D point clouds and high-resolution DEM reconstruction (centimetric/decimetric). For almost 20 years, airborne LiDAR technology offers several application possibilities for landslide investigation presented in Jaboyedoff et al. (2012). Today, there are more than 10 000 references in Google Scholar since 2016. This technology provides major information on topography, especially with the full-waveform laser scanning systems that are able to record the entire emitted and backscattered signal of each laser pulse even for vegetated areas with a high-point density possibility throughout vegetation. The theoretical principles of full-waveform LiDAR are presented in Mallet and Bretar (2009). This technology has highly advanced the accuracy of landslide inventory maps (Schulz 2004; Ardizzone et al. 2007), the monitoring of surface displacement, and provides essential data for landslide susceptibility (i.e., DEM derivatives with slope, surface roughness, curvature calculation in Van Den Eeckhaut et al. 2012). The gain of information by LiDAR also concerns detailed morphological features investigation at the sub-meter scale (Lissak et al. 2014; Bunn et al. 2019). But exploration of this kind of Very High-Resolution (VHR) data can be limited by the cost of surveys. Consequently, UAVs can be an alternative to acquire high-resolution data (3D point cloud, DEM and orthophotos). But once again, it depends on the parameter studied that can be too large to be overflown by UAV and by boat-based mobile laser scanning (Michoud et al. 2014).

2.4 Ground Systems

Ground techniques can be complementary to other techniques (Stumpf et al. 2015; Wilkinson et al. 2016) mentioned above. Because airborne data acquisition depends on flying conditions, and/or can be expensive, spaceborne data availability depends on slope orientation and on the temporal/spatial resolution. Three main categories of ground-based remote-sensing techniques are used in landslide monitoring: 1) Terrestrial Optical Photogrammetry (TOP), 2) Terrestrial Laser Scanning (TLS), 3) Ground-Based Synthetic Aperture Radar Interferometry (GB-InSAR).

2.4.1 Terrestrial Optical Photogrammetry (TOP) Technique

This technique is reputed in geoscience for the 3D-textured restitution, but also for the construction of high-resolution DEMs at high spatial and temporal resolution (centimeter to sub-decimetric accuracy). Due to the fact this technique is simple to use and can be easily repeated close in time, TOP with SfM (Structure from Motion) technique has been increasingly used in recent years (Abellan et al. 2016; James et al. 2019). SfM photogrammetry is based on an improved principle of stereoscopy, i.e., the reproduction of a relief perception from two flat images in the same way as human vision. The objective is to model a real object or environment in 3D from a multitude of 2D images using algorithms that have ability to detect and identify similar elements between two pictures (i.e., "Scale Invariant Feature Transform" in Lowe 1999, 2004). Through a number of photographs, algorithms will use common pixels of each picture to reconstruct the geometry of the object which must be modelled.

Several publications highlight the application of this technique to study the soil erosion (Gudino-Elizondo et al. 2018; Heindel et al. 2018; Di Stefano et al. 2019), volcano hazard (Carr et al. 2018; Gomez and Kennedy 2018; Biass et al. 2019), glacier or ice sheet evolution (Brun et al. 2016; Rossini et al. 2018; Groos et al. 2019), rivers geometry and their dynamics (Marteau et al. 2016; Jugie et al. 2018; Rusnák et al. 2018). Although analogue photogrammetry has existed since the second part of the 19th century (Aimé Laussedat in 1849, Albrecht Meydenbauer who first defined the concept in 1867), the progress in the field of computing and the democratization of computers over time have allowed the digital development of this technique. It is a fast, inexpensive and universally accessible modelling technique that is currently used in a wide range of scientific fields (Westoby et al. 2012; Jaud et al. 2019; Valkaniotis et al. 2018). For landslide studies, TOP can be applied for permanent monitoring with a fixed digital automatic camera in front of the landslide to assess displacement rates (in pixels.day^{-1} in Travelletti et al. 2012; Gance et al. 2014) or by multitemporal data acquisition for the 3D geometry reconstruction (Rossi et al. 2018; Ma et al. 2019), characterization of the kinematics (Chanut et al. 2017; Warrick et al. 2019).

2.4.2 Multitemporal Terrestrial Laser Scanning (TLS)

TLS is the second ground-based remote-sensing technique widely used for landslide assessment (Bitelli et al. 2004) with more than 4000 references in Google Scholar since 2016. The measurement principles are presented in Petrie and Toth (2008), Shan and Toth (2018) and Jaboyedoff et al. (2012) prepared a review of laser scanner (airborne and terrestrial) applications for landslide studies. The TLS is a measuring instrument based on laser technology that can measure distance to a high degree of accuracy between the instrument and an object to be measured. It is based on a point cloud using distance measurement by the delay between the sending of an infrared laser pulse and the return of the reflected pulse (Slob and Hack 2004; Teza et al. 2007). The laser instrument is able to measure the precise time interval between the pulse emitted by the laser beam located at point A and its return after reflection from the object to be measured (e.g., slope, river bank...) located at point B (Petrie and Toth 2008). Compared to terrestrial photogrammetry, this instrument provides a high density point cloud (per m^2) measuring all elements of the landscape. Consequently, the reflected pulse can be processed to distinguish the vegetation from the soil to extract it. However, TLS is more expensive from a financial point of view than TOP. It also requires more operators on the field. Campaign measurement can be laborious for difficult to access locations such as mountains or coastal areas with high tides (Medjkane et al. 2018). Because TLS has been developed for precision surveying applications, combined with field measurement, this instrument is now commonly employed to produce highly detailed 3D point clouds in geoscience (Telling et al. 2017; Piégay et al. 2020). Over the last two decades, TLS has proven to be an increasingly practical option for landslide assessment (Delacourt et al. 2007; Jaboyedoff et al. 2012). This technique is employed at various scales ranging from a fixed sector (i.e., a specific part of the landslide such as landslide toe), to large scale (i.e., entire landslide area, Tyszkowski and Cebulski 2019) and for largest areas, TLS with mobile platforms can be used for coverage of several kilometres (Michoud et al. 2014). Application of TLS for the characterization of the landslide kinematics or for the reconstruction of 3D geometric models is significant.

2.4.3 Ground-Based Radar for SAR Interferometry

Ground-based SAR (GBSAR, Tarchi et al. 2003; Corsini et al. 2006, 2013; Herrera et al. 2009; Barla et al. 2010; Monserrat et al. 2014) uses a Radar sensor (in most cases working in Ku band) in a moving configuration for enabling Synthetic Aperture processing in a similar way as for spaceborne imagery. Currently, two concepts are most used to control the motion of such a sensor. The first, known as linear SAR, is to install the radar on a rail (typically 2–3 m long) allowing a translation motion (e.g., Tarchi et al. 2003). On the second, the radar is on a tripod with a mechanism allowing a rotation motion of the tool (e.g., Werner et al. 2008). The choice between both configurations depends on the context of the motion to be observed. In particular, for given sensor's characteristics, the configurations are not equivalent in terms of range and swath—use of rail is generally better for longer ranges but has a reduced swath compared to the tripod—and ease of installation—installed on a tripod is generally a more portable device.

For slope instabilities (notably landslides, but such GBSAR systems are also widely used for monitoring active open pit mines) monitoring, GBSAR is used in an interferometric configuration. The tool can be installed in front of the slope to be monitored and acquires data with a repeat cycle up to about one minute. Typically, the tool is adapted to monitor slopes in a range between about 100 m and few kilometers with a resolution of the order of 0.1 m—depending on the distance to the sensor. It therefore allows to monitor a wide spectrum of landslides, in terms of size and kinematics, to be monitored. With respect to spaceborne interferometric techniques, GBSAR is suitable to monitor slopes with previously known motion or high estimated susceptibility and having a specific interest in terms of risk management (e.g., that could represent a threat for identified assets/persons). If spaceborne interferometric techniques cover wider areas and give information on past motions they cannot—with the current missions—provide a high temporal resolution comparable with GBSAR. Both techniques having different domains of application can therefore be used in a complementary way. In addition, due to their characteristics, GBSAR tools with adapted communications systems can be used in early warning systems.

Finally, noteworthy is the fact that—in a similar way as for spaceborne SAR imagery—offset tracking techniques on the amplitude measurements can be applied to GBSAR data as a complement to interferometric processing (Crosetto et al. 2014). This non-interferometric approach to estimate slope deformation can be useful for monitoring very fast motions (several m day^{-1}) where GB-InSAR is not reliable.

3 Earth Observation Data and Methods for Landslide Detection and Inventory

In risk assessment, there are two main issues: (1) hazard identification and (2) mapping. Both are essential to avoid the exposure of goods and people to hazard. In a context of crisis management, it is often time pressure to detect landslides in specific areas in order to assist people and lead rescue operations. Inventory maps are useful tools for authorities for risk management and to gain knowledge on hazard extension. But most of the time they are only available for limited areas (Guzzetti et al. 2012). Moreover, inventories should be regularly updated, complemented by historical databases to consider the geographical distribution of past (Svennevig 2019) and recent landslides in different time periods (location of the hazard initiation and extension, age…). The regular updating of inventories is also complicated and requires large effort (Bell et al. 2012; Burns and Madin 2009; Burns et al. 2012; Galli et al. 2008; Guzzetti et al. 2012) especially for inaccessible high altitude areas (Du et al. 2020). But they are essential for Landslide Susceptibility Mapping (LSM) for risk mitigation and planning; the accurate detection of landslide locations highly influence the landslide susceptibility analysis (Galli et al. 2008, Song et al. 2012).

Traditionally, for event-based inventories, field reconnaissance approaches by scientists can be conducted (Brunsden 1993). The aim is to identify and delineate landslides, but field-based approaches are time-consuming and can be laborious and tedious for large areas (Yu and Chen 2017) and especially when the area is inaccessible or covered by dense vegetation (Ardizzone et al. 2007). Furthermore, for regularly updating inventories, field reconnaissance approaches are almost impossible, especially after high intensity hazard occurrence, when multiple instabilities trigger simultaneously (Xu et al. 2019). Thus, aerial and space borne data analysis are good alternatives for landslide investigation, especially over large areas (>200 km^2). With remotely sensed data, temporal sequences of images can accurately indicate spectral changes based on surface physical condition variations and induced by landslide triggering. In this way, the use of airborne and spaceborne data has gradually complemented field surveys with a simplified acquisition of multi-resolution images (multispectral or panchromatic, radar), and models (Digital Terrain Model DTM, Digital Elevation Model DEM, Digital Surface Model DSM) with increasing resolution degrees. In this way, various types of remotely sensed data exist and their choice in their use depends on the study site properties and funding.

3.1 Data Preprocessing

Data preparation consists of extraction of metrics from satellite and aerial images and from DEMs. The preprocessing data will be useful for (1) landslide identification, inventory based on visual interpretation of images or based on image classification, (2) for landslide susceptibility mapping (LSM) with the production of geospatial data to define the conditioning factors of landslide triggering, (3) for image comparison for the characterization of the landslide kinematics. Two different types of preprocessing are highlighted. One, focused on radiometric information to create additional geospatial raster layers, is mainly based on optical satellite images. The other considers the spatial relationship of pixels and is focused on the radiometric information to create additional geospatial raster layers. This analysis is mainly based on optical satellite images. The second type of preprocessing considers the spatial relationship of pixels and is mainly based on DEM analysis.

Fig. 1 Examples of radiometric, spatial and textural indicators calculated with R libraries for landslide ▶ analysis from UAV derived-DSM (flight on 31 July 2017, images provided by Kobe University) and multispectral image (Pléiades image, 30 September 2017) above Kyushu island (Japan). **a** RGB drone image, **b** true colour Pleiades image (2 m), **c** true colour Pan-sharpened Pleiades image (0.5 m), **d** NDVI index, **e** ATSAVI index, **f** SAVI index, **g** local relief model by low-pass filter in LRM toolbox ArcGIS ® (Novák 2014), **h** curvature, **i** openness

3.1.1 Radiometric Analysis

In the selected papers for this review, the radiometric analysis of optical images mainly consists of index calculations to discriminate areas covered by vegetation from exposed bare soils (Du et al. 2020) and to detect anomalies in vegetation cover (Fig. 1). To facilitate the distinction between vegetation and bare soil several spectral indices can be calculated using Red Green Blue (RGB) bands of orthophotos (Comert et al. 2019) in Eq. (1)–(5). But for landslide modelling or detection, most studies considering radiometric analysis rely on satellite images to calculate the Normalized Difference Vegetation Index (NDVI) with near-infrared (NIR) and Red bands in Eq. (6) (Yang and Chen 2010; Song et al. 2012; Yang et al. 2013; Behling et al. 2014; Moosavi et al. 2014; Achour and Pourghasemi 2019; Arabameri et al. 2019; Ghorbanzadeh et al. 2019; Wang et al. 2019; Bui et al. 2020; Du et al. 2020; Fang et al. 2020; Hong et al. 2015; Hu et al. 2020; Huang et al. 2020).

$$\text{Normalized Green Blue Difference Band Index (NGBDI)} : (G - B)/(G + B) \quad (1)$$

$$\text{Red Band Ratio (RBR)} : R/(R + G + B) \quad (2)$$

$$\text{Green Band Ratio (GBR)} : G/(R + G + B) \quad (3)$$

$$\text{Excess Greenness Index (EGI)} : 2 * G - R - B \quad (4)$$

$$\text{Green-Red Vegetation Index (GRVI) (Rau et al. 2011)} : (G - R)/(G + R) \quad (5)$$

$$\text{Normalized Difference Vegetation Index (NDVI)} : (\text{NIR} - R)/(\text{NIR} + R) \quad (6)$$

$$\text{Soil-adjusted vegetation index (SAVI)} : (1 + L)(\text{NIR} - R)/(\text{NIR} + R + L) \quad (7)$$

$$\text{Normalized Difference Blue-Red Band Index (NDBRBI)} : (B - R)/(B + R) \quad (8)$$

But other metrics can be calculated to distinguish specific features related to landslides (e.g., Normalized Difference Blue–Red Band Index (NDBRBI) in Eq. (8) in Comert et al. 2019 was considered as effective for extracting the shadow areas on orthophotos). Brightness for RGB images is also significant to distinguish landslides (Rau et al. 2011) because landslide areas have higher intensity than the other image objects.

3.1.2 Spatial Analysis

The spatial analysis of images (Fig. 1) consists here to study the spatial relationship of pixels in the image to gain knowledge on (1) topographic and morphometric features, (2)

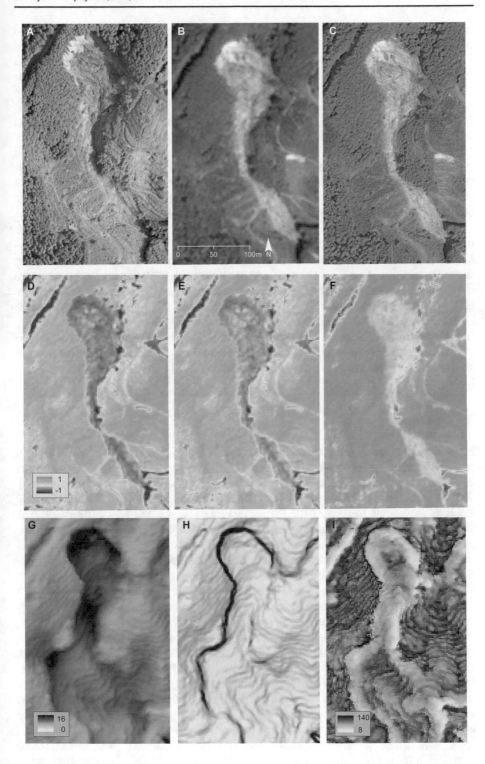

hydrological environment, and (3) texture and roughness of the terrain. Several metrics and derivatives based on statistical methods and filtering are usually calculated to gain knowledge on geographical context of the study area. Among all existing DEM derivatives some conventional ones are almost systematically calculated such as slope, altitude, curvature. Pawluszek and Borkowski (2016), and Van den Eeckhaut et al. (2012) summarize the main DEM derivatives information used for landslide detection with topographic metrics (e.g., openness, roughness, morphological gradient, curvature..), and hydromorphological metrics (e.g., Stream Power Index SPI to measure the erosion power of the stream, Topographic Wetness Index—TWI to measure the degree of accumulation of water at a site and thus influences the occurrence of landslides in Catani et al. (2013). The Sky View Factor (SVF) is another relief visualization technique that represents the ratio between the visible sky and a hemisphere centred over the study area in a given point. This technique is based on the direct illumination of relief to intuitively recognize features (Kokalj et al. 2016).

Statistical methods focused on texture analysis can be also exploited to consider the spatial relationship of pixels and emphase the relief features. In Mezaal et al. (2018) the Grey-Level Co-occurrence Matrix (GLCM) texture features were calculated on airborne laser scanning data with eCognition software, and in Comert et al. (2019) on the red band image to highlight specific patterns features for detecting and differentiating landslides. The GLCM function (Package 'glcm' in R software) is used to characterize the texture images by calculating how often pairs of pixels with specific values and in a specified spatial relationship occur in an image, to create a matrix. DEM filtering combined with conventional derivatives is also a useful technique to detect features associated with landslide, such as convolution filtering, low-pass filtering in Chen et al. (2014).

3.2 Data Interpretation

Interpretation of aerial photos or satellite images are approaches commonly used to identify past and recent mass movements (Chigira et al. 2004; Catani et al. 2005; Ardizzone et al. 2007; Galli et al. 2008; Yang and Chen 2010; Song et al. 2012; Chen et al. 2014; Xu et al. 2014; Zhang et al. 2014; Ciampalini et al. 2015; Fressard et al. 2016; Fan et al. 2017, 2018; Roulland et al. 2019; Bui et al. 2019; Görüm 2019; Lewkowicz and Way 2019; Pánek et al. 2019; Pham et al. 2019; Wang et al. 2019; Du et al. 2020). Based on morphological features of the landscape and visible 'anomalies', visual interpretation of images can be faster than the ground survey approach to identify mass movement. Nevertheless, ground investigation is meaningful in a second step of the inventory process to validate interpretations. Nevertheless, the quality of the visual interpretation highly depends on the complexity of the terrain, the vegetation cover, and on the acquisition procedures. For areas of dense vegetation, the use of a LiDAR-derived elevation model (3D models: pointcloud, 3D meshes, DEMs) helps to identify undercovered features (Chigira et al. 2004; Mckean and Roering 2004; Ardizzone et al. 2007; Van Den Eeckhaut et al. 2012; Razak et al. 2013; Lissak et al. 2014; Pawluszek and Borkowski 2016; Bunn et al. 2019; Görüm 2019). DEM derivatives such as slope values, aspect, roughness, orientation, openness, and Sky View Factor indicators can be calculated to highlight morphological features induced by landslides and extract hazard boundaries. Thus, statistical differences between field-based inventory and image/models interpretation can exist. For example, the size of landslide mapped by field recognition can be larger than landslide mapped using LiDAR-derived DEM (Ardizzone et al. 2007). It can easily be explained by the accessibility and visibility of the study area.

3.3 Normalized Difference Vegetation Index (NDVI)

In some specific contexts, morphological changes induced by landslide triggering can be linked to land cover changes. According to the postulate that a region is character-ized by a relatively consistent vegetation cover between 2 years, a multitemporal optical remote-sensing approach based on radiometric analysis with NDVI values can be a solu-tion to detect landslide occurrence (Lin et al. 2004). For large areas inventories, most studies are based on post-failure images because major changes in the reflectance char-acteristics are linked to landslide surface (Yang et al. 2013). For systematic spatiotem-poral mapping of landslides, an approach based on NDVI trajectory over time seems to be efficient (Behling et al. 2014). NDVI index can be generated at various temporal intervals (several days, during one year, or bitemporal) to identify post-event landslides or to define evolution patterns of reactivated one. NDVI times series are analysed in Yang et al. 2013, considering the potential disturbance induced by cloud or atmosphere to compare NDVI value before co-seismic landslide event and after 2008 Wenchuan earthquake. But variation in phenology states due to seasonal vegetation cover evolution should also be considered before associating NDVI values and landsliding.

NDVI values can be a useful indicator to detect landslides in various environments. For example, tundra is in most parts of the Arctic characterized by vegetation cover-age. Any environmental disturbance results in removal of vegetation and soils are then exposed. Mass movements are abundant in these areas, more common than in other regions around the world (several tens of thousand have been documented). Their occur-rence is conditioned by permafrost and thus they are sensitive to temperature changes. Progressive rise in mean summer air temperature due to climate change is therefore expected to trigger specifically retrogressive thaw slumps according to Lewkowicz and Way (2019). Approximately 22% of the Northern Hemisphere are underlain by perma-frost based on a recent account (Obu et al. 2019). Such mass movements are a promi-nent example for the need of automatic mapping procedures. A major role is played by deposits of former glaciations and marine terraces specifically the presence of ground ice (Lewkowicz and Way 2019; Leibman et al. 2015). Ice melt causes thermokarst (top-ographic depression generated by thawing ground ice) which results in various specific surface features including landslides. Unconsolidated sediments are currently frozen but can be mobilized under exceptionally warm conditions. The depth of the seasonal thaw layer (active layer thickness—ALT) exceeds normal conditions in exceptionally warm years (Bartsch et al. 2019). Ice lenses at the base of the active layer melt leading to high porewater pressures, a reduction in effective shear strength, and eventually slope failure. Retrogressive Thaw Slumps (RTS) are a common type of cryogenic landslides which are caused by this mechanism (Lantz et al. 2009). They are therefore more likely to be initiated under unusually warm conditions. This has been described for sites in Canada (Lewkowicz and Way 2019; Jones et al. 2019) and in Russia (Babkina et al. 2019). As an example, more than 4000 thaw slumps have been initiated since 1984 over an area of 70,000 km^2 (Lewkowicz and Way 2019), covering an area of 64 km^2. Clusters of RTS have been reported for different regions representing a range of climate conditions span-ning from -19.7 °C to -7° Mean Annual Air Temperature (Jones et al. 2019; Babkina et al. 2019). Headwall retreat after initiation is depending more on local conditions, especially terrain factors (Jones et al. 2019) but reactivation is also triggered by high temperatures (Babkina et al. 2019). Retrogressive thaw slumps continue to grow over several years until they stabilize. The lifespan of a RTS is determined by the ratio of the slope of the slump floor compared to (and running parallel to) the slope of undisturbed terrain (Jones et al. 2019). It can be up to 50 years in extreme cases (French and Williams 2017). RTS enlarge by retrogression at typical rates of 5–27 m yr^{-1} (Lewkowicz and Way 2019; Jones et al. 2019). Extension is limited to temperatures above 0 °C, so growth is taking place within a few months per year only. They are on average smaller than 2 ha, but can be larger. Mega slumps are defined as features larger than 20 ha.

Further features in this context are active layer detachment slides (Lewkowicz 2007; Rudy et al. 2016). They are in general smaller than thaw slumps. They can occur on slopes as low as 3° (French and Williams 2017). Active layer detachment slides result in the formation of bare mineral scar (to of frozen ground) and depositional areas, where an earth mass shifts with vegetation.

To detect the occurrence of thaw slumps and detachment slides and their changes automatically, multispectral images such as available from Landsat are usually applied for analyses of trends (Nitze 2018; Lewkowicz and Way 2019; Jones et al. 2019), especially to quantify changes of vegetation indices such as NDVI in areas which have been mapped as thaw slumps leading to thermocirques (amphitheatrical hollows in Fig. 2). Nevertheless, a major constraint in this case is spatial resolution, as features are comparably small and multispectral regular acquisitions which go back to the 1990s are of comparably coarse spatial resolution. Landsat resolution (30 m) prevents the identification of thaw slump areas in many cases as they have a width of few pixels only (example of two-pixel width in Fig. 2). Lewkowicz and Way (2019) therefore could not fully apply automatic detection and eventually relied on large scale manual post-processing utilizing crowdsourcing. In this context, Sentinel-2 with its 10 m resolution provides an important step forward in monitoring of retrogressive thaw slumps. Consequently, recent features can be mapped with Sentinel-2, also revealing changes within the season and the re-establishment of vegetation starting in the lower part.

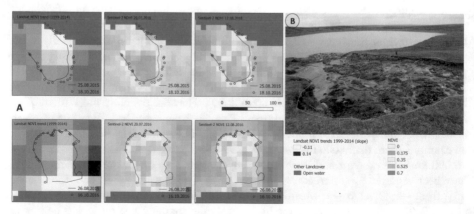

Fig. 2 Thaw slump vegetation properties from Landsat and Sentinel-2 for two sites located on the Yamal peninsula, Russia. **a** From left to right: NDVI trends from Landsat (Nitze et al. 2018), NDVI from two Sentinel-2 acquisitions in 2016 for feature #1 (top) and #2 (bottom). Lines and dots represent outlines based on GPS surveys. **b** photograph of feature #2 (thermocirque, viewing direction from NE to SW) (Picture: Bartsch 26 August 2015)

3.4 Landslide Identification Using Machine Learning Approaches

The identification of landslides for risk assessment and management is possible through a visual interpretation of multi-source data. Although effective, this technique is time-consuming (especially for large areas) and very difficult to apply for diachronic studies. Consequently, many studies aim to automate or semi-automate detection methods. For this purpose, Machine Learning (ML) and Deep Learning (DL) approaches can be useful. Whereas these approaches were few employed for landslide assessment until the early 2000s (Bui et al. 2019), today they are developed for various applications: landslide triggering prediction (Farahmand and AghaKouchak 2013), landslide displacement prediction (Lian et al. 2013; Zhao and Du 2016), landslide detection (Stumpf and Kerle 2011a, b; Chen et al. 2014; Moosavi et al. 2014; Bunn et al. 2019; Ghorbanzadeh et al. 2019).

3.4.1 Pixel/Object-Based Techniques

For landslide detection, two groups of techniques can be suggested. The first technique, and the most frequent, is the Pixel-Based Image Analysis (PBIA). This technique considers image pixels as fundamental units of analysis. The second is the Object-Based Image Analysis (OBIA). This technique is based on the creation of image objects, or segments used for image analysis.

In pixel-based approaches, each pixel is classified without considering neighbouring pixels and all pixels are considered as spatially independent from each other. Consequently, pixel-based approaches can be sensitive to noise (Van den Eeckhaut et al. 2012), especially with Very High-Resolution (VHR) images that provide numerous information with high spatial resolution but low spectral domain (Lv et al. 2020).

Object-based approach is a good alternative to detect landslides (Stumpf and Kerle 2011a, b; Kurtz et al. 2014; Moosavi et al. 2014; Li et al. 2015; Casagli et al. 2016, 2017; Bunn et al. 2019) from various data sources (as well as Very High-Resolution optical data than LiDAR-derived DEM). This technique is based on 2 steps: the image segmentation and the image classification. The image segmentation relies on various pixels in groups into homogeneous objects or regions, considering their similarities between neighbours (Fig. 3). Several algorithms of segmentation exist. In eCognition® software, the most commonly used algorithm is "Multi-Resolution Segmentation" (MRS). This method is based on the pairwise region-merging technique and provides good results for landslide inventories (Moosavi et al. 2014; Mezaal et al. 2018). But for an optimal segmentation, the structure of the segmentation into several levels of segmentation must be a possibility to cluster image pixels according to their homogeneity in spectral, spatial and textural characteristics (Anders et al. 2011). Objects are merged or distinguished according to three parameters: colour, scale, and shape (Fig. 3).

For better segmentation, an a priori topographic information can be integrated in the process (e.g., landslide morphology from DEM analysis with location of main scarp, deposits, secondary scarps...). It can considerably influence the result of segmentation (Van Den Eeckhaut et al. 2012; Li et al. 2015). Spectral bands of orthophotos or multispectral images can be used as input layers to create image objects. But various studies employed other layers for segmentation (Rau et al. 2011; Stumpf and Kerle 2011a; Chen et al. 2014). Van Den Eeckhaut et al. (2012) use 45 segmentation layers of LiDAR derivative maps (e.g., altitude, slope, aspect, curvature, Sky View Factor).

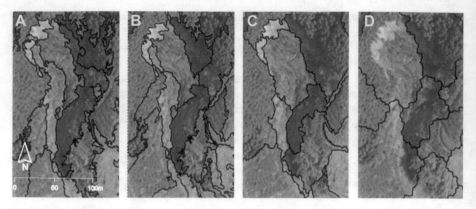

Fig. 3 RGB drone image segmentation using eCognition software ®. At this step, the segmentation parameters are focused on the shape of objects. Several parameters are tested here for a better pixel merging: compactness value (0.5) and shape value A)0.1, B)0.3, D) 0.5, C) 0.9

3.4.2 From Conventional Classification Algorithms to Artificial Neural Network

Machine Learning (ML) methods are effective for image classification and landslide detection based on various ML methods and classifiers using supervised and unsupervised algorithms. Automatic landslide detection by ML depends on the algorithm used, training samples (for the supervised classification) and validation data. While in pixel-based approaches a class is attributed for each pixel according to spectral information; in object-based approaches each segment or object is classified according to spectral, geometric, contextual, and textural information of the image object. All mentioned classification presented below can be applied for supervised pixel-based or object-based classification. But regarding several references, landslide detection and inventory using object-based approach seems to provide better results than a pixel-based approach (Comert et al. 2019). For example, in Bunn et al. (2019) the recognition of landslide features is 70% accurate. But with OBIA, the result of the classification highly depends on the segmentation quality which one also depends on (1) the image resolution (Stumpf and Kerle 2011a; Kurtz et al. 2014), (2) the number of available bands, (3) the segmentation scale (under or over segment images in Moosavi et al. 2014), and/or the training sample quality for supervised methods. Training samples can be expert-based (Van Den Eeckhaut et al. 2012) or random-based (Chen et al. 2014; Pawluszek and Borkowski 2016). Despite the good results with object-based approaches, pixel-based approaches are still the predominant methods (Casagli et al. 2016). Pixel-based approaches can provide results with several misclassified pixels, especially with high spectral variance data in High-Resolution images, and if classification is only focused on spectral characteristics (Moosavi et al. 2014) with threshold-based approaches and classification (Li et al. 2016).

Regardless of the technique (pixel/object-based), several classification algorithms exist and can be applied separately or jointly (e.g., Support Vector Machine algorithm—SVM in Van Den Eeckhaut et al. 2012; Random Forest—RF algorithm in Stumpf and Kerle 2011a, both in Li et al. 2015). Both techniques have been used in a wide range of remote-sensing applications including landslide detection and susceptibility mapping (Ballabio and Sterlacchini 2012; Catani et al. 2013; Achour and Pourghasemi 2019; Arabameri et al. 2019; Bui et al. 2020; Fang et al. 2020). For a good integration of the spectral and spatial information several classification algorithms (Chunhui et al. 2018) and processes (Fig. 4) can be useful to detect landslides (Comert et al. 2019).

Concerning the support data, classification processes to detect landslides are mainly based on spectral information from orthophotos or multispectral images (Ghorbanzadeh et al. 2019). But additional information from DEM (Kurtz et al. 2014), DEM derivatives (Van den Eeckhaut et al. 2012), or radiometric analysis (NDVI) can be used to improve the classification accuracy (Comert et al. 2019; Fang et al. 2020).

Random Forest (RF) and Support Vector Machine (SVM) algorithms are popular powerful supervised learning techniques (Cortes and Vapnik 1995; Pal 2005) based on a set of training samples for image classification and regression analysis. Huang and Zhao 2018 underline the interest of these modelling techniques, especially the SVM method which seems to be more effective than other methods (Moosavi et al. 2014).

The RF technique is based on multiple decision trees to train and predict samples (Breiman 2001). RF is considered as less sensitive to the over-fitting problem caused by complex datasets than other decision trees. This technique can be considered as the most effective non-parametric ensemble learning methods (Ghorbanzadeh et al. 2019). But both methods (SVM and RF) are generally used simultaneously (Table 2, Fig. 5). The advantage of the SVM method is the possibility to classify each pixel according to a hyperplane and separate classes that cannot be split with a linear classifier. Thus, different Kernel functions can be specified (Hong et al. 2016) to perform the SVM classification (e.g., polynomial, sigmoid, and Radial Basis Function—RBF). Some authors have proposed to merge classifications to improve results using fusion techniques (i.e., Dempster–Shafer theory—DST and variants) on results issued from various classifiers such as SVM, K-nearest neighbour (KNN) and RF in Mezaal et al. (2018). The optimization of the both methods is also possible. To illustrate, in Bui et al. (2019) the Least Squares Support Vector Machine (LSSVM) technique has been employed to label pixels to be either "non landslide" (negative class) or "landslide" (positive class) for landslide prediction modelling. The optimization of classifiers can be performed by using boosting algorithms like AdaBoost (Li et al. 2008; Kadavi et al. 2018), LogitBoost, Multiclass Classifier, Bagging models (Bui et al. 2019), and multi-boost models (Pham et al. 2019). Adaboost (Freund and Schapire 1995) is one of the most used machine learning ensemble algorithms to create a series of individual classifiers to classify training data. The interest of the method is that Adaboost is based on an adaptive resampling

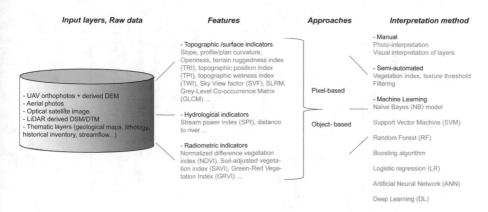

Fig. 4 Flowchart of the various methods and data used for landslide detection

Table 2 Listing of various remote sensing methods used for landslide identification and associated references

Approach	Method	Study cases
Manual	Visual interpretation	Catani et al. (2005), Ardizzone et al. (2007), Galli et al. (2008), Yang and Chen (2010), Song et al. (2012), Chen et al. (2014), Xu et al. (2014), Ciampalini et al. (2015), Li et al. (2015), Fan et al. (2018), Huang and Zhao (2018), Zhou et al. (2018), Bui et al. (2019), Görüm (2019), Lewkowicz and Way (2019), Pánek et al. (2019), Pham et al. (2019)
Semi-automated	Thresholding, filtering	Lin et al. (2004), McKean and Roering (2004), Yang et al. (2013), Behling et al. (2014), Li et al. (2016)
Machine learning	Naïve Bayes (NB) model	Hu et al. (2020)
	Support vector machine (SVM)	Ballabio and Sterlacchini (2012), Van den Eeckhaut et al. (2012), Moosavi et al. (2014), Li et al. (2015), Hong et al. (2016), Pawluszek and Borkowski (2016), Chen et al. (2017), Huang and Zhao (2018), Kalantar et al. (2018), Mezaal et al. (2018), Zhou et al. (2018), Achour and Pourghasemi (2019), Bui et al. (2019, 2020), Ghorbanzadeh et al. (2019), Fang et al. (2020), Huang et al. (2020)
	Random forest (RF)	Stumpf and Kerle (2011a, b), Catani et al. (2013), Chen et al. (2014), Aditian et al. (2018), Huang and Zhao (2018), Mezaal et al. (2018), Achour and Pourghasemi (2019), Ghorbanzadeh et al. (2019), Mohammady et al. (2019), Bui et al. (2020), Fang et al. (2020), Hu et al. (2020), Huang et al. (2020)
	Boosting algorithm	Kadavi et al. (2018), Pham et al. (2019)
	Logistic regression (LR)	Huang and Zhao (2018), Kalantar et al. (2018), Zhou et al. (2018), Du et al. (2020), Huang et al. (2020)
	Artificial neural network (ANN)	Song et al. (2012), Moosavi et al. (2014), Aditian et al. (2018), Huang and Zhao (2018), Kalantar et al. (2018), Zhou et al. (2018), Ghorbanzadeh et al. (2019), Du et al. (2020), Huang et al. (2020)
	Deep learning (DL)	Gong et al. (2015), Ghorbanzadeh et al. (2019), Lei et al. (2019a, b), Wang et al. (2019), Bui et al. (2020), Du et al. (2020), Fang et al. (2020), Lv et al. (2020)

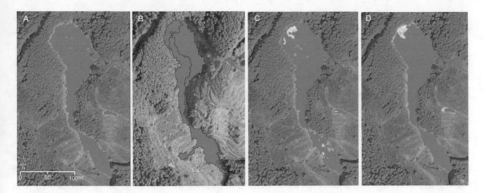

Fig. 5 Extract of landslide inventory map on Kyushu island after 2017 landslide event. Multi-algorithms were tested with **a** K-means classification, **b** Object-based classification, **c** support vector machine classification (SVM), **d** random forest (RF) classification

technique for which classifiers can be progressively adjusted according to misclassified dataset.

To illustrate, several classifiers have been performed (Fig. 5) with two supervised (RF, SVM) and one unsupervised method (K-means) for pixel-based and object-based approaches (Fig. 5b). The OBIA approach has been tested in eCognition software® for UAV images because of their low spectral resolution (with slope, Local Relief Model, R-G-B bands, vegetation index GRVI (Eq. 4) and EGI (Eq. 5)). The pixel-based approach was performed on Pléiades image with 'Caret' library in R software (with NDVI, ATSAVI, Slope, curvature, Sky View Factor, R-G-B-NIR bands). Several datasets have been tested (Fig. 1) to define the most relevant layers. Evidently, an important preprocessing work is necessary with radiometric analysis [several indexes tested potentially redundant (Eq. (1)–(8))], DEM derivatives calculation (see below: LRM, SVF, various slope classifications…). Some layers are a priori considered as essential for landslide assessment (e.g., the topographic layer, Soil-Adjusted Vegetation Index (SAVI) values to consider the soil on brightness influence, openness…). Regarding the results in Ghorbanzadeh et al. 2019, despite the removal of the topographic layers in their datasets no difference has been seen in the classification (RF and SVM). The pixel approach with Very High-Resolution data seems to be sensitive to noise and consequently post-processing operations are necessary to improve the results (data filtering). In our study, OBIA approach has provided good results with 70% of landslide detection and less influenced by image noise. The performance of RF is better than SVM in several studies mentioned in Huang and Zhao (2018).

4 Artificial Neural Network (ANN) and Deep Learning (DL) Algorithms for Landslide Identification

Another family of classification algorithms exploits Artificial Neural Networks (ANN). Over the past 10 years, performances of related approaches generally systematically outperform conventional classification methods (Pakhale and Gupta 2010). The interest lies in the possibility of ANN with deep layers to automatically compute efficient spatial features and classify them in a single framework. Artificial Neural Networks can consist of deep structure and refer to 'Deep Learning' technique (Table 2). Among ANNs, deep learning architecture provides an opportunity to jointly optimize several related tasks together considers multi-layers processing in the model (to get more information on deep neural networks refer to LeCun et al. 2015; Schmidhuber 2015). Deep neural network approaches using Earth Observation have been particularly used over the early years, mainly thanks to the power of Convolutional Neural Networks—CNNs (Zhu et al. 2017; Ma et al. 2019). Zhao et al. (2019) present a review of deep learning-based object detection frameworks with a specific presentation of CNNs and their applications in landslide studies. CNN is a biologically-inspired DL technique that has shown powerful capabilities in feature extraction (Girshick 2015). This type of network is considered as the most popular model of DL, even in the field of landslide researches (Table 2). The main idea is to apply, in each layer of a deep neural network, spatial convolutions where the weights are learned by the model during the training stage. In each layer only a subset of the information embedded in it is kept (pooling). It results in images of decreasing sizes but with more and more features since many convolutions are applied. This enables to automatically compute many spatial features adapted to the observed data (see, for example, Zhao et al. (2019)). DL with CNN already have been applied for landslide detection with multispectral, aerial or SAR images (Gong et al. 2015; Lei et al. 2019a, b; Lv et al. 2020) and can be employed for features extraction and then combined with another classifier (i.e., CNN + SVM, CNN + RF, … in Fang et al. 2020).

As a matter of fact, the deep architecture of CNN provides an exponentially increased expressive capability for object detection such as landslides (Table 2). Regarding the bibliography, the results of the image classification, and consequently the landslide detection, will depend on image accuracy, input data selected, chosen algorithm and CNN architecture. Initially the CNN approaches aimed to compute features and then to classify entire images (i.e., one label per image). Based on the idea of CNN, several extensions have been proposed with *deconvolution* layers able to spatially relocalize the extracted features. For example, objects can be detected with bounding boxes (R-CNN) or each pixel of the image can be classified (*semantic segmentation*, FCN and variants) based on numerous input datasets (multispectral images, various radiometrics indice layers, DEM derivatives layers… and also training samples for supervised methods). Deep learning techniques with CNN, R-CNN or FCN provide great potential in the feature extraction process (Chen et al. 2017; Zhao and Du 2016; Lei et al. 2019a, b; Fang et al. 2020) for landslide detection. These approaches are briefly described below.

(1) Region-based Convolutional Network (R-CNN) and fast R-CNN (Girshick 2015) are region-based methods able to extract several bounding boxes in images. These techniques do not seem to be the most adequate for risk assessment, because one of the main challenges in risk assessment is to define with accuracy the limits of landslides in landscape. A region-based approach is consequently not sufficient for hazard assessment.

(2) Fully Convolutional Networks (FCN) techniques and their extensions predict a class for each pixel of the input image instead of classify region (Long et al. 2015). Numerous recent work has shown the effectiveness of FCN principles for semantic segmentation (FCN-Fast-FCN, U-Net, Res-Net, etc.). Although considered to be a robust approach, Ghorbanzadeh et al. 2019 consider CNN methods to be as effective (if not less) as conventional learning algorithms (e.g., RF, SVM, ANN). Particularly when modelling is based on spectral information only and training samples for supervised methods randomly selected. However, other studies (Lei et al. 2019a, b; Liu et al. 2020) highlight

the performance of FCN to learn better image features to improve landslide inventory and Landslide Susceptibility Mapping (Zhao and Du 2016; Fang et al. 2020).

5 Mass Movement and Change Detection by Remote Sensing

To characterize and quantify accurately surface and depth deformation several monitoring techniques are possible (GNSS receiver, benchmark, extensometers, inclinometers…, López-Davalillo et al. 2014; Jiang et al. 2016; Carlà et al. 2019). Numerous field monitoring techniques can provide admittedly high accuracy 3D data of slope dynamics but with a limited spatial coverage. Field investigation over large landslides is highly dependent on field accessibility and costs. Consequently, a dense multi-sensors coverage is hardly conceivable. In this context, remote-sensed techniques can be employed as additional support to field monitoring based on sensors to increase the size of investigated areas and the spatial resolution of data. The choice of the appropriate method mainly depends on (1) landslide velocity, (2) size, (3) location orientation, or its morphostructural context (high coastal mountain north–south orientation…), (4) and vegetation cover. Thus, three approaches frequently emerge. While for large landslides undercovered by vegetation, radar imaging (Crosetto et al. 2016) will be favoured, lasergrammetry or photogrammetry, image correlation (Casson et al. 2003; Ayoub et al. 2009; Travelletti et al. 2012) will be privileged in poorly vegetated and accessible areas. For large landslide detection and displacement measurement (Berardino et al. 2002; Colesanti and Wasowski 2006) the most widely used method is Synthetic Aperture Radar Interferometry (InSAR) based on scattering properties of Earth surface (Bamler and Hartl 1998). This technique measures the phase and the amplitudes of the backscattered microwaves signals, comparing the phase information between signals acquired at different epochs, phase differences being proportional to ground motion. Persistent Scatterer Interferometry (PS-InSAR, Crosetto et al. 2016), Differential SAR Interferometry (DInSAR, Rudy et al. 2018), and Small Baseline interferometry (SBAS) are some examples of how the use of SAR has evolved with improved precision. These methods make it possible to detect and quantify surface deformations (landslides including thaw slumps and active layer detachment slides, rock glacier, solifluction detection (Barboux et al. 2013, 2014; Echelard et al. 2013; Zwieback et al. 2018; Rouyet et al. 2019, Paquette et al. 2020) with repeat pass InSAR, by measuring, field for example, the phase difference between two radar images acquired in a satellite, airborne or terrestrial context.

5.1 SAR and Optical Cross-Correlation for Measuring Landslide Kinematics

To quantify surface deformation, radar-amplitude images are frequently combined with optical images (Crippen 1992; Michel et al. 1999; Van Puymbroeck et al. 2000). It has been proven complementary to InSAR in a number of geophysical studies (Klinger et al. 2006; de Michele and Briole 2007; de Michele et al. 2010). The use of cross-correlation of optical spaceborne imagery to measure displacement fields of the Earth surface was first conceptualized by Robert Crippen (1992) and applied to landslide motion measurement from SPOT data. The method relies on the fact that spaceborne images, acquired at different times, can be resampled to the same geometry with the use of a DEM and a robust camera model. Residual offsets that remain within the resampled images might then be due to terrain motion within the footprint of the images. Surely, other (non-geophysical)

residual distortions can exist within the offset field. These can be due to CCD misalign-
ment, the jitter of the satellite, roll, pitch and yaw motion of the sensor. In theory, these latter distortions can be modelled and removed. Offsets are commonly calculated by differentiating the phase of the Fourier transform calculated on a subset of the images, by means of
a moving window. The correlation peak is then interpolated to achieve sub-pixel precision.
It is generally proven that this methodology can be as precise as 1/10th of the pixel size.
The method has been applied to aerial photos and to satellite optical data (Delacourt et al.
2004, 2009; Stumpf et al. 2014; Le Bivic et al. 2017; Lacroix et al. 2019). In the optical
domain the offset fields measured by cross-correlation is in two directions: lines and columns of the image matrix. These fields are commonly regarded as horizontal displacement
fields. Sometimes this approximation is incorrect; the measured offset field is the apparent
horizontal offset, which is the projection of the downslope motion induced by the gravitational movements, on the image plane.

In the radar domain, Synthetic Aperture Radar (SAR) data can be used, along with the
sub-pixel offset method (Fig. 6), to measure displacements fields of the Earth surface due
to gravitational movements. The SAR sub-pixel correlation method, today called "offset
tracking", has been developed for earthquake studies by Michel et al. (1999). It demonstrated useful for landslides motion detection in a number of studies (Debella-Gilo and
Kääb 2011; Li et al. 2011; Raucoules et al. 2013; Singleton et al. 2014; Wang et al. 2016).
This technique exploits the amplitude channel of the SAR system. It can be applied in
the observation of fast landslides movements—as opposed to InSAR, since InSAR signal
decorrelates if the ground motion gradient is higher than half an interferometric fringe per
pixel—even in scarcely temporally coherent areas. The SAR instrument records the SAR
echoes in two directions, the Line of Sight (LOS) and the Azimuth directions (i.e., the
orbit direction). The LOS direction has an angle with respect to the vertical. Therefore, by
combining ascending and descending correlograms, one can retrieve the 3D vectors of the
landslide displacement field over time, yielding a spatiotemporal distribution of landslides
kinematics as described in Raucoules et al. 2013. Today, new generations of optical and
radar satellites, with improved repeat frequency, can be used along with the sub-pixel offset

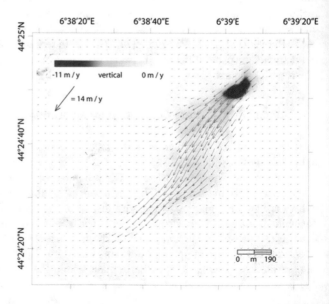

Fig. 6 3D surface displacement field of La Valette landslide (France) from sub-pixel offset of TERRA-SAR X data

method to extract landslide velocity fields from space and create time series. For instance, Li et al. (2011) and Sun and Muller (2016) used multitemporal Terra-SARX data to derive landslide motion rate at a sub-pixel level. Valkaniotis et al. (2018) used multiple sensors, SAR and optical, to study a co-seismic landslide in Iran; Lacroix et al. (2019) estimated the ground displacement from time series analysis of Landsat 8 images, spanning a 5.7-year period. They show systematic patterns that correlate with topography and seasonal variations. They finally show complex nonlinear interannual displacement patterns.

A special case in this context are Arctic coasts, which are highly affected by erosion, up to 10 m per year (Lantuit et al. 2012). Coastal retrogressive thaw slumps will for example cause the loss of about 50% of cultural sites for 2100 along the Beaufort Sea coast (Canada; Irrgang et al. 2019). Many sites have been lost every year since the 1950s. Usually, aerial photos, sporadic high-resolution satellite data and, in some extreme cases, Landsat data can be used to manually digitize coastlines and quantify their change over time. In general, this technique allows only the detection of year to year changes or over several decades and the monitoring of seasonal behaviour is impeded by frequent cloud cover across the Arctic. In this context, SAR data could be a solution to overcome these constraints but the spatial resolution of SAR data is insufficient in case of most available sensors. However, Stettner et al. (2018) demonstrated the utility of X-band SAR (2.35 m nominal resolution) for retrogressive thaw slumps in association with river bank erosion (Lena Delta, Russia). The rate of about 2.5 m over three weeks barely matches the resolution of the sensors and can only be retrieved by analysing the progression over the whole season. Stettner et al. (2018) proposed a method which is, however, only applicable for slopes facing directly towards the sensor as the detection principle relies on the foreshortening effect of radar data. In such cases slopes appear brighter and can be therefore easily distinguished from surrounding tundra and river banks. The usually wet (and vegetation free) surfaces add to the magnitude of backscatter as the response is conditioned by dielectric properties. A further disadvantage is, however, that actual positioning of the cliff-top requires the existence of an elevation model valid for the time of acquisition. Rates are therefore relative but can give nevertheless valuable insight into seasonality and enable to identify the driving factors in these environments. Further developments are needed to extend the use of high-resolution SAR data to further coastlines, also not facing the sensor.

5.2 Terrestrial Laser Scanner (Repeated Surveys)

Common landslide monitoring techniques with inclinometers, GNSS receivers, or InSAR can be difficult to apply with adequate spatial or temporal resolution; specifically, in forested and steep slope environments. In various geographical contexts, such as coastal (Conner and Olsen 2014; Costa et al. 2019), volcanic (Pesci et al. 2011), or mountainous areas (Travelletti et al. 2012, 2014; Kenner et al. 2014), repeated campaigns of terrestrial laser scanner (TLS) have proven to be an effective way to analyse patterns of mass movement displacements (Jaboyedoff et al. 2012; Telling et al. 2017). Indeed, application of TLS for displacements measurement can be advantaged for very fast and very slow moving landslides. These techniques provide high-resolution 3D points clouds and infra-centimetric resolution models. In some cases, multitemporal point clouds from TLS can be combined with ALS surveys to assess vertical/horizontal displacement fields at various scales, to maximize spatial coverage and point density (Fig. 7). But according to the study area application of TLS can be spatially limited because of the field accessibility, vegetation

Fig. 7 TLS combined with ALS survey and ground control points (GCPs) for velocity measurement in coastal landslide in France

cover and laser range (Niethammer et al. 2012). In this context, the characterization of the landslide kinematics is challenging and requires the combination of tools.

Normandy (France) coastal landslides (Costa et al. 2019) can easily illustrate the necessity and the difficulties to combine different sources of data to reduce measurement uncertainties inherent to the geographical context of the study area. Villerville landslide is affected by slow and complex kinematics for which monitoring has been performed by conventional techniques (inclinometers) and GNSS surveys since the 1980s. Villerville landslide is affected by complex movement patterns with deformations ranging from a few millimetres, to several centimetres per year. These displacement values are often close to the detection limit of conventional monitoring equipment. Due to the complex nature of its dynamics, and for early warning strategies, various techniques of investigation have been implemented with discrete measurements and continuous monitoring. A network of two permanent GNSS stations and 17 GNSS stations for campaign measurements was deployed on the landslide. Because of a dense vegetation cover and the landslide dimension, surveys are limited (mainly because of accessibility, logistical and economic constraints). In 2011, to detect failures under vegetation cover a first airborne full-waveform LiDAR was performed (Fig. 7). Another LiDAR survey was performed by IGN in 2015 (Litto3D®) and used to study the deformation pattern of the landslide between 2010 and 2015. Although Airborne LiDAR modelling accuracy can reach few decimetres, the result of LiDAR-derived DEM differencing (DoD) between 2010 and 2015 was not sufficient (i.e., the displacement field was below the laser model accuracy. To address this issue, TLS was deployed since 2018 for yearly campaigns at the foot of the landslide to generate very high-resolution model of this part of the landslide. Data were taken by a RIEGL VZ-400 instrument equipped with 1550 nm laser wavelength and unique echo digitization (RIEGL Laser Measurement Systems 2014). For the best reconstruction of the surface geometry and avoiding occlusion the TLS station must be located in the most appropriate positions (especially in complex geometry areas) involving multi-scan approach. But a multi-scan, a multi-station approach can induce several sources of error affecting the 3D modelling and as a result the estimation of displacement values (Barbarella et al. 2017). The surveys of the landslide toe were carried out where the vegetation cover is sparse. But the study area is located in a coastal environment, consequently the field of view is limited by the sea. The generation of accurate multitemporal models of the landslide deformation is now carried out with both airborne laser scanning for the largest and undercovered area and terrestrial laser scanning modelling with very high resolution (<cm) for the landslide toe. Besides, the protocol to be implemented with TLS can be difficult, especially for irregular terrain and coastal areas, Terrestrial Optical Photogrammetry (TOP) with Structure of Motion (SfM) has recently emerged as an alternative and competing technology to provide high-resolution 3D point clouds and HR models for landslide studies.

5.3 Terrestrial Optical Photogrammetry (Repeated Survey)

For landslide assessment, Terrestrial Optical Photogrammetry (TOP) provides a low-cost system for high-resolution monitoring in various environment: continental areas (Gance et al. 2014; Stumpf et al. 2015; Fernández et al. 2016; Kromer et al. 2019) and coastal areas (Francioni et al. 2018; Westoby et al. 2018; Gilham et al. 2019; Jaud et al. 2019; Warrick et al. 2019). As example, the monitoring network of the Vaches Noires cliffs (Normandy, France) can be presented to illustrate the monitoring of landslides using photogrammetric techniques (Medjkane et al. 2018; Roulland et al. 2019). These cliffs form a 4.5 km coastal line. Composed of marls, limestones and chalks layers, they have a badland morphology that evolve under combined action of subaerial, continental and marine processes. Hydrogravity processes on these cliffs are multiple and interlock (landslides, rockfalls, mudflows, toe cliff erosion…). While the upper part can only be removed by ablation, the lower part (i.e., toe cliff) alternates between periods of progradation (by feeding materials from upstream) and periods of erosion (by sea erosion). The nonlinear functioning in time and space of these coastal slopes is the result of hydrogravity processes relays that are defined and quantified on a test site of the Vaches Noires cliffs with the help of SfM photogrammetry. Six or seven models are carried out each year on these cliffs to determine their seasonal activity. The data acquisition is as follows: using a reflex camera (Nikon D810, 35 mm Sigma lens), 450–700 shots are taken along four photographic lines located at the bottom and top of the beach, on the top of basal scarp and at the foot of the gullies. Photographs must have a recovery rate of 60% between them. In order to set all models on the same geographical reference (RGF93—Lambert 93), fifteen targets are distributed over the cliff studied portion, then surveyed using a Trimble differential GPS that has a centimetric accuracy in longitude, latitude and altitude. Once the acquisition in the field is completed, pictures are inserted into the Agisoft Photoscan ® software, then the three-dimension models are built according to the different steps.

On each 3D model produced, a DTM is extracted and then integrated into a GIS software. Each DTM is compared with the one acquired previously, by subtracting the altitude values between both. This allows mapping eroded areas (represent a loss between -0.05 and -3 m in red) and accumulation areas (represent a deposit between $+0.05$ and $+3$ m in blue) (Fig. 8). Each mass movement is then identified, digitized and integrated into a database where is integrated the type of movement, its surface area and also the volume of materials mobilized. The repeated use over time of photogrammetry SfM, as well as its centimetric accuracy, allow to improve the understanding of mass movements of the Vaches Noires cliffs, but also to determine the rates and rhythms of evolution in relation to marine, hydrological, and meteorological conditions.

It is necessary to keep in mind that SfM photogrammetry is one of the many spatial remote-sensing tools available for mass movement analysis. It must be always supported by observations and field measurements or monitoring. It has several advantages such as the 3D processing speed (from the field to the laboratory step) and also the possibility of quickly mobilizing the equipment on the field during major morphogenetic events (storms, floods, …). The centimetric accuracy obtained with quality measuring instruments (differential GPS, total station) or the textured 3D model obtained that facilitates the reading and analysis of the modelled geographic objects for geomorphologists. However, there are limits due to photographic protocol (Fig. 8). For example, a uniform light is essential on each picture, and the camera configurations should not be changed during the photographic acquisition. Nowadays, it is also difficult to remove totally vegetation from 3D models. Hence, others 3D modelling tools such as LiDAR are used to counter these difficulties. Several papers try a comparative analysis of the performance of these two methods (TLS and SfM in Salvini et al. 2013; Ouédraogo et al. 2014). Various studies underline that these two techniques provide very high-resolution topographic data with heterogeneous point spacing and density. The resolution of the data will mainly depend on the protocol of survey and on the specificities of the study area. The steeper and more vegetated the study area is, the less accurate the point cloud will be.

6 Discussion

Landslides detection is still a challenging task due to many forms and sizes landslides can take and the context of their occurrence. Remotely sensed data and associated tools have become essential for landslide detection and monitoring. In recent decades, several systems have been developed with the possibility of free of charge satellite solutions with high-resolution images such as Copernicus Sentinel-1 and Sentinel-2, and low-cost airborne solutions such as UAVs equipped with different types of sensors (i.e.,

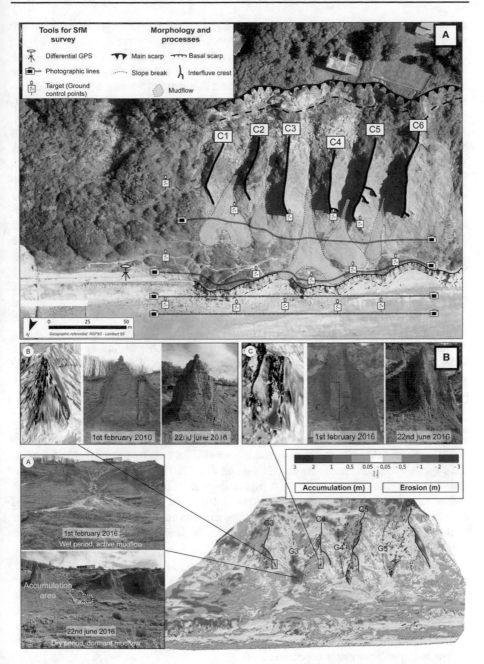

Fig. 8 Presentation of the photogrammetric SfM technique uses on the Vaches Noires cliffs (Normandy, France). **a** Data acquisition strategy on the field, **b** difference elevation model with pictures of actives areas

multi- hyperspectral, LiDAR sensors). Several applications use remote sensing for landslide mapping and monitoring (Tofani et al. 2013) and these approaches can be considered today as important as field surveys. For risk assessment, the amount of data used is always important. The challenge is to define the most appropriate spatial and temporal scale of analysis and the most appropriate support (spatial, airborne, ground-based).

With the increase of the number of remotely sensed data and their increase in resolution, automatic processing techniques have become more and more widespread. The landscape change detection is usually performed automatically using algorithms developed to compare 3D point clouds, DEMs or multispectral images. But the manual interpretation of data (spaceborne, airborne, terrestrial images) is still effective.

For example, in permafrost regions, monitoring of retrogressive thaw slumps has been so far mostly based on manual interpretation, especially with Landsat data. Machine learning has been rarely used to date for land cover classification tasks in high latitudes (see also Bartsch et al. 2016). This is attributed to the size of the features (mixed pixel effect) and ambiguities in reflectance patterns in tundra landscapes. Many further bare tundra surface types are existing. Many thousands of landslides are initiated and reactivated related to air temperature fluctuations and subsequent permafrost thaw across the entire Arctic. Visual interpretation is therefore insufficient to obtain a complete picture for the Arctic. Moreover, it is time-consuming and labour-intensive. Sentinel-2 with its 10 m resolution and high revisit time provides an important step forward in monitoring of these retrogressive thaw slumps and in general landslides. Consequently, inventories can be frequently updated with recent features and seasonal changes. Obviously, automatic detection of these features based on high-resolution data will be of high value for climate change impact assessment in such regions.

The application of machine learning and especially deep learning approaches with VHR images is expected to ameliorate both landslides detection and their evolution in various environments in the near future. Automation approach with convolutional neural networks (CNNs), has made a series of improvement in image classification and object detection. Although conventional deep learning architectures are frequently applied (Table 2) for Landslide Susceptibility Mapping, with application of CNN or derived methods (object detection, semantic segmentation) for landslide detection are still limited. Conventional machine learning techniques (e.g., SVM, RF) are suitable for landslide assessment with small dataset (Huang and Zhao 2018), but the distinction of landslide type remains difficult and results highly depend on the chosen algorithm, network architecture and dataset (Ghorbanzadeh et al. 2019). Most of the studies today are based on supervised approaches that require training samples for training the network. These methods are robust but labelling training samples is time-consuming. Today, the open challenges are:

- VHR satellite images can be acquired timely after a major landslide event and/or with daily temporal resolution at nearly global coverage. In combination with the potentiality of deep learning algorithms, one of the major challenges is to provide robust solutions for near real-time hazard detection along the lines of what is being done for flood management (e.g., European Flood Awareness System from Copernicus in https://emergency.copernicus.eu/).
- Reduce costs in terms of data handling and processing and the technical skills for near real-time hazard detection.
- Combine SAR offset tracking and InSAR analysis to improve landslides inventories on a national scale level.
- Foster knowledge transfer from scientific community to stakeholders and populations at risk.

7 Conclusions

Numerous fatalities and thousands of deaths result from landslides each year. In a context of global changes, numerous uncertainties concern landslide occurrence. Consequently, it is necessary to identify areas affected by landslides. On the one hand, they caused structural, physical and economic damages when they occur. On the other hand, these processes may cause major environmental damages, as is the case in the Arctic. The Arctic is one of the areas which has recently gained more and more attention in the context of mass movements due to their abundance and relationship to climate change. They are not only a hazard for people and infrastructure in the Arctic but are also relevant on global scale. As soils are carbon-rich in the Arctic, they also play a role for the carbon cycle. Mass movements play an important role for carbon transport into streams and the ocean, leading to acidification. It has been estimated that CO_2 efflux in rill runoff thaw streams (runoff) within retrogressive thaw slumps (RTSs) is four times greater than in adjacent streams. The quantification of overall transport into the oceans, by streams and coastal erosion still remains to be quantified. Satellite data are expected to support such analyses by combination of marine (ocean colour) and terrestrial observations (land surface features as proxy for soil properties). A combination of multispectral, Lidar and radar information together with advanced analyses techniques are needed to fully capture their occurrence and impact. The presented examples illustrate the applicability as well as gaps of the various types of remote-sensing techniques (InSAR, sub-pixel correlation, photogrammetry…) and highlight the necessity of automatization of the processing especially for landslide detection, mapping and surface deformation assessment. This paper presents numerous remote-sensing techniques and highlights the difficulties related to these methods, both in terms of spatial and temporal resolution and sometimes their difficulty to implement because of the specificity of the terrain (accessibility, vegetation cover, landslide velocity and size). Consequently, it may be necessary to integrate data from different sources of investigations (TLS associated to SfM or TLS associated to ALS) to overcome the limitations of each remote-sensed technique and to cross-validate the result with conventional techniques. Remotely sensed data can be considered as powerful and well-established information sources for landslide mapping, monitoring and hazard analysis and a wide range of available techniques and supports can be useful depending on the size and velocity of the hazard.

Acknowledgements This paper is a result of the international workshop on "Natural and man-made hazards monitoring by the Earth Observation missions: current status and scientific gaps" held at the International Space Science Institute (ISSI), Bern, Switzerland, on April 15–18, 2019. We thank all the authors for their contribution in this article with the result of their researches. The repeated surveys in Normandy, are supported by the ANR Project "RICOCHET: multi-RIsk assessment on Coastal territory in a global CHange context" funded by the French Research National Agency (ANR-16-CE03-0008). Permafrost researches benefited from Nunataryuk Project (H2020 Research and Innovation Programme under Grant Agreement No. 773421), ESA DUE GlobPermafrost Project (4000116196/15/I–NB) as well as ESA Climate Change Initiative Project on Permafrost (4000123681/18/I-NB).

References

Abellan A, Derron MH, Jaboyedoff M (2016) Use of 3D points clouds in geohazards. Special issue: current challenges and future trends. Remote Sens 8(2):130. https://doi.org/10.3390/rs8020130
Achour Y, Pourghasemi HR (2019) How do machine learning techniques help in increasing accuracy of landslide susceptibility maps? Geosci Front 11(3):871–883. https://doi.org/10.1016/j.gsf.2019.10.001

Aditian A, Kubota T, Shinohara Y (2018) Comparison of GIS-based landslide susceptibility models using frequency ratio, logistic regression, and artificial neural network in a tertiary region of Ambon, Indonesia. Geomorphology 318:101–111. https://doi.org/10.1016/j.geomorph.2018.06.006

Anders NS, Seijmonsbergen AC, Bouten W (2011) Segmentation optimization and stratified object-based analysis for semi-automated geomorphological mapping. Remote Sens Environ 115(12):2976–2985. https://doi.org/10.1016/j.rse.2011.05.007

Arabameri A, Pradhan B, Rezaei K, Lee CW (2019) Assessment of landslide susceptibility using statistical- and artificial intelligence-based FR–RF integrated model and multiresolution DEMs. Remote Sens 11(9):999. https://doi.org/10.3390/rs11090999

Ardizzone F, Cardinali M, Galli M, Guzzetti F, Reichenbach P (2007) Identification and mapping of recent rainfall-induced landslides using elevation data collected by airborne Lidar. Nat Hazards Earth Syst Sci 7(6):637–650

Ayoub F, Leprince S, Avouac JP (2009) Co-registration and correlation of aerial photographs for ground deformation measurements. ISPRS J Photogramm Remote Sens 64(6):551–560. https://doi.org/10.1016/j.isprsjprs.2009.03.005

Babkina EA, Leibman MO, Dvornikov YA, Fakashchuk NY, Khairullin RR, Khomutov AV (2019) Activation of cryogenic processes in Central Yamal as a result of regional and local change in climate and thermal state of permafrost. Russ Meteorol Hydrol 44(4):283–290. https://doi.org/10.3103/S1068373919040083

Ballabio C, Sterlacchini S (2012) Support vector machines for landslide susceptibility mapping: the Staffora river basin case study, Italy. Math Geosci 44:47–70. https://doi.org/10.1007/s11004-011-9379-9

Bamler R, Hartl P (1998) Synthetic aperture radar interferometry. Inverse Prob 14(4):R1

Barbarella M, Fiani M, Lugli A (2017) Uncertainty in terrestrial laser scanner surveys of landslides. Remote Sens 9(2):113. https://doi.org/10.3390/rs9020113

Barboux C, Delaloye R, Lambiel C, Strozzi T, Collet C, Raetzo H (2013) Surveying the activity of landslides and rock glaciers above the tree line with InSAR. In: Graf C (ed) Mattertal-ein Tal in Bewegung. Jahrestagung der Schweizerischen Geomorphologischen Gesellschaft 29. Juni–1. Juli 2011, St. Niklaus, Birmensdorf, Eidg. Forschungsanstalt WSL, pp 7–19

Barboux C, Delaloye R, Lambiel C (2014) Inventorying slope movements in an Alpine environment using DInSAR. Earth Surf Proc Land 39(15):2087–2099. https://doi.org/10.1002/esp.3603

Barla G, Antolini F, Barla M, Mensi E, Piovano G (2010) Monitoring of the Beauregard landslide (Aosta Valley, Italy) using advanced and conventional techniques. Eng Geol 116(3–4):218–235. https://doi.org/10.1016/j.enggeo.2010.09.004

Bartsch A, Höfler A, Kroisleitner C, Trofaier AM (2016) Land cover mapping in northern high latitude permafrost regions with satellite data: achievements and remaining challenges. Remote Sens 8:979. https://doi.org/10.3390/rs8120979

Bartsch A, Leibman M, Strozzi T, Khomutov A, Widhalm B, Babkina E et al (2019) Seasonal progression of ground displacement identified with satellite radar interferometry and the impact of unusually warm conditions on permafrost at the Yamal Peninsula in 2016. Remote Sens 11(16):1865. https://doi.org/10.3390/rs11161865

Behling R, Roessner S, Kaufmann H, Kleinschmit B (2014) Automated spatiotemporal landslide mapping over large areas using rapideye time series data. Remote Sens 6(9):8026–8055. https://doi.org/10.3390/rs6098026

Bell R, Petschko H, Röhrs M, Dix A (2012) Assessment of landslide age, landslide persistence and human impact using airborne laser scanning digital terrain models. Geografiska Annaler Ser A Phys Geogr 94(1):135–156. https://doi.org/10.1111/j.1468-0459.2012.00454.x

Berardino P, Fornaro G, Lanari R, Sansosti E (2002) A new algorithm for surface deformation monitoring based on small baseline differential SAR interferograms. IEEE Trans Geosci Remote Sens 40(11):2375–2383. https://doi.org/10.1109/TGRS.2002.803792

Biass S, Orr TR, Houghton BF, Ratrick MR, James M, Turner N (2019) Insights into Pāhoehoe lava emplacement using visible and thermal structure-from-motion photogrammetry. J Geophys Res Solid Earth 124(6):5678–5695. https://doi.org/10.1029/2019JB017444

Bitelli G, Dubbini M, Zanutta A (2004) Terrestrial laser scanning and digital photogrammetry techniques to monitor landslide bodies. Int Arch Photogramm Remote Sens Spatial Inf Sci 35(B5):246–251

Breiman L (2001) Random forests. Mach Learn 45(1):5–32

Brun F, Buri P, Miles E, Wagnon P, Steiner J, Berthier E, Ragettli S, Kraaijenbrink P, Immerzeel W, Pellicciotti F (2016) Quantifying volume loss from ice cliffs on debris-covered glaciers using high-resolution terrestrial and aerial photogrammetry. J Glaciol 62(234):684–695. https://doi.org/10.1017/jog.2016.54

Brunsden D (1993) Mass movement; the research frontier and beyond: a geomorphological approach. Geomorphology 7(1–3):85–128

Bui DT, Hoang ND, Nguyen H, Tran XL (2019) Spatial prediction of shallow landslide using Bat algorithm optimized machine learning approach: a case study in Lang Son Province, Vietnam. Adv Eng Inf 42:100978. https://doi.org/10.1016/j.aei.2019.100978

Bui DT, Tsangaratos P, Nguyen VT, Van Liem N, Trinh PT (2020) Comparing the prediction performance of a deep learning neural network model with conventional machine learning models in landslide susceptibility assessment. CATENA 188:104426. https://doi.org/10.1016/j.catena.2019.104426

Bunn MD, Leshchinsky BA, Olsen MJ, Booth A (2019) A simplified, object-based framework for efficient landslide inventorying using LIDAR digital elevation model derivatives. Remote Sens 11(3):303. https://doi.org/10.3390/rs11030303

Burns WJ, Madin I (2009) Protocol for inventory mapping of landslide deposits from light detection and ranging (LiDAR) imagery. Oregon Department of Geology and Mineral Industries, Portland, pp 1–30

Burns WJ, Duplantis S, Jones CB, English JT (2012) Lidar data and landslide inventory maps of the North Fork Siuslaw River and Big Elk Creek watersheds. Lane, Lincoln, and Benton Counties, Oregon

Carlà T, Tofani V, Lombardi L, Raspini F, Bianchini S, Bertolo D et al (2019) Combination of GNSS, satellite InSAR, and GBInSAR remote sensing monitoring to improve the understanding of a large landslide in high alpine environment. Geomorphology 335:62–75. https://doi.org/10.1016/j.geomorph.2019.03.014

Carr BB, Clarke BA, Arrowsmith JR, Vanderkluysen L, Eko Dhanu B (2018) The emplacement of the active lava flow at Sinabung Volcano, Sumatra, Indonesia, documented by structure-from-motion photogrammetry. J Volcanol Geoth Res 382:164–172. https://doi.org/10.1016/j.jvolgeores.2018.02.004

Casagli N, Cigna F, Bianchini S, Hölbling D, Füreder P, Righini G et al (2016) Landslide mapping and monitoring by using radar and optical remote sensing: examples from the EC-FP7 project SAFER. Remote Sens Appl Soc Environ 4:92–108. https://doi.org/10.1016/j.rsase.2016.07.001

Casagli N, Frodella W, Morelli S, Tofani V, Ciampalini A, Intrieri E et al (2017) Spaceborne, UAV and ground-based remote sensing techniques for landslide mapping, monitoring and early warning. Geoenviron Disasters 4(9):1–23. https://doi.org/10.1186/s40677-017-0073-1

Casson B, Baratoux D, Delacourt D, Allemand P (2003) Seventeen years of the "La Clapière" landslide evolution analysed from ortho-rectified aerial photographs. Eng Geol 68(1–2):123–139

Casson B, Delacourt C, Allemand P (2005) Contribution of multi-temporal remote sensing images to characterize landslide slip surface? Application to the La Clapière landslide (France). NHESS 5:425–437

Catani F, Casagli N, Ermini L, Righini G, Menduni G (2005) Landslide hazard and risk mapping at catchment scale in the Arno River basin. Landslides 2(4):329–342. https://doi.org/10.1007/s10346-005-0021-0

Catani F, Lagomarsino D, Segoni S, Tofani V (2013) Exploring model sensitivity issues across different scales in landslide susceptibility. NHESS 1(2):583–623

Chanut MA, Kasperski J, Dubois L, Dauphin S, Duranthon JP (2017) Quantification des déplacements 3D par la méthode PLaS: application au glissement du Chambon (Isère). Rev Fr Géotech 150(4):1–14. https://doi.org/10.1051/geotech/2017009

Chen W, Li X, Wang Y, Chen G, Liu S (2014) Forested landslide detection using LiDAR data and the random forest algorithm: a case study of the Three Gorges, China. Remote Sens Environ 152:291–301. https://doi.org/10.1016/j.rse.2014.07.004

Chen W, Pourghasemi HR, Kornejady A, Zhang N (2017) Landslide spatial modeling: introducing new ensembles of ANN, MaxEnt, and SVM machine learning techniques. Geoderma 305:314–327. https://doi.org/10.1016/j.geoderma.2017.06.020

Chigira M, Duan F, Yagi H, Furuya T (2004) Using an airborne laser scanner for the identification of shallow landslides and susceptibility assessment in an area of ignimbrite overlain by permeable pyroclastics. Landslides 1(3):203–209. https://doi.org/10.1007/s10346-004-0029-x

Chigira M, Wu X, Inokuchi T, Wang G (2010) Landslides induced by the 2008 Wenchuan earthquake, Sichuan, China. Geomorphology 118(3–4):225–238. https://doi.org/10.1016/j.geomorph.2010.01.003

Chunhui Z, Bing G, Lejun Z, Xiaoqing W (2018) Classification of Hyperspectral Imagery based on spectral gradient, SVM and spatial random forest. Infrared Phys Technol 95:61–69. https://doi.org/10.1016/j.infrared.2018.10.012

Ciampalini A, Raspini F, Bianchini S, Frodella W, Bardi F, Lagomarsino D et al (2015) Remote sensing as tool for development of landslide databases: the case of the Messina Province (Italy) geodatabase. Geomorphology 249:103–118. https://doi.org/10.1016/j.geomorph.2015.01.029

Colesanti C, Wasowski J (2006) Investigating landslides with space-borne Synthetic Aperture Radar (SAR) interferometry. Eng Geol 88(3–4):173–199. https://doi.org/10.1016/j.enggeo.2006.09.013

Comert R, Avdan U, Gorum T, Nefeslioglu HA (2019) Mapping of shallow landslides with object-based image analysis from unmanned aerial vehicle data. Eng Geol 260:105264. https://doi.org/10.1016/j.enggeo.2019.105264

Conner JC, Olsen MJ (2014) Automated quantification of distributed landslide movement using circular tree trunks extracted from terrestrial laser scan data. Comput Geosci 67:31–39. https://doi.org/10.1016/j.cageo.2014.02.007

Corsini A, Farina P, Antonello G, Barbieri M, Casagli N, Coren F et al (2006) Space-borne and ground-based SAR interferometry as tools for landslide hazard management in civil protection. Int J Remote Sens 27(12):2351–2369. https://doi.org/10.1080/01431160600554405

Corsini A, Berti M, Monni A, Pizziolo M, Bonacini F, Cervi F et al (2013) Rapid assessment of landslide activity in Emilia Romagna using GB-InSAR short surveys. Landslide Sci Pract. https://doi.org/10.1007/978-3-642-31445-2_51

Cortes C, Vapnik V (1995) Support-vector networks. Mach Learn 20(3):273–290

Costa S, Maquaire O, Letortu P, Thirard G, Compain V, Roulland T et al (2019) Sedimentary Coastal cliffs of Normandy: modalities and quantification of retreat. J Coast Res 88(SI):46–60. https://doi.org/10.2112/SI88-005.1

Crippen RE (1992) Measurement of subresolution terrain displacements using SPOT panchromatic imagery. Report 15(1):56–61

Crosetto M, Monserrat O, Luzi G, Cuevas-González M, Devanthéry N (2014) A non interferometric procedure for deformation measurement using GB-SAR imagery. IEEE Geosci Remote Sens Lett 11(1):34–38. https://doi.org/10.1109/LGRS.2013.2245098

Crosetto M, Monserrat O, Cuevas-González M, Devanthéry N, Crippa B (2016) Persistent scatterer interferometry: a review. ISPRS J Photogramm Remote Sens 115:78–89. https://doi.org/10.1016/j.isprs.2015.10.011

Dai K, Li Z, Tomás R, Liu G, Yu B, Wang X, Cheng H, Chen J, Stockamp J (2016) Monitoring activity at the Daguangbao mega-landslide (China) using Sentinel-1 TOPS time series interferometry. Remote Sens Environ 186:501–513. https://doi.org/10.1016/j.rse.2016.09.009

De Michele M, Briole P (2007) Deformation between 1989 and 1997 at Piton de la Fournaise volcano retrieved from correlation of panchromatic airborne images. Geophys J Int 169(1):357–364. https://doi.org/10.1111/j.1365-246X.2006.03307.x

De Michele M, Raucoules D, De Sigoyer J, Pubellier M, Chamot-Rooke N (2010) Three-dimensional surface displacement of the 2008 May 12 Sichuan earthquake (China) derived from Synthetic Aperture Radar: evidence for rupture on a blind thrust. Geophys J Int 183(3):1097–1103. https://doi.org/10.1111/j.1365-246X.2010.04807.x

Debella-Gilo M, Kääb A (2011) Sub-pixel precision image matching for measuring surface displacements on mass movements using normalized cross-correlation. Remote Sens Environ 115(1):130–142. https://doi.org/10.1016/j.rse.2010.08.012

Delacourt C, Allemand P, Casson B, Vadon H (2004) Velocity field of the "La Clapière" landslide measured by the correlation of aerial and QuickBird satellite images. Geophys Res Lett 31(15):1–5. https://doi.org/10.1029/2004GL020193

Delacourt C, Allemand P, Berthier E, Raucoules D, Casson B, Grandjean P, Pambrun C, Varel E (2007) Remote-sensing techniques for analysing landslide kinematics: a review. Bulletin de la Société Géologique de France 178(2):89–100. https://doi.org/10.2113/gssgfbull.178.2.89

Delacourt C, Raucoules D, Le Mouélic S, Carnec C, Feurer D, Allemand P, Cruchet M (2009) Observation of a large landslide on La Reunion Island using differential SAR interferometry (JERS and Radarsat) and correlation of optical (Spot5 and Aerial) images. Sensors 9(1):616–630. https://doi.org/10.3390/s90100616

Di Stefano C, Palmeri V, Pampalone V (2019) An automatic approach for rill network extraction to measure rill erosion by terrestrial and low-cost unmanned aerial vehicle photogrammetry. Hydrol Process 33(13):1883–1895. https://doi.org/10.1002/hyp.13444

Du J, Glade T, Woldai T, Chai B, Zeng B (2020) Landslide susceptibility assessment based on an incomplete landslide inventory in the Jilong Valley, Tibet, Chinese Himalayas. Eng Geol. https://doi.org/10.1016/j.enggeo.2020.105572

Echelard T, Krysiecki JM, Gay M, Schoeneich P (2013) Détection des mouvements de glaciers rocheux dans les Alpes françaises par interférométrie radar différentielle (D-InSAR) dérivée des archives satellitaires ERS (European Remote Sensing). Géomorphologie Relief Processus Environnement 19(3):231–242. https://doi.org/10.4000/geomorphologie.10264

Fan X, Xu Q, Scaringi G, Dai L, Li W, Dong X, Zhu X, Pei X, Dai K, Havenith HB (2017) Failure mechanism and kinematics of the deadly June 24th 2017 Xinmo landslide, Maoxian, Sichuan, China. Landslides 14(6):2129–2146. https://doi.org/10.1007/s10346-017-0907-7

Fan X, Domènech G, Scaringi G, Huang R, Xu Q, Hales TC et al (2018) Spatio-temporal evolution of mass wasting after the 2008 Mw 7.9 Wenchuan earthquake revealed by a detailed multi-temporal inventory. Landslides 15(12):2325–2341. https://doi.org/10.1007/s10346-018-1054-5

Fang Z, Wang Y, Peng L, Hong H (2020) Integration of convolutional neural network and conventional machine learning classifiers for landslide susceptibility mapping. Comput Geosci. https://doi.org/10.1016/j.cageo.2020.104470

Farahmand A, AghaKouchak A (2013) A satellite-based global landslide model. Nat Hazards Earth Syst Sci 13(5):1259–1267. https://escholarship.org/uc/item/0p905918

Fernández T, Pérez J, Cardenal J, Gómez J, Colomo C, Delgado J (2016) Analysis of landslide evolution affecting olive groves using UAV and photogrammetric techniques. Remote Sens 8(10):837. https://doi.org/10.3390/rs8100837

Francioni M, Coggan J, Eyre M, Stead D (2018) A combined field/remote sensing approach for characterizing landslide risk in coastal areas. Int J Appl Earth Obs Geoinf 67:79–95. https://doi.org/10.1016/j.jag.2017.12.016

French HM, Williams P (2017) The periglacial environment, 4th edn. Longman, London. https://doi.org/10.1002/9781119132820

Fressard M, Maquaire O, Thiery Y, Davidson R, Lissak C (2016) Multi-method characterisation of an active landslide: case study in the Pays d'Auge plateau (Normandy, France). Geomorphology 270:22–39. https://doi.org/10.1016/j.geomorph.2016.07.001

Freund Y, Schapire RE (1995) A desicion-theoretic generalization of on-line learning and an application to boosting. Eur Conf Comput Learn Theory. https://doi.org/10.1007/3-540-59119-2_166

Galli M, Ardizzone F, Cardinali M, Guzzetti F, Reichenbach P (2008) Comparing landslide inventory maps. Geomorphology 94(3–4):268–289. https://doi.org/10.1016/j.geomorph.2006.09.023

Gance J, Malet JP, Dewez T, Travelletti J (2014) Target Detection and Tracking of moving objects for characterizing landslide displacements from time-lapse terrestrial optical images. Eng Geol 172:26–40. https://doi.org/10.1016/j.enggeo.2014.01.003

Ghorbanzadeh O, Blaschke T, Gholamnia K, Meena SR, Tiede D, Aryal J (2019) Evaluation of different machine learning methods and deep-learning convolutional neural networks for landslide detection. Remote Sens 11(2):196. https://doi.org/10.3390/rs11020196

Gilham J, Barlow J, Moore R (2019) Detection and analysis of mass wasting events in chalk sea cliffs using UAV photogrammetry. Eng Geol 250:101–112. https://doi.org/10.1016/j.enggeo.2019.01.013

Girshick R (2015) Fast r-cnn. Proc IEEE Int Conf Comput Vis. https://doi.org/10.1109/ICCV.2015.169

Gomez C, Kennedy B (2018) Capturing volcanic plumes in 3D with UAV-based photogrammetry at Yasur Volcano Vanuatu. J Volcanol Geoth Res 350:84–88. https://doi.org/10.1016/j.jvolgeores.2017.12.007

Gong M, Zhao J, Liu J, Miao Q, Jiao L (2015) Change detection in synthetic aperture radar images based on deep neural networks. IEEE Trans Neural Netw Learn Syst 27(1):125–138. https://doi.org/10.1109/TNNLS.2015.2435783

Görüm T (2019) Landslide recognition and mapping in a mixed forest environment from airborne LiDAR data. Eng Geol 258:105155. https://doi.org/10.1016/j.enggeo.2019.105155

Groos AR, Bertschinger TJ, Kummer CM, Erlwein S, Munz L, Philipp A (2019) The potential of low-cost UAVs and open-source photogrammetry software for high-resolution monitoring of Alpine glaciers: a case study from the Kanderfirn (Swiss Alps). Geosciences 9(8):356. https://doi.org/10.3390/geosciences9080356

Gudino-Elizondo N, Biggs TW, Castillo C, Bingner RL et al (2018) Measuring ephemeral gully erosion rates and topographical thresholds in an urban watershed using unmanned aerial systems and structure from motion photogrammetric techniques. Land Degrad Dev 29(6):1896–1905. https://doi.org/10.1002/ldr.2976

Guzzetti F, Mondini AC, Cardinali M, Fiorucci F, Santangelo M, Chang KT (2012) Landslide inventory maps: new tools for an old problem. Earth Sci Rev 112(1–2):42–66. https://doi.org/10.1016/j.earscirev.2012.02.001

Heindel RC, Chipman JW, Dietrich JT, Virginia RA (2018) Quantifying rates of soil deflation with structure-from-motion photogrammetry in west Greenland. Arct Antarct Alp Res 50(1):S100012. https://doi.org/10.1080/15230430.2017.1415852

Herrera G, Fernández-Merodo JA, Mulas J, Pastor M, Luzi G, Monserrat O (2009) A landslide forecasting model using ground based SAR data: the Portalet case study. Eng Geol 105(3–4):220–230. https://doi.org/10.1016/j.enggeo.2009.02.009

Hong H, Pradhan B, Xu C, Bui DT (2015) Spatial prediction of landslide hazard at the Yihuang area (China) using two-class kernel logistic regression, alternating decision tree and support vector machines. CATENA 133:266–281. https://doi.org/10.1016/j.catena.2015.05.019

Hong H, Pradhan B, Jebur MN, Bui DT, Xu C, Akgun A (2016) Spatial prediction of landslide hazard at the Luxi area (China) using support vector machines. Environ Earth Sci 75(1):40. https://doi.org/10.1007/s12665-015-4866-9

Hu Q, Zhou Y, Wang S, Wang F (2020) Machine learning and fractal theory models for landslide susceptibility mapping: case study from the Jinsha River Basin. Geomorphology 351:106975. https://doi.org/10.1016/j.geomorph.2019.106975

Huang RQ, Li AW (2009) Analysis of the geo-hazards triggered by the 12 May 2008 Wenchuan Earthquake, China. Bull Eng Geol Environ 68(3):363–371. https://doi.org/10.1007/s10064-009-0207-0

Huang Y, Zhao L (2018) Review on landslide susceptibility mapping using support vector machines. CATENA 165:520–529. https://doi.org/10.1016/j.catena.2018.03.003

Huang F, Cao Z, Guo J, Jiang SH, Li S, Guo Z (2020) Comparisons of heuristic, general statistical and machine learning models for landslide susceptibility prediction and mapping. CATENA 191:104580. https://doi.org/10.1016/j.catena.2020.104580

Hungr O, Leroueil S, Picarelli L (2014) The Varnes classification of landslide types, an update. Landslides 11(2):167–194. https://doi.org/10.1007/s10346-013-0436-y

Ingles J, Darrozes J, Soula JC (2006) Effects of the vertical component of ground shaking on earthquake-induced landslide displacements using generalized Newmark analysis. Eng Geol 86(2–3):134–147. https://doi.org/10.1016/j.enggeo.2006.02.018

Irrgang AM, Lantuit H, Gordon RR, Piskor A, Manson GK (2019) Impacts of past and future coastal changes on the Yukon coast: threats for cultural sites, infrastructure, and travel routes. Arct Sci 5(2):107–126. https://doi.org/10.1139/as-2017-0041

Jaboyedoff M, Oppikofer T, Abellán A, Derron MH, Loye A, Metzger R, Pedrazzini A (2012) Use of LIDAR in landslide investigations: a review. Nat Hazards 61(1):5–28. https://doi.org/10.1007/s11069-010-9634-2

Jaboyedoff M, Del Gaudio V, Derron MH, Grandjean G, Jongmans D (2019) Characterizing and monitoring landslide processes using remote sensing and geophysics. Eng Geol 259:105167

James MR, Chandler JH, Eltner A, Fraser C, Miller PE, Mills JP, Noble T, Robson S, Lane SN (2019) Guidelines on the use of structure-from-motion photogrammetry in geomorphic research. Earth Surf Process Landf 44:2081–2084. https://doi.org/10.1002/esp.4637

Jaud M, Letortu P, Théry C, Grandjean P, Costa S, Maquaire O, Davidson R, Le Dantec N (2019) UAV survey of a coastal cliff face: selection of the best imaging angle. Measurement 139:10–20. https://doi.org/10.1016/j.measurement.2019.02.024

Jiang S, Wen BP, Zhao C, Li RD, Li ZH (2016) Kinematics of a giant slow-moving landslide in Northwest China: constraints from high-resolution remote sensing imagery and GPS monitoring. J Asian Earth Sci 123:34–46. https://doi.org/10.1016/j.jseaes.2016.03.019

Jones MKW, Pollard WH, Jones BM (2019) Rapid initialization of retrogressive thaw slumps in the Canadian high Arctic and their response to climate and terrain factors. Environ Res Lett 14(5):055006. https://doi.org/10.1088/1748-9326/ab12fd

Jorgenson MT, Grosse G (2016) Remote sensing of landscape change in permafrost regions. Permafrost Periglac Process 27(4):324–338. https://doi.org/10.1002/ppp.1914

Jugie M, Gob F, Virmoux C, Brunstein D, Tamisier V, Le Cœur C, Grancher D (2018) Characterizing and quantifying the discontinuous bank erosion of a small low energy river using structure-from-motion photogrammetry and erosion pins. J Hydrol 563:418–434. https://doi.org/10.1016/j.jhydrol.2018.06.019

Kadavi PR, Lee CW, Lee S (2018) Application of ensemble-based machine learning models to landslide susceptibility mapping. Remote Sens 10(8):1252. https://doi.org/10.3390/rs10081252

Kalantar B, Pradhan B, Naghibi SA, Motevalli A, Mansor S (2018) Assessment of the effects of training data selection on the landslide susceptibility mapping: a comparison between support vector machine (SVM), logistic regression (LR) and artificial neural networks (ANN). Geomat Nat Hazards Risk 9(1):49–69. https://doi.org/10.1080/19475705.2017.1407368

Kenner R, Bühler Y, Delaloye R, Ginzler C, Phillips M (2014) Monitoring of high alpine mass movements combining laser scanning with digital airborne photogrammetry. Geomorphology 206:492–504. https://doi.org/10.1016/j.geomorph.2013.10.020

Klinger Y, Michel R, King GCP (2006) Evidence for an earthquake barrier model from Mw~7.8 Kokoxili (Tibet) earthquake slip-distribution. Earth Planet Sci Lett 242(3–4):354–364. https://doi.org/10.1016/j.epsl.2005.12.003

Ko FW, Lo FL (2018) From landslide susceptibility to landslide frequency: a territory-wide study in Hong Kong. Eng Geol 242:12–22. https://doi.org/10.1016/j.enggeo.2018.05.001

Kokalj Ž, Zakšek K, Oštir K, Pehani P, Čotar K, Somrak M et al (2016) Relief Visualization Toolbox, ver. 2.2.1 Manual. Remote Sens 3(2):398–415

Kromer R, Walton G, Gray B, Lato M, Group R (2019) Development and optimization of an automated fixed-location time-lapse photogrammetric rock slope monitoring system. Remote Sens 11:1890–1908. https://doi.org/10.3390/rs11161890

Kurtz C, Stumpf A, Malet JP, Gançarski P, Puissant A, Passat N (2014) Hierarchical extraction of landslides from multiresolution remotely sensed optical images. ISPRS J Photogramm Remote Sens 87:122–136. https://doi.org/10.1016/j.isprsjprs.2013.11.003

Lacroix P, Berthier E, Maquerhua ET (2015) Earthquake-driven acceleration of slow-moving landslides in the Colca valley, Peru, detected from Pléiades images. Remote Sens Environ 165:148–158. https://doi.org/10.1016/j.rse.2015.05.010

Lacroix P, Bièvre G, Pathier E, Kniess U, Jongmans D (2018) Use of Sentinel-2 images for the detection of precursory motions before landslide failures. Remote Sens Environ 215:507–516. https://doi.org/10.1016/j.rse.2018.03.042

Lacroix P, Araujo G, Hollingsworth J, Taipe E (2019) Self-entrainment motion of a slow-moving landslide inferred from landsat-8 time series. J Geophys Res Earth Surf 124(5):1201–1216. https://doi.org/10.1029/2018JF004920

Lanaras C, Bioucas-Dias J, Galliani S, Baltsavias E, Schindler K (2018) Super-resolution of Sentinel-2 images: learning a globally applicable deep neural network. ISPRS J Photogramm Remote Sens 146:305–319. https://doi.org/10.1016/j.isprsjprs.2018.09.018

Lang S, Füreder P, Rogenhofer E (2018) Earth observation for humanitarian operations. In: Yearbook on space policy 2016, Springer, Cham, pp 217–229. https://doi.org/10.1007/978-3-319-72465-2_10

Lantuit H, Pollard WH, Couture N, Fritz M, Schirrmeister L, Meyer H, Hubberten HW (2012) Modern and late Holocene retrogressive thaw slump activity on the Yukon coastal plain and Herschel Island, Yukon Territory, Canada. Permafr Periglac Process 23(1):39–51. https://doi.org/10.1002/ppp.1731

Lantz TC, Kokelj SV, Gergel SE, Henry GH (2009) Relative impacts of disturbance and temperature: persistent changes in microenvironment and vegetation in retrogressive thaw slumps. Glob Change Biol 15(7):1664–1675. https://doi.org/10.1111/j.1365-2486.2009.01917.x

Le Bivic R, Allemand P, Quiquerez A, Delacourt C (2017) Potential and limitation of SPOT-5 ortho-image correlation to investigate the cinematics of landslides: the example of "Mare à Poule d'Eau" (Réunion, France). Remote Sens 9(2):106. https://doi.org/10.3390/rs9020106

LeCun Y, Bengio Y, Hinton G (2015) Deep learning. Nature 521(7553):436–444. https://doi.org/10.1038/nature14539

Lei T, Zhang Y, Lv Z, Li S, Liu S, Nandi AK (2019a) Landslide inventory mapping from bitemporal images using deep convolutional neural networks. IEEE Geosci Remote Sens Lett 16(6):982–986. https://doi.org/10.1109/LGRS.2018.2889307

Lei T, Zhang Q, Xue D, Chen T, Meng H, Nandi AK (2019b) End-to-end change detection using a symmetric fully convolutional network for landslide mapping. In: ICASSP 2019-2019 IEEE international conference on acoustics, speech and signal processing (ICASSP), pp 3027–3031. https://doi.org/10.1109/ICASSP.2019.8682802

Leibman MO, Khomutov AV, Gubarkov AA, Dvornikov YA, Mullanurov DR (2015) The research station" Vaskiny Dachi", Central Yamal, West Siberia, Russia–a review of 25 years of permafrost studies. Fennia-Int J Geogr 193(1):3–30

Letortu P, Costa S, Cador JM, Coinaud C, Cantat O (2015a) Statistical and empirical analyses of the triggers of coastal chalk cliff failure. Earth Surf Process Landf 40:1371–1386. https://doi.org/10.1002/esp.3741

Letortu P, Costa S, Maquaire O, Delacourt C, Augereau E, Davidson R, Suanez S, Nabucet J (2015b) Retreat rates, modalities and agents responsible for erosion along the coastal chalk cliffs of Upper Normandy: the contribution of terrestrial laser scanning. Geomorphology 245:3–14. https://doi.org/10.1016/j.geomorph.2015.05.007

Letortu P, Costa S, Maquaire O, Davidson R (2019) Marine and subaerial controls of coastal chalk cliff erosion in Normandy (France) based on a 7-year laser scanner monitoring. Geomorphology 335:75–91. https://doi.org/10.1016/j.geomorph.2019.03.005

Lewkowicz AG (2007) Dynamics of active-layer detachment failures, Fosheim peninsula, Ellesmere Island, Nunavut, Canada. Permaf Periglac Process 18(1):89–103. https://doi.org/10.1002/ppp.578

Lewkowicz AG, Way RG (2019) Extremes of summer climate trigger thousands of thermokarst landslides in a High Arctic environment. Nat Commun 10(1):1329. https://doi.org/10.1038/s41467-019-09314-7

Li X, Wang L, Sung E (2008) AdaBoost with SVM-based component classifiers. Eng Appl Artif Intell 21(5):785–795. https://doi.org/10.1016/j.engappai.2007.07.001

Li X, Muller JP, Fang C, Zhao Y (2011) Measuring displacement field from TerraSAR-X amplitude images by subpixel correlation: an application to the landslide in Shuping. Three Gorges Area Acta Petrologica Sinica 27(12):3843–3850

Li X, Cheng X, Chen W, Chen G, Liu S (2015) Identification of forested landslides using LiDar data, object-based image analysis, and machine learning algorithms. Remote Sens 7(8):9705–9726. https://doi.org/10.3390/rs70809705

Li Z, Shi W, Myint SW, Lu P, Wang Q (2016) Semi-automated landslide inventory mapping from bitemporal aerial photographs using change detection and level set method. Remote Sens Environ 175:215–230. https://doi.org/10.1016/j.rse.2016.01.003

Lian C, Zeng Z, Yao W, Tang H (2013) Displacement prediction model of landslide based on a modified ensemble empirical mode decomposition and extreme learning machine. Nat Hazards 66(2):759–771. https://doi.org/10.1007/s11069-012-0517-6

Lin CY, Lo HM, Chou WC, Lin WT (2004) Vegetation recovery assessment at the Jou–Jou Mountain landslide area caused by the 921 Earthquake in Central Taiwan. Ecol Model 176(1–2):75–81. https://doi.org/10.1016/j.ecolmodel.2003.12.037

Lissak C, Maquaire O, Malet JP, Bitri A, Samyn K, Grandjean G et al (2014) Airborne and ground-based data sources for characterizing the morpho-structure of a coastal landslide. Geomorphology 217:140–151. https://doi.org/10.1016/j.geomorph.2014.04.019

Liu P, Wei Y, Wang Q, Chen Y, Xie J (2020) Research on post-earthquake landslide extraction algorithm based on improved u-net model. Remote Sens 12(5):894. https://doi.org/10.3390/rs12050894

Long J, Shelhamer E, Darrell T (2015) Fully convolutional networks for semantic segmentation. In Proceedings of the IEEE conference on computer vision and pattern recognition, pp 3431–3440

López-Davalillo JG, Monod B, Alvarez-Fernandez MI, Garcia GH et al (2014) Morphology and causes of landslides in Portalet area (Spanish Pyrenees): probabilistic analysis by means of numerical modelling. Eng Fail Anal 36:390–406. https://doi.org/10.1016/j.engfailanal.2013.10.015

Lowe D (1999) Object recognition from local scale-invariant features. In: Proceedings of the international conference of computer vision, Corfu, Greece

Lowe D (2004) Distinctive image features from scale-invariant keypoints. Int J Comput Vis 60:91–110

Lv Z, Liu T, Kong X, Shi C, Benediktsson JA (2020) Landslide inventory mapping with bitemporal aerial remote sensing images based on the dual-path full convolutional network. IEEE J Sel Top Appl Earth Observ Remote Sens 14(8)

Ma S, Xu C, Chao X, Zhang P, Liang X, Tian Y (2019) Geometric and kinematic features of a landslide in Mabian Sichuan, China, derived from UAV photography. Landslides 16:373–381. https://doi.org/10.1007/s10346-018-1104-z

Mallet C, Bretar F (2009) Full-waveform topographic lidar: state-of-the-art. ISPRS J Photogramm Remote Sens 64(1):1–16. https://doi.org/10.1016/j.isprsjprs.2008.09.007

Marc O, Meunier P, Hovius N (2017) Prediction of the area affected by earthquake-induced landsliding based on seismological parameters. Nat Hazards Earth Syst Sci 17(7):1159–1175. https://doi.org/10.5194/nhess-17-1159-2017

Marteau B, Vericat D, Gibbins C, Batalla RJ, Green DR (2016) Application of structure-from-motion photogrammetry to river restoration. Earth Surf Proc Land 42(3):503–515. https://doi.org/10.1002/esp.4086

McKean J, Roering J (2004) Objective landslide detection and surface morphology mapping using high-resolution airborne laser altimetry. Geomorphology 57(3–4):331–351. https://doi.org/10.1016/S0169-555X(03)00164-8

Medjkane M, Maquaire O, Costa S, Roulland T, Letortu P, Fauchard C, Antoine R, Davidson R (2018) High-resolution monitoring of complex coastal morphology changes: cross-efficiency of SfM and TLS-based survey (Vaches Noires cliffs, Normandy, France). Landslides 15(6):1097–1108. https://doi.org/10.1007/s10346-017-0942-4

Mezaal MR, Pradhan B, Rizeei HM (2018) Improving landslide detection from airborne laser scanning data using optimized Dempster-Shafer. Remote Sens 10(7):1029. https://doi.org/10.3390/rs10071029

Michel R, Avouac JP, Taboury J (1999) Measuring ground displacements from SAR amplitude images: application to the Landers earthquake. Geophys Res Lett 26(7):875–878. https://doi.org/10.1029/1999GL900138

Michoud C, Carrea D, Costa S, Derron MH, Jaboyedoff M, Davidson R, Delacourt C, Letortu P, Maquaire O (2014) Landslide detection and monitoring capability of boat-based mobile laser scanning along Dieppe coastal cliffs Normandy. Landslides 12(2):403–418. https://doi.org/10.1007/s10346-014-0542-5

Mohammady M, Pourghasemi HR, Amiri M (2019) Land subsidence susceptibility assessment using random forest machine learning algorithm. Environ Earth Sci 78(16):503. https://doi.org/10.1007/s12665-019-8518-3

Monserrat O, Crosetto M, Luzi G (2014) A review of ground-based SAR interferometry for deformation measurement. ISPRS J Photogramm Remote Sens 93:40–48. https://doi.org/10.1016/j.isprsjprs.2014.04.001

Moosavi V, Talebi A, Shirmohammadi B (2014) Producing a landslide inventory map using pixel-based and object-oriented approaches optimized by Taguchi method. Geomorphology 204:646–656. https://doi.org/10.1016/j.geomorph.2013.09.012

Nichol J, Wong MS (2005a) Detection and interpretation of landslides using satellite images. Land Degrad Dev 16(3):243–255. https://doi.org/10.1002/ldr.648

Nichol J, Wong MS (2005b) Satellite remote sensing for detailed landslide inventories using change detection and image fusion. Int J Remote Sens 26(9):1913–1926. https://doi.org/10.1080/01431160512331314047

Niethammer U, James MR, Rothmund S, Travelletti J, Joswig M (2012) UAV-based remote sensing of the Super-Sauze landslide: evaluation and results. Eng Geol 128:2–11. https://doi.org/10.1016/j.enggeo.2011.03.012

Nitze I (2018 Trends of land surface change from Landsat time-series 1999-2014, Transect T4, Eastern Canada. Alfred Wegener Institute, Helmholtz Centre for Polar and Marine Research, Bremerhaven, PANGAEA. https://doi.org/10.1594/PANGAEA.884276

Nitze I, Grosse G, Jones BM, Romanovsky VE, Boike J (2018) Remote sensing quantifies widespread abundance of permafrost region disturbances across the Arctic and Subarctic. Nat Commun 9(1):5423. https://doi.org/10.1038/s41467-018-07663-3

Novák D (2014) Local relief model (LRM) toolbox for ArcGIS. Czech Academy of Science, Staré Město

Obu J, Westermann S, Bartsch A, Berdnikov N, Christiansen HH, Dashtseren A et al (2019) Northern Hemisphere permafrost map based on TTOP modelling for 2000–2016 at 1 km2 scale. Earth Sci Rev 193:299–316. https://doi.org/10.1016/j.earscirev.2019.04.023

Ouédraogo MM, Degré A, Debouche C, Lisein J (2014) The evaluation of unmanned aerial system-based photogrammetry and terrestrial laser scanning to generate DEMs of agricultural watersheds. Geomorphology 214:339–355. https://doi.org/10.1016/j.geomorph.2014.02.016

Pakhale GK, Gupta PK (2010) Comparison of advanced pixel based (ANN and SVM) and object-oriented classification approaches using landsat-7 Etm+ data. Int J Eng Technol 2(4):245–251

Pal M (2005) Random forest classifier for remote sensing classification. Int J Remote Sens 26(1):217–222. https://doi.org/10.1080/01431160412331269698

Pánek T, Březný M, Kapustová V, Lenart J, Chalupa V (2019) Large landslides and deep-seated gravitational slope deformations in the Czech Flysch Carpathians: new LiDAR-based inventory. Geomorphology 346:106852. https://doi.org/10.1016/j.geomorph.2019.106852

Paquette M, Rudy AC, Fortier D, Lamoureux SF (2020) Multi-scale site evaluation of a relict active layer detachment in a High Arctic landscape. Geomorphology 359(15):107159. https://doi.org/10.1016/j.geomorph.2020.107159

Pawłuszek K, Borkowski A (2016) Landslides identification using airborne laser scanning data derived topographic terrain attributes and support vector machine classification. In: The international archives of the photogrammetry, remote sensing and spatial information sciences, XXIII ISPRS Congress, vol 8

Pesci A, Teza G, Casula G, Loddo F, De Martino P, Dolce M, Obrizzo F, Pingue F (2011) Multitemporal laser scanner-based observation of the Mt. Vesuvius crater: characterization of overall geometry and recognition of landslide events. ISPRS J Photogramm Remote Sens 66(3):327–336. https://doi.org/10.1016/j.isprsjprs.2010.12.002

Petley DN, Crick WDO, Hart AB (2002) The use of satellite imagery in landslide studies in high mountain areas. In: Proceedings of the 23rd Asian conference on remote sensing (ACRS'2002), Kathmandu. http://www.gisdevelopment.net/aars/acrs/2002/hdm/48.pdf

Petrie G, Toth CK (2008) Introduction to laser ranging, profiling, and scanning. Topographic laser ranging and scanning: principles and processing, pp 1–28

Pham BT, Jaafari A, Prakash I, Bui DT (2019) A novel hybrid intelligent model of support vector machines and the MultiBoost ensemble for landslide susceptibility modeling. Bull Eng Geol Env 78(4):2865–2886. https://doi.org/10.1007/s10064-018-1281-y

Piégay H, Arnaud F, Belletti B, Bertrand M, Bizzi S, Carbonneau P et al (2020) Remotely sensed rivers in the Anthropocene: state of the art and prospects. Earth Surf Proc Land 45(1):157–188. https://doi.org/10.1002/esp.4787

Proy C, Tinel C, Fontannaz D (2013) Pleiades in the context of the International Charter "space and major disasters". In: 2013 IEEE international geoscience and remote sensing symposium-IGARSS, pp 4530–4533. https://doi.org/10.1109/IGARSS.2013.6723843

Rau JY, Jhan JP, Lo CF, Lin YS (2011) Landslide mapping using imagery acquired by a fixed-wing UAV. Int Arch Photogramm Remote Sens Spat Inf Sci 38(1/C22):195–200

Raucoules D, De Michele M, Malet JP, Ulrich P (2013) Time-variable 3D ground displacements from high-resolution synthetic aperture radar (SAR). Application to La Valette landslide (South French Alps). Remote Sens Environ 139:198–204. https://doi.org/10.1016/j.rse.2013.08.006

Razak KA, Santangelo M, Van Westen CJ, Straatsma MW, de Jong SM (2013) Generating an optimal DTM from airborne laser scanning data for landslide mapping in a tropical forest environment. Geomorphology 190:112–125. https://doi.org/10.1016/j.geomorph.2013.02.021

RIEGL Laser Measurement Systems (2014) Datasheet VZ-400 (RIEGL Laser Measurement Systems GmbH). Austria

Roessner S, Wetzel HU, Kaufmann H, Sarnagoev A (2001) Satellite remote sensing for regional assessment of landslide hazard in Kyrgyzstan (Central Asia). In: Proceedings of second symposium on Katastrophenvorsorge, Leipzig, pp 24–25

Rossi G, Tanteri L, Tofani V, Vannocci P, Moretti S, Casagli N (2018) Multitemporal UAV surveys for landslide mapping and characterization. Landslides 15:1045–1052. https://doi.org/10.1007/s10346-018-0978-0

Rossini M, Di Mauro B, Garzonio R, Baccolo G, Cavallini G, Mattavelli M, De Amicis M, Colombo R (2018) Rapid melting dynamics of an alpine glacier with repeated UAV photogrammetry. Geomorphology 304:159–172. https://doi.org/10.1016/j.geomorph.2017.12.039

Roulland T, Maquaire O, Costa S, Compain V, Davidson R, Medjkane M (2019) Dynamique des falaises des Vaches Noires: analyse diachronique historique et récente à l'aide de documents multi-sources (Normandie, France). Géomorphologie Relief Processus Environnement 25(1):37–55. https://doi.org/10.4000/geomorphologie.12989

Rouyet L, Lauknes TR, Christiansen HH, Strand SM, Larsen Y (2019) Seasonal dynamics of a permafrost landscape, Adventdalen, Svalbard, investigated by InSAR. Remote Sens Environ 231:111236. https://doi.org/10.1016/j.rse.2019.111236

Rudy AC, Lamoureux SF, Treitz P, Ewijk KV, Bonnaventure PP, Budkewitsch P (2016) Terrain controls and landscape-scale susceptibility modelling of active-layer detachments, Sabine Peninsula, Melville Island, Nunavut. Permaf Periglac Process 28(1):79–91. https://doi.org/10.1002/ppp.1900

Rudy AC, Lamoureux SF, Treitz P, Short N, Brisco B (2018) Seasonal and multi-year surface displacements measured by DInSAR in a high Arctic permafrost environment. Int J Appl Earth Obs Geoinf 64:51–61. https://doi.org/10.1016/j.jag.2017.09.002

Rusnák M, Sládek J, Pacina J, Kidov A (2018) Monitoring of avulsion channel evolution and river morphology changes using UAV photogrammetry: case study of the gravel bed Ondava river in outer western carpathians. Area 51(3):549–560. https://doi.org/10.1111/area.12508

Salvini R, Francioni M, Riccucci S, Bonciani F, Callegari I (2013) Photogrammetry and laser scanning for analyzing slope stability and rockfall runout along the Domodossola-Iselle railway, the Italian Alps. Geomorphology 185:110–122. https://doi.org/10.1016/j.geomorph.2012.12.020

Sato HP, Harp EL (2009) Interpretation of earthquake-induced landslides triggered by the 12 May 2008, M7.9 Wenchuan earthquake in the Beichuan area, Sichuan Province, China using satellite imagery and Google Earth. Landslides 2:153–159. https://doi.org/10.1007/s10346-009-0147-6

Schmidhuber J (2015) Deep learning in neural networks: an overview. Neural Netw 61:85–117. https://doi.org/10.1016/j.neunet.2014.09.003

Schulz WH (2004) Landslides mapped using LIDAR imagery, Seattle, Washington. US Geological Survey Open-File Report, 1396(11)

Shafique M, van der Meijde M, Khan MA (2016) A review of the 2005 Kashmir earthquake-induced landslides; from a remote sensing perspective. J Asian Earth Sci 118:68–80. https://doi.org/10.1016/j.jseaes.2016.01.002

Shan J, Toth CK (2018) Topographic laser ranging and scanning: principles and processing. Taylor & Francis Group CRC Press, Boca Raton

Singleton A, Li Z, Hoey T, Muller JP (2014) Evaluating sub-pixel offset techniques as an alternative to D-InSAR for monitoring episodic landslide movements in vegetated terrain. Remote Sens Environ 147:133–144. https://doi.org/10.1016/j.rse.2014.03.003

Slob S, Hack R (2004) 3D terrestrial laser scanning as a new field measurement and monitoring technique. In: Engineering geology for infrastructure planning in Europe. Springer, Berlin, Heidelberg, pp 179–189. https://doi.org/10.1007/978-3-540-39918-6_22

Song KY, Oh HJ, Choi J, Park I, Lee C, Lee S (2012) Prediction of landslides using ASTER imagery and data mining models. Adv Space Res 49(5):978–993. https://doi.org/10.1016/j.asr.2011.11.035

Stettner S, Beamish AL, Bartsch Heim B, Grosse G, Roth A, Lantuit H (2018) Monitoring inter-and intra-seasonal dynamics of rapidly degrading ice-rich permafrost riverbanks in the Lena Delta with TerraSAR-X time series. Remote Sens 10(1):51. https://doi.org/10.3390/rs10010051

Stumpf A, Kerle N (2011a) Object-oriented mapping of landslides using Random Forests. Remote Sens Environ 115(10):2564–2577. https://doi.org/10.1016/j.rse.2011.05.013

Stumpf A, Kerle N (2011b) Combining random forests and object-oriented analysis for landslide mapping from very high resolution imagery. Proc Environ Sci 3:123–129. https://doi.org/10.1016/j.proenv.2011.02.022

Stumpf A, Malet JP, Allemand P, Ulrich P (2014) Surface reconstruction and landslide displacement measurements with Pléiades satellite images. ISPRS J Photogramm Remote Sens 95:1–12. https://doi.org/10.1016/j.isprsjprs.2014.05.008

Stumpf A, Malet JP, Allemand A, Pierrot-Deseilligny M, Skupinski G (2015) Ground-based multiview photogrammetry for the monitoring of landslide deformation and erosion. Geomorphology 231:130–145. https://doi.org/10.1016/j.geomorph.2014.10.039

Stumpf A, Malet JP, Delacourt C (2017) Correlation of satellite image time-series for the detection and monitoring of slow-moving landslides. Remote Sens Environ 189:40–55. https://doi.org/10.1016/j.rse.2016.11.007

Sun L, Muller JP (2016) Evaluation of the use of sub-pixel offset tracking techniques to monitor landslides in densely vegetated steeply sloped areas. Remote Sens 8(8):659. https://doi.org/10.3390/rs8080659

Svennevig K (2019) Preliminary landslide mapping in Greenland. Geol Surv Den Greenl Bull. https://doi.org/10.34194/GEUSB-201943-02-07

Tarchi D, Casagli N, Fanti R, Leva DD, Luzi G, Pasuto A et al (2003) Landslide monitoring by using ground-based SAR interferometry: an example of application to the Tessina landslide in Italy. Eng Geol 68(1–2):15–30. https://doi.org/10.1016/S0013-7952(02)00196-5

Telling J, Lyda A, Hartzell P, Glennie C (2017) Review of Earth science research using terrestrial laser scanning. Earth Sci Rev 169:35–68. https://doi.org/10.1016/j.earscirev.2017.04.007

Teza G, Galgaro A, Zaltron N, Genevois R (2007) Terrestrial laser scanner to detect landslide displacement fields: a new approach. Int J Remote Sens 28(16):3425–3446. https://doi.org/10.1080/01431160601024234

Tofani V, Segoni S, Agostini A, Catani F, Casagli N (2013) Use of remote sensing for landslide studies in Europe. Nat Hazards Earth Syst Sci. https://doi.org/10.5194/nhess-13-299-2013

Travelletti J, Delacourt C, Allemand P, Malet JP, Schmittbuhl J, Toussaint R, Bastard M (2012) Correlation of multi-temporal ground-based optical images for landslide monitoring: application, potential and limitations. ISPRS J Photogramm Remote Sens 70:39–55. https://doi.org/10.1016/j.isprsjprs.2012.03.007

Travelletti J, Malet JP, Delacourt C (2014) Image-based correlation of laser scanning point cloud time series for landslide monitoring. Int J Appl Earth Obs Geoinf 32:1–18. https://doi.org/10.1016/j.jag.2014.03.022

Tyszkowski S, Cebulski J (2019) Practical aspects of landslides surveys using terrestrial laser scanning in diverse geomorphological terrains: case studies from Polish Carpathians and Lower Vistula Valley. Zeitschrift für Geomorphologie 62(2):107–124. https://doi.org/10.1127/zfg/2019/0500

Valkaniotis S, Papathanassiou G, Ganas A (2018) Mapping an earthquake-induced landslide based on UAV imagery; case study of the 2015 Okeanos landslide, Lefkada, Greece. Eng Geol 245:141–152. https://doi.org/10.1016/j.enggeo.2018.08.010

Van Den Eeckhaut M, Kerle N, Poesen J, Hervás J (2012) Object-oriented identification of forested landslides with derivatives of single pulse LiDAR data. Geomorphology 173:30–42. https://doi.org/10.1016/j.geomorph.2012.05.024

Van Puymbroeck N, Michel R, Binet R, Avouac JP, Taboury J (2000) Measuring earthquakes from optical satellite images. Appl Opt 39(20):3486–3494. https://doi.org/10.1364/AO.39.003486

Van Westen CJ, Castellanos E, Kuriakose SL (2008) Spatial data for landslide susceptibility, hazard, and vulnerability assessment: an overview. Eng Geol 102(3–4):112–131. https://doi.org/10.1016/j.enggeo.2008.03.010

Voigt S, Giulio-Tonolo F, Lyons J, Kučera J, Jones B, Schneiderhan T et al (2016) Global trends in satellite-based emergency mapping. Science 353(6296):247–252. https://doi.org/10.1126/science.aad8728

Wang D, Jiang Y, Wang W, Wang Y (2016) Bias reduction in sub-pixel image registration based on the antisymmetric feature. Meas Sci Technol 27(3):035206. https://doi.org/10.1088/0957-0233/27/3/035206

Wang Y, Fang Z, Hong H (2019) Comparison of convolutional neural networks for landslide susceptibility mapping in Yanshan County, China. Sci Total Environ 666:975–993. https://doi.org/10.1016/j.scitotenv.2019.02.263

Warrick JA, Ritchie AC, Schmidt KM, Reid ME, Logan J (2019) Characterizing the catastrophic 2017 Mud Creek landslide, California, using repeat structure-from-motion (SfM) photogrammetry. Landslides 16(6):1201–1219. https://doi.org/10.1007/s10346-019-01160-4

Wasowski J, Bovenga F (2014) Investigating landslides and unstable slopes with satellite multi temporal interferometry: current issues and future perspectives. Eng Geol 174:103–138. https://doi.org/10.1016/j.enggeo.2014.03.003

Werner C, Strozzi T, Wiesmann A, Wegmüller U (2008) GAMMA's portable radar interferometer. In: Proceedings of 13th FIG symposium of deformation measurements and analysis, pp 1–10

Westoby MJ, Brasington J, Glasser NF, Hambrey MJ, Reynolds JM (2012) Structure-from-Motion photogrammetry: a low-cost, effective tool for geoscience applications. Geomorphology 179:300–314. https://doi.org/10.1016/j.geomorph.2012.08.021

Westoby M, Lim M, Hogg MJ, Pound M, Dunlop L, Woodward J (2018) Cost-effective erosion monitoring of coastal cliffs. Coast Eng 138:152–164. https://doi.org/10.1016/j.coastaleng.2018.04.008

Wilkinson MW, Jones RR, Woods CE, Gilment SR, McCaffrey KJW, Kokkalas S, Long JJ (2016) A comparison of terrestrial laser scanning and structure-from-motion photogrammetry as methods for digital outcrop acquisition. Geosphere 12(6):1865–1880. https://doi.org/10.1130/GES01342.1

Xu C (2015) Preparation of earthquake-triggered landslide inventory maps using remote sensing and GIS technologies: principles and case studies. Geosci Front 6(6):825–836. https://doi.org/10.1016/j.gsf.2014.03.004

Xu C, Xu X, Yao X, Dai F (2014) Three (nearly) complete inventories of landslides triggered by the May 12, 2008 Wenchuan Mw 7.9 earthquake of China and their spatial distribution statistical analysis. Landslides 11(3):441–461. https://doi.org/10.1007/s10346-013-0404-6

Xu Q, Ouyang C, Jiang T, Fan X, Cheng D (2019) DFPENet-geology: a deep learning framework for high precision recognition and segmentation of co-seismic landslides. arXiv:1908.10907

Yamaguchi YS, Tanaka S, Odajima T, Kamai T, Tsuchida S (2003) Detection of a landslide movement as geometric misregistration in image matching of SPOT HRV data of two different dates. Int J Remote Sens 24(18):3523–3534. https://doi.org/10.1080/01431160110111063

Yang X, Chen L (2010) Using multi-temporal remote sensor imagery to detect earthquake-triggered landslides. Int J Appl Earth Obs Geoinf 12(6):487–495. https://doi.org/10.1016/j.jag.2010.05.006

Yang W, Wang M, Shi P (2013) Using MODIS NDVI time series to identify geographic patterns of landslides in vegetated regions. IEEE Geosci Remote Sens Lett 10(4):707–710. https://doi.org/10.1109/LGRS.2012.2219576

Yin Y, Wang F, Sun P (2009) Landslide hazards triggered by the 2008 Wenchuan earthquake, Sichuan, China. Landslides 6(2):139–152. https://doi.org/10.1007/s10346-009-0148-5

Yu B, Chen F (2017) A new technique for landslide mapping from a large-scale remote sensed image: a case study of Central Nepal. Comput Geosci 100:115–124. https://doi.org/10.1016/j.cageo.2016.12.007

Zhang S, Zhang LM, Glade T (2014) Characteristics of earthquake-and rain-induced landslides near the epicenter of Wenchuan earthquake. Eng Geol 175:58–73. https://doi.org/10.1016/j.enggeo.2014.03.012

Zhao W, Du S (2016) Learning multiscale and deep representations for classifying remotely sensed imagery. ISPRS J Photogramm Remote Sens 113:155–165. https://doi.org/10.1016/j.isprsjprs.2016.01.004

Zhao ZQ, Zheng P, Xu ST, Wu X (2019) Object detection with deep learning: a review. IEEE Trans Neural Netw Learn Syst 30(11):3212–3232. https://doi.org/10.1109/TNNLS.2018.2876865

Zhou C, Yin K, Cao Y, Ahmed B, Li Y, Catani F, Pourghasemi HR (2018) Landslide susceptibility modeling applying machine learning methods: a case study from Longju in the Three Gorges Reservoir area, China. Comput Geosci 112:23–37. https://doi.org/10.1016/j.cageo.2017.11.019

Zhu XX, Tuia D, Mou L, Xia GS, Zhang L, Xu F, Fraundorfer F (2017) Deep learning in remote sensing: a comprehensive review and list of resources. IEEE Geosci Remote Sens Mag 5(4):8–36. https://doi.org/10.1109/MGRS.2017.2762307

Zolkos S, Tank SE, Striegl RG, Kokelj SV (2019) Thermokarst effects on carbon dioxide and methane fluxes in streams on the Peel Plateau (NWT, Canada). J Geophys Res Biogeosci 124(7):1781–1798. https://doi.org/10.1029/2019JG005038

Zwieback S, Kokelj SV, Günthe F, Boike J, Grosse G, Hajnsek I (2018) Sub-seasonal thaw slump mass wasting is not consistently energy limited at the landscape scale. Cryosphere 12(2):549–564. https://doi.org/10.3929/ethz-b-000244496

Publisher's Note Springer Nature remains neutral with regard to jurisdictional claims in published maps and institutional affiliations.

Affiliations

Candide Lissak[1] · Annett Bartsch[2,3] · Marcello De Michele[4] · Christopher Gomez[5,6] · Olivier Maquaire[1] · Daniel Raucoules[4] · Thomas Roulland[1]

[1] Normandie Université, UNICAEN, CNRS, LETG, Esplanade de la Paix, 14000, Caen, France

[2] b.geos, Industriestrasse 1, 2100 Korneuburg, Austria

[3] Austrian Polar Research Institute, c/o Universität Wien, Universitätsstraße 7, 1010 Vienna, Austria

[4] BRGM, French Geological Survey, 3 Av. Claude Guillemin, 45000 Orléans, France

[5] Graduate School of Maritime Sciences, Laboratory of Sediment Hazards and Disaster Risk, Kobe University, 5-1-1 Fukaeminamimachi Higashinadaku, Kobe 658-0022, Japan

[6] Faculty of Geography, University Gadjah Mada, Sekip Utara, Bulaksumur, 55281 Yogyakarta, Indonesia

Surveys in Geophysics (2020) 41:1437–1459
https://doi.org/10.1007/s10712-020-09610-8

Fire Danger Observed from Space

M. Lucrecia Pettinari[1] · **Emilio Chuvieco**[1]

Received: 18 December 2019 / Accepted: 6 August 2020 / Published online: 19 August 2020
© Springer Nature B.V. 2020

Abstract

Biomass burning is one of the critical components of the Earth system, significantly affecting atmospheric emissions and carbon budgets. Fires occurring in the interface between wildland and urban areas also have important socioeconomic effects, affecting people's lives and resources. Even though fires are natural in many ecosystems, climate and societal changes have recently caused particularly severe fire seasons (Australia, California, Amazonia, Portugal…). Mitigating the negative impacts of fire requires further efforts to assess fire danger conditions. Satellite Earth observation provides considerable capabilities to evaluate the different variables involved in fire danger. Data obtained from remote sensors offer information on possible sources of fire ignition, on fuel status and abundance, and on the topography and the meteorological conditions that will affect fire spread. Satellite observations also provide near-real-time information on fire occurrence for early response teams. This article describes the different variables affecting fire danger and illustrates how satellite data can offer useful information to estimate these variables, focusing on global and continental fire danger systems.

Keywords Fire hazard · Fire danger · Remote sensing · Fire causes · Active fires · Fire danger information systems

1 Fire in the Earth system

Fire has been present in the Earth system since the origin of plants, playing a key role in plant adaptation and ecosystem distribution (Pausas and Keeley 2009). With the emergence of humans on Earth, the causes of fire ignition increased and fire regimes changed, as vegetation burning became an integral form of agriculture, traditional pastoralism and hunting practices (Pyne 1995; Pyne and Goldammer 1997). Even today, fires are used for a variety of purposes, such as the clearing of forests and savannahs for agriculture and grazing, removing unpalatable grasses, stubble and waste removal after harvest, shifting agriculture, hunting, and production of charcoal (Levine et al. 1995; Shlisky et al. 2007).

✉ M. Lucrecia Pettinari
  mlucrecia.pettinari@uah.es

1 Environmental Remote Sensing Research Group, University of Alcala, Calle Colegios 2,
   Alcalá de Henares 28801, Spain

Wildland fires are an important driver of environmental transformation, affecting the Earth system in many different and, at the same time, interlinked ways. Even though fires have many positive impacts on biodiversity and plant succession (Archibald et al. 2018), they also imply increasing soil erosion and runoff, water quality degradation, and in many cases end in deforestation and land cover change (Cano-Crespo et al. 2015; Eva and Lambin 2000). They are also recognized as a critical factor affecting carbon budgets and greenhouse gas emissions (Jones et al. 2019; Thonicke et al. 2010; van der Werf et al. 2010). At the same time, the changes in the climate in the past decades are also affecting fire occurrence and effects (Achard et al. 2008; de Groot et al. 2013).

Moreover, fires are a major natural hazard for societies (Bowman et al. 2009; Knorr et al. 2016; McCaffrey 2004), causing loss of lives, properties, and infrastructures, and these effects are more evident when fire occurs in the wildland–urban interface (WUI) (Collins 2005; Meldrum et al. 2018; Syphard et al. 2007). Fires also disturb the services provided by different natural ecosystems (Chuvieco et al. 2014a, 2014b) and can even affect intangible values such as sacred environments and life views (Moritz et al. 2014). In addition, human exposure to fire is expected to increase in the future, due to projected population growth in areas with frequent wildfires (Knorr et al. 2016). Humans are not only the most important source of fire ignition (Balch et al. 2017; Syphard et al. 2017), but also play an important role in fire propagation, directly through fire suppression or prescribed burning, and indirectly through land use management (Archibald et al. 2010; Bistinas et al. 2013).

Recent assessments based on coarse-resolution Earth observation satellites estimate that between 3.5 and 5 million km^2 are burned globally each year (Giglio et al. 2018; Lizundia-Loiola et al. 2020). Most probably, these estimations are conservative, as comparisons of global with local products have observed a tendency of the former to underestimation of burned area (Hawbaker et al. 2017; Roteta et al. 2019). In addition, expected climate and societal changes will likely imply the extension of wildfire-prone areas to new regions, particularly Northern latitudes and evergreen tropical forests (Evangeliou et al. 2019).

The objective of this article is to describe the different variables affecting fire danger and illustrate how satellite data can provide useful information to estimate them. The analysis is focused on sensors and methods that can provide data for global or continental fire danger assessment.

2 Fire Danger Assessment

For a fire to start, three elements must be present: a fuel that can burn (in this case live or dead vegetation), a source of ignition (either natural or anthropogenic), and the oxygen of the air that acts as an oxidizing agent. For this spark to become a wildland fire, other elements are required: a chain reaction to maintain the fire, fuel continuity, and meteorological and topographical conditions to spread the fire to the surroundings.

It is currently accepted in the emergency management community that hazard refers to the possible future occurrence of an event that may have adverse effects on vulnerable and exposed elements (Cardona et al. 2012). This term differentiates the hazard component from the risk itself (which also includes the inventory of the elements that can be affected by the hazard—the exposure—and the level of affection of those elements—the vulnerability). But in a wildfire context, the term "hazard" has frequently been used to account only for the fuel conditions independently of the weather state (Hardy 2005), and the term "danger" has been the common way to describe all the variables affecting the probability of fire occurrence (Chuvieco et al. 2003; FAO and GFMC 1999). For that reason, the term fire danger will be used in this article as the assessment of the conditions under which a fire can be ignited and will spread.

Fire danger assessment and early warning systems allow fire managers, emergency responders and even the public to identify periods of extreme fire danger, implement fire prevention measures, establish early response mechanisms and reduce fire consequences in terms of cost to properties, natural values and even human lives (de Groot et al. 2006). Fire danger systems and indices have been in use in different parts of the world for many years. These models are usually non-spatial, but are the core of currently operational geographical early warning systems. Such is the case of the Canadian Forest Fire Danger Rating System (CFFDRS) developed by the Canadian Forest Service (Stocks et al. 1989), with its sub-system Fire Weather Index (FWI, Van Wagner 1987), which is the basis of the Canadian Wildland Fire Information System (CWFIS, https://cwfis.cfs.nrcan.gc.ca/home, accessed June 2020). Another example is the United States Forest Service (USFS) Wildland Fire Assessment System (WFAS, Burgan et al. 1997), available at https://www.wfas.net/index .php (accessed June 2020), which uses the equations of the National Fire-Danger Rating System (Bradshaw et al. 1983; Deeming et al. 1977). Australia uses, among others, the McArthur Fire Danger Rating System (FDRS, Luke and McArthur 1978; McArthur 1967), one of its major components being the Forest Fire Danger Index (FFDI, also called Mark 5). The near-real-time (NRT) fire danger rating maps for each Australian region can be accessed through the Australian Bureau of Meteorology (https://www.bom.gov.au/weather-services/fire-weather-centre/fire-weather-services/index.shtml, accessed June 2020).

The European Commission, through its Joint Research Centre (JRC), developed the European Forest Fire Information System (EFFIS, European Forest Fire Information System 2020; San Miguel-Ayanz et al. 2012), available at https://effis.jrc.ec.europa.eu/ (accessed June 2020), which is currently part of the Copernicus Emergency Management Service (CMES, https://emergency.copernicus.eu/, accessed June 2020). This system provides a forecast of fire danger using different fire danger models based on meteorological information: FWI and its intermediate components, the Burning Index (BI) and its components of the NFRDS, the components of Mark 5, and the Keetch–Byram Drought Index (KBDI, Keetch and Byram 1968). The JRC also developed the Global Wildfire Information System (GWIS, https://gwis.jrc.ec.europa.eu/static/gwis_current_situation/public/index .html, accessed June 2020) as an extension of EFFIS, providing global estimates of these indices, as well as other fire related information, in NRT.

Many of the components and indices described above, and used in fire danger systems, rely solely on weather information. For example, the inputs for the KBDI are only the maximum temperature (or dry-bulb temperature) and the total precipitation (P) of the previous 24 h (Keetch and Byram 1968). FFDI, besides the dry-bulb temperature (T), also needs for its computation the relative humidity (H), wind speed (W) and a drought factor dependent on the time since last rain, P, and the dryness of the soil as calculated by the KBDI (Noble et al. 1980). The inputs for the calculation of FWI and it components are T, H, W, P, and the day length to compute the slow-reacting moisture codes (Dowdy et al. 2009; Van Wagner 1987). In the case of the components of the BI in the NFDRS, calculating them requires both weather information (T, H, dew point, W, wind direction, P, precipitation duration, and solar radiation) and other inputs such as fuel model and slope (Bradshaw et al. 1983; Schlobohm and Brain 2002).

A general framework to evaluate fire danger, considering all the different variables involved in fire ignition and spread, is shown in Fig. 1. It requires weather data to retrieve

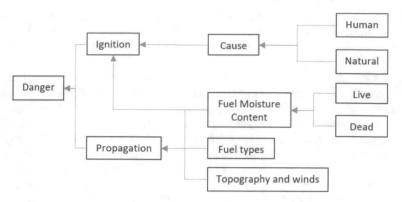

Fig. 1 Fire danger framework. Adapted from Chuvieco et al. (2014a)

fuel moisture and wind conditions, but also information on potential fire causes, the topography of the terrain being assessed for fire danger, and the fuel characteristics of that area. These different variables, their role in fire danger, and the remote sensing (RS) sources available to obtain them are described in the following sections.

3 Fire Danger Variables Estimation from Remote Sensing

Earth observation from space is a basic tool to retrieve information useful for fire danger estimation, especially at global or continental scale, due to its global coverage, high temporal resolution, and free availability of images when working at spatial resolutions of 10 m or lower. Using RS data for fire danger requires a balance between the amount of data to be handled, with its corresponding cost in processing time and hardware requirements, and the scale at which fire danger is being assessed. For global or continental scales, as is the focus of this article, the use of coarse to medium resolution (250–1000 m) is a good trade between data processing and product resolution. For finer scales (e.g., states), a medium spatial resolution of 10–100 m would be desirable, while for even finer scales (e.g., counties or private fields), information at 1–5 m would be the most appropriate option. The latter will require the acquisition of satellite images from private companies, with their associated cost. Some examples of these last kinds of products are included in the article for illustration. The amount of data to be processed does not depend only on spatial resolution, but also on temporal resolution. Geostationary satellites providing information every 5–15 min are very useful to follow the progress of rapidly changing events such as lightning strikes, fire spread, etc., but at the same time increase greatly the amount of data to be processed.

3.1 Fire Causes

Fire ignition can have a natural or anthropogenic origin. The natural sources of ignition are few, mostly comprising lightning strikes (Barroso Ramos-Neto and Pivello 2000; Zhao et al. 2007), but can also include volcanic eruptions, meteorites, or other natural sources of sparks. Human causes of fire, on the other hand, are numerous and more difficult to study.

Lightning strikes were the main fire cause when fire regimes were natural (Bowman et al. 2011), and even though that is no longer the case, lightning is still an important source of ignitions in remote areas (Conedera et al. 2006; Zhao et al. 2007). Remote sensing can provide valuable data regarding lightning strikes that could be used for early warning of probable fire ignitions. Lightning detectors have primarily been ground-based networks of multiple antennas or mobile systems aboard aircraft, but they have extended since the nineties to include satellite sensors.

There are several sensors that have been designed or used for satellite lightning detection. The first one was the optical transient detector (OTD), installed on the Orbview-1 satellite, which collected data of lightning activity around the globe for the period 1995–2000 (Christian et al. 2003, 1996). The lightning imaging sensor (LIS), installed on the Tropical Rainfall Measuring Mission (TRMM) satellite, stayed on orbit collecting data from 1997 to 2015 (Albrecht et al. 2011). A second LIS instrument was placed on the International Space Station (ISS) in February 2017 (Blakeslee et al. 2014). From the OTD and LIS data, a Climatology Dataset was developed in 2015 for the years 1995–2014 (Cecil 2015; Cecil et al. 2014).

Newer satellites have also included instruments for lightning detection, this time located in geostationary platforms to assure a high temporal resolution. The Geostationary Lightning Mapper (GLM, Goodman et al. 2013) has been placed on the Geostationary Operational Environmental Satellites GOES-16 and GOES-17 and has a spatial resolution of approximately 10 km. These satellites were launched in 2016 and 2018, respectively, and collect data for the Western Hemisphere, covering the Americas and the Pacific Ocean. The China Meteorological Administration (CMA) and National Remote Sensing Centre of China (NRSCC) also included the Lightning Mapper Instrument (LMI, Yang et al. 2017), with a spatial resolution of 7.8 km at nadir, on the Feng-Yung-4A satellite launched in 2016. Although the satellite provides coverage for Asia and Australia, the current LMI field of view is restricted to China and neighbouring territories, due to the characteristics of the instrument. The European Space Agency (ESA) is also planning to include the Lightning Imager (LI) on board the Meteosat Third-Generation (MTG) geostationary satellite that will be launched in 2021 (Lorenzini et al. 2012; Stuhlmann et al. 2005). It will provide information over Europe, the Mediterranean Sea, Africa, and the Atlantic Ocean, with a spatial resolution at nadir of 4.5 km.

Lightning information from space is rapidly becoming an input for fire danger assessment at global scale. For example, the European Centre for Medium-Range Weather Forecasts (ECMWF) has developed a model to predict lightning strikes calibrated with simulations with OTD and LIS data (Lopez 2016) that has been incorporated to the GWIS. The coarse spatial resolution of the data can only provide a general location of potential ignition sources. Still, this information can be very useful to emergency responders, as it can pinpoint locations where vigilance for possible fire occurrence needs to be strengthened after lightning episodes.

Fire regimes based on lightning as a source of ignition have switched to anthropogenic regimes in most of the world (Bowman et al. 2009; Pyne and Goldammer 1997), and human factors have superseded natural conditions in their influence on fire regimes (Syphard et al. 2017). It is estimated that around 90% of the fires throughout the world are human-caused, either deliberately or due to negligence or accidents (Balch et al. 2017; Ganteaume et al. 2013; Lauk and Erb 2009; Vilar et al. 2016).

Some typical variables that are analysed to model human-caused ignitions are population density, density of transport networks, distance to transport networks and to urban or recreational areas, or the WUI and the wildland–agriculture interface (WUA) (Costafreda-Aumedes et al. 2017; Chuvieco et al. 2014a; Chuvieco and Justice 2010; Martínez-Fernández et al. 2013; Martínez et al. 2009). Socioeconomic variables are also often included in the analysis, such as level of industrialization, agricultural practises, unemployment rate, gross domestic product per capita, etc. (Chuvieco and Justice 2010; Lauk and Erb 2009). To obtain these variables for fire danger assessment, existing databases are compiled in geographic information systems, and indices are developed to assess how these variables can increase fire danger (Chuvieco et al. 2014a; Chuvieco and Justice 2010; Rodrigues et al. 2014, 2016). When dealing with anthropogenic variables, remote sensing alone cannot provide all the information that is needed, but it can be used to derive some of those variables. For instance, land use—land cover maps can be used for interface determination and to calculate distances to networks (Martín et al. 2019; Vilar et al. 2016), and these datasets are usually based on RS data. One example of this type of maps, in the case of Europe, is the Corine Land Cover (CLC) (Martín et al. 2019; Vilar et al. 2016) product (Büttner 2014). This dataset was developed with the use of different RS inputs (Büttner et al. 2017): the CLC 2012 (Büttner et al. 2017) used data from the Indian Space Research Organization (ISRO) IRS-P6 and Resourcesat-2 satellites, the Satellite Pour l'Observation de la Terre (SPOT) of the French Space Agency (CNES, Maliet 2013) and RapidEye (Tyc et al. 2005), while the CLC 2018 is based on data from the Sentinel-2 MultiSpectral Instrument (MSI, European Space Agency 2015), with Landsat-8 Operational Land Imager (OLI, United States Geological Survey 2019) data for filling. The CLC has a minimum mapping unit of 25 ha for areal phenomena, and a minimum width of 100 m for linear elements.

Population density is also an important variable to evaluate fire ignition potential, and different global- and continental-extent gridded population datasets currently exist (Leyk et al. 2019). Even if early efforts to create these datasets relied only on a regular distribution of population census information within political units (e.g., the Gridded Population of the World (GPW, Doxsey-Whitfield et al. 2015)), RS information was later used to distribute the population data in a more realistic way. To mention some current examples of global population products, the Global Human Settlement Population Grid (GHS-POP) at a spatial resolution of 1 km (Freire et al. 2018) allocated the census data according to built-up areas obtained from Landsat sensors Multispectral Scanner (MSS), Thematic Mapper (TM) and Enhanced Thematic Mapper (ETM) information (Pesaresi et al. 2016). The WorldPop dataset (Tatem 2017) with a resolution of 100 m is based on the GPW version 3 and used information from different RS sensors to distribute the population dasymmetrically. Nowadays, the population datasets are moving towards higher resolution products. The most prominent example is the Global Urban Footprint (GUF) dataset developed by the German Aerospace Centre (DLR) at a resolution of approximately 12 m using TanDEM-X and TerraSAR-X radar images collected in 2011–2012 (Esch et al. 2017).

Most of the different variables related to potential anthropogenic fire ignitions (e.g., population density, roads, WUI) have a slower rate of change than natural phenomena such as lightning, and for that reason, an annual or quinquennial update of the databases could be sufficient to show the changes in human distribution and its characteristics related to fire danger. Still, some seasonal variables, such as tourism or agricultural practices, might require a more frequent update. The optimum spatial resolution of these variables, on the other hand, will depend on the scale of analysis, as commented at the beginning of Sect. 3.

3.2 Fuel Moisture Content

Fuel moisture content (FMC) is a decisive in the capability of the vegetation to ignite and to propagate the fire (Albini 1976; Rothermel 1983; Viegas et al. 1992), thus becoming fundamental in fire danger assessment. The dryer the vegetation is, the more easily it will ignite, and the higher rate of spread of the fire will be. FMC is defined as the ratio between the mass of water contained in a plant and its total dry mass, in both live (LFMC) or dead (DFMC) fuels.

The most accurate method to obtain LFMC values is through fieldwork, but it is costly and labour-intensive and therefore unfeasible for monitoring large areas. LFMC can also be retrieved from satellite observations, profiting from the direct effect of water absorption on the vegetation reflectance in the near-infrared (NIR) and short-wave-infrared (SWIR) spectral regions (Yebra et al. 2013). One technique for LFMC retrieval is the use of spectral vegetation indices to obtain information that is later statistically fitted to field-measured values. Yebra et al. (2013) summarized the spectral indices usually used to estimate LFMC. Another method to obtain LFMC from remote sensing is to use physically based techniques, such as the inversion of radiative transfer models (RTMs, Quan et al. 2016; Wang et al. 2013; Yebra et al. 2018). These models are based on physical relationships between water content and vegetation reflectance and are independent of sensor or site conditions, making them more robust, but their selection and parameterization are complex (Yebra et al. 2008). Recent studies have also shown the potential of Synthetic Aperture Radar (SAR) data to estimate LFMC, as the backscatter measurements are affected by the moisture of the vegetation (Konings et al. 2019).

Coarse-to-medium-resolution optical sensors have been used for LFMC estimation, since they provide a good temporal resolution and the necessary bands to calculate typical vegetation indices or invert RTMs through the use of lookup tables. Some examples of the use of optical sensors for LFMC estimation include the Advanced Very High Resolution Radiometer (AVHRR, https://www.bou.class.noaa.gov/data_available/avhrr/index.htm, accessed June 2020) (Chuvieco et al. 2004; García et al. 2008) located on the polar-orbiting National Oceanic and Atmospheric Administration (NOAA) and most recently on the Meteorological Operation (MetOP) satellites, the Terra and Aqua Moderate Resolution Imaging Spectroradiometer (MODIS, https://modis.gsfc.nasa.gov/about/specifications.php, accessed June 2020) (Argañaraz et al. 2016; Luo et al. 2019; Mendiguren et al. 2015; Myoung et al. 2018), and the Visible/Infrared Imager Radiometer Suite (VIIRS, Hillger et al. 2013) on board the Suomi National Polar-orbiting Partnership (NPP) and most recently the NOAA-20 satellite (Whitney et al. 2019). Higher spatial resolution is provided by the Landsat sensors (Chuvieco et al. 2002; Quan et al. 2016, 2017), and the Sentinel-2 MultiSpectral Instrument (MSI, Shu et al. 2019), but they have lower temporal resolution. Regarding the use of SAR data, Sentinel-1A and the Soil Moisture Active Passive (SMAP) radiometer have been used for LFMC estimation (Jia et al. 2019; Wang et al. 2019a). However, in all cases, field measurements are necessary to constrain and validate the results. The importance of accurate values of LFMC has derived in the creation of the Globe-LFMC database (Yebra et al. 2019), a global database of field samples used to calibrate and validate RS algorithms used to predict live fuel moisture content.

Plants have different physiological traits to control their total water content, but once they die, the FMC is only controlled by the meteorological conditions. In addition, the small dead fuels (dead leaves and needles, small fallen branches, etc.) constitute the fuelbed that most easily dries and becomes the primary substrate for fire to ignite and start spreading (Rothermel 1983). DFMC is calculated from weather information: temperature, relative humidity, precipitations, days without rain, and for fine fuels, even the cloudiness and the length of the day are important factors, as insolation will rapidly dry them. Dead fuels being the primary substrate for fire are one of the reasons why many fire danger models rely only on weather data to estimate fire danger.

Traditionally, this information was obtained from weather stations, and they are still widely used, but information from RS is also becoming a main input for fire danger estimation, especially at continental or global scale. These types of applications estimate DFMC through the use of meteorological indices (Camia et al. 1999), such as the KBDI or FWI. For example, EFFIS and GWIS calculate FWI from numerical weather predictions of the ERA5 reanalysis from ECMWF at a spatial resolution of approximately 8 km (San-Miguel-Ayanz et al. 2018). ERA5 is a climate reanalysis dataset, covering the period 1950 to present (Hersbach et al. 2018), and developed from in situ data and information from more than 45 satellite sensors (the documentation of ERA5 including the complete list of sensors is available at https://confluence.ecmwf.int/display/CKB/ERA5%3A+data+documentation, accessed June 2020). Figure 2 shows an example of the GWIS FWI forecast.

Recently, datasets with long time series of FWI have been developed. Vitolo et al. (2019) created a global fire danger dataset, using the components of FWI, for the period 1980 to present. This dataset was developed using ERA5 weather information at global scale (Vitolo et al. 2019) and complemented with the KBDI, and some components of the FDRS and NFDRS. This dataset is available at the Copernicus Climate Change Service (https://cds.climate.copernicus.eu/cdsapp#!/dataset/cems-fire-historical, accessed June 2020). Another example is the Global Fire WEather Database (GFWED) developed by Field et al. (2015). This dataset calculates the different components of FWI since 1981 to present using weather information from the Modern Era Retrospective Analysis for Research and Applications version 2 (MERRA-2, Rienecker et al. 2011). This reanalysis dataset, in turn, uses information from different RS sensors, which are detailed in Appendix B of Rienecker et al. (2011). The GFWED dataset can be downloaded from https://data.giss.nasa.gov/impacts/gfwed (accessed June 2020).

Fig. 2 Screen capture of the GWIS Current Situation viewer (https://gwis.jrc.ec.europa.eu/static/gwis_current_situation/public/index.html, accessed July 2020) showing the global Fire Weather Index (FWI) of 1 July 2020

One drawback of these types of reanalysis data is the coarse spatial resolution of the information they provide. This resolution is useful for global applications, but it can be too coarse for realistic fire danger assessment at local scales, especially in mountainous areas or regions with steep climate gradients (Hijmans et al. 2005). Current efforts to solve this limitation are underway, with the development of climate datasets a 1 km spatial resolution (Fick and Hijmans 2017).

3.3 Fuel Types Classification

Fuels (i.e. vegetation components) are classified in different types, according to their fire behaviour (Pyne et al. 1996). Different fuel types classifications exist (Cohen and Deeming 1985; Ottmar et al. 2007; Rothermel 1972, 1983; Scott and Burgan 2005), each of which incorporate the necessary variables to run different fire danger or behaviour models. These variables include from dead fuel loads at different fuel sizes, live fuel loads, fuel depths (depth of the fuelbed on the ground), etc., to percentage of fuel cover, canopy base height (CBH), total height and number of trees. Traditionally, these characteristics were extracted from field campaigns, and although some of the required variables still need to be determined on-site (e.g., small dead fuels on the ground), many others can be obtained using satellite RS information.

Multispectral optical data have been used for many years to retrieve information of vegetation characteristics to develop fuel type maps (Chuvieco et al. 2009). Keane et al. (2001) described the advantages, limitations, and examples of different-resolution optical sensors used for fuel type mapping. These sensors allow to obtain information related to fuel types such as land cover types (Arino et al. 2007; Friedl et al. 2010) or vegetation cover (Hansen et al. 2003, 2007).

Some sensors widely used for this purpose include the Landsat TM, ETM and OLI (He et al. 2019; Stefanidou et al. 2018; Van Wagtendonk and Root 2003), the Terra and Aqua MODIS (Bajocco et al. 2015; Lanorte and Lasaponara 2008), the Advanced Spaceborne Thermal Emission and Reflection Radiometer (ASTER, Yamaguchi et al. 1998) on board Terra (Lasaponara and Lanorte 2007), and even AVHRR (Nadeau et al. 2005). More recently, Sentinel-2 data have also been used for forest characterization and fuel mapping (Franke et al. 2018; Hościło and Lewandowska 2019; Yankovich et al. 2019). High-resolution sensors, such as QuickBird, have also been used for fuel type classification in small areas (Arroyo et al. 2006).

Many fuel type maps have been developed based on optical RS information, from local to continental scales. Some of the largest fuel maps include the one developed by EFFIS for Europe based on CLC information (European Forest Fire Information System 2017), the United States LANDFIRE maps (Nelson et al. 2016; Rollins 2009) created from Landsat information, and the Canadian fuel types map (Nadeau et al. 2005) from SPOT-VEGETATION data (Latifovic et al. 2004). There have also been attempts to develop global fuel maps using data derived from different sources, including RS derived products such as vegetation cover, canopy height, and land cover classes (Pettinari and Chuvieco 2016) to be used for coarse-resolution fire danger and behaviour estimations (Pettinari and Chuvieco 2017).

Still, optical sensors are limited in the sense that they can only provide information at the top of the canopy, and hence they are inadequate to describe the vertical distribution of the fuels or the surface fuels, due to obstruction problems where canopies are dense. These limitations have been overcome in recent years with the use of active sensors, such as Light Detection and Ranging (LiDAR) and SAR. LiDAR allows obtaining information on fuel load, canopy height, CBH, bulk density and canopy cover (García et al. 2010, 2017; Hermosilla et al. 2014; Huesca et al. 2019; Liu et al. 2019). Some approaches for fuel mapping also use hybrid methods with a combination of LiDAR and optical information (Alonso-Benito et al. 2016; García et al. 2011; Marino et al. 2016; Mutlu et al. 2008; Sánchez Sánchez et al. 2018).

LiDAR information for fuel mapping is normally retrieved from airborne sensors, but satellite sensors have also been used for this purpose (García et al. 2012). Such is the example of the Geoscience Laser Altimeter System (GLAS) on board the Ice, Cloud, and land Elevation Satellite (ICESat, Schutz et al. 2005): data from this sensor were used to develop global canopy height maps (Lefsky 2010; Simard et al. 2011) and biomass maps (Baccini et al. 2012; Saatchi et al. 2011) aided by ancillary data. The launch in September 2018 of the ICESat-2 (Markus et al. 2017), carrying the Advanced Topographic Laser Altimeter System (ATLAS), as well as the Global Ecosystem Dynamics Investigation (GEDI) sensor (Dubayah et al. 2020) located on board the ISS since the end of December 2018, provides a new opportunity to retrieve fuel information from space and improve the existing canopy height products at a global scale (Narine et al. 2019; Neuenschwander and Magruder 2016; Popescu et al. 2018; Schneider et al. 2020).

SAR data have also been used for fuel characteristics retrieval at regional and global scale. For example, recently Sentinel-1 data have been used to develop a biomass map of herbaceous vegetation in the Kruger National Park (Berger et al. 2019), and the Advanced Land Observing Satellite (ALOS) Phased Array type L-Band Synthetic Aperture Radar (PALSAR, Rosenqvist et al. 2007) was used to create an aboveground biomass (AGB) map of Africa (Bouvet et al. 2018). The ESA DUE (Data User Element) Biomass project developed a global biomass map for 2010 based on the Envisat Advanced Synthetic Aperture Radar (ASAR) sensor, ALOS PALSAR, and Landsat-7 ETM + data (Quegan et al. 2017; Santoro 2018).

The upcoming ESA Biomass mission, programmed to be launched in 2022, will carry a polarimetric P-band SAR designed specifically for the determination of forest AGB and forest height (Quegan et al. 2019) and will provide further data for fuel type classification.

The information on vegetation characteristics retrieved from satellite instruments, due to their coarse resolution, is useful to broadly estimate fuel properties for continental or global fire danger assessment, but is inadequate when the scale of analysis is higher. At the moment, only airborne sensors can provide the level of detail necessary to adequately characterize fuels with high resolution in small areas.

3.4 Topography and Wind

Topography is an important variable influencing fire behaviour once a fire has started. First, the slope of the terrain will directly affect fire spread, as a fire moving upwards preheats the fuels ahead of the fire front, hence drying them and accelerating their ignition (Rothermel 1972, 1983). What is more, the aspect of the terrain will also influence the fuel moisture conditions, as a slope facing south (in the northern hemisphere) will receive more solar radiation and increase the drying of the fuels (Holden and Jolly 2011). Topography will also indirectly affect fire danger influencing other variables such as the distribution of the fuels, wind behaviour, and differences in temperature, relative humidity, and precipitation due to elevation (Schunk et al. 2013).

All these topographic characteristics can be obtained from digital elevation models (DEMs) at different scales. Between the different methods for DEM generation (Florinsky 2016), satellite RS plays an important role. Obtaining information on topography from satellites can involve the use of passive sensors, using stereo pairs of visible or infrared images, or active sensors, especially SAR.

There are currently several DEMs available generated with the use of satellite data, most of which are summarized in Florinsky (2016). Some examples of widely used products include the freely available SRMT, ASTER GDEM, and AW3D30, and the commercial WorldDEM and AW3D. The Shuttle Radar Topography Mission (SRTM) was a joint endeavour of the United States National Aeronautics and Space (NASA), the National Geospatial-Intelligence Agency, and the German and Italian Space Agencies and flew for 11 days in February 2000 on the Space Shuttle Endeavour. It acquired interferometric radar data using dual C-band radar antennas and produced a near-global 1 arc-second (approximately 30 m at the equator) product covering approximately 80% of the Earth landmass, from 60° N to 56° S (Farr et al. 2007). The ASTER sensor, launched on board the Terra satellite on 1999, collects in-track stereo pairs using nadir- and backward-looking infrared cameras. Its data were used to generate the Global DEM (GDEM) in 2009, as a collaboration between NASA and the Ministry of Economy, Trade, and Industry (METI) from Japan. The current product is on its third version, released in 2016. The GDEM V3 is a 1 arc-second grid covering the planet from 83° N to 83° S (Abrams 2016). Neither of the two sources of data for DEM generation is perfect. While the use of visible or infrared stereo pairs results in artefacts created by clouds obscuring the surface, SAR data can look through clouds. But at the same time, data from SAR can include voids due to steep slopes facing away from the radar (shadowing) or towards the radar (foreshortening) (Farr et al. 2007) that ASTER, with its nadir view, can circumvent. For this reason, the last release of the SRTM DEM, called SRTM Plus or SRTM NASA V3, includes data from the GDEM Version 2 to fill voids of the SRTM, and some small areas were also filled using United States Geological Survey national DEMs (SRTM 2015). Other researchers have also combined ASTER and SRTM DEMs to improve their results in different regions (Pham et al. 2018).

The ALOS World 3D - 30m product (also called AW3D30) was generated by the Japan Aerospace Exploration Agency (JAXA) using the data of the Panchromatic Remote-Sensing Instrument for Stereo Mapping (PRISM) on board ALOS, which operated from 2006 to 2011 (Takaku et al. 2014). The product at 1 arc-second spatial resolution was released free of charge on 2016, based on the commercial AWS3D dataset at 0.15 arc-seconds (approximately 5 m resolution) (Tadono et al. 2016, 2017). Finally, the TanDEM-X mission, operated between 2010 and 2015 by DLR and Airbus, produced a global 0.4 arc-second (approximately 12 m) WorldDEM product, using two satellites (TerraSAR-X and TanDEM-X) flying in twin formation (Wessel et al. 2018; Zink et al. 2014).

For regional or local effects of topography on fire danger, high-resolution DEMs can also be generated from high-resolution sensors, for example using stereo imagery from Pleiades-1A and 1B (Bagnardi et al. 2016; Nasir et al. 2015), or from Worldview-2 (Wang et al. 2019b). LiDAR data are also usually used for the generation of DEM maps (Hubacek et al. 2016) or to fill voids in global DEMs (Schenk et al. 2014), although at the moment these sensors are airborne and not located on satellites.

Wind speed and direction will directly affect fire behaviour, modifying the rate of spread of the fire and its direction (Andrews 2012). Wind conditions vary in time in the order of hours, minutes, and even seconds and are also highly heterogeneous in space. The dynamic nature of the winds requires information with a high temporal resolution to quickly identify changes in wind direction and velocity and assess how these changes will affect fire danger. Topography also affects surface winds, influencing their speed and direction (Abd-Elaal et al. 2018; Jackson and Hunt 1975; Taylor et al. 1987), and for this reason, a spatial resolution as high as possible is desirable to be able to detect these changes.

Wind vector information has been derived from satellite information for years (Velden et al. 2005), but mostly focused on winds over the ocean. Some current geostationary sensors provide this information also over land, such as the Advanced Baseline Imager (ABI, Schmit et al. 2017) on board GOES-16 and GOES-17, which has a spatial resolution of 2 km and a temporal resolution of 5–15 min. The Advanced Himawari Imager (AHI, Bessho et al. 2016) on the Himawari-8 and -9 satellites of the Japan Meteorological Agency (JMA) has a spatial resolution of 2 km and a temporal resolution of 10 min, while the Spinning Enhanced Visible Infra-Red Imager (SEVIRI) on board the Meteosat Second-Generation (MSG, Schmetz et al. 2002) satellites has a spatial resolution of 3 km and a temporal resolution of 5–15 min. The Aeolus satellite, launched in 2018 and with a polar orbit, is the first satellite designed specifically to provide profiles of wind, using the Atmospheric Laser Doppler Instrument (ALADIN), and providing information with a maximum of 10 km spatial resolution (Kanitz et al. 2019), orbiting the Earth every 90 min and with a revisit time of 7 days. Wind vector information can also be obtained from the ERA5 reanalysis (Hersbach et al. 2018), with a spatial resolution of approximately 30 km.

But even though some fire danger models, such as FWI, NFDRS and FDRS, use wind speed as one of the models' input parameters, the coarse resolution of the information based on RS (1–30 km) limits considerably the usefulness of wind information for realistic fire danger assessment. Still, the improvement in the spatial and temporal resolution of the new generation of geostationary sensors is starting to provide wind vector information that could be useful to estimate shifts in wildfire direction and pyroconvection (Apke et al. 2020).

4 Early Detection of Fire Occurrence from Space

The early detection of fires could be considered a final step in the assessment of fire danger, as early warning systems allow fire managers and emergency responders to act promptly when a fire occurs, reducing in many cases the total fire size and/or its consequences. This information is also useful to validate fire danger maps. Information on fire activity can be obtained from middle and thermal infrared bands of passive sensors, comparing the temperature of a potential fire with the temperature of the land surrounding it: if the difference is above a given threshold, the potential fire is confirmed as an active fire (AF) or "hot spot" (Giglio et al. 2016; Schroeder et al. 2014; Wooster et al. 2015). Most recently, new algorithms have been developed including the temporal dimension in their classification process (Schroeder et al. 2016), in order to discriminate between permanent sources of heat (bright surfaces, gas flares, etc.) and fire events (Lin et al. 2018).

The first sensors used for AF detection were the AVHRR (Flannigan and Haar 1986) with a spatial resolution of approximately 1 km, supplemented afterwards by MODIS, whose thermal channels have the same resolution as AVHRR (Giglio et al. 2018, 2003). More recently, the NPP, launched in 2011, incorporated the VIIRS sensor that detects active fires at 375 and 750 m and provides information twice a day (Oliva and Schroeder 2015). The VIIRS sensor was incorporated to the NOAA-20 satellite, launched in 2017, and it is programmed to be included in the Joint Polar Satellite System (JPSS)-2, -3, and

-4 satellites, the heirs of the NOAA satellites. Other polar satellites that also incorporate thermal channels for AF detection are the FengYun-3 B and C with the Visible and Infra-Red Radiometer (VIIR, Lin et al. 2018). The combination of these satellites allows information on active fires to be obtained several times each day.

Still, while these satellites provide a higher resolution compared to geostationary satellites, they have a much lower temporal resolution in comparison with the 5–15-min image interval of geostationary sensors, which allow to detect short-lived fires and to track fire spread (Wooster et al. 2015). The reference geostationary sensors for this purpose are, as in the case of the wind detection, the GOES ABI, Meteosat SEVIRI, and Himawari AHI. These sensors have a spatial resolution of 2–3 km, which can be a disadvantage in the detection of small fires, as their radiative power can be below the threshold for AF detection.

Higher resolution sensors such as Landsat-8 OLI (Schroeder et al. 2016) or Sentinel-2 MSI (Dell'Aglio et al. 2019) have also been proposed for active fire detection, but the lower temporal resolution of these sensors limits their usefulness for NRT applications.

Currently, several systems exist to display AF in NRT globally or at continental level. An example of a global system is the NASA Fire Information for Resource Management System (FIRMS), which displays the active fires from MODIS and VIIRS (Fig. 3). EFFIS and GWIS also show active fires from those sensors in their systems (see left panel in the screen capture of Fig. 2), for Europe and northern Africa and globally, respectively, as well as the Global Forest Watch Fires website https://fires.globalforestwatch.org/map (accessed June 2020). The Australian My Fire Watch map service (https://myfirewatch.landgate.wa.gov.au/, accessed June 2020) displays the active fires for Australia and South East Asia. In addition, the Brazilian Instituto Nacional de Pesquisas Espaciais (INPE) provides global information on active fires at NRT (https://queimadas.dgi.inpe.br/queimadas/portal-static/situacao-atual/, accessed June 2020), complementing MODIS and VIIRS information with AVHRR and SEVIRI data.

Fig. 3 Screen capture of the FIRMS website (https://firms.modaps.eosdis.nasa.gov/map/, accessed July 2020) showing the active fires of 1 July 2020

5 Conclusions

As shown in the previous sections, satellite remote sensing has become an invaluable source of information for fire danger assessment, especially at continental and global scales. Extensive research has been performed to incorporate RS information to model diverse variables that influence fire occurrence, and that can be used to predict fire probability and spread. Some of these models have become operational products that provide information on the probability of fire occurrence to guide public campaigns for fire prevention and for early response management.

The new sensors launched in the past few years, along with the upcoming satellites, will increase the spatial and temporal resolution of the remote sensing information offering a larger volume of available data and improving the mechanisms for fire danger assessment.

Acknowledgements The authors thank the International Space Science Institute (ISSI) for organizing the "Natural and man-made hazards monitoring by the Earth Observation missions: current status and scientific gaps" that led to this publication.

References

Abd-Elaal E-S, Mills JE, Ma X (2018) Numerical simulation of downburst wind flow over real topography. J Wind Eng Ind Aerodyn 172:85–95. https://doi.org/10.1016/j.jweia.2017.10.026

Abrams M (2016) ASTER global DEM version 3, and new ASTER water body dataset. Int Arch Photogramm Remote Sens Spatial Inf Sci XLI-B4:107–110. https://doi.org/10.5194/isprs-archives-XLI-B4-107-2016

Achard F, Eva HD, Mollicone D, Beuchle R (2008) The effect of climate anomalies and human ignition factor on wildfires in Russian boreal forests. Philos Trans R Soc Lond B Biol Sci 363:2331–2339. https://doi.org/10.1098/rstb.2007.2203

Albini FA (1976) Estimating wildfire behavior and effects, General Technical Report INT-30. USDA Forest Service, Intermountain Forest and Range Experiment Station, Odgen, UT

Albrecht RI, Goodman SJ, Petersen WA, Buechler DE, Bruning EC, Blakeslee RJ, Christian HJ (2011) The 13 years of TRMM lightning imaging sensor: from individual flash characteristics to decadal tendencies. In: XIV international conference on atmospheric electricity, Rio de Janeiro, Brasil.

Alonso-Benito A, Arroyo LA, Arbelo M, Hernández-Leal P (2016) Fusion of WorldView-2 and LiDAR data to map fuel types in the Canary Islands. Remote Sens 8:669

Andrews PL (2012) Modeling wind adjustment factor and midflame wind speed for Rothermel's surface fire spread model, Gen. Tech. Rep. RMRS-GTR-266. USDA Forest Service, Rocky Mountain Research Station, Fort Collins, CO

Apke JM, Hilburn KA, Miller SD, Peterson DA (2020) Towards objective identification and tracking of convective outflow boundaries in next-generation geostationary satellite imagery. Atmos Meas Tech 13:1593–1608. https://doi.org/10.5194/amt-13-1593-2020

Archibald S et al (2018) Biological and geophysical feedbacks with fire in the Earth system. Environ Res Lett 13:033003. https://doi.org/10.1088/1748-9326/aa9ead

Archibald S, Scholes RJ, Roy DP, Roberts G, Boschetti L (2010) Southern African fire regimes as revealed by remote sensing. Int J Wildland Fire 19:861–878

Argañaraz JP, Landi MA, Bravo SJ, Gavier-Pizarro GI, Scavuzzo CM, Bellis LM (2016) Estimation of live fuel moisture content from MODIS images for fire danger assessment in Southern Gran Chaco. IEEE J Sel Top Appl Earth Observ Remote Sens 9:5339–5349. https://doi.org/10.1109/jstars.2016.2575366

Arino O et al. GlobCover: ESA service for global land cover from MERIS. In: International geoscience and remote sensing symposium, IGARSS 2007, Barcelona, Spain, 2007. IEEE- Inst Electrical Electronics Engineer Inc., pp 2412–2415. https://doi.org/10.1109/IGARSS.2007.4423328

Arroyo LA, Healey SP, Cohen WB, Cocero D, Manzanera JA (2006) Using object-oriented classification and high-resolution imagery to map fuel types in a Mediterranean region. J Geophys Res. https://doi.org/10.1029/2005JG000120,2006

  Springer

Baccini A et al (2012) Estimated carbon dioxide emissions from tropical deforestation improved by carbon-density maps Nat. Clim Change 2:182–185. https://doi.org/10.1038/nclimate1354

Bagnardi M, González PJ, Hooper A (2016) High-resolution digital elevation model from tri-stereo Pleiades-1 satellite imagery for lava flow volume estimates at Fogo Volcano. Geophys Res Lett 43:6267–6275. https://doi.org/10.1002/2016gl069457

Bajocco S, Dragoz E, Gitas I, Smiraglia D, Salvati L, Ricotta C (2015) Mapping forest fuels through vegetation phenology: the role of coarse-resolution satellite time-series. PLoS ONE 10:e0119811–e0119811. https://doi.org/10.1371/journal.pone.0119811

Balch JK, Bradley BA, Abatzoglou JT, Nagy RC, Fusco EJ, Mahood AL (2017) Human-started wildfires expand the fire niche across the United States. Proc Natl Acad Sci 114:2946. https://doi.org/10.1073/pnas.1617394114

Barroso Ramos-Neto M, Pivello VR (2000) Lightning fires in a brazilian savanna National Park: rethinking management strategies. Environ Manag 26:675–684. https://doi.org/10.1007/s002670010124

Berger C, Werner S, Wigley-Coetsee C, Smit I, Schmullius C (2019) Multi-temporal sentinel-1 data for wall-to-wall herbaceous biomass mapping in Kruger National Park, South Africa—first results. In: IGARSS 2019-2019 IEEE international geoscience and remote sensing symposium, 28 July–2 Aug. 2019, pp 7358–7360. https://doi.org/10.1109/igarss.2019.8898045

Bessho K et al (2016) An Introduction to Himawari-8/9— Japan’s new-generation geostationary. Meteorol Satellites J Meteorol Soc Jpn Ser II 94:151–183. https://doi.org/10.2151/jmsj.2016-009

Bistinas I, Oom D, Sá ACL, Harrison SP, Prentice IC, Pereira JMC (2013) Relationships between human population density and burned area at continental and global scales. PLoS ONE 8:e81188–e81188. https://doi.org/10.1371/journal.pone.0081188

Blakeslee RJ et al (2014) Lightning imaging sensor (LIS) for the international space station (ISS): missio n description and science goals. In: XV international conference on atmospheric electricity, Norman, Oklahoma

Bouvet A, Mermoz S, Le Toan T, Villard L, Mathieu R, Naidoo L, Asner GP (2018) An above-ground biomass map of African savannahs and woodlands at 25m resolution derived from ALOS PALSAR. Remote Sens Environ 206:156–173. https://doi.org/10.1016/j.rse.2017.12.030

Bowman DMJS et al (2011) The human dimension of fire regimes on Earth. J Biogeogr 38:2223–2236. https://doi.org/10.1111/j.1365-2699.2011.02595.x

Bowman DMJS et al (2009) Fire in the earth system. Science 324:481–484. https://doi.org/10.1126/science.1163886

Bradshaw LS, Deeming JE, Burgan RE, Cohen JD (1983) The 1978 National fire-danger rating system: technical documentation, GTR INT-169. USDA Forest Service, Intermountain Forest and Range Experiment Station, Ogden, UT

Burgan RE, Andrews PL, Bradshaw LS, Chase CH, Hartford RA, Latham DJ (1997) WFAS: wildland fire assessment system. Fire Manag Notes 57:14–17

Büttner G (2014) CORINE Land Cover and Land Cover Change Products. In: Manakos I, Braun M (eds) Land use and land cover mapping in Europe: practices & trends. Springer, Dordrecht, pp 55–74. https://doi.org/10.1007/978-94-007-7969-3_5

Büttner G, Kosztra B, Soukup T, Sousa A, Langanke T (2017) CLC2018 Technical Guidelines, Service Contract No 3436/R0-Copernicus/EEA.56665. European Topic Centre on Urban, Land and Soil Systems,

Camia A, Bovio G, Aguado I, Stach N (1999) Meteorological fire danger indices and remote sensing. In: Chuvieco E (ed) Remote sensing of large wildfires in the european mediterranean basin. Springer, Berlin, pp 39–59

Cano-Crespo A, Oliveira PJC, Boit A, Cardoso M, Thonicke K (2015) Forest edge burning in the Brazilian Amazon promoted by escaping fires from managed pastures. J Geophys Res Biogeosci 120:2095–2107. https://doi.org/10.1002/2015jg002914

Cardona OD et al. (2012) Determinants of risk: exposure and vulnerability. Cambridge, UK, and New York, NY, USA

Cecil DJ (2015) LIS/OTD gridded lightning climatology data collection, version2.3.2015. NASA EOSDIS globalhydrology resource center distributed active archive center, Huntsville, Alabama, U.S.A. https://dx.doi.org/10.5067/LIS/LIS-OTD/DATA311

Cecil DJ, Buechler DE, Blakeslee RJ (2014) Gridded lightning climatology from TRMM-LIS and OTD: dataset description. Atmos Res 135–136:404–414. https://doi.org/10.1016/j.atmosres.2012.06.028

Cohen JD, Deeming JE (1985) The National fire-danger rating system: basic equations, general technical report PSW-82. USDA Forest Service, Pacific Southwest Forest and Range Experiment Station, Berkeley, CA

Collins TW (2005) Households, forests, and fire hazard vulnerability in the American West: a case study of a California community. Environ Hazards 6:23–37

Conedera M, Cesti G, Pezzatti GB, Zumbrunnen T, Spinedi F (2006) Lightning-induced fires in the Alpine region: an increasing problem. In: Viegas DX (ed) V International conference on forest fire research, Coimbra, Portugal, 2006. Coimbra University Press

Costafreda-Aumedes S, Comas C, Vega-Garcia C (2017) Human-caused fire occurrence modelling in perspective: a review. Int J Wildland Fire 26:983–998. https://doi.org/10.1071/WF17026

Christian HJ et al (2003) Global frequency and distribution of lightning as observed from space by the Optical Transient Detector. J Geophys Res Atmos 108:ACL41–ACL415. https://doi.org/10.1029/2002jd002347

Christian HJ, Driscoll K, Goodman S, Blakeslee R, Mach DDB (1996) The optical transient detector (OTD). In: Proceedings of the 10th international conference on atmospheric electricity, Osaka, Japan, June 10–14 1996. pp 368–371

Chuvieco E et al (2014a) Integrating geospatial information into fire risk assessment. Int J Wildland Fire 23:606–619. https://doi.org/10.1071/WF12052

Chuvieco E, Allgöwer B, Salas J (2003) Integration of physical and human factors in fire danger assessment. In: Chuvieco E (ed) Wildland fire danger estimation and mapping, series in remote sensing. World Scientific Pub Co Inc, Singapore, pp 197–218. https://doi.org/10.1142/9789812791177_00074

Chuvieco E, Cocero D, Riaño D, Martin P, Martínez-Vega J, de la Riva J, Pérez F (2004) Combining NDVI and surface temperature for the estimation of live fuel moisture content in forest fire danger rating. Remote Sens Environ 92:322–331. https://doi.org/10.1016/j.rse.2004.01.019

Chuvieco E, Justice C (2010) Relations between human factors and global fire activity. In: Chuvieco E, Li J, Yang X (eds) Advances in earth observation of global change. Springer, London, pp 187–199

Chuvieco E, Martínez S, Román MV, Hantson S, Pettinari ML (2014b) Integration of ecological and socio-economic factors to assess global vulnerability to wildfire. Global Ecol Biogeogr 23:245–258. https://doi.org/10.1111/geb.12095

Chuvieco E, Riaño D, Aguado I, Cocero D (2002) Estimation of fuel moisture content from multitemporal analysis of landsat thematic mapper reflectance data: applications in fire danger assessment. Int J Remote Sens 23:2145–2162. https://doi.org/10.1080/01431160110069818

Chuvieco E, Wagtendonk J, Riaño D, Yebra M, Ustin SL (2009) Estimation of fuel conditions for fire danger assessment. In: Chuvieco E (ed) Earth observation of wildland fires in Mediterranean ecosystems. Springer, Berlin, pp 83–96. https://doi.org/10.1007/978-3-642-01754-4

de Groot WJ, Flannigan MD, Cantin AS (2013) Climate change impacts on future boreal fire regimes. For Ecol Manag 294:35–44. https://doi.org/10.1016/j.foreco.2012.09.027

de Groot WJ et al (2006) Developing a global early warning system for wildland fire. In: Viegas DX (ed) V International conference on forest fire research, Coimbra, Portugal, 27–30 Nov 2006. p 12

Deeming JE, Burgan RE, Cohen JD (1977) The national fire-danger rating system—1978, General Technical Report INT-39. Ogden, Utah

Dell'Aglio DAG, Gargiulo M, Iodice A, Riccio D, Ruello G (2019) Active fire detection in multispectral super-resolved sentinel-2 images by means of sam-based approach. In: 2019 IEEE 5th international forum on research and technology for society and industry (RTSI), 9–12 Sept 2019. pp 124–127. https://doi.org/10.1109/rtsi.2019.8895538

Dowdy AJ, Mills GA, Finkele K, de Groot WJ (2009) Australian fire weather as represented by the McArthur forest fire danger index and the canadian forest fire weather index, CAWCR technical report No. 10. Centre for Australian Weather and Climate Research,

Doxsey-Whitfield E, MacManus K, Adamo SB, Pis-tolesi L, Squires J, Borkovska O, Baptista SR (2015) Taking advantage of the improved availability of census data: a first look at the gridded population of the world. Appl Geogr 1:226–234. https://doi.org/10.1080/23754931.2015.1014272

Dubayah R et al (2020) The Global Ecosystem Dynamics Investigation: high-resolution laser ranging of the Earth's forests and topography. Sci Remote Sens 1:100002. https://doi.org/10.1016/j.srs.2020.100002

Esch T et al (2017) Breaking new ground in mapping human settlements from space—the global urban footprint ISPRS. J Photogramm 134:30–42. https://doi.org/10.1016/j.isprsjprs.2017.10.012

European Forest Fire Information System (2017) European Forest Map. JRC Contract No 384347 on the "Development of a European Fuel Map". European Commission

European Forest Fire Information System (2020) User Guide to EFFIS applications, version 2.5.

European Space Agency (2015) Sentinel-2 User Handbook, Issue 1, Rev 2.

Eva H, Lambin EF (2000) Fires and land-cover change in the tropics: a remote sensing analysis at the landscape scale. J Biogeogr 27:765–776

Evangeliou N et al (2019) Open fires in Greenland in summer 2017: transport, deposition and radiative effects of BC, OC and BrC emissions. Atmos Chem Phys 19:1393–1411. https://doi.org/10.5194/acp-19-1393-2019

FAO, GFMC (1999) 1999 Revision of the FAO Wildland Fire Management Terminology, by the Global Fire Monitoring Center (GFMC)

Farr TG et al (2007) The shuttle radar topography mission. Rev Geophys. https://doi.org/10.1029/2005rg000183

Fick SE, Hijmans RJ (2017) WorldClim 2: new 1-km spatial resolution climate surfaces for global land areas. Int J Climatol 37:4302–4315. https://doi.org/10.1002/joc.5086

Field RD et al (2015) Development of a global fire weather database. Nat Hazards Earth Syste Sci. https://doi.org/10.5194/nhess-15-1407-2015

Flannigan MD, Haar THV (1986) Forest fire monitoring using NOAA satellite AVHRR. Can J Forest Res 16:975–982. https://doi.org/10.1139/x86-171

Florinsky IV (2016) Chapter 3-digital elevation models. In: Florinsky IV (ed) Digital terrain analysis in soil science and geology, 2nd edn. Academic Presss, Cambridge, pp 77–108. https://doi.org/10.1016/B978-0-12-804632-6.00003-1

Franke J et al (2018) Fuel load mapping in the Brazilian Cerrado in support of integrated fire management. Remote Sens Environ 217:221–232. https://doi.org/10.1016/j.rse.2018.08.018

Freire S et al (2018) Enhanced data and methods for improving open and free global population grids: putting 'leaving no one behind' into practice. Int J Digital Earth. https://doi.org/10.1080/17538947.2018.1548656

Friedl MA, Sulla-Menashe D, Tan B, Schneider A, Ramankutty N, Sibley A, Huang X (2010) MODIS Collection 5 global land cover: algorithm refinements and characterization of new datasets. Remote Sens Environ 114:168–182. https://doi.org/10.1016/j.rse.2009.08.016

Ganteaume A, Camia A, Jappiot M, San-Miguel-Ayanz J, Long-Fournel M, Lampin C (2013) A review of the main driving factors of forest fire ignition over. Europe Environ Manag 51:651–662. https://doi.org/10.1007/s00267-012-9961-z

García M, Chuvieco E, Nieto H, Aguado I (2008) Combining AVHRR and meteorological data for estimating live fuel moisture content. Remote Sens Environ 112:3618–3627. https://doi.org/10.1016/j.rse.2008.05.002

García M, Popescu S, Riaño D, Zhao K, Neuenschwander A, Agca M, Chuvieco E (2012) Characterization of canopy fuels using ICESat/GLAS data. Remote Sens Environ 123:81–89. https://doi.org/10.1016/j.rse.2012.03.018

García M, Riaño D, Chuvieco E, Danson FM (2010) Estimating biomass carbon stocks for a Mediterranean forest in central Spain using LiDAR height and intensity data. Remote Sens Environ 114:816–830. https://doi.org/10.1016/j.rse.2009.11.021

García M, Riaño D, Chuvieco E, Salas J, Danson FM (2011) Multispectral and LiDAR data fusion for fuel type mapping using Support Vector Machine and decision rules. Remote Sens Environ 115:1369–1379. https://doi.org/10.1016/j.rse.2011.01.017

García M, Saatchi S, Casas A, Koltunov A, Ustin SL, Ramirez C, Balzter H (2017) Extrapolating forest canopy fuel properties in the california rim fire by combining airborne LiDAR and landsat OLI data. Remote Sens 9:394

Giglio L, Boschetti L, Roy DP, Humber ML, Justice CO (2018) The Collection 6 MODIS burned area mapping algorithm and product. Remote Sens Environ 217:72–85. https://doi.org/10.1016/j.rse.2018.08.005

Giglio L, Descloitres J, Justice CO, Kaufman YJ (2003) An enhanced contextual fire detection algorithm for MODIS. Remote Sens Environ 87:273–282. https://doi.org/10.1016/S0034-4257(03)00184-6

Giglio L, Schroeder W, Justice CO (2016) The collection 6 MODIS active fire detection algorithm and fire products. Remote Sens Environ 178:31–41. https://doi.org/10.1016/j.rse.2016.02.054

Goodman SJ et al (2013) The GOES-R geostationary lightning mapper (GLM). Atmos Res 125126:34–49. https://doi.org/10.1016/j.atmosres.2013.01.006

Hansen MC, DeFries RS, Townsend JRG, Carroll M, Dimiceli C, Sohlberg RA (2003) Global percent tree cover at a spatial resolution of 500 meters: first results of the MODIS vegetation continuous fields algorithm. Earth Interact 7:1–15

Hardy CC (2005) Wildland fire hazard and risk: problems, definitions, and context. Forest Ecol Manag 211:73–82. https://doi.org/10.1016/j.foreco.2005.01.029

Hawbaker TJ et al (2017) Mapping burned areas using dense time-series of landsat data. Remote Sens Environ 198:504–522. https://doi.org/10.1016/j.rse.2017.06.027

He J, Loboda TV, Jenkins L, Chen D (2019) Mapping fractional cover of major fuel type components across Alaskan tundra. Remote Sens Environ 232:111324. https://doi.org/10.1016/j.rse.2019.111324

Hermosilla T, Ruiz LA, Kazakova AN, Coops NC, Moskal LM (2014) Estimation of forest structure and canopy fuel parameters from small-footprint full-waveform LiDAR data. Int J Wildland Fire 23:224–233. https://doi.org/10.1071/WF13086

Hersbach H et al (2018) Operational global reanalysis: progress, future directions and synergies with NWP, ERA Report Series 27. European Centre for Medium Range Weather Forecasts, Reading, UK. https://doi.org/10.21957/tkic6g3wm

Hijmans RJ, Cameron SE, Parra JL, Jones PG, Jarvis A (2005) Very high resolution interpolated climate surfaces for global land areas. Int J Climatol 25:1965–1978. https://doi.org/10.1002/joc.1276

Hillger D et al (2013) First-Light Imagery from Suomi NPP VIIRS. Bull Am Meteor Soc 94:1019–1029. https://doi.org/10.1175/bams-d-12-00097.1

Holden ZA, Jolly WM (2011) Modeling topographic influences on fuel moisture and fire danger in complex terrain to improve wildland fire management decision support. For Ecol Manag 262:2133–2141. https://doi.org/10.1016/j.foreco.2011.08.002

Hościło A, Lewandowska A (2019) Mapping forest type and tree species on a regional scale using multitemporal sentinel-2 data. Remote Sens 11:929

Hubacek M, Kovarik V, Kratochvil V (2016) Analysis of influence of terrain relief roughness on DEM accuracy generated from LIDAR in the Czech Republic territory. Int Arch Photogramm Remote Sens Spatial Inf Sci 4XLI-B4:25–30. https://doi.org/10.5194/isprs-archives-XLI-B4-25-2016

Huesca M, Riaño D, Ustin SL (2019) Spectral mapping methods applied to LiDAR data: application to fuel type mapping. Int J Appl Earth Obs Geoinf 74:159–168. https://doi.org/10.1016/j.jag.2018.08.020

Jackson PS, Hunt JCR (1975) Turbulent wind flow over a low hill. Q J R Meteorol Soc 101:929–955. https://doi.org/10.1002/qj.49710143015

Jia S, Kim SH, Nghiem SV, Kafatos M (2019) Estimating live fuel moisture using SMAP L-band radiometer soil moisture for Southern California, USA. Remote Sens 11:1575

Jones MW, Santín C, van der Werf GR, Doerr SH (2019) Global fire emissions buffered by the production of pyrogenic carbon. Nat Geosci 12:742–747. https://doi.org/10.1038/s41561-019-0403-x

Kanitz T et al (2019) Aeolus first light: first glimpse, vol 11180. In: International conference on space optics-ICSO 2018. SPIE

Keane RE, Burgan RE, van Wagtendonk J (2001) Mapping wildland fuels for fire management across multiple scales: integrating remote sensing, GIS, and biophysical modeling. Int J Wildland Fire 10:301–319

Keetch JJ, Byram GM (1968) A drought index for forest fire control, Research Paper SE-38 (revised 1988). United States Department of Agriculture - Forest Service, Ashville, NC

Knorr W, Arneth A, Jiang L (2016) Demographic controls of future global fire risk. Nat Clim Change 6:781–785. https://doi.org/10.1038/nclimate2999

Konings AG, Rao K, Steele-Dunne SC (2019) Macro to micro: microwave remote sensing of plant water content for physiology and ecology. New Phytol 223:1166–1172. https://doi.org/10.1111/nph.15808

Lanorte A, Lasaponara R (2008) Fuel type characterization based on coarse resolution MODIS satellite data. Forest Biogeosci For 1:60–64. https://doi.org/10.3832/ifor0451-0010060

Lasaponara R, Lanorte A (2007) Remotely sensed characterization of forest fuel types by using satellite ASTER data. Int J Appl Earth Obs Geoinf 9:225–234. https://doi.org/10.1016/j.jag.2006.08.001

Latifovic R, Zhu Z-L, Cihlar J, Giri C, Olthof I (2004) Land cover mapping of North and Central America-Global Land Cover 2000. Remote Sens Environ 89:116–127. https://doi.org/10.1016/j.rse.2003.11.002

Lauk C, Erb K-H (2009) Biomass consumed in anthropogenic vegetation fires: global patterns and processes. Ecol Econ 69:301–309. https://doi.org/10.1016/j.ecolecon.2009.07.003

Lefsky MA (2010) A global forest canopy height map from the moderate resolution imaging spectroradiometer and the geoscience laser altimeter system. Geophys Res Lett 37:5. https://doi.org/10.1029/2010GL043622

Levine JS, Cofer WR, Cahoon DR, Winstead EL (1995) Biomass burning: a driver for global change. Environ Sci Technol 29:120A–125A. https://doi.org/10.1021/es00003a746

Leyk S et al (2019) The spatial allocation of population: a review of large-scale gridded population data products and their fitness for use. Earth Syst Sci Data 11:1385–1409. https://doi.org/10.5194/essd-11-1385-2019

Lin Z, Chen F, Niu Z, Li B, Yu B, Jia H, Zhang M (2018) An active fire detection algorithm based on multitemporal FengYun-3C VIRR data. Remote Sens Environ 211:376–387. https://doi.org/10.1016/j.rse.2018.04.027

Liu L, Lim S, Shen X, Yebra M (2019) A hybrid method for segmenting individual trees from airborne lidar data. Comput Electron Agric 163:104871. https://doi.org/10.1016/j.compag.2019.104871

Lizundia-Loiola J, Otón G, Ramo R, Chuvieco E (2020) A spatio-temporal active-fire clustering approach for global burned area mapping at 250 m from MODIS data. Remote Sens Environ 236:111493. https ://doi.org/10.1016/j.rse.2019.111493

Lopez P (2016) A lightning parameterization for the ECMWF integrated forecasting system. Mon Weather Rev 144:3057–3075. https://doi.org/10.1175/mwr-d-16-0026.1

Lorenzini S, Bardazzi R, Giampietro MD, Feresin F, Taccola M, Cuevas LP (2012) Optical design of the lightning imager for MTG. In: Cugny B, Armandillo E, Karafolas N (eds) International conference on space optics 2012, Ajaccio, Corsica, France, 2012. SPIE. https://doi.org/10.1117/12.2309091

Luke RH, McArthur AG (1978) Bushfires in Australia. Australian Government Publishing Service, Canberra

Luo K, Quan X, He B, Yebra M (2019) Effects of live fuel moisture content on wildfire occurrence in fire-prone regions over Southwest China. Forests 10:887

Maliet E (2013) SPOT 6 and SPOT 7: offering SPOT data continuity. In: 64th International astronautical congress (IAC 2013), Beijing, China

Marino E, Ranz P, Tomé JL, Noriega MÁ, Esteban J, Madrigal J (2016) Generation of high-resolution fuel model maps from discrete airborne laser scanner and landsat-8 OLI: a low-cost and highly updated methodology for large areas. Remote Sens Environ 187:267–280. https://doi.org/10.1016/j.rse.2016.10.020

Markus T et al (2017) The ice, cloud, and land elevation satellite-2 (ICESat-2): science requirements, concept, and implementation. Remote Sens Environ 190:260–273. https://doi.org/10.1016/j.rse.2016.12.029

Martín Y, Zúñiga-Antón M, Rodrigues Mimbrero M (2019) Modelling temporal variation of fire-occurrence towards the dynamic prediction of human wildfire ignition danger in northeast Spain Geomatics. Nat Hazards Risk 10:385–411. https://doi.org/10.1080/19475705.2018.1526219

Martínez-Fernández J, Chuvieco E, Koutsias N (2013) Modelling long-term fire occurrence factors in Spain by accounting for local variations with geographically weighted regression. Nat Hazards Earth Syst Sci 13:311–327. https://doi.org/10.5194/nhess-13-311-2013

Martínez J, Vega-Garcia C, Chuvieco E (2009) Human-caused wildfire risk rating for prevention planning in Spain. J Envinon Manag 90:1241–1252

McArthur AG (1967) Fire behaviour in eucalypt forests, Leaflet N. 107. Department of National Development, Forestry and Timber Bureau, Canberra

McCaffrey S (2004) Thinking of wildfire as a natural hazard. Soc Nat Resour 17:509–516. https://doi.org/10.1080/08941920490452445

Meldrum JR, Brenkert-Smith H, Champ PA, Falk L, Wilson P, Barth CM (2018) Wildland-urban interface residents' relationships with wildfire: variation within and across communities. Soc Nat Resour 31:1132–1148. https://doi.org/10.1080/08941920.2018.1456592

Mendiguren G, Pilar Martín M, Nieto H, Pacheco-Labrador J, Jurdao S (2015) Seasonal variation in grass water content estimated from proximal sensing and MODIS time series in a Mediterranean Fluxnet site. Biogeosciences 12:5523–5535. https://doi.org/10.5194/bg-12-5523-2015

Moritz MA et al (2014) Learning to coexist with wildfire. Nature 515:58–66. https://doi.org/10.1038/nature13946

Mutlu M, Popescu SC, Stripling C, Spencer T (2008) Mapping surface fuel models using lidar and multispectral data fusion for fire behavior. Remote Sens Environ 112:274–285. https://doi.org/10.1016/j.rse.2007.05.005

Myoung B, Kim HS, Nghiem VS, Jia S, Whitney K, Kafatos CM (2018) Estimating live fuel moisture from MODIS satellite data for wildfire danger assessment in Southern California USA. Remote Sens. https://doi.org/10.3390/rs10010087

Nadeau LB, McRae DJ, Jin J-Z (2005) Development of a national fuel-type map for Canada using fuzzy logic, information report NOR-X-406. Canadian Forest Service, Edmonton

Narine LL, Popescu SC, Malambo L (2019) Synergy of ICESat-2 and landsat for mapping forest aboveground biomass with deep learning. Remote Sens 11:1503

Nasir S, Iqbal IA, Ali Z, Shahzad A (2015) Accuracy assessment of digital elevation model generated from pleiades tri stereo-pair. In: 2015 7th international conference on recent advances in space technologies (RAST), 16–19 June 2015, pp 193–197. https://doi.org/10.1109/rast.2015.7208340

Nelson KJ, Long DG, Connot JA (2016) LANDFIRE 2010—Updates to the national dataset to support improved fire and natural resource management, 2016–1010. Reston, VA. https://doi.org/10.3133/ofr20161010

Neuenschwander AL, Magruder LA (2016) The potential impact of vertical sampling uncertainty on ICESat-2/ATLAS terrain and canopy height retrievals for multiple ecosystems. Remote Sens 8:1039

Noble IR, Gill AM, Bary GAV (1980) McArthur's fire-danger meters expressed as equations Australian. J Ecol 5:201–203. https://doi.org/10.1111/j.1442-9993.1980.tb01243.x

Oliva P, Schroeder W (2015) Assessment of VIIRS 375m active fire detection product for direct burned area mapping. Remote Sens Environ 160:144–155. https://doi.org/10.1016/j.rse.2015.01.010

Ottmar RD, Sandberg DV, Riccardi CL, Prichard SJ (2007) An overview of the fuel characteristic classification system—Quantifying, classifying, and creating fuelbeds for resource planning. Can J Forest Res 37:2383–2393

Pausas JG, Keeley JE (2009) A burning story: the role of fire in the history of life. Bioscience 59:593–601. https://doi.org/10.1525/bio.2009.59.7.10

Pesaresi M, Ehrlich D, Florczyk AJ, Freire S, Julea A, Kemper T, Syrris V (2016) The global human settlement layer from landsat imagery. In: 2016 IEEE international geoscience and remote sensing symposium (IGARSS), 10–15 July 2016, pp 7276–7279. https://doi.org/10.1109/igarss.2016.7730897

Pettinari ML, Chuvieco E (2017) Fire behavior simulation from global fuel and climatic information. Forests 8:179. https://doi.org/10.3390/f8060179

Pettinari ML, Chuvieco E (2016) Generation of a global fuel data set using the fuel characteristic classification system. Biogeosciences 13:2061–2076. https://doi.org/10.5194/bg-13-2061-2016

Pham HT, Marshall L, Johnson F, Sharma A (2018) A method for combining SRTM DEM and ASTER GDEM2 to improve topography estimation in regions without reference data. Remote Sens Environ 210:229–241. https://doi.org/10.1016/j.rse.2018.03.026

Popescu SC, Zhou T, Nelson R, Neuenschwander A, Sheridan R, Narine L, Walsh KM (2018) Photon counting LiDAR: an adaptive ground and canopy height retrieval algorithm for ICESat-2 data. Remote Sens Environ 208:154–170. https://doi.org/10.1016/j.rse.2018.02.019

Pyne SJ (1995) World fire. The culture of fire on earth. Henry Colt and Company Inc, New York

Pyne SJ, Andrews PL, Laven RD (1996) Introduction to wildland fire. Wiley, New York

Pyne SJ, Goldammer JG (1997) The culture of fire: an introduction to anthropogenic fire history. Sediment records of biomass burning and global change. Springer, Berlin, pp 71–114

Quan X, He B, Li X, Liao Z (2016) Retrieval of grassland live fuel moisture content by parameterizing radiative transfer model with interval estimated LAI. IEEE J Sel Top Appl Earth Observ Remote Sens 9:910–920. https://doi.org/10.1109/jstars.2015.2472415

Quan X, He B, Yebra M, Yin C, Liao Z, Li X (2017) Retrieval of forest fuel moisture content using a coupled radiative transfer model. Environ Model Softw 95:290–302. https://doi.org/10.1016/j.envsoft.2017.06.006

Quegan S et al (2019) The European space agency BIOMASS mission: measuring forest above-ground biomass from space. Remote Sens Environ 227:44–60. https://doi.org/10.1016/j.rse.2019.03.032

Quegan S et al. (2017) D6–Global Biomass Map: Algorithm Theoretical Basis Document.

Rienecker MM et al (2011) MERRA: NASA's modern-era retrospective analysis for research and applications. J Clim 24:3624–3648. https://doi.org/10.1175/jcli-d-11-00015.1

Rodrigues M, de la Riva J, Fotheringham S (2014) Modeling the spatial variation of the explanatory factors of human-caused wildfires in Spain using geographically weighted logistic regression. Appl Geogr 48:52–63. https://doi.org/10.1016/j.apgeog.2014.01.011

Rodrigues M, Jiménez A, de la Riva J (2016) Analysis of recent spatial–temporal evolution of human driving factors of wildfires in Spain. Nat Hazards 84:2049–2070. https://doi.org/10.1007/s11069-016-2533-4

Rollins MG (2009) LANDFIRE: a nationally consistent vegetation, wildland fire, and fuel assessment. Int J Wildland Fire 18:235–249

Rosenqvist A, Shimada M, Ito N, Watanabe M (2007) ALOS PALSAR: a pathfinder mission for global-scale monitoring of the environment. IEEE Trans Geosci Remote 45:3307–3316. https://doi.org/10.1109/tgrs.2007.901027

Roteta E, Bastarrika A, Padilla M, Storm T, Chuvieco E (2019) Development of a sentinel-2 burned area algorithm: generation of a small fire database for sub-Saharan Africa. Remote Sens Environ 222:1–17. https://doi.org/10.1016/j.rse.2018.12.011

Rothermel RC (1972) A mathematical model for predicting fire spread in wildland fuels, Research Paper INT-115. USDA Forest Service, Intermountain Forest and Range Experiment Station, Odgen, UT

Rothermel RC (1983) How to predict the spread and intensity of forest and range fires, INT-143. National Wildfire Coordinating Group, USDA Forest Service Intermountain Research Station, Boise, ID

Saatchi SS et al (2011) Benchmark map of forest carbon stocks in tropical regions across three continents. PNAS 108:9899–9904. https://doi.org/10.1073/pnas.1019576108

San-Miguel-Ayanz J et al. (2018) Basic Criteria to assess wildfire risk at the pan-European level. EUR 29500 EN. https://doi.org/10.2760/052345

 Springer

San Miguel-Ayanz J et al. (2012) Comprehensive monitoring of wildfires in Europe: the European Forest Fire Information System (EFFIS). In: Tiefenbacher J (ed) Approaches to managing disaster—assessing hazards, emergencies and disaster impacts. InTech, Rijeka, Croatia, pp 87–108. https://doi.org/10.5772/28441

Sánchez Sánchez Y, Martínez-Graña A, Santos Francés F, Mateos Picado M (2018) Mapping wildfire ignition probability using sentinel 2 and LiDAR (Jerte Valley, Cáceres, Spain). Sensors 18:826

Santoro M (2018) GlobBiomass-global datasets of forest biomass. PANGAEA. https://doi.org/10.1594/pangaea.894711

Scott JH, Burgan RE (2005) Standard fire behavior fuel models: a comprehensive set for use with Rothermel's Surface Fire Spread Model, RMRS-GTR-153. USDA Forest Service, Rocky Mountain Research Station, Fort Collins, CO

Schenk T, Csatho B, van der Veen C, McCormick D (2014) Fusion of multi-sensor surface elevation data for improved characterization of rapidly changing outlet glaciers in Greenland. Remote Sens Environ 149:239–251. https://doi.org/10.1016/j.rse.2014.04.005

Schlobohm P, Brain J (2002) Gaining an understanding of the National Fire Danger Rating System, PMS 932. National Wildfire Coordinating Group, Fire Danger Working Team, Boise, ID

Schmetz J, Pili P, Tjemkes S, Just D, Kerkmann J, Rota S, Ratier A (2002) An introduction to meteosat second generatio (MSG). Bull Am Meteor Soc 83:977–992. https://doi.org/10.1175/1520-0477(2002)083

Schmit TJ, Griffith P, Gunshor MM, Daniels JM, Goodman SJ, Lebair WJ (2017) A closer look at the ABI on the GOES-R series. Bull Am Meteor Soc 98:681–698. https://doi.org/10.1175/bams-d-15-00230.1

Schneider FD, Ferraz AA, Hancock S, Duncanson LI, Dubayah RO, Pavlick RP, Schimel DS (2020) Towards mapping the diversity of canopy structure from space with GEDI. Environ Res Lett. https://doi.org/10.1088/1748-9326/ab9e99

Schroeder W, Oliva P, Giglio L, Csiszar IA (2014) The New VIIRS 375m active fire detection data product: algorithm description and initial assessment. Remote Sens Environ 143:85–96. https://doi.org/10.1016/j.rse.2013.12.008

Schroeder W, Oliva P, Giglio L, Quayle B, Lorenz E, Morelli F (2016) Active fire detection using Landsat-8/OLI data. Remote Sens Environ 185:210–220. https://doi.org/10.1016/j.rse.2015.08.032

Schunk C, Wastl C, Leuchner M, Schuster C, Menzel A (2013) Forest fire danger rating in complex topography—Results from a case study in the Bavarian Alps in autumn 2011. Nat Hazards Earth Syst Sci 13:2157–2167. https://doi.org/10.5194/nhess-13-2157-2013

Schutz BE, Zwally HJ, Shuman CA, Hancock D, DiMarzio JP (2005) Overview of the ICESat mission geophysical. Res Lett. https://doi.org/10.1029/2005gl024009

Shlisky A et al (2007) Fire, ecosystems and people: threats and strategiesfor global biodiversity conservation. The Nature Conservancy, Arlington, VA

Shu Q, Quan X, Yebra M, Liu X, Wang L, Zhang Y (2019) Evaluating the sentinel-2a satellite data for fuel moisture content retrieval. In: IGARSS 2019-2019 IEEE international geoscience and remote sensing symposium, 28 July–2 Aug 2019, pp 9416–9419. https://doi.org/10.1109/igarss.2019.8900104

Simard M, Pinto N, Fisher JB, Baccini A (2011) Mapping forest canopy height globally with spaceborne lidar. J Geophys Res 116:12. https://doi.org/10.1029/2011JG001708

SRTM (2015) The Shuttle Radar Topography Mission (SRTM) Collection User Guide. https://lpdaac.usgs.gov/documents/179/SRTM_User_Guide_V3.pdf.

Stefanidou A, Dragozi E, Stavrakoudis D, Gitas IZ (2018) Fuel type mapping using object-based image analysis of DMC and Landsat-8 OLI imagery. Geocarto Int 33:1064–1083. https://doi.org/10.1080/10106049.2017.1333532

Stocks BJ, Lawson BD, Alexander ME, Van Wagner CE, McAlpine RS, Lynham TJ, Dubé DE (1989) Canadian forest fire danger rating system: an overview. For Chron 65:258–265

Stuhlmann R et al (2005) Plans for EUMETSAT's third generation meteosat geostationary satellite programme. Adv Space Res 36:975–981. https://doi.org/10.1016/j.asr.2005.03.091

Syphard AD, Keeley JE, Pfaff AH, Ferschweiler K (2017) Human presence diminishes the importance of climate in driving fire activity across the United States. Proc Natl Acad Sci 114:13750–13755. https://doi.org/10.1073/pnas.1713885114

Syphard AD, Radeloff VC, Keeley JE, Hawbaker TJ, Clayton MK, Stewart SI, Hammer RB (2007) Human influence on California fire regimes. Ecol Appl 17:1388–1402

Tadono T, Nagai H, Ishida H, Oda F, Naito S, Minakawa K, Iwamoto H (2016) Generation of the 30 m-mesh global digital surface model. Int Arch Photogramm Remote Sens Spatial Inf Sci XLI-B4:157–162. https://doi.org/10.5194/isprs-archives-XLI-B4-157-2016

Tadono T, Takaku J, Ohgushi F, Doutsu M, Kobayashi K (2017) Updates of 'AW3D30' 30 M-MESH global digital surface model dataset. In: 2017 IEEE international geoscience and remote sensing symposium (IGARSS), 23–28 July 2017, pp 5656–5657. https://doi.org/10.1109/igarss.2017.8128290

Takaku J, Tadono T, Tsutsui K (2014) Generation of high resolution global DSM from ALOS PRISM. Int Arch Photogramm Remote Sens Spatial Inf Sci XL4:243–248. https://doi.org/10.5194/isprsarchives-XL-4-243-2014

Tatem AJ (2017) WorldPop, open data for spatial demography. Sci Data 4:170004. https://doi.org/10.1038/sdata.2017.4

Taylor PA, Mason PJ, Bradley EF (1987) Boundary-layer flow over low hills. Bound-Layer Meteorol 39:107–132. https://doi.org/10.1007/bf00121870

Thonicke K, Spessa A, Prentice IC, Harrison SP, Dong L, Carmona-Moreno C (2010) The influence of vegetation, fire spread and fire behaviour on biomass burning and trace gas emissions: results from a process-based model. Biogeosciences 7:1991–2011. https://doi.org/10.5194/bg-7-1991-2010

Tyc G, Tulip J, Schulten D, Krischke M, Oxfort M (2005) The RapidEye mission design. Acta Astronaut 56:213–219. https://doi.org/10.1016/j.actaastro.2004.09.029

United States Geological Survey (2019) Landsat 8 (L8) Data Users Handbook, version 5.0, LSDS-1574.

van der Werf GR et al (2010) Global fire emissions and the contribution of deforestation, savanna, forest, agricultural, and peat fires (1997–2009). Atmos Chem Phys 10:11707–11735. https://doi.org/10.5194/acp-10-11707-2010

Van Wagner CE (1987) Development and structure of the Canadian Forest Fire Weather Index System. Canadian Forestry Service, Ottawa

Van Wagtendonk JW, Root RR (2003) The use of multi-temporal landsat normalized difference vegetation index (NDVI) data for mapping fuel models in Yosemite National Park, USA. Int J Remote Sens 24:1639–1651. https://doi.org/10.1080/01431160210144679

Vegetation Continuous Field MOD44B, Collection 4, Version 3 (2007) University of Maryland, College Park, Maryland. https://www.glcf.umd.edu/data/vcf/. Accessed last accessed January 2012

Velden C et al (2005) Recent innovations in deriving tropospheric winds from meteorological satellites. Bull Am Meteor Soc 86:205–224. https://doi.org/10.1175/bams-86-2-205

Viegas D, Viegas M, Ferreira A (1992) Moisture content of fine forest fuels and fire occurrence in Central Portugal. Int J Wildland Fire 2:69–86. https://doi.org/10.1071/WF9920069

Vilar L, Camia A, San-Miguel-Ayanz J, Martín MP (2016) Modeling temporal changes in human-caused wildfires in Mediterranean Europe based on Land Use-Land Cover interfaces. For Ecol Manag 378:68–78. https://doi.org/10.1016/j.foreco.2016.07.020

Vitolo C, Di Giuseppe F, Krzeminski B, San-Miguel-Ayanz J (2019) A 1980–2018 global fire danger reanalysis dataset for the Canadian Fire Weather Indices. Sci Data 6:190032. https://doi.org/10.1038/sdata.2019.32

Wang L, Hunt ER, Qu JJ, Hao X, Daughtry CST (2013) Remote sensing of fuel moisture content from ratios of narrow-band vegetation water and dry-matter indices. Remote Sens Environ 129:103–110. https://doi.org/10.1016/j.rse.2012.10.027

Wang L, Quan X, He B, Yebra M, Xing M, Liu X (2019a) Assessment of the dual polarimetric sentinel-1A data for forest fuel moisture content estimation. Remote Sens 11:1568

Wang S et al (2019b) DEM generation from Worldview-2 stereo imagery and vertical accuracy assessment for its application in active tectonics. Geomorphology 336:107–118. https://doi.org/10.1016/j.geomorph.2019.03.016

Wessel B, Huber M, Wohlfart C, Marschalk U, Kosmann D, Roth A (2018) Accuracy assessment of the global TanDEM-X digital elevation model with GPS data ISPRS. J Photogramm 139:171–182. https://doi.org/10.1016/j.isprsjprs.2018.02.017

Whitney KL, Kim SH, Kafatos M (2019) Modeling live fuel moisture content with MODIS and VIIRS satellite data in Los Angeles County, California. In: American Geophysical Union, Fall Meeting 2019, San Francisco

Wooster M et al (2015) LSA SAF Meteosat FRP products - Part 1: algorithms, product contents, and analysis tmospheric. Chem Phys 15:13217–13239. https://doi.org/10.5194/acp-15-13217-2015

Yamaguchi Y, Kahle AB, Tsu H, Kawakami T, Pniel M (1998) Overview of advanced spaceborne thermal emission and reflection radiometer (ASTER). IEEE Trans Geosci Remote 36:1062–1071. https://doi.org/10.1109/36.700991

Yang J, Zhang Z, Wei C, Lu F, Guo Q (2017) Introducing the new generation of chinese geostationary weather satellites, Fengyun-4. Bull Am Meteor Soc 98:1637–1658. https://doi.org/10.1175/bams-d-16-0065.1

Yankovich EP, Yankovich KS, Baranovskiy NV, Bazarov AV, Sychev RS, Badmaev NB (2019) Mapping of vegetation cover using Sentinel-2 to estimate forest fire danger vol 11152. SPIE Remote Sensing. SPIE

Yebra M, Chuvieco E, Riaño D (2008) Estimation of live fuel moisture content from MODIS images for fire risk assessment. Agric For Meteorol 148:523–536. https://doi.org/10.1016/j.agrformet.2007.12.005

Yebra M et al (2013) A global review of remote sensing of live fuel moisture content for fire danger assessment: moving towards operational products. Remote Sens Environ 136:455–468. https://doi.org/10.1016/j.rse.2013.05.029

Yebra M, Quan X, Riaño D, Rozas Larraondo P, van Dijk AIJM, Cary GJ (2018) A fuel moisture content and flammability monitoring methodology for continental Australia based on optical remote sensing. Remote Sens Environ 212:260–272. https://doi.org/10.1016/j.rse.2018.04.053

Yebra M et al (2019) Globe-LFMC, a global plant water status database for vegetation ecophysiology and wildfire applications. Sci Data 6:155. https://doi.org/10.1038/s41597-019-0164-9

Zhao J et al (2007) Spatial and temporal distributions of lightning activities in Northeast China from satellite observation and analysis for lightning fire. In: Gao W, Ustin SL (eds) Remote sensing and modeling of ecosystems for sustainability IV, vol 6679 San Diego, CA, p 66790M. https://doi.org/10.1117/12.729349

Zink M et al (2014) TanDEM-X: the new global DEM takes shape. IEEE Geosci Remote Sens Mag 2:8–23. https://doi.org/10.1109/mgrs.2014.2318895

Publisher's Note Springer Nature remains neutral with regard to jurisdictional claims in published maps and institutional affiliations.

Surveys in Geophysics (2020) 41:1461–1487
https://doi.org/10.1007/s10712-020-09618-0

On the Use of Satellite Remote Sensing to Detect Floods and Droughts at Large Scales

T. Lopez[1,2] · A. Al Bitar[3] · S. Biancamaria[4] · A. Güntner[5,6] · A. Jäggi[7]

Received: 23 March 2020 / Accepted: 10 September 2020 / Published online: 10 October 2020
© Springer Nature B.V. 2020

Article highlights

- Each component of the terrestrial water storage is a key hydrological variable to understand floods and drought events
- Their monitoring at river basin scale and over long periods of time is facilitated by largescale sensors.
- The combination of Earth observations with other datasets can be an asset for the prediction ofhydrological events and for monitoring.

✉ T. Lopez
   tlopez.science@protonmail.com

   A. Al Bitar
   ahmad.albitar@ccsbio.cnes.fr

   S. Biancamaria
   sylvain.biancamaria@legos.obs-mip.fr

   A. Güntner
   andreas.guentner@gfz-potsdam.de

   A. Jäggi
   adrian.jaeggi@aiub.unibe.ch

[1] Institut de Recherche Technologique (IRT) Saint-Exupéry - Fondation STAE, GET, 14 avenue Edouard Belin, 31400 Toulouse, France

[2] International Space Science Institute (ISSI), Hallerstrasse 6, 3012 Bern, Switzerland

[3] CESBIO, Université de Toulouse, CNES, CNRS, INRAE, IRD, UPS, 13 avenue du Colonel Roche, 31400 Toulouse, France

[4] LEGOS, Université de Toulouse, CNES, CNRS, IRD, UPS, 14 avenue Edouard Belin, Toulouse 31400, France

[5] Helmholtz Centre Potsdam GFZ German Research Centre for Geosciences, 14473 Telegrafenberg, Germany

[6] Institute of Environmental Science and Geography, University of Potsdam, 14476 Potsdam, Germany

[7] Astronomical Institute, University of Bern, Sidlerstrasse 5, 3012 Bern, Switzerland

Surveys in Geophysics (2020) 41:1461–1487

Abstract

Hydrological extremes, in particular floods and droughts, impact all regions across planet Earth. They are mainly controlled by the temporal evolution of key hydrological variables like precipitation, evaporation, soil moisture, groundwater storage, surface water storage and discharge. Precise knowledge of the spatial and temporal evolution of these variables at the scale of river basins is essential to better understand and forecast floods and droughts. In this article, we present recent advances on the capability of Earth observation (EO) satellites to provide global monitoring of floods and droughts. The local scale monitoring of these events which is traditionally done using high-resolution optical or SAR (synthetic aperture radar) EO and in situ data will not be addressed. We discuss the applications of moderate- to low-spatial-resolution space-based observations, e.g., satellite gravimetry (GRACE and GRACE-FO), passive microwaves (i.e. SMOS) and satellite altimetry (i.e. the JASON series and the Copernicus Sentinel missions), with supporting examples. We examine the benefits and drawbacks of integrating these EO datasets to better monitor and understand the processes at work and eventually to help in early warning and management of flood and drought events. Their main advantage is their large monitoring scale that provides a "big picture" or synoptic view of the event that cannot be achieved with often sparse in situ measurements. Finally, we present upcoming and future EO missions related to this topic including the SWOT mission.

Keywords Floods · Droughts · Large scale · Terrestrial water storage · GRACE · SMOS · Satellite altimetry · SWOT

1 Introduction

Hydro-meteorological extreme events count among the costliest natural disasters affecting human societies. They produce the largest cumulative total of lives lost and of socio-economic costs. In the context of global warming, these extreme events are intensifying and becoming more frequent in recent years (Stocker et al. 2013; Yamazaki et al. 2018 and references therein). This review paper focuses on floods and droughts and is part of a special issue on the benefits of integrating spaced-based or airborne observations in order to better predict, monitor and help in post-disaster management of natural hazards in which several themes as storm surges (Melet et al. 2020), tsunamis (Hébert et al. 2020), landslides (Lissak et al. 2020) and fire (Pettinari and Chuvieco 2020) are investigated. Space-based or airborne Earth observation (EO) have several benefits: (1) their instruments are not affected by the events, (2) they collect consistent data at different wavelengths and over different spatio-temporal scales and (3) they cover dangerous/inaccessible areas. Overall, EO data allow better understanding of the relationship between the hydrosphere, atmosphere, biosphere and solid Earth, and provide a global view of the phenomena and are complementary to in situ measurements (Tralli et al. 2005; Petiteville et al. 2015). Moreover, EO data can be used synergistically with demographic and socio-economic data to understand how hazards impact human societies, enhance our knowledge of the human influence on risks and thus elaborate mitigation, disaster and post-disaster management (Tralli et al. 2005).

In this article, the focus is put on recent advances on the monitoring of floods and droughts at the scale of large river basins through the observation of specific components of the hydrological cycle via space-based EO. At the end of last century, droughts counted for one-fifth of the damage caused by natural hazards and has steadily increased in recent

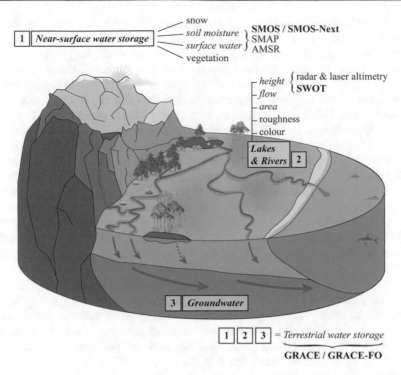

Fig. 1 Sketch of the different water storage compartments investigated in this study. Terms in italic are state variables considered, and bold terms are satellite missions presented in this article

decades. On the other hand, the cost of floods is expected to increase by tenfold in 2030 (World Resources Institute 2019). It is important to note that this article does not address the local scale monitoring of these events which is traditionally done using high-resolution optical or SAR (synthetic aperture radar) EO (e.g., Domeneghetti et al. 2019) and in situ data. We review the innovative applications on the use of moderate- to low-spatial-resolution missions to provide information facilitating the understanding and thus the forecasting of floods and droughts and discuss the future of these Earth observation data.

2 Water Storage on the Continents: General Remarks

Freshwater represents less than 3% of the total amount of water on Earth. On land, freshwater is stored in various reservoirs such as ice caps, snow, glaciers, groundwater, soil moisture (in the unsaturated soil and root zone, i.e. in the upper few metres of the soil (e.g., Hillel 1998)) and surface water bodies (rivers, lakes, man-made reservoirs, wetlands and inundated areas) (Fig. 1). These different storage compartments are in direct interactions with the atmosphere. For example, in the tropical Pacific, long-term droughts and floods are under the influence of the El Niño-Southern Oscillation (ENSO) events (e.g., Ward et al. 2014; Fok et al. 2018 and references therein).

Short-time hydrological events such as droughts and floods (i.e. events of water surplus or deficit that deviate from the average conditions over days to several months) are controlled by natural phenomena and anthropogenic activities. Key hydrological variables that drive and characterise these events are precipitation, soil moisture, groundwater storage and river discharge (e.g., Niu et al. 2014). Precipitation is considered as the main variable for the assessment of meteorological droughts and it is the main driver of flood events and thus the main input for hydrological flood models. The World Meteorological Organization (WMO) recommends the use of the Standardized Precipitation Index (SPI) for the assessment of wet and dry conditions (Svoboda et al. 2012). While its importance is not debated, complementary indices have emerged in the recent decades for characterising of droughts and floods at large scale. Here, we address the monitoring of flood and drought events in terms of soil moisture, inundated areas, groundwater, total water storage change and river discharge. Soil moisture is a key variable of the water cycle that links sub-surface and surface to atmospheric processes (e.g., Robinson et al. 2008; Niu et al. 2014; Grillakis et al. 2016; Babaeian et al. 2019). It is identified as an Essential Climate Variable (ECV) by the Global Climate Observing System (GCOS) (Bojinski et al. 2014) and is a major part of the so-called Green Water (Hoekstra et al. 2011). As stated by Babaeian et al. (2019) "soil moisture is an effective indicator for drought conditions and flood risks and thus plays a unique role in their prediction". In fact, soil moisture availability in the root zone is a direct indicator for agricultural droughts and can also inform on the wetness status in advance of strong rainfall events as an indicator of possible flood generation. Soil moisture accounts for about 0.79% of the available freshwater on Earth, but it is the main interface for the atmosphere/biosphere/sub-surface interactions.

Open water bodies include lakes, rivers, wetlands, peatlands, floodplains and manmade reservoirs. Wetlands cover about 4% of the Earth's land surface (Pekel et al. 2016) and are mostly present in tropical areas, for example, up to 20% of the Amazon Basin (Parrens et al. 2017). They play an important role in the hydrological regimes of large river basins as transition zones. Wetlands and floodplains are also the place of ecosystems with rich biodiversity (Costa et al. 2013), but they are very vulnerable to extreme changes.

Rivers account for 0.015% of the available freshwater globally, but like soil moisture they are the transfer medium for a large amount of the available freshwater. Since more than 50% of the world's population lives closer than 3 km to a surface freshwater body, and in majority near small to large rivers (74%) (Kummu et al. 2011), most of the world population is impacted by hydrological droughts and floods.

Groundwater is the main reservoir of available freshwater globally (96.7%) and accounts for 80% of the water extracted by humans. Groundwater is becoming an important supply of freshwater (~50% of the drinking water) (Foster and Loucks 2006) in different regions of the world as the surface water is becoming less reliable (e.g., Richey et al. 2015). Arid and semi-arid regions represent ~30% of the Earth's surface and groundwater constitutes a significant resource of freshwater in these areas; poorly renewable as the water input is often very localised, but exploitable (e.g., Margat and Van der Gun 2013). The importance of groundwater supply is known in drought conditions, but only few EO-based studies have addressed its contribution to flood events. In karstic regions, groundwater directly contributes to the generation of floods. However, they remain less common than river flooding and generates less damage (e.g., Yu et al. 2019). Runoff from groundwater storage tends to be delayed relative to the faster-reacting near-surface storage compartments (Yu et al. 2019 and references therein) and can thus have a major contribution to the later stages of a flood event (Huntingford et al. 2014).

Precise knowledge of the spatial and temporal evolution of the aforementioned storage variables and that of their sum as an integrative measure of the water storage in a river basin, i.e. total terrestrial water storage (TWS) change which is one of the four fundamental components of the terrestrial water balance besides precipitation, evapotranspiration and runoff, are important to better understand and forecast flood and drought events at the scale of river basins.

3 Space-Based Observations of the Water Cycle Components

3.1 Monitoring of Soil Moisture

The observation of surface soil moisture can be obtained using active or passive microwaves (Njoku et al. 2003; Kerr et al. 2012; Tomer et al. 2015). Optical remote sensing, on the other hand, will provide proxy information of the impact of soil moisture on soil emissivity and is more applied for soil texture characterisation (Gomez et al. 2019). The retrieval of soil moisture from active microwave is generally applied using SAR data from ASCAT (Wagner et al. 2013), RadarSat (Tomer et al. 2015) or Sentinel-1 (Paloscia et al. 2013). The retrieval of soil moisture from passive microwave has been applied to C-Band (Njoku et al. 2003) and L-Band (Entekhabi et al. 2010; Kerr et al. 2012). The more recent L-Band data from the SMOS (ESA) and SMAP (NASA) missions are considered as the most adapted for soil moisture retrieval considering radiative transfer physics (Ulaby 1981), but they can be hampered by their coarse resolution (~ 50 km and 36 km, respectively). Downscaling algorithms combining the microwave to optical or radar have been applied to enhance the resolution (Merlin et al. 2010; Tomer et al. 2016). The computation of root zone soil moisture from surface soil moisture can be provided through parsimonious models (Albergel et al. 2008; Al Bitar et al. 2013) or assimilation into land surface models (LSM) (Reichle 2018). The advantage of the parsimonious models is their ability to reduce the number of inputs mainly due to their independence of precipitation data. This is a major advantage in large irrigated areas. The advantage of data assimilation (DA) systems is their coherent retrieval or prediction of the ensemble of state variables (e.g., root zone soil moisture, surface and/or vegetation temperatures); still, one must keep in mind that errors on one variable will impact the other ones. Nevertheless, SMOS and SMAP missions provide a high temporal revisit (3 days for ascending and descending orbits). They gives access to consistent information over the globe as the acquisition configuration is similar across ecosystems. Yet several drawbacks can be mentioned as (1) the sensitivity of L-Band acquisition to radio-frequency interference from illegal emissions in the observation bandwidth from civil and military applications, (2) the retrieval using parsimonious models requires one to take into consideration the transpiration of the vegetation that is ill characterised and, in many cases, bypassed and (3) the resolution of the original data often needs to be enhanced using error-prone disaggregation approaches.

3.2 Monitoring Inundated Areas using Passive Microwaves

Flooded areas are commonly monitored using high-resolution thermal, visible or radar observations, as they can provide information at local scale. For example, Pekel et al. (2016) provided a 30-year database of surface water changes from LANDSAT (NASA) dataset. Mueller et al. (2016) presented a similar analysis over the Australian continent while providing the details for an operational continental water detection suite. These three techniques present major drawbacks in specific conditions. For optical (thermal and visible) remote sensing, the cloud cover can be a major issue during flood events and over tropical areas. Moreover, thermal remote sensing is highly sensible to the presence of water vapour in the atmosphere, while relative humidity over 80% is frequent in tropical regions and during wet weather. SAR is widely used to monitor the flood extents (Bonn and Dixon 2005; Matgen et al. 2011; Twele et al. 2016) and is an essential data source used during the activation of the International Disasters Charter. The main drawbacks of the SAR data are the impact of dense vegetation and soil roughness in dry conditions. In this paper, we are addressing the monitoring of inundated areas at the scale of river basins. In these conditions, the use of passive microwaves has been demonstrated in several studies (Prigent et al. 2007; Schroeder et al. 2015; Al Bitar et al. 2016).

Recently, the multi-angular full-polarisation brightness temperatures from SMOS (Al Bitar et al. 2017) have been used in a contextual approach to retrieve the flooded water fractions at 25 km resolution (Parrens et al. 2017). The application to the Amazonian basin shows that flood and drought events related to water surface changes can be monitored at 10 days temporal resolution. To reduce the inconvenience of coarse resolution, disaggregation algorithms to sub-kilometric resolutions have been applied to such datasets using auxiliary data related to flood probabilities, digital elevation models (DEMs) and basin hydromorphological information (Aires et al. 2017; Parrens et al. 2019). At these resolutions, the dataset is highly dependent on the input auxiliary data quality, mainly the DEM. The main advantage of microwave-based datasets is their ability to provide large-scale information. For example, the SMOS swath is about 1400 km. Another advantage related to the large scale is the ability through the frequent revisits (3 days global coverage for ascending and descending orbits) to provide coherent and consistent information at global scale. Finally, the use of low-frequency passive acquisition (e.g., L-Band at 1.4 GHz) is less prone to screening effect from the vegetation compared to passive C-Band (AMSR-E). Low-frequency passive sensors are also less sensitive to vegetation structure and surface roughness than in active microwaves (ASCAT). Still the data have several drawbacks, e.g., (1) the low resolution of the acquisition and (2) their strong vulnerability to radio-frequency interference (RFI) (Soldo et al. 2016) because of the swath width and the low measured surface emissions (10^{-4} W m^{-2}).

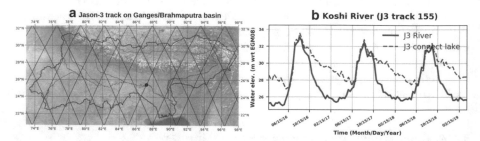

Fig. 2 **a** Jason-3 satellite tracks (red lines) over the Ganges/Brahmaputra basin (magenta boundaries, which come from GRDC Major River Basins database). The background image comes from the NASA MODIS "Blue Marble Next Generation" image (Stöckli et al. 2005). **b** Jason-3 water elevation time series on the Koshi river (red dot on panel a.) and on a nearby lake connected to the river during seasonal floods (red and blue lines, respectively)

3.3 River Elevation and Discharge

Nadir radar and SAR altimeters (e.g., Jason-1,2,3, Envisat, Saral/AltiKa and Sentinel-3A/B) provide measurement of water elevations (i.e. the distance between water body surface and a reference surface, ellipsoid or geoid). However, they can provide neither water depth nor any information on water body bathymetry below the lowest water level observed. Initially conceived to measure ocean surface topography, they have been used since the 1990s to also measure inland water elevations with an accuracy of few decimetres to a metre over continents (Santos da Silva et al. 2010), depending on the observation configuration, previous measurements and instrument characteristics. The main limitations of such instruments are their spatial and temporal sampling. They only provide measurements along the satellite track, missing all water bodies that are not beneath by the satellite. Current missions on a repetitive orbit (the satellite passes over the same point at a constant time step) have an inter-track distance (i.e. the distance between two adjacent ascending or descending tracks) at the equator between 315 km (the Jason series) to 52 km (when considering the two satellites of the Sentinel-3 series). The lower the inter-track distance, the higher the repeat period. For example, the Jason satellites have a 10-day revisit time, whereas it is 27 days for the Sentinel-3 satellites. In contrast, in situ gauges provide daily or better time sampling. These characteristics restrict nadir altimeter observation capacity to large-scale floods (Coe and Birkett 2004; Frappart et al. 2005; Biancamaria et al. 2011; Boergens et al. 2019). Nevertheless, satellite altimetry provides free measurements over multiple points of the river network (see Fig. 2a for an example of Jason-3 space sampling over the Ganges/Brahmaputra basin). They have thus been quite useful for large and poorly gauged basins, especially in a context of transboundary basins, to provide measurements both on the river and on the floodplain (Fig. 2b).

To overcome the space and time sampling issue, many studies have combined observations from the same mission at different locations along the river network or even multiple observations from different altimetry missions, in order to increase water level time series to observe as many events as possible (Hossain et al. 2014; Tourian et al. 2016, Boergens et al. 2019). For example, Tourian et al. (2017) have been able to compute daily water level and discharge on the Niger River mainstream using Envisat, SARAL/AltiKal and Jason-2 data and the few in situ data available for this basin.

Combining altimetry observations with measurements from other space sensors is also a promising perspective. Combination of altimetry data with optical sensors, especially the Moderate Resolution Imaging Spectroradiometer (MODIS), has been the most investigated (Tarpanelli et al. 2015, 2017, 2019; Ovando et al. 2018; Pham et al. 2018). In these studies, relationships between MODIS images for different wavebands and altimetry-based water elevations are computed to complement altimetry time series with the high temporal resolution of optical sensors enabled by their large swaths.

3.4 Monitoring of Terrestrial Water Storage Including Groundwater

In spite of its fundamental role for the global water cycle, for a long time there was no technique available that allowed for monitoring variations in total terrestrial water storage in an integrative way, including all relevant storage compartments on and below the Earth surface. Also, the lack of in situ data with the ability to measure long-term changes in the groundwater storage explains why the groundwater variability was poorly constrained (e.g.,

Famiglietti et al. 2011; Famiglietti 2014; Chen et al. 2016; Bonsor et al. 2018) before the launch of new types of sensors. Indeed, since the beginning of the century, a new generation of low-Earth-orbit (LEO) satellites has been operated to precisely determine the Earth's static gravity field, such as the GOCE (Gravity field and steady-state Ocean Circulation Explorer) mission (Drinkwater et al. 2006). The most important one for hydrological applications is the Gravity Recovery and Climate Experiment (GRACE) mission (Tapley et al. 2004) and its successor GRACE-FO (Landerer et al. 2020), which is dedicated to map the static gravity field and its time variations. Since its launch in March 2002, with the co-orbiting twin vehicles, it has measured the time variations of the Earth's gravity field for 15 years and with unprecedented precision. These two satellites are in the same orbit at 450–500 km altitude with a separation of about 200 km. The high-precision measurements of the inter-satellite distance change were enabled by the on-board K-Band Range Rate (KBRR at a precision of $0.1 \ \mathrm{m \ s^{-1}}$) system (Tapley et al. 2004). Together with Global Positioning System (GPS) observations and on-board accelerometer and star camera observations, the estimation of the gravity variations is indirectly obtained by inverting the ultra-precise KBRR observations between the two vehicles (Houborg et al. 2012). GRACE and GRACE-FO yield crucial information on the temporal and spatial variations of TWS for the period 2002 to today, with a gap between GRACE and GRACE-FO from June 2017 to June 2018. TWS is a unique integrative observation that notably includes groundwater storage variations down to any aquifer depth and the mass loss of ice caps and glaciers, typically with a temporal resolution of one month and spatial resolution of ~300 km. From the TWS variations, water storage variations in individual storage compartments such as groundwater can be singled out by the subtraction of water mass variations in other components if available from complementary observations and models or in a model-based data assimilation framework (see, for example, Frappart and Ramillien 2018). Both approaches are at the expense of additional errors introduced in the final product by the uncertainties of the separation approaches themselves.

4 Progress in Flood Applications

Flood is defined by the World Meteorological Organization International Glossary of Hydrology (WMO 2012) as: "(1) Rise, usually brief, in the water level of a stream or water body to a peak from which the water level recedes at a slower rate. (2) Relatively high flow as measured by stage height or discharge". In this review study, we focus on long-lasting floods with large coverage in space which can be captured by moderate- to low-spatial-resolution space-based observations. We do not consider flash floods (defined as "flood of short duration with a relatively high peak discharge" in WMO (2012)) which can last a few hours and be very intense locally and cannot be observed with such satellite instruments. The following subsections present how the space-based observations presented in Sect. 3 can observe and provide meaningful measurements for these large-scale floods.

4.1 Added Value of Satellite Altimetry

Daily discharge has been computed using data assimilation (DA) techniques by updating hydraulic/hydrological model state variables with altimetry-derived observations (Schumann et al. 2010; Hirpa et al. 2013; Michailovsky et al. 2013; Finsen et al. 2014; Emery et al. 2018). These improved time series help to get better estimates of the dynamics and the magnitude of flood events at the river basin scale. A particular advantage compared to classical approaches based on often scarce in situ data is that the altimetry-based reanalyses may allow for a spatially distributed discharge estimation for large basins. In a few studies, the calibration of hydraulic model parameters with altimetry time series has been investigated (e.g., Schneider et al. 2017; Jiang et al. 2019). As highlighted by Jiang et al. (2019), for parameter estimation, contrarily to water elevation/discharge time series densification by higher temporal resolution, a better spatial coverage is more important than revisit times with high frequency. The benefit of a better calibrated model is the possibility to simulate more accurately past events and to eventually forecast them with lower uncertainty in future. Bates et al. (2014) provided a more in-depth review of satellite data used to observe floods and how they have been combined with models.

However, very few studies implemented a proof-of-concept of flood forecasting system using altimetry observations. Based on the work from Biancamaria et al. (2011), Hossain et al. (2014) implemented a real flood forecasting system of daily water levels with a 5-day lead time over Bangladesh (Ganges/Brahmaputra/Meghna basin) using Jason-2 data and a hydraulic model. This system has then been used operationally by Bangladesh Institute of Water Modeling as a decision-making tool (Hossain et al. 2014). Tarpanelli et al. (2017) also developed two proof-of-concept systems to forecast discharge with a 4-day lead time over the Niger basin, using either MODIS images or altimetry data. They found that altimetry data provide more accurate forecasts, but lack sufficiently frequent revisit times. On the contrary, the MODIS-based forecast system is somewhat less accurate, but enables daily forecasts which are more suitable for operational agencies. However, to the authors' knowledge, this system has not yet been used operationally.

One must keep in mind that nadir radar altimeters due to the vertical accuracy and time and space samplings cannot bring as much information as in situ data. However, due to their large spatial coverage, especially for big transboundary basins and where there are few gauges or where the gauge data are not shared with downstream countries, altimetry measurements could help to provide more knowledge on large floods.

4.2 Monitoring and Predicting Flood Events With Satellite Gravimetry

The mass changes observed by GRACE and GRACE-FO encompass water storage changes associated with the evolution of hydrological events. Satellite gravimetry thus can be used to assess the magnitude of large-scale flood events in terms of their storage amplitude, duration and frequency in river basins worldwide. Furthermore, monitoring catchment wetness conditions opens up the opportunity to use GRACE/GRACE-FO in a predictive mode for early warning applications, i.e. for indicating river basin states that favour the generation of flood events. Added value for flood monitoring and forecasting relative to other observation data can be expected as gravimetry is the only comprehensive observation technique that delivers the wetness state in an integrated form for all TWS compartments.

4.2.1 Monitoring Flood Events With GRACE/GRACE-FO

Monthly GRACE products from various processing centres and with different solution strategies (see, for example, Jean et al. 2018 for an overview of available solutions)

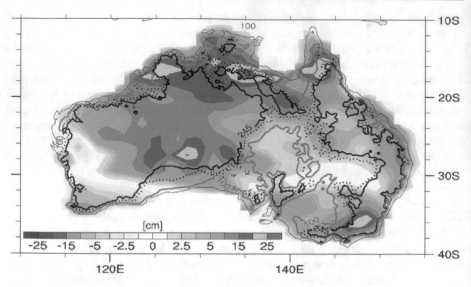

Fig. 3 Water storage anomaly (expressed in cm of equivalent water height) for Australia based on GRACE for the period June 2010 to February 2011 (modified from Fasullo et al. 2013)

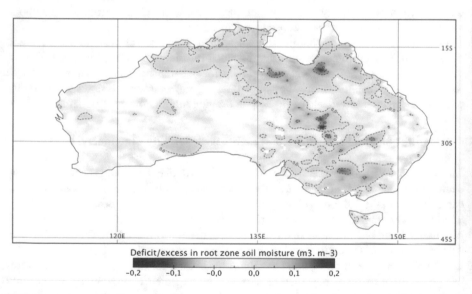

Fig. 4 Root zone soil moisture deficit and excess over Australia in $m^3 \, m^{-3}$ for February 2011 based on SMOS root zone soil moisture

have frequently been used to describe flood events. For example, Seitz et al. (2008) reported evidence of observed extreme weather fluctuations in Central Europe for the period 2003–2008 including the associated flood events by performing a regional analysis of GRACE data over Europe based on spherical wavelet/B-spline and global spherical harmonic solutions. Chen et al. (2010) studied the exceptional 2009 Amazon flood and inter-annual terrestrial water storage change observed by GRACE and concluded that the TWS increase in the lower Amazon basin in the first half of 2009 were clearly associated with the exceptional flood season in that region.

Abelen et al. (2015) related soil moisture from the WaterGAP Global Hydrology Model (WGHM) (Döll et al. 2003) and from the satellite sensors Advanced Microwave Scanning Radiometer—Earth Observing System (AMSR-E) and Advanced Scatterometer (ASCAT) to GRACE-based total water storage variations. For the La Plata basin in South America, they found that GRACE water storage dynamics represented the ENSO-dependent sequence of drought and flood events with a temporal delay of few months compared to near-surface soil moisture. Furthermore, Boening et al. (2012) and Fasullo et al. (2013) reported heavy precipitation over the Australian continent during the 2010–2011 La Niña. Based on GRACE observations, they highlighted regional spots of much higher wetness than normal, particularly in northern Australia. This is illustrated in Fig. 3 showing GRACE-based water storage anomalies in Australia over a period of 8 months from June 2010 to February 2011. This information can be associated with the anomaly in total water storage including surface water and groundwater with respect to long-term average. Figure 4 shows the excess and deficit information from the SMOS-based root zone soil moisture during February 2011 compared to the seasonal averages. The maps in Figs. 3 and 4 show similar patterns in several regions, but also discrepancies that can be explained by the associated representative time period for each observed component by the two satellites that are complementary.

Zhou et al. (2017) exploited TWS observations from the temporal gravity field model HUST-Grace2016, which was developed by a new low-frequency noise processing strategy, for identifying flood events in the Yangtze River basin and its sub-basins. "Universal" floods were found in 2010, 2012 and 2016, while a "regional" flood was observed in 2003. Similarly, Sun et al. (2017) showed for the Yangtze River basin that GRACE water storage anomalies and a GRACE-based flood potential index could effectively monitor large-scale flood events.

Following a probabilistic approach of the occurrence frequency and the expected return levels of hydrological extremes (i.e. the expected anomalous flux or storage values once in N years), Kusche et al. (2016) mapped these statistics on a global scale. They identified hot spots of anomalously high water storage that can be related to flood events with good statistical significance despite a quite short time period of 144 months of GRACE data. Moreover, they predicted that with the continuation of the GRACE mission by GRACE-FO, it would be possible by around the year 2020 to detect changes in the frequency of total fluxes for at least 10–20% of the continental area.

While all flood analyses mentioned before used standard monthly GRACE products, Gouweleeuw et al. (2018) were the first to assess daily GRACE data for flood monitoring. They analysed two daily gravity field solutions based on GRACE observations and evaluated them against daily river runoff data for major flood events in the Ganges–Brahmaputra Delta (GBD) in 2004 and 2007. They found that variations over periods of a few days of the daily GRACE data reflect temporal variations in daily river runoff during major flood events. This confirmed in particular the potential of daily GRACE gravity field solutions based on a Kalman filter approach for gravity-based large-scale flood monitoring. They concluded that the release of daily GRACE gravity field solutions in near real time may enable flood monitoring for large events. Gruber and Gouweleeuw (2019) further detailed how GRACE dynamic measurements and empirical external covariance information can be combined in the Kalman filter approach for the daily, short-latency mass changes products and reported on an evaluation of the product for describing flood dynamics in the Mekong delta.

Instead of using GRACE data after conversion to a dataset of surface mass variations, Han et al. (2009) studied the terrestrial water storage in the Amazon basin and its surrounding areas by exploring directly the instantaneous measurements of distance changes between two satellites from the GRACE mission. These measurements are directly influenced by large water masses in the river network and adjacent floodplains and are thus potentially sensitive to flood dynamics. Han et al. (2009) anticipated that the assimilation of GRACE inter-satellite range-rate data will improve the surface water models by tuning the effective flow velocities within the large basins. Following a similar concept, Ghobadi-Far et al. (2018) developed a transfer function to determine the in situ line-of-sight gravity difference (LGD) directly from GRACE range–acceleration measurements. They demonstrated the use of LGD data based on forward computation of the gravity effects from hydrological models and comparison with the GRACE LGD data and concluded that the approach could serve as a tool suitable to study mass changes at high temporal resolution, including near-real-time monitoring of floods.

4.2.2 GRACE-Based Flood Indicators and Flood Potential Assessment

In the studies mentioned above, GRACE provided information on water storage anomalies in flood-affected regions for the event period itself. This is a valuable contribution for assessing the impact of the hydrological event on the water cycle in terms of magnitude and duration. Several studies, however, also explored the potential of GRACE information in a predictive way for forecasting flood events. The fundamental concept behind this is that the GRACE mass anomalies describe the wetness state of a river basin, which in turn is a key factor for flood generation. Wet preconditions in the sense of high, i.e. near-surface groundwater levels, close to saturated soils or filled surface water bodies indicate a reduced ability of the river basin to store additional water that comes with the next rainfall event. Thus, the probability that much of this rainfall becomes runoff and eventually generates a flood is increased. On this basis, Reager and Famiglietti (2009) estimated the flood potential of a region by inferring its storage capacity from the repeated maxima of the monthly time series of GRACE water storage anomalies. The flood potential is then defined as the actual water storage anomaly relative to the storage capacity. Molodtsova et al. (2016) evaluated this flood potential index for a large number of floods in the USA and demonstrated its skill in particular for predicting large-area, long-duration floods in the summer season that are rainfall driven. Idowu and Zhou (2019) confirmed the value of the flood potential index for indicating flood events in the Niger basin. Jäggi et al. (2019) presented a wetness index based on daily GRACE water storage data. They showed for the example of the Danube basin that the index exceeded threshold values several weeks before the flood peak, making it a potentially useful candidate for early flood warning.

Reager et al. (2014) further developed the flood potential concept towards predicting flood discharge by an autoregressive model, driven with water storage anomalies. They showed that the GRACE-based information on elevated total water storage could improve the prediction with lead times of several months, for example, of the Mississippi flood in 2011. With a similar approach for a Ganges sub-basin, Chinnasamy (2017) confirmed that adding GRACE-based water storage anomalies improved flood discharge prediction with long lead times in comparison with a model that relies on precipitation only. Also, Chao and Wang (2017), for the Yangtze River basin, showed with a time-lagged autoregressive model that TWS changes from GRACE were effective for characterising flood potential and for flood early warning with lead times of several months.

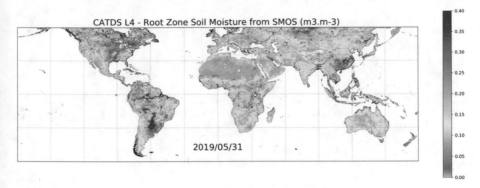

Fig. 5 Global root zone soil moisture in $m^3 \ m^{-3}$ obtained from SMOS surface soil moisture at 25 km resolution using a parsimonious model for 31 May 2019

Beyond statistical forecast models as used in the studies mentioned before, integration of GRACE-based water storage anomalies into physically based hydrological and flood forecasting models may allow for additional early warning applications. As a first example, Reager et al. (2015) assimilated GRACE water storage anomalies into a hydrological model. For the 2011 Missouri flood, flood generation understanding and modelling could be improved by resulting in wetter pre-event conditions and by providing more detailed information on the contributions of different storage compartments to the pre-event flood potential.

5 Progress in Drought Applications

Droughts are the costliest hydrological events in terms of economic, environmental and human life losses, and they impact a broad variety of regions in the world (e.g., Zhang and Jia 2013; Bayissa et al. 2017 and references therein). They are generally controlled by a deficit of groundwater, water available in the unsaturated zone or in surface water bodies. They can be classified (e.g., Wilhite 2000; Zhang and Jia 2013; Du et al. 2019 and references therein) as meteorological droughts that are mainly associated with a deficit of precipitation and/or an increase in the potential evapotranspiration. They can also be caused by sea surface temperature anomalies that lead to persistent continental drought, while a warm temperature with low precipitation over a short period of time generates flash drought. Agricultural droughts are driven by a deficit of soil moisture associated or not with a deficiency of surface/sub-surface water supply, i.e. hydrological droughts. The latter can also be described as socio-economic droughts where the available water is insufficient to cover the demand of some socio-economic goods and/or habits. This may induce a slower-progressing drought along with a deficiency of TWS and more particularly of groundwater availability. These different types of drought can occur alone, but most of the time, a meteorological drought triggers the other ones.

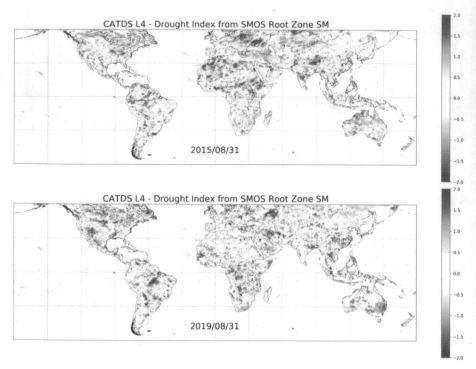

Fig. 6 Global drought index obtained from the SMOS root zone soil moisture for August 2015 (upper panel) and August 2019 (lower panel)

5.1 Agricultural Droughts of 2019 from L-Band Radiometers

Major information related to agricultural drought assessment is the water availability in the root zone. The definition of the water availability and root zone depth depends on the observed cropping type, phase of development and irrigation practices, but it is generally admitted that the root zone definition of crops at large scale is related to the first meter of soil and the water availability is defined as percentage of available water between the wilting point and the field capacity associated with the soil texture. The root zone soil moisture is not directly accessible currently by remote sensing technologies. Some studies addressed through local radiometers the capacity of future P-Band mission from active or passive remote sensing to measure root zone soil moisture (Garrison et al. 2018), but there is currently no operational satellite in this bandwidth. Still, root zone soil moisture can be indirectly quantified from remote sensing through vegetation stress condition (Swain et al. 2013), TWS or the link between precipitation, surface and root zone soil moisture. The vegetation stress condition can be monitored using thermal remote sensing (Anderson et al. 2007) which provides a posteriori information about drought as the vegetation starts to show water stress after available water depletes. As mentioned in previous sections, TWS information can be acquired using gravimetric data at a coarse spatial resolution and root zone soil moisture may be singled out by removing the contributions from other water storage compartments if available or by

🌀 Springer

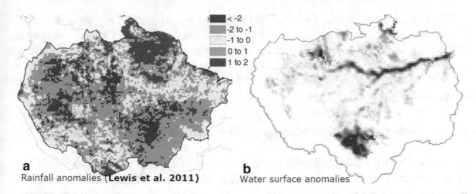

a Rainfall anomalies **(Lewis et al. 2011)**

b Water surface anomalies

Fig. 7 Standardised anomalies of **a** rainfall (adapted from Lewis et al. 2011) and **b** water surfaces based on the surface water areal fraction (SWAF) (Parrens et al. 2017) over the Amazon River basin during the 2010 drought event

modelling via TWS data assimilation. The link between surface soil moisture and root zone soil moisture is more direct. Figure 5 shows the root zone soil moisture obtained from the surface soil moisture SMOS data used in a parsimonious model. The model is based on two steps. First an exponential function is used to update the root zone soil moisture based on the observed one. Second a soil model is used to compute the infiltration and vegetation extraction through computation of transpiration. In this case, the MODIS NDVI data are used to compute the FAO56 crop index KC (Er-Raki et al. 2009).

Once a root zone soil moisture is obtained, agricultural drought indicators can be determined. Al Bitar et al. (2013) and Kerr et al. (2016) defined the drought based on the standard anomaly index where an empirical cumulative density function is fitted to the time series statistics at each node and the drought is defined on probability thresholds. Figure 6 shows the drought index defined as root zone soil moisture anomalies for August 2015 and August 2019. The figure clearly shows the increase in droughts in 2019 to an already dry condition in 2015 leading to the massive forest fires in Australia. The figure also shows the dramatic conditions in large parts of Amazonia and in South Africa in 2019.

The main advantage of the use of passive L-Band remote sensing for agricultural drought monitoring is that, when combined with a parsimonious depth-scaling model, it provides the closest information to the state variable of interest, i.e. the root zone soil moisture.

5.2 Droughts of Amazon Basin of 2010 observed by SMOS Water Fractions

Anomalies of the extent of water surfaces may represent droughts that have an impact on wetlands and floodplains and can thus be classified as indicators of hydrological drought. Figure 7 shows two complementary information layers on meteorological and hydrological drought conditions over the Amazonian basin for the 2010 drought event. Figure 7a shows the anomalies obtained from rainfall data (adapted from Lewis et al. 2011), while Fig. 7b shows the anomalies of water surface extent during the drought event. These water surface anomalies, expressed as anomalies of the surface water areal fraction (SWAF) in 25-km grid cells, have been derived from SMOS by Parrens et al. (2017). They show two facets of this drought event that impacted the rainforest and the floodplains. The two components of this drought event had dramatic consequences on the ecosystem. While the meteorological drought had a direct impact on biomass and thus the carbon budget, the hydrological drought had an impact on the Amazon River floodplains and Bolivian wetlands ecosystem in the south of the basin.

5.3 Monitoring and Predicting Drought Events With GRACE-Based Gravimetry

TWS estimations from GRACE give useful information about the water content variations through all the compartments at large spatial scale and over more than a decade. In order to better characterise droughts, information concerning water variations of all compartments is crucial and thus justifies the use of GRACE TWS estimations for droughts monitoring (e.g., Rodell 2012). Moreover, the drought event that impacted Texas in 2011 demonstrated that the depletion of water observed by GRACE is dominated by changes in soil moisture storage (Long et al. 2013). Since its launch, in different regions of the world, droughts have been observed directly with GRACE TWS data as, for example, in the Amazon River basin (Chen et al. 2009) and in Texas (Long et al. 2013). It has also demonstrated its ability to monitor droughts that affected several tropical basins during an El Niño event. For example, the 2015–2016 El Niño led to one of the most intense droughts ever recorded over southern Africa (e.g., Siderius et al. 2018). This event prevented groundwater recharge and hence produced groundwater decline over two consecutive years as reported from an analysis of GRACE data (Kolusu et al. 2019).

Some methods have been proposed to better characterise droughts, by detecting anomalies in the TWS series (Thomas et al. 2014) and with the normalised GRACE-derived groundwater storage deviation that permit quantification of the groundwater storage deficit (Thomas et al. 2017). Meanwhile, various drought indices obtained from GRACE have been used, e.g., the Drought Severity Index (DSI) and the Total Storage Deficit Index (Cao et al. 2015; Zhao et al. 2017; Nie et al. 2018). These partial examples demonstrate

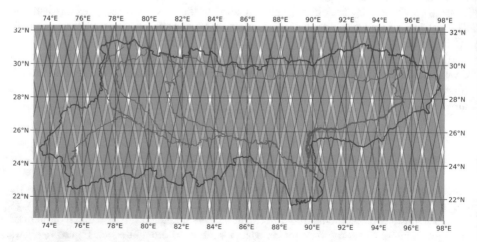

Fig. 8 SWOT coverage (green polygons) over the Ganges/Brahmaputra basins (magenta boundaries). Light green corresponds to regions with one SWOT observation during a satellite repeat period (21 days), green regions correspond to two observations during a repeat period, and white regions will never be observed by SWOT. Blue lines represent the Ganges river and its major tributaries, and the Brahmaputra river

the interest and the ability of GRACE to allow globally consistent and effective drought monitoring. Two disadvantages of GRACE data for the monitoring of droughts are: (1) the short period of time acquisition regarding the monitoring of drought events over several decades and (2) the spatial resolution. Concerning the short time period of acquisitions, Long et al. (2014) and Zhang et al. (2016) developed a method based on artificial neural network approach to reconstruct longer TWS series from which it is possible to observe past drought events. In order to compensate for the low spatial resolution of GRACE, some studies have demonstrated the benefits of GRACE data assimilation into land surface models, allowing spatial and temporal downscaling and providing information about groundwater and soil moisture by vertical decomposition (Houborg et al. 2012; Li et al. 2012). Despite their limitations, GRACE TWS estimations are the only information that permits the monitoring of droughts in regions where in situ data are sparse or non-existent. On the contrary, complementing in situ measurements and modelling, these estimations are extremely useful for groundwater resource management, especially during ENSO-driven events. Moreover, assimilation of GRACE and other data into regional/global hydrological models may be of great value to forecast flood and drought events affecting river basins.

6 Future of Earth Observations for Large-Scale Flood and Drought Surveys

6.1 Future Altimetry Missions

Jason-3 (launched in 2016), Sentinel-3A (launched in 2016) and Sentinel-3B (launched in 2018) are the nadir radar altimetry missions currently in operation at the time of writing. A new nadir altimetry mission (Sentinel-6A) will be launched in 2020 on the same nominal orbit as the Jason series. However, to alleviate nadir altimetry important observation gaps, the Surface Water and Ocean Topography (SWOT) mission will be launched in early 2022. SWOT will provide 2-D images of water surface topography, which should enhance hydrological hazard observation at least spatially. SWOT will observe all rivers wider than 100 m (with a goal to go down to 50 m) and all water bodies with an area above 250 m×250 m, i.e. 62,500 m^2 (with a goal to go down to 100 m×100 m) (Desai 2018). The main payload will measure water topography over two 50 km wide swaths. It will observe 96.5% of continents between 78°S and 78°N (Biancamaria et al. 2016), providing an almost global coverage of all water bodies on continental surfaces (see Fig. 8 for SWOT coverage over the Ganges/Brahmaputra basin). If the repeat period of the orbit is 21 days, thanks to swaths overlaps, it will observe many regions multiple times during one repeat period (Fig. 8). Compared to nadir or even SAR altimeters, SWOT will provide unprecedented spatial observations of water elevation, along with water area and river surface slope. From these measurements, global estimates of water storage change and river discharge will be derived (Durand et al. 2016; Desai 2018), which are essential for water management.

Despite its almost global coverage, SWOT will not be able to observe floods lasting only a few days or less, because of its still coarse time sampling. Frasson et al. (2019) investigated specifically the question of flood observability with SWOT. Comparing the 4,664 past flood events reported in the Dartmouth Flood Observatory database with SWOT orbit, they concluded that "SWOT would have seen 55% of these, with higher probabilities associated with more extreme events and with those that displaced more than 10,000 people". More specifically over the Cumberland River basin within continental USA, Yoon et al. (2016) estimated that SWOT has "only a 5% chance of direct observations of the 2-day flash flood event". Nonetheless, even if such new observation could help to forecast floods, SWOT uneven time sampling might be an issue. Therefore, SWOT will definitely improve large-scale flood observation, but not smaller-scale flood events. This issue could be somewhat alleviated using some DA techniques, currently under development, to derive enhanced space and time sampling discharge time series by taking advantages of the unprecedented SWOT spatial coverage over large to medium river basins (Brisset et al. 2018; Oubanas et al. 2018; Yang et al. 2019).

It is also expected that SWOT will be better suited than nadir altimeter to calibrate/ correct model parameters thanks to its spatial coverage (Yoon et al. 2012; Pedinotti et al. 2014; Boergens et al. 2019), which then could provide better forecasts.

Benefits of SWOT DA into hydraulic or hydrological models have also been investigated to optimise reservoir control for flood mitigation in the Cumberland River basin (Yoon et al. 2016), to guarantee minimum flow within the upper Niger River basin (Munier et al. 2015) and to forecast some flood events within the Ohio River basin (Andreadis and Schumann 2014). In particular, this latter study estimated that assimilating SWOT data into a hydraulic model should help to forecast water height up to lead times of 11 days. However, it is still unclear how SWOT data will be effectively used into operational forecast systems, as SWOT wide swath products might not be delivered in real time (the time latency should be less than 45 days, according to SWOT Product Description Documents for hydrology surfaces) and its required nominal lifetime is only 3 years, with a goal to go up to 5 years (Desai 2018).

Space and time samplings of nadir altimetry missions limit the use of these sensors to observe and even more to forecast floods that last few days or less. In comparison, SWOT wide swath altimetry will provide global coverage, but its time sampling also restricts flood observation to major floods (Frasson et al. 2019). To improve samplings, a constellation of satellites seems to be a good option, as no single orbit will allow both frequent time and space samplings. The Centre National d'Etudes Spatiales (CNES) in France is currently proposing a constellation of two satellites with a wide swath altimeter, as potential candidate to the Copernicus—Next Generation Sentinel-3 Topography program (after 2030). It should observe similar water bodies as SWOT, but with a better time sampling (~ 10 days) and a longer lifetime more suitable for operational use (Cheymol et al. 2019). To increase temporal sampling, CNES is also studying the feasibility of a constellation of nadir altimeters optimised for inland water bodies, called SMall Altimetry Satellites for Hydrology (SMASH) (Blumstein et al. 2019). It should provide daily measurements of water elevation with 10 cm accuracy, a data latency of less than 6 h and a long mission lifetime (> 10 years), particularly well suited for operational applications.

6.2 Future of L-Band Microwave EO

L-Band EO data for flood and drought applications currently are provided by the SMOS and SMAP missions which are active since 2010 and 2015, respectively, with both missions beyond their nominal lifetime of 3 years. The continuity of L-Band observation is under study in the context of the ESA future Copernicus Imaging Microwave Radiometer (CIMR). If L-Band observation is confirmed aboard, this will ensure the continuity of the L-Band acquisitions, but will not meet the needs expressed by the scientific community (Escorihuela and Kerr 2018) with respect to the spatial resolution. Actually, the L-Band radiometer on board CIMR is foreseen to have a coarser resolution than SMOS due to the mesh antenna size. Other initiatives for high resolution like SMOS-HR are under investigation at CNES (Rodríguez-Fernández et al. 2019). It is worth mentioning that the future NISAR (NASA/ISRO) mission which will include an S-Band and L-Band SAR will provide interesting opportunities for soil moisture and groundwater subsidence monitoring.

6.3 Future of Gravimetry Missions

Space-based TWS estimates by gravity data currently rely only on GRACE-FO mission launched in 2018 (Flechtner et al. 2014), thus extending the 15-year-long record of GRACE, in spite of a close to 1-year gap (GRACE ended in October 2017). But only by ensuring continuous and further improved TWS measurements from space, numerous questions regarding the changes and dynamic processes in land hydrology, cryosphere, ocean, atmosphere and solid Earth can be addressed (Pail et al. 2015). The NASA Earth Science Decadal Survey Report (National Academies of Sciences, Engineering, and Medecine 2018) thus highlighted monitoring spaced-based TWS as one of the five top priorities in Earth observation for the next decade (National Academies of Sciences, Engineering, and Medecine 2018). Based on this prioritisation, NASA is currently running various studies how a Next Generation Gravity Mission (NGGM) should be realised, e.g., in terms of the number of satellites (or satellite pairs), orbit configuration and instrumentation. Also, ESA launched a feasibility study on NGGM with the use of Laser interferometer for ranging measurements (Dionisio et al. 2018). A NGGM based on the innovative observational concept of a high–low tracking formation with micrometre ranging accuracy (Pail et al. 2019) is currently being studied by CNES. Eventually also, the reversing of GNSS-based kinematic trajectories of low-Earth-orbit constellations might be able to monitor groundwater evolution (e.g., Bezděk et al. 2016 and references therein).

7 Conclusions

This article presents a review on the major scientific and technical developments made over more than 20 years with low- and moderate-resolution satellite-based datasets in order to promote their use for flood and drought prediction and monitoring. Satellite gravimetry, L-Band passive microwave Earth observation and nadir altimetry provide unique opportunities to monitor total terrestrial water storage and its individual compartments, soil moisture, surface water level and extent, and river discharge, i.e. key hydrological variables to understand, predict and monitor flood and drought events. In fact, GRACE/GRACE-FO, SMOS and nadir altimeters missions have considerably improved our knowledge on the evolution at large spatial and long temporal scale of these variables. New methods developed along with the use of these sensors offer a better estimation of the dynamics and magnitudes of flood events, as well as a better spatially distributed discharge estimation in river basins with sparse in situ observations. Furthermore, a better estimation of the long-term evolution of groundwater storage and that of the total wetness state of river basins as represented by gravity-based TWS have been an asset to study the storage amplitude, duration and frequency of large-scale floods and to highlight important groundwater withdrawals or severe droughts. The most important benefit of using moderate- to low-spatial-resolution sensors is their ability to provide a synoptic view of these events from the river basin to the global scale. Very often, this synoptic view cannot be achieved with often sparse in situ measurements or with remote sensing techniques of high resolution, but low spatial coverage. Finally, the most promising perspective to increase the understanding, forecasting and mitigation of hydrological events will be to assimilate the moderate- to low-resolution satellite data into LSMs, to combine them with higher-spatial-resolution data, either space- or UAV-based (Antoine et al. 2020), and to use new methodologies, involving Artificial Intelligence, for instance.

Acknowledgements This paper arose from the international workshop on "Natural and man-made hazards monitoring by the Earth Observation missions: current status and scientific gaps" held at the International Space Science Institute (ISSI), Bern, Switzerland, on April 15–18, 2019. The authors wish to thank the two anonymous reviewers for their constructive suggestions which significantly improved this article; Dr. Anny Cazenave, for her invitation to prepare this paper and Anne-Marie Cousin for her help in producing Fig. 1.

References

Abelen S, Seitz F, Abarca-del-Rio R, Güntner A (2015) Droughts and floods in the La Plata Basin in soil moisture data and GRACE. Remote Sensing 7:7324–7349. https://doi.org/10.3390/rs70607324

Aires F, Miolane L, Prigent C et al (2017) A global dynamic long-term inundation extent dataset at high spatial resolution derived through downscaling of satellite observations. J Hydrometeor 18:1305–1325. https://doi.org/10.1175/JHM-D-16-0155.1

Al Bitar A, Kerr Y, Merlin O et al (2013) Root zone soil moisture and drought index from SMOS. In: Satellite soil moisture validation and application workshop. 2013-07-012013-07-03, Frascati, ITA.

Al Bitar A, Mialon A, Kerr Y et al (2017) The Global SMOS Level 3 daily soil moisture and brightness temperature maps. Earth Syst Sci Data 9:293–315. https://doi.org/10.5194/essd-9-293-2017

Al Bitar A, Parrens M, Frappart F et al (2016) How are the wetlands over tropical basins impacted by the extreme hydrological events? AGU Fall Meeting Abstracts 51.

Albergel C, Rüdiger C, Pellarin T et al (2008) From near-surface to root-zone soil moisture using an exponential filter: an assessment of the method based on in-situ observations and model simulations. Hydrol Earth Syst Sci 12:1323–1337. https://doi.org/10.5194/hess-12-1323-2008

Anderson MC, Norman JM, Mecikalski JR et al (2007) A climatological study of evapotranspiration and moisture stress across the continental United States based on thermal remote sensing: 2. Surface moisture climatology. J Geophys Res Atmospheres 112. https://doi.org/10.1029/2006JD007507

Andreadis KM, Schumann GJ-P (2014) Estimating the impact of satellite observations on the predictability of large-scale hydraulic models. Adv Water Resources 73:44–54. https://doi.org/10.1016/j.advwatres.2014.06.006

Antoine R, Fauchard C, Gaillet L et al (2020) Geoscientists in the sky: unmanned aerial Vehicles for geohazards response. Surv Geophys. https://doi.org/10.1007/s10712-020-09611-7

Babaeian E, Sadeghi M, Jones SB et al (2019) Ground, proximal, and satellite remote sensing of soil moisture. Rev Geophys 57:530–616. https://doi.org/10.1029/2018RG000618

Bates PD, Neal JC, Alsdorf D, Schumann GJ-P (2014) Observing global surface water flood dynamics. Surv Geophys 35:839–852. https://doi.org/10.1007/s10712-013-9269-4

Bayissa Y, Tadesse T, Demisse G, Shiferaw A (2017) Evaluation of satellite-based rainfall estimates and application to monitor meteorological drought for the Upper Blue Nile Basin. Ethiopia Remote Sensing 9:669. https://doi.org/10.3390/rs9070669

Bezděk A, Sebera J, Teixeira da Encarnação J, Klokočník J (2016) Time-variable gravity fields derived from GPS tracking of Swarm. Geophys J Int 205:1665–1669. https://doi.org/10.1093/gji/ggw094

Biancamaria S, Hossain F, Lettenmaier DP (2011) Forecasting transboundary river water elevations from space. Geophys Res Lett. https://doi.org/10.1029/2011GL047290

Biancamaria S, Lettenmaier DP, Pavelsky TM (2016) The SWOT mission and its capabilities for land hydrology. Surv Geophys 37:307–337. https://doi.org/10.1007/s10712-015-9346-y

Blumstein D, Guérin A, Lamy A, et al (2019) SMASH: a constellation of small altimetry satellites dedicated to hydrology. In: 6th Workshop on advanced RF sensors and remote sensing instruments (ARSI'19) and 4th Ka-band Earth Observation Radar Missions Workshop (KEO'19). ESA-ESTEC, Noordwijk, The Netherlands

Boening C, Willis JK, Landerer FW et al (2012) The 2011 La Niña: So strong, the oceans fell. Geophysical Res Lett. https://doi.org/10.1029/2012GL053055

Boergens E, Dettmering D, Seitz F (2019) Observing water level extremes in the Mekong River Basin: the benefit of long-repeat orbit missions in a multi-mission satellite altimetry approach. J Hydrol 570:463–472. https://doi.org/10.1016/j.jhydrol.2018.12.041

Bojinski S, Verstraete M, Peterson TC et al (2014) The concept of essential climate variables in support of climate research, applications, and policy. Bull Am Meteor Soc 95:1431–1443. https://doi.org/10.1175/BAMS-D-13-00047.1

Bonn F, Dixon R (2005) Monitoring flood extent and forecasting excess runoff risk with RADARSAT-1 data. Nat Hazards 35:377–393. https://doi.org/10.1007/s11069-004-1798-1

Bonsor H, Shamsudduha M, Marchant B et al (2018) Seasonal and decadal groundwater changes in African sedimentary aquifers estimated using GRACE products and LSMs. Remote Sensing 10:904. https://doi.org/10.3390/rs10060904

Brisset P, Monnier J, Garambois P-A, Roux H (2018) On the assimilation of altimetric data in 1D Saint-Venant river flow models. Adv Water Resour 119:41–59. https://doi.org/10.1016/j.advwatres.2018.06.004

Cao Y, Nan Z, Cheng G (2015) GRACE gravity satellite observations of terrestrial water storage changes for drought characterization in the arid land of Northwestern China. Remote Sensing 7:1021–1047. https://doi.org/10.3390/rs70101021

Chao N, Wang Z (2017) Characterized flood potential in the Yangtze River Basin from GRACE gravity observation, hydrological model, and in-situ hydrological station. J Hydrol Eng 22:05017016. https://doi.org/10.1061/(ASCE)HE.1943-5584.0001547

Chen JL, Wilson CR, Tapley BD et al (2016) Long-term groundwater storage change in Victoria, Australia from satellite gravity and in situ observations. Global Planet Change 139:56–65. https://doi.org/10.1016/j.gloplacha.2016.01.002

Chen JL, Wilson CR, Tapley BD (2010) The 2009 exceptional Amazon flood and interannual terrestrial water storage change observed by GRACE. Water Resour Res. https://doi.org/10.1029/2010WR009383

Chen JL, Wilson CR, Tapley BD et al (2009) 2005 drought event in the Amazon River basin as measured by GRACE and estimated by climate models. J Geophys Res Solid Earth 114. https://doi.org/10.1029/2008JB006056

Cheymol C, Coutin-Faye S, Le Traon P-Y, et al (2019) WiSA: a wide Swath Altimetry Mission for Operational Oceanography and Hydrology A good candidate for Copernicus-NG Sentinel 3-Topo program. Ocean Surface Topography Science Team Meeting (OSTST)

Chinnasamy P (2017) Inference of basin flood potential using nonlinear hysteresis effect of basin water storage: case study of the Koshi basin. Hydrol Res 48:1554–1565. https://doi.org/10.2166/nh.2016.268

Coe MT, Birkett CM (2004) Calculation of river discharge and prediction of lake height from satellite radar altimetry: Example for the Lake Chad basin. Water Resour Res. https://doi.org/10.1029/2003WR002543

Costa MPF, Silva TS, Evans TL (2013) Wetland classification. Remote sensing of natural resources. CRC Press, Boca Raton.

Desai S (2018) Surface Water and Ocean Topography Mission (SWOT) Project, Science Requirements Document. NASA/JPL technical document D–61923 Revision B:29

Dionisio S, Anselmi A, Bonino L et al (2018) The "Next Generation Gravity Mission": challenges and consolidation of the system concepts and technological innovations. In: 15th International conference on space operations. https://doi.org/10.2514/6.2018-2495

Döll P, Kaspar F, Lehner B (2003) A global hydrological model for deriving water availability indicators: model tuning and validation. J Hydrol 270:105–134. https://doi.org/10.1016/S0022-1694(02)00283-4

Domeneghetti A, Schumann GJ-P, Tarpanelli A (2019) Preface: remote sensing for flood mapping and monitoring of flood dynamics. Remote Sensing 11:943. https://doi.org/10.3390/rs11080943

Drinkwater M, Haagmans R, Muzi D et al (2006) The GOCE gravity mission: ESA'sfirst core explorer. In: Proceedings 3rd GOCE User Workshop. Frascati, Italy, pp 1–7

Du J, Kimball JS, Velicogna I et al (2019) Multicomponent satellite assessment of drought severity in the contiguous United States From 2002 to 2017 using AMSR-E and AMSR2. Water Resour Res 55:5394–5412. https://doi.org/10.1029/2018WR024633

Durand M, Gleason CJ, Garambois PA et al (2016) An intercomparison of remote sensing river discharge estimation algorithms from measurements of river height, width, and slope. Water Resour Res 52:4527–4549. https://doi.org/10.1002/2015WR018434

Emery CM, Paris A, Biancamaria S et al (2018) Large-scale hydrological model river storage and discharge correction using a satellite altimetry-based discharge product. Hydrol Earth Syst Sci 22:2135–2162. https://doi.org/10.5194/hess-22-2135-2018

Entekhabi D, Njoku EG, O'Neill PE et al (2010) The soil moisture active passive (SMAP) mission. Proc IEEE 98:704–716. https://doi.org/10.1109/JPROC.2010.2043918

Er-Raki S, Chehbouni A, Guemouria N et al (2009) Citrus orchard evapotranspiration: comparison between eddy covariance measurements and the FAO-56 approach estimates. Plant Biosyst Int J Deal Aspects Plant Biol 143:201–208. https://doi.org/10.1080/11263500802709897

Escorihuela MJ, Kerr Y (2018) Low frequency passive microwave user requirement consolidation study: D-02 white paper on L-band radiometry for earth observation: status and achievements

Famiglietti JS (2014) The global groundwater crisis. Nat Climate Change 4:945–948. https://doi.org/10.1038/nclimate2425

Famiglietti JS, Lo M, Ho SL et al (2011) Satellites measure recent rates of groundwater depletion in California's Central Valley. Geophys Res Lett 38:L03403. https://doi.org/10.1029/2010GL046442

Fasullo JT, Boening C, Landerer FW, Nerem RS (2013) Australia's unique influence on global sea level in 2010–2011. Geophys Res Lett 40:4368–4373. https://doi.org/10.1002/grl.50834

Finsen F, Milzow C, Smith R et al (2014) Using radar altimetry to update a large-scale hydrological model of the Brahmaputra river basin. Hydrol Res 45:148–164. https://doi.org/10.2166/nh.2013.191

Flechtner F, Morton P, Watkins M, Webb F (2014) Status of the GRACE Follow-On mission. Gravity, geoid and height systems,pp 117–121. https://doi.org/10.1007/978-3-319-10837-7_15

Fok H, He Q, Chun K et al (2018) Application of ENSO and drought indices for water level reconstruction and prediction: a case study in the Lower Mekong River Estuary. Water 10:58. https://doi.org/10.3390/w10010058

Foster S, Loucks DP (2006) Non-renewable groundwater resources: a guidebook on socially-sustainable management for water-policy makers, United Nations Educational. Scientific and Cultural Organization, UNESCO, Paris, France

de Frasson RP, Schumann GJ -P, Kettner AJ et al (2019) Will the surface water and ocean topography (SWOT) satellite mission observe floods? Geophys Res Lett 46:10435–10445. https://doi.org/10.1029/2019GL084686

Frappart F, Ramillien G (2018) Monitoring groundwater storage changes using the gravity recovery and climate experiment (GRACE) satellite mission: a review. Remote Sensing 10:829. https://doi.org/10.3390/rs10060829

Frappart F, Seyler F, Martinez J-M et al (2005) Floodplain water storage in the Negro River basin estimated from microwave remote sensing of inundation area and water levels. Remote Sens Environ 99:387–399. https://doi.org/10.1016/j.rse.2005.08.016

Garrison JL, Kurum M, Nold B et al (2018) Remote sensing of root-zone soil moisture using I- and P-band signals of opportunity: Instrument Validation Studies. IGARSS 2018 - 2018 IEEE International Geoscience and Remote Sensing Symposium, pp 8305–8308. https://doi.org/10.1109/IGARSS.2018.8518772

Ghobadi-Far K, Han S-C, Weller S et al (2018) A transfer function between line-of-sight gravity difference and GRACE intersatellite ranging data and an application to hydrological surface mass variation. J Geophys Res Solid Earth 123:9186–9201. https://doi.org/10.1029/2018JB016088

Gomez C, Dharumarajan S, Féret J-B et al (2019) Use of Sentinel-2 time-series images for classification and uncertainty analysis of inherent biophysical property: case of soil texture mapping. Remote Sensing 11:565. https://doi.org/10.3390/rs11050565

Gouweleeuw BT, Kvas A, Gruber C et al (2018) Daily GRACE gravity field solutions track major flood events in the Ganges-Brahmaputra Delta. Hydrol Earth Syst Sci 22:2867–2880. https://doi.org/10.5194/hess-22-2867-2018

Grillakis MG, Koutroulis AG, Komma J et al (2016) Initial soil moisture effects on flash flood generation— a comparison between basins of contrasting hydro-climatic conditions. J Hydrol 541:206–217. https://doi.org/10.1016/j.jhydrol.2016.03.007

Gruber C, Gouweleeuw B (2019) Short-latency monitoring of continental, ocean- and atmospheric mass variations using GRACE intersatellite accelerations. Geophys J Int 217:714–728. https://doi.org/10.1093/gji/ggz042

Han S-C, Kim H, Yeo I-Y et al (2009) Dynamics of surface water storage in the Amazon inferred from measurements of inter-satellite distance change. Geophys Res Lett 36. https://doi.org/10.1029/2009GL037910

Hébert H, Gailler A, Gupta H, et al (2020) Contribution of space missions to a better tsunami science: observations, models and warning. Surv Geophys 41 (in press)

Hillel D (1998) Environmental soil physics: fundamentals, applications, and environmental considerations. Elsevier, Amsterdam

Hirpa FA, Hopson TM, De Groeve T et al (2013) Upstream satellite remote sensing for river discharge forecasting: application to major rivers in South Asia. Remote Sens Environ 131:140–151. https://doi.org/10.1016/j.rse.2012.11.013

Hoekstra AY, Chapagain AK, Mekonnen MM, Aldaya MM (2011) The water footprint assessment manual: Setting the global standard. Routledge

Hossain F, Siddique-E-Akbor AH, Mazumder LC et al (2014) Proof of concept of an Altimeter-based river forecasting system for transboundary flow inside Bangladesh. IEEE J Selected Topics Appl Earth Observ Remote Sensing 7:587–601. https://doi.org/10.1109/JSTARS.2013.2283402

Houborg R, Rodell M, Li B et al (2012) Drought indicators based on model-assimilated Gravity Recovery and Climate Experiment (GRACE) terrestrial water storage observations. Water Resour Res 48. https://doi.org/10.1029/2011WR011291

Huntingford C, Marsh T, Scaife AA et al (2014) Potential influences on the United Kingdom's floods of winter 2013/14. Nature Clim Change 4:769–777. https://doi.org/10.1038/nclimate2314

Idowu Z (2019) Performance evaluation of a potential component of an early flood warning system—a case study of the 2012 flood, Lower Niger River Basin. Nigeria Remote Sensing 11:1970. https://doi.org/10.3390/rs11171970

Jäggi A, Weigelt M, Flechtner F et al (2019) European Gravity Service for Improved Emergency Management (EGSIEM)—from concept to implementation. Geophys J Int 218:1572–1590. https://doi.org/10.1093/gji/ggz238

Jean Y, Meyer U, Jäggi A (2018) Combination of GRACE monthly gravity field solutions from different processing strategies. J Geodesy 92:1313–1328. https://doi.org/10.1007/s00190-018-1123-5

Jiang L, Madsen H, Bauer-Gottwein P (2019) Simultaneous calibration of multiple hydrodynamic model parameters using satellite altimetry observations of water surface elevation in the Songhua River. Remote Sens Environ 225:229–247. https://doi.org/10.1016/j.rse.2019.03.014

Kerr YH, Waldteufel P, Richaume P et al (2012) The SMOS soil moisture retrieval algorithm. IEEE Trans Geosci Remote Sens 50:1384–1403. https://doi.org/10.1109/TGRS.2012.2184548

Kerr YH, Wigneron J-P, Al Bitar A et al (2016) Soil moisture from space: techniques and limitations. Satellite Soil Moisture Retrieval, pp 3–27. https://doi.org/10.1016/B978-0-12-803388-3.00001-2

Kolusu SR, Shamsudduha M, Todd MC et al (2019) The El Niño event of 2015–2016: climate anomalies and their impact on groundwater resources in East and Southern Africa. Hydrol Earth Syst Sci 23:1751–1762. https://doi.org/10.5194/hess-23-1751-2019

Kummu M, De Moel H, Ward PJ, Varis O (2011) How close do we live to water? A global analysis of population distance to freshwater bodies. PLoS ONE 6:e20578. https://doi.org/10.1371/journal.pone.0020578

Kusche J, Eicker A, Forootan E et al (2016) Mapping probabilities of extreme continental water storage changes from space gravimetry. Geophys Res Lett 43:8026–8034. https://doi.org/10.1002/2016GL069538

Landerer FW, Flechtner FM, Save H et al (2020) Extending the Global Mass Change Data Record: GRACE Follow-On Instrument and Science Data Performance. Geophys Res Lett 47:e2020GL088306. https://doi.org/10.1029/2020GL088306

Lewis SL, Brando PM, Phillips OL et al (2011) The 2010 Amazon drought. Science 331:554–554. https://doi.org/10.1126/science.1200807

Li B, Rodell M, Zaitchik BF et al (2012) Assimilation of GRACE terrestrial water storage into a land surface model: evaluation and potential value for drought monitoring in western and central Europe. J Hydrol 446–447:103–115. https://doi.org/10.1016/j.jhydrol.2012.04.035

Lissak C, De Michele M, Bartsch A et al (2020) Remote sensing for mass movement assessment. Surv Geophys. https://doi.org/10.1007/s10712-020-09609-1

Long D, Scanlon BR, Longuevergne L et al (2013) GRACE satellite monitoring of large depletion in water storage in response to the 2011 drought in Texas. Geophys Res Lett 40:3395–3401. https://doi.org/10.1002/grl.50655

Long D, Shen Y, Sun A et al (2014) Drought and flood monitoring for a large karst plateau in Southwest China using extended GRACE data. Remote Sens Environ 155:145–160. https://doi.org/10.1016/j.rse.2014.08.006

Margat J, Van der Gun J (2013) Groundwater around the world: a geographic synopsis. CRC Press, Taylor & Francis Group

Matgen P, Hostache R, Schumann G et al (2011) Towards an automated SAR-based flood monitoring system: lessons learned from two case studies. Phys Chem Earth A/B/C 36:241–252. https://doi.org/10.1016/j.pce.2010.12.009

Melet A, Teatini P, Le Cozannet G et al (2020) Earth observations for monitoring Marine Coastal Hazards and their drivers. Surv Geophys. https://doi.org/10.1007/s10712-020-09594-5

Merlin O, Al Bitar A, Walker JP, Kerr Y (2010) An improved algorithm for disaggregating microwave-derived soil moisture based on red, near-infrared and thermal-infrared data. Remote Sens Environ 114:2305–2316. https://doi.org/10.1016/j.rse.2010.05.007

Michailovsky CI, Milzow C, Bauer-Gottwein P (2013) Assimilation of radar altimetry to a routing model of the Brahmaputra River: Radar Altimetry to a Routing Model. Water Resour Res 49:4807–4816. https://doi.org/10.1002/wrcr.20345

Molodtsova T, Molodtsov S, Kirilenko A et al (2016) Evaluating flood potential with GRACE in the United States. Natural Hazards Earth Syst Sci 16:1011–1018. https://doi.org/10.5194/nhess-16-1011-2016

Mueller N, Lewis A, Roberts D et al (2016) Water observations from space: mapping surface water from 25 years of Landsat imagery across Australia. Remote Sens Environ 174:341–352. https://doi.org/10.1016/j.rse.2015.11.003

Munier S, Polebistki A, Brown C et al (2015) SWOT data assimilation for operational reservoir management on the upper Niger River Basin. Water Resour Res 51:554–575. https://doi.org/10.1002/2014WR016157

National Academies of Sciences, Engineering, and Medecine (2018) Thriving on our changing planet: a decadal strategy for earth observation from space. https://doi.org/10.17226/24938

Nie N, Zhang W, Chen H, Guo H (2018) A global hydrological drought index dataset based on Gravity Recovery and Climate Experiment (GRACE) Data. Water Resour Manage 32:1275–1290. https://doi.org/10.1007/s11269-017-1869-1

Niu J, Chen J, Sivakumar B (2014) Teleconnection analysis of runoff and soil moisture over the Pearl River basin in southern China. Hydrol Earth Syst Sci 18:1475–1492. https://doi.org/10.5194/hess-18-1475-2014

Njoku EG, Jackson TJ, Lakshmi V et al (2003) Soil moisture retrieval from AMSR-E. IEEE Trans Geosci Remote Sens 41:215–229. https://doi.org/10.1109/TGRS.2002.808243

Oubanas H, Gejadze I, Malaterre P-O et al (2018) Discharge estimation in ungauged basins through variational data assimilation: the potential of the SWOT mission. Water Resour Res 54:2405–2423. https://doi.org/10.1002/2017WR021735

Ovando A, Martinez JM, Tomasella J et al (2018) Multi-temporal flood mapping and satellite altimetry used to evaluate the flood dynamics of the Bolivian Amazon wetlands. Int J Appl Earth Obs Geoinf 69:27–40. https://doi.org/10.1016/j.jag.2018.02.013

Pail R, Bamber J, Biancale R et al (2019) Mass variation observing system by high low inter-satellite links (MOBILE)—a new concept for sustained observation of mass transport from space. J Geodetic Sci 9:48–58. https://doi.org/10.1515/jogs-2019-0006

Pail R, Braitenberg C, Dobslaw H et al (2015) Science and user needs for observing global mass transport to understand global change and to benefit society. Surv Geophys 36:743–772. https://doi.org/10.1007/s10712-015-9348-9

Paloscia S, Pettinato S, Santi E et al (2013) Soil moisture mapping using Sentinel-1 images: algorithm and preliminary validation. Remote Sens Environ 134:234–248. https://doi.org/10.1016/j.rse.2013.02.027

Parrens M, Al Bitar A, Frappart F et al (2017) Mapping dynamic water fraction under the tropical rain forests of the Amazonian Basin from SMOS brightness temperatures. Water 9:350. https://doi.org/10.3390/w9050350

Parrens M, Bitar AA, Frappart F et al (2019) High resolution mapping of inundation area in the Amazon basin from a combination of L-band passive microwave, optical and radar datasets. Int J Appl Earth Obs Geoinf 81:58–71. https://doi.org/10.1016/j.jag.2019.04.011

Pedinotti V, Boone A, Ricci S et al (2014) Assimilation of satellite data to optimize large-scale hydrological model parameters: a case study for the SWOT mission. Hydrol Earth Syst Sci 18:4485–4507. https://doi.org/10.5194/hess-18-4485-2014

Pekel J-F, Cottam A, Gorelick N, Belward AS (2016) High-resolution mapping of global surface water and its long-term changes. Nature 540:418–422. https://doi.org/10.1038/nature20584

Petiteville I, Ishida C, Danzeglocke J et al (2015) WCDRR and the CEOS activites on disasters. Int Arch Photogrammetry Remote Sensing Spatial Information Sci 76: https://doi.org/10.5194/isprsarchives-XL-7-W3-845-2015

Pettinari ML, Chuvieco E (2020) Fire danger observed from space. Surv Geophys. https://doi.org/10.1007/s10712-020-09610-8

Pham HT, Marshall L, Johnson F, Sharma A (2018) Deriving daily water levels from satellite altimetry and land surface temperature for sparsely gauged catchments: a case study for the Mekong River. Remote Sens Environ 212:31–46. https://doi.org/10.1016/j.rse.2018.04.034

Prigent C, Papa F, Aires F et al (2007) Global inundation dynamics inferred from multiple satellite observations, 1993–2000. J Geophys Res Atmospheres 112. https://doi.org/10.1029/2006JD007847

Reager J, Thomas A, Sproles E et al (2015) Assimilation of GRACE terrestrial water storage observations into a land surface model for the assessment of regional flood potential. Remote Sensing 7:14663–14679. https://doi.org/10.3390/rs71114663

Reager JT, Famiglietti JS (2009) Global terrestrial water storage capacity and flood potential using GRACE. Geophys Res Lett 36. https://doi.org/10.1029/2009GL040826

Reager JT, Thomas BF, Famiglietti JS (2014) River basin flood potential inferred using GRACE gravity observations at several months lead time. Nat Geosci 7:588–592. https://doi.org/10.1038/ngeo2203

Reichle RHA (2018) Soil Moisture Active Passive (SMAP) Mission Level 4 Surface and Root Zone Soil Moisture (L4_SM) Product Specification Document

Richey AS, Thomas BF, Lo M-H et al (2015) Uncertainty in global groundwater storage estimates in a total groundwater sress framework. Water Resour Res 51:5198–5216. https://doi.org/10.1002/2015WR017351

Robinson DA, Campbell CS, Hopmans JW et al (2008) Soil moisture measurement for ecological and hydrological watershed-scale observatories: a review. Vadose Zone J 7:358–389. https://doi.org/10.2136/vzj2007.0143

Rodell M (2012) Satellite Gravimetry applied to drought monitoring. Remote Sensing of Drought Innovative Monitoring Approaches 261. https://doi.org/10.1201/b11863-18

Rodríguez-Fernández NJ, Anterrieu E, Rougé B, et al (2019) SMOS-HR: a high resolution L-Band passive Radiometer for Earth Science and Applications. IGARSS 2019–2019 IEEE International Geoscience and Remote Sensing Symposium. https://doi.org/10.1109/IGARSS.2019.8897815

Santos da Silva J, Calmant S, Seyler F et al (2010) Water levels in the Amazon basin derived from the ERS 2 and ENVISAT radar altimetry missions. Remote Sens Environ 114:2160–2181. https://doi.org/10.1016/j.rse.2010.04.020

Schneider R, Godiksen PN, Villadsen H et al (2017) Application of CryoSat-2 altimetry data for river analysis and modelling. Hydrol Earth Syst Sci 21:751–764. https://doi.org/10.5194/hess-21-751-2017

Schroeder R, McDonald KC, Chapman BD et al (2015) Development and evaluation of a multi-year fractional surface water data set derived from active/passive microwave remote sensing data. Remote Sensing 7:16688–16732. https://doi.org/10.3390/rs71215843

Schumann G, Di Baldassarre G, Alsdorf D, Bates PD (2010) Near real-time flood wave approximation on large rivers from space: application to the River Po. Italy Water Resources Res 46. https://doi.org/10.1029/2008WR007672

Seitz F, Schmidt M, Shum CK (2008) Signals of extreme weather conditions in Central Europe in GRACE 4-D hydrological mass variations. Earth Planet Sci Lett 268:165–170. https://doi.org/10.1016/j.epsl.2008.01.001

Siderius C, Gannon KE, Ndiyoi M et al (2018) Hydrological response and complex impact pathways of the 2015/2016 El Niño in Eastern and Southern Africa. Earth's Future 6:2–22. https://doi.org/10.1002/2017EF000680

Soldo Y, Khazaal A, Cabot F, Kerr YH (2016) An RFI index to quantify the contamination of SMOS data by radio-frequency interference. IEEE J Selected Topics Appl Earth Observ Remote Sensing 9:1577–1589. https://doi.org/10.1109/JSTARS.2015.2425542

Stocker TF, Qin D, Plattner G-K, et al (2013) Climate change 2013: The physical science basis. Contribution of working group I to the fifth assessment report of the intergovernmental panel on climate change 1535.

Stöckli R, Vermote E, Saleous N, et al (2005) Blue Marble Next Generation—a true color earth dataset including seasonal dynamics from MODIS. In: NASA Earth Observatory. https://earthobservatory.nasa.gov/features/BlueMarble. Accessed 29 Jan 2020

Sun Z, Zhu X, Pan Y, Zhang J (2017) Assessing terrestrial water storage and flood potential using GRACE data in the Yangtze River Basin. China Remote Sensing 9:1011. https://doi.org/10.3390/rs9101011

Svoboda M, Hayes M, Wood D (2012) Standardized precipitation index user guide. World Meteorological Organization Geneva, Switzerland

Swain S, Wardlow BD, Narumalani S et al (2013) Relationships between vegetation indices and root zone soil moisture under maize and soybean canopies in the US Corn Belt: a comparative study using a close-range sensing approach. Int J Remote Sens 34:2814–2828. https://doi.org/10.1080/01431161.2012.750020

Tapley BD, Bettadpur S, Ries JC et al (2004) GRACE measurements of mass variability in the earth system. Science 305:503–505. https://doi.org/10.1126/science.1099192

Tarpanelli A, Amarnath G, Brocca L et al (2017) Discharge estimation and forecasting by MODIS and altimetry data in Niger-Benue River. Remote Sens Environ 195:96–106. https://doi.org/10.1016/j.rse.2017.04.015

Tarpanelli A, Brocca L, Barbetta S et al (2015) Coupling MODIS and Radar altimetry data for discharge estimation in poorly gauged river basins. IEEE J Selected Topics Appl Earth Observ Remote Sensing 8:141–148. https://doi.org/10.1109/JSTARS.2014.2320582

Tarpanelli A, Santi E, Tourian MJ et al (2019) Daily river discharge estimates by merging satellite optical sensors and radar altimetry through artificial neural network. IEEE Trans Geosci Remote Sens 57:329–341. https://doi.org/10.1109/TGRS.2018.2854625

Thomas AC, Reager JT, Famiglietti JS, Rodell M (2014) A GRACE-based water storage deficit approach for hydrological drought characterization. Geophys Res Lett 41:1537–1545. https://doi.org/10.1002/2014GL059323

Thomas BF, Famiglietti JS, Landerer FW et al (2017) GRACE groundwater drought index: evaluation of California Central Valley groundwater drought. Remote Sens Environ 198:384–392. https://doi.org/10.1016/j.rse.2017.06.026

Tomer SK, Al Bitar A, Sekhar M et al (2015) Retrieval and multi-scale validation of soil moisture from multi-temporal SAR data in a semi-arid tropical region. Remote Sensing 7:8128–8153. https://doi.org/10.3390/rs70608128

Tomer SK, Al Bitar A, Sekhar M et al (2016) MAPSM: a spatio-temporal algorithm for merging soil moisture from active and passive microwave remote sensing. Remote Sensing 8:990. https://doi.org/10.3390/rs8120990

Tourian MJ, Schwatke C, Sneeuw N (2017) River discharge estimation at daily resolution from satellite altimetry over an entire river basin. J Hydrol 546:230–247. https://doi.org/10.1016/j.jhydrol.2017.01.009

Tourian MJ, Tarpanelli A, Elmi O et al (2016) Spatiotemporal densification of river water level time series by multimission satellite altimetry. Water Resour Res 52:1140–1159. https://doi.org/10.1002/2015WR017354

Tralli DM, Blom RG, Zlotnicki V et al (2005) Satellite remote sensing of earthquake, volcano, flood, landslide and coastal inundation hazards. ISPRS J Photogrammetry Remote Sensing 59:185–198. https://doi.org/10.1016/j.isprsjprs.2005.02.002

Twele A, Cao W, Plank S, Martinis S (2016) Sentinel-1-based flood mapping: a fully automated processing chain. Int J Remote Sens 37:2990–3004. https://doi.org/10.1080/01431161.2016.1192304

Ulaby FTM (1981) Microwave remote sensing: active and passive. Volume 1—Microwave remote sensing fundamentals and radiometry

Wagner W, Hahn S, Kidd R et al (2013) The ASCAT soil moisture product: a review of its specifications, validation results, and emerging applications. Meteorol Z 22:5–33. https://doi.org/10.1127/0941-2948/2013/0399

Ward PJ, Jongman B, Kummu M et al (2014) Strong influence of El Nino Southern Oscillation on flood risk around the world. Proc Natl Acad Sci 111:15659–15664. https://doi.org/10.1073/pnas.1409822111

Wilhite DA (2000) Drought as a natural hazard: Concepts and Definitions, 22

WMO (2012) International glossary of Hydrology, UNESCO

World Resources Institute (2019) Aqueduct Floods. In: Aqueduct Floods. https://www.wri.org/applications/aqueduct/floods/. Accessed 28 Jul 2020

Yamazaki D, Watanabe S, Hirabayashi Y (2018) Global flood risk modeling and projections of climate change impacts. Global Flood Hazard Appl Model Mapping Forecasting 233:254. https://doi.org/10.1002/9781119217886.ch11

Yang Y, Lin P, Fisher CK et al (2019) Enhancing SWOT discharge assimilation through spatiotemporal correlations. Remote Sens Environ 234:111450. https://doi.org/10.1016/j.rse.2019.111450

Yoon Y, Beighley E, Lee H et al (2016) Estimating flood discharges in reservoir-regulated river basins by integrating synthetic SWOT satellite observations and hydrologic modeling. J Hydrol Eng 21:05015030. https://doi.org/10.1061/(ASCE)HE.1943-5584.0001320

Yoon Y, Durand M, Merry CJ et al (2012) Estimating river bathymetry from data assimilation of synthetic SWOT measurements. J Hydrol 464–465:363–375. https://doi.org/10.1016/j.jhydrol.2012.07.028

Yu X, Moraetis D, Nikolaidis NP et al (2019) A coupled surface-subsurface hydrologic model to assess groundwater flood risk spatially and temporally. Environ Modell Softw 114:129–139. https://doi.org/10.1016/j.envsoft.2019.01.008

Zhang A, Jia G (2013) Monitoring meteorological drought in semiarid regions using multi-sensor microwave remote sensing data. Remote Sens Environ 134:12–23. https://doi.org/10.1016/j.rse.2013.02.023

Zhang D, Zhang Q, Werner AD, Liu X (2016) GRACE-based hydrological drought evaluation of the Yangtze River Basin, China. J Hydrometeor 17:811–828. https://doi.org/10.1175/JHM-D-15-0084.1

Zhao L, Dong H, Edelmann RE et al (2017) Coupling of Fe(II) oxidation in illite with nitrate reduction and its role in clay mineral transformation. Geochim Cosmochim Acta 200:353–366. https://doi.org/10.1016/j.gca.2017.01.004

Zhou H, Luo Z, Tangdamrongsub N et al (2017) Characterizing drought and flood events over the Yangtze River Basin using the HUST-Grace2016 solution and Ancillary data. Remote Sensing 9:1100. https://doi.org/10.3390/rs9111100

Publisher's Note Springer Nature remains neutral with regard to jurisdictional claims in published maps and institutional affiliations.

Surveys in Geophysics (2020) 41:1489–1534
https://doi.org/10.1007/s10712-020-09594-5

Earth Observations for Monitoring Marine Coastal Hazards and Their Drivers

A. Melet[1] · P. Teatini[2] · G. Le Cozannet[3] · C. Jamet[4] · A. Conversi[5] · J. Benveniste[6] · R. Almar[7]

Received: 4 December 2019 / Accepted: 2 May 2020 / Published online: 5 June 2020
© The Author(s) 2020

Abstract

Coastal zones have large social, economic and environmental values. They are more densely populated than the hinterland and concentrate large economic assets, critical infra-structures and human activities such as tourism, fisheries, navigation. Furthermore, coastal oceans are home to a wealth of living marine resources and very productive ecosystems. Yet, coastal zones are exposed to various natural and anthropogenic hazards. To reduce the risks associated with marine hazards, sustained coastal zone monitoring programs, fore-casting and early warning systems are increasingly needed. Earth observations (EO), and in particular satellite remote sensing, provide invaluable information: satellite-borne sensors allow an effective monitoring of the quasi-global ocean, with synoptic views of large areas, good spatial and temporal resolution, and sustained time-series covering several years to decades. However, satellite observations do not always meet the precision required by users, in particular in dynamic coastal zones, characterized by shorter-scale variability. A variety of sensors are used to directly monitor the coastal zone and their observations can also be integrated into numerical models to provide a full 4D monitoring of the ocean and forecasts. Here, we review how EO, and more particularly satellite observations, can moni-tor coastal hazards and their drivers. These include coastal flooding, shoreline changes, maritime security, marine pollution, water quality, and marine ecology shifts on the one hand, and several physical characteristics (bathymetry, topography, vertical land motion) of coastal zones, meteorological and oceanic (metocean) variables that can act as forcing fac-tors for coastal hazards on the other hand.

Keywords Coastal zone · Earth observation · Monitoring · Hazards · Water quality · Flooding

---

✉ A. Melet
   amelet@mercator-ocean.fr

Extended author information available on the last page of the article

1 Introduction

Coastal zones are the transition areas that connect the terrestrial and marine environment. The global coastlines exceed 1.6 million kilometers (Burke et al. 2001) and 84% of the countries of the world have a coastline (either with the open oceans, inland seas or both).

Globally, coastal zones are more densely populated than the hinterland (Small and Nicholls 2003), exhibit higher rates of population growth and urbanisation, and are concentrating economic assets and critical infrastructures. The low-elevation coastal zone, defined as the contiguous and hydrologically connected zone of land along the coast with an elevation above sea level of less than 10 m, covers only 2% of the world's land area but~10% of the world population lives there (McGranahan et al. 2007, Neumann et al. 2015). The growth of coastal population is projected to continue over the coming years, in response to population growth and coastward migration associated with the global trend of urbanisation. The world population is expected to reach 8.5 billion in 2030 and 9.7 billion in 2050 (projections of United Nations 2015) and most of the world's megacities are located in the coastal zone, many in large deltas (McGranahan et al. 2007). Coastal migration is also driven by the combinations of specific economic, geographic and historical conditions, concentration of densely settled agricultural areas in well-watered, fertile deltas and coastal plains (Hugo 2011; McGranahan et al. 2007).

Coastal zones represent a huge economic value. For instance, maritime transport is essential to the world economy as over 90% of the world trade is carried by sea, with large portions of maritime routes in the coastal ocean (International Maritime Organization). In 2010, the global ocean economy represented USD 1.5 trillion in added value, with a strong contribution from offshore oil and gas, maritime and coastal tourism, ports and maritime equipment, and ocean-based industries employment dominated by fisheries and maritime and coastal tourism (OECD 2016). In addition to these established activities, emerging ones are projected to grow in the coming decades, including marine aquaculture, ocean renewable energy, marine safety and surveillance (OECD 2016). By 2030, conservative estimates assess that ocean economy will grow to more than USD 3 trillion (in constant 2010 USD) much of which will rely on coastal tourism, offshore oil and gas and port activities. Marine aquaculture is estimated to grow at an annual rate of 5.7% between 2010 and 2030.

Coastal ecosystems are home to a wealth of terrestrial and marine flora and fauna. Marine coastal ecosystems are among the most productive on Earth and provide a range of social and economic benefits to humans. They yield 90% of global fisheries and almost 80% of known species of marine fish (Cicin-Sain et al. 2002). Reefs, mangroves and sand dunes also play a regulating role and protect the shoreline (Wells et al. 2006), for instance by largely attenuating the energy of wind-generated waves. Coral reefs are among the most biodiverse ecosystems on Earth, and together with river estuaries, have a high biomass productivity.

Therefore, coastal zones have a tremendous social, economic and biological value (Martínez et al. 2007). They are providing many services to human society, including food, energy and other resources, shoreline protection, ocean recreation, tourism and coastal livelihoods, water quality maintenance, waste treatment, biogeochemical cycling, and regulating services, support of the green and blue economy and importantly, the maintenance of the basic global life support systems.

However, coastal zones are exposed to different hazards of natural or anthropogenic origins (Fig. 1). Natural hazards can be driven by extreme states of the natural environment

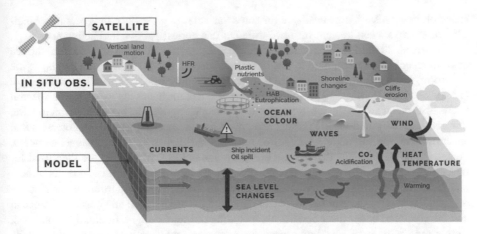

Fig. 1 Schematic representation (non-exhaustive) of the coastal zone, hazards (in normal font), and metocean variables (in bold font) that are relevant for coastal marine hazards and their monitoring [HFR: high-frequency radar; HAB: harmful algae bloom]

such as marine heat waves causing coral bleaching and fish mortality, large waves and extreme sea levels for coastal flooding, erosion and salinization of aquifers, low oxygen and acidification for degradation of ecosystems and habitats, etc. Several of them are exacerbated by climate change and induced ocean warming, acidification, and deoxygenation, sea level rise, increase in harmful algae blooms and invasive species (Gattuso et al. 2015; Bindoff et al. 2019).

The increasing development and utilisation of coastal zones (on land and ocean) over recent decades have led them to experience high pressures and have increased the vulnerability of coastal ecosystems (e.g., Crossland et al. 2005). Coastal hazards related to anthropogenic activities include maritime pollution, unsafe maritime conditions (e.g., for shipping and activities at sea), poor water quality, eutrophication, overfishing, degradation or loss of marine and coastal ecosystems and habitats, over-exploitation of subsurface fluids (groundwater, hydrocarbon) causing land subsidence. Forcing factors for marine coastal hazards are therefore widespread (Fig. 1).

Marine coastal hazards can lead to high risks for coastal zones due to the combination of high exposure (dense population, living marine resources, activities at sea, large economic assets and critical infrastructure) and vulnerability to these hazards in coastal areas (e.g., low adaptation capabilities of a given marine ecosystem to adapt to extreme or changing conditions, low capacity of a ship facing extreme wave conditions, etc.). Monitoring, understanding and predicting marine coastal hazards are therefore of increasing importance. A variety of sensors is used to monitor coastal zone characteristics, various metocean variables that are relevant for different coastal hazards, and some coastal hazards themselves. Metocean is the abbreviation of meteorology and oceanography and encompasses atmospheric and physical oceanographic variables such as characterizing wind, waves, sea level, bathymetry, currents, sea-ice (thickness, extent), seawater properties (salinity, temperature, stratification), water quality parameters.

Satellite-borne sensors allow an effective monitoring of the quasi-global ocean, with synoptic views of large areas in conjunction with good spatial and temporal resolution. Space observations can also allow to monitor multi-decadal changes as long time-series of space observations are now available for some variables (e.g., surface temperature). Satellite-borne sensors can be split into two different categories: passive and active ones (e.g., Le Traon et al. 2018). Passive techniques measure the natural radiation emitted from the sea or from reflected solar radiation in one or several spectral bands (wavelength bands) of the electromagnetic spectrum. Multispectral imagery generally has 3–25 bands of wavelength (13 for Sentinel-2), while hyperspectral imagery relies on hundreds of narrower spectral bands distributed across the electromagnetic spectrum. Passive sensors include microwave, infrared (thermal), and visible radiometers, imaging radiometers, spectro-radiometers and spectrometers. On the other hand, active sensors emit their own electromagnetic energy, potentially at wavelengths that are not energetic enough in sunlight and are then used for measuring the signal reflected, refracted or scattered by the Earth's surface or its atmosphere. As the majority of active sensors operate in the microwave portion of the electromagnetic spectrum, the emitted wave can penetrate the atmosphere and observe the ocean and the coastland under most conditions, including during the night, or cloudy conditions. Active sensors of particular importance for metocean and coastland monitoring include lidars, radar altimeters, synthetic or real aperture radars (SAR, RAR), and scatterometers.

The main aim of this article is to review how Earth Observations (EO), and more particularly satellite observations, can monitor several coastal zone characteristics that are relevant to hazard assessment (bathymetry, topography, vertical land motion, land cover and land use, in Sect. 2), metocean variables that can act as forcing factors for coastal hazards (Sect. 3) and coastal hazards themselves (Sect. 4). Among the various existing coastal hazards, attention is focused in Sect. 4 on coastland flooding and coastline dynamics, maritime security, marine pollution, water quality, and marine ecology shifts. Although of importance, coastal hazards related to ice (e.g., for navigation safety on sea-ice or ice-infested coastal waters, permafrost thaw, etc.) and plastic pollution (Viatte et al. 2020) are left out of this review, and so do the aspects related to the assessment of exposure and vulnerability, such as urbanisation and land management practices (Le Cozannet et al. 2020). Current gaps and perspectives or recommendations on using EO for a reliable evaluation and prediction of coastal hazards are discussed in the final section.

2 Earth Observations to Characterize Coastal Zones

Characterization of physical properties of the coastal zone is needed for coastal hazard assessments. In this section, we focus on EO monitoring of different essential physical properties of the coastal zone: land topography and ocean bathymetry, vertical land motion and land cover/land use.

2.1 Bathymetry

A great demand exists nowadays for up-to-date bathymetric (seafloor) and topographic (over land) maps of shallow water areas and adjacent emerged zones (e.g., beaches and sand barriers, lagoons, estuaries and mud coasts, cliffs). Indeed, estimates of the coastal zone topography and bathymetry with high spatial and temporal resolution and a good accuracy is a prerequisite to model or estimate different drivers of coastal hazards or hazards themselves (Cazenave et al. 2017). Inaccurate or outdated coastal Digital Elevation Models (DEMs) (including topography and bathymetry) lead to flooding modeling errors that can ultimately mislead coastal risk management (Neumann et al. 2015).

Bathymetric measurements by ship- or air-borne echo-sounders and lidars are tedious and expensive. As a result, only a rather small fraction ($< 15\%$) of the ocean depth has been determined at a horizontal resolution of at least 1 km (Mayer et al. 2018). In a large number of coastal areas of interest, bathymetric information is unavailable or is often decades old. Satellite observations are therefore an important complement to existing in situ observation systems, with the advantage of providing a more systematic monitoring (Benveniste et al. 2019). Satellite altimetry has allowed estimating the world's ocean bathymetry in deep water, but with an average achievable resolution of 8 km. Satellite altimetry measures the sea surface height, which is notably affected by the gravitational effects of seafloor topographic features. The horizontal resolution of altimetry-derived bathymetry is much lower than direct measurements with echo-sounders, but it allowed mapping the seafloor over extended areas. Using satellite altimetry-derived bathymetry from Geosat and ERS-1 to guide the interpolation between direct depth soundings, Smith and Sandwell (1997) provided maps of the world's ocean bathymetry with a horizontal resolution of 1–12 km. These maps have been refined over time using new altimeter missions (e.g., Jason-1 and 2, SARAL/AltiKa geodetic missions, CryoSat-2) and will be further improved thanks to the future SWOT mission (Tozer et al. 2019).

In the coastal zone, satellite-derived bathymetry (SDB) can reach much finer horizontal resolutions thanks to two other methodologies (Pleskachevsky et al. 2011): methods based on the modeling of radiative transfer of light in water for the processing of optical images, and methods based on the influence of topography on hydrodynamic processes (e.g., current variations and wave characteristics); the latter ones can use optical or SAR images. Each method/sensor comes with its own strengths, limitations, and scope of applications (Gao 2009; Jawak et al. 2015). Specifically, methods based on the radiative transfer perform better in clear and calm waters, whereas techniques based on water depth inversion requires waves and can perform over turbid waters.

The pioneering technique using aerial multispectral photographs (Lyzenga 1978) was expanded to multi-spectral optical satellite images with first attempts made using Landsat data and then a wider application to other satellites such as Sentinel-2 (Caballero and Stumpf 2019; Evagorou et al. 2019; Sagawa et al. 2019). SDB is calculated based on the attenuation of radiance as a function of depth and wavelength in the water column, using analytical or empirical imaging methods. Analytical models are based on radiative transfer models and optical properties of the sea water, such as the attenuation coefficient and backscattering, the spectral signatures of suspended and dissolved matter, and bottom reflectance (Lee et al. 1998). Empirical methods are based on statistical relationships between image pixel values and ground truth depth measurements. Bathymetries can be estimated down to depths from a few meters (e.g., northern Baltic Sea) to 20–30 m (e.g., Mediterranean Sea) depending on the maximum penetration depth of sunlight, which varies with seasons, locations, turbidity, bottom reflectance, etc. The horizontal resolution of such bathymetries depends on the optical sensor, from coarse with for instance Landsat (~ 30 m resolution), to medium with SPOT and Sentinel-2 (10 m resolution, Fig. 2), and to high with the use of commercial satellites such as IKONOS (1 m resolution) and QuickBird or WorldView (50–60 cm resolution) (Monteys et al. 2015). The accuracy of this method is equal to or is lower than with lidars or echo-sounders (~ 0.5 m), but is an effective solution to map the nearshore bathymetry over large areas.

Other techniques for estimating bathymetry from optical images use wave characteristics (Abileah 2006; Danilo and Binet 2013; Poupardin et al. 2015), capitalizing on the methodologies developed by the coastal community (Holman et al. 2016) for video imagery and drones. This allows depths of up to 40–50 m to be resolved. Bathymetric

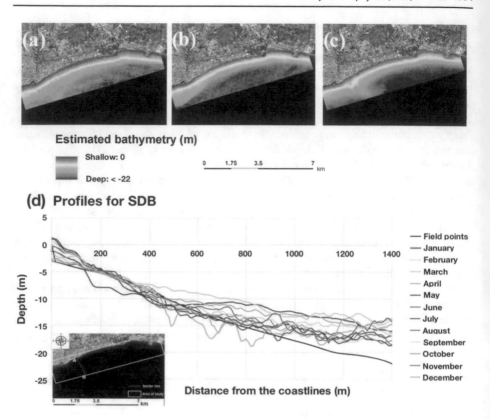

Estimated bathymetry (m)

Shallow: 0

Deep: < -22

(d) Profiles for SDB

Fig. 2 Satellite-derived bathymetry (SDB) from Sentinel-2 multispectral optical data in **a** January 2018, **b** February 2018 and **c** March 2018 for depth shallower than 22 m over a coastal area of southern Cyprus. The monthly evolution of a bathymetric profile (shown in the lower inset) is shown in panel **d** from April 2017 to March 2018 with Sentinel-2 data. Field points correspond to LIDAR data acquired in 2014. Adapted from Evagorou et al. (2019)

inversion code developed from the wave dispersion relationship in intermediate to shallow water make use of the temporal information contained in the spatial images and were applied to different satellite observations: IKONOS (Abileah 2006), WorldView-2 (McCarthy 2010), SPOT5/6 (Poupardin et al. 2016), Sentinel-2 (Bergsma et al. 2019) and Pleiades (Danilo and Binet 2013; Almar et al. 2019). One of the great advantages of this approach is that it is autonomous and does not require additional wave information from observation or models besides those acquired by means of satellite remote sensing. Nearshore bathymetry can also be derived from SAR sensors, such as those on-board Sentinel-1 satellites. Similar to optical methods, shallow water bathymetry is derived from directional wave spectrum (e.g., Wiehle and Pleskachevsky 2018). At larger regional scale and for deeper waters, Alpers and Hennings (1984) proposed the first theoretical model to map bathymetry based on sea surface features induced by current variations over bottom topography. The range of validity of wave-based SDB does not depend on local calibration, water turbidity and bottom type, as color-based methods do, and this method can cover most shelves to depth up to 40–50 m (Bergsma and Almar 2020). However, color-based SDB seems to better resolve small-scale features in very shallow waters, so that a combination of methods is beneficial.

It is noteworthy that a recently increasing number of works make use of deep learning (Sagawa et al. 2019) that brings great expectations to solve satellite-based bathymetry issues of complex physics and environments, method fusion and computational costs (Danilo and Melgani 2019).

Coastal morphology changes over a wide range of timescales (from storm events, seasonal and interannual variability to longer-term adaptation to changing environmental conditions), in particular in response to changing incoming wave regimes (e.g., Karunarathna et al. 2016; Bergsma et al. 2019, Fig. 2d) and human interventions. Despite its high potential, SDB has only been applied to limited space domains, and efforts remain to be done to map nearshore bathymetry and its time-evolution at global scale (e.g., Mayer et al. 2018; Wölfl et al. 2019; Benveniste et al. 2019).

2.2 Coastal Land Topography

Several methods exist for mapping the coastal topography (see recent review by Salameh et al. 2019) using various passive (SPOT, Landsat-8/OLI, Sentinel-2/MSI, WorldView, Quickbird, IKONOS, Pléiades, etc.) and active (ERS-1&2, ENVISAT, TerraSAR-X, Sentinel-1, etc.) sensors. Satellite remote sensing techniques now offer a good alternative for digital elevation model (DEM) construction over large spatial areas with sufficient horizontal resolution (< 100 m). Global DEMs such as the 90 m resolution SRTM (Farr et al. 2007) (widely used for broad scale flood modeling, e.g., Ettritch et al. 2018; Neumann et al. 2015; Kulp and Strauss 2018), 30 m resolution ALOS AW3D30 (JAXA, Tadono et al. 2016, recently released and used incipiently by Zhang et al. 2009) and 8 m resolution WorldDEM (Airbus, on-demand) are available. However, they are often years old and do not reflect the rapid evolution of dynamic coastal areas.

By analyzing stereoscopic pairs of satellite optical images (Tateishi and Akutsu 1992), it is possible to study large coastal areas with vertical errors of less than 0.5 m. Examples using Pléiades are given in Collin et al. (2018) and Almeida et al. (2019) (Fig. 3). Intertidal areas between spring tides also benefit from the wide range of satellite data available. The waterline method for intertidal topography mapping introduced by Mason et al. (1995) is one of the most widely adopted techniques. The method consists in determining the horizontal position of the shoreline using SAR (Heygster et al. 2010; Li et al. 2014) or optical images (Bergmann et al. 2018; Khan et al. 2019), and then superimposing the heights above mean sea level obtained from the knowledge of tide and surge on this boundary. For multiple images obtained over a range of different waterline levels, a set of elevated shorelines can be constructed and from there, a raster DEM can be interpolated.

Over the last decade, other methods have also proven reliable for monitoring the topography of coastal flat areas. Specifically, highly accurate DEMs have been obtained by processing SAR images acquired by Tandem missions. ESA conducted between 27 September 2007 and 12 February 2008 a dedicated ERS-2–ENVISAT Tandem mission. A unique opportunity offered by these two SAR instruments operated in the same orbital configuration was ERS–ENVISAT cross-interferometry. ENVISAT was operated in the same orbits as ERS-2, preceding ERS-2 by approximately 28 min. The almost simultaneous acquisition of SAR images by the two satellites and a large perpendicular baseline of approximately 2 km between the two satellites for mid to high northern latitudes allowed the generation of precise DEMs in relatively flat areas (Wegmüller et al. 2009). Similarly, cross satellite SAR interferometry based on two TerraSAR-X radar satellites flying in close formation has been made available by TanDEM-X (TerraSAR add-on for Digital Elevation Measurement—TDX) mission. The

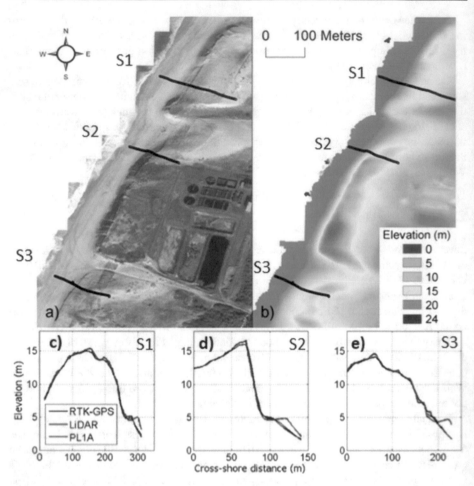

Fig. 3 **a** Orthophoto and **b** digital Surface Model from Pleiades stereo pair showing the locations of the transects S1, S2, S3 for comparison of beach profiles between the RTK-GPS, Pleiades (PL1A) and LiDAR data shown in panel **c** for S1, **d** for S2 and **e** for S3 (from Almeida et al. 2019)

TDX datasets acquired using a short-term or no temporal baseline (10 s), a large spatial baseline (3 to 4 km), and a high-resolution imaging mode (5 to 7 m) provided a great opportunity to generate DEMs with significantly improved vertical accuracy (height error in the range of 0.10–0.15 m) over intertidal environments (Lee and Ryu 2017; Choi and Kim 2018). Overall, vertical errors of 0.2 m or less correspond to current requirements for local high-resolution flood modeling (Le Roy et al. 2015; Ablain et al. 2016). While such accuracies in urbanized zones can only be achieved with lidar today, other products are useful for broad scale modeling, as well as in areas characterized by flooding by overflow and lack of data.

2.3 Vertical Land Motion

The capability of monitoring the vertical movements of the Earth surface, and specifically land subsidence in coastal areas, has increased significantly over the last two decades with the availability of satellite SAR acquisitions and the simultaneous development of interferometric processing chains (e.g., Teatini et al. 2005; Chaussard et al. 2013; Higgins et al. 2014; Tosi et al. 2016; Da Lio et al. 2018). Several SAR-borne satellites have been in operation since 1991 (ERS-1/2; ENVISAT; JERS-1; RadarSAT-1/2, ALOS, TerraSAR-X, Cosmo-SkyMed, and Sentinel-1A&B from the mid-2014), thus a large satellite SAR data archive exists over many areas. Differential SAR Interferometry—DInSAR (Gabriel et al. 1989), Permanent Scatterer InSAR—PSInSAR (Ferretti et al. 2001), Small Baseline Subset—SBAS (Berardino et al. 2002), Interferometric Point Target Analysis—IPTA (Wegmüller et al. 2004), and "Squeezed" SAR—SqueeSAR (Ferretti et al. 2011) are only the most well-known and widely used SAR processing chains among a continuously increasing variety of algorithms.

The quantification of land subsidence in coastlands has always been challenging for traditional ground-based methods (e.g., levelling, GNSS) because of the peculiarity of low-lying coastal zones. On the one hand, in natural conditions coastlands are characterized by difficult access due to the presence of marshlands, lagoons, and watercourses. On the other hand, the use of in situ techniques in large urban settlements and megacities (which are continuously growing in coastal zones worldwide) usually provides an over-simplified representation of the actual pattern of the land displacements due to the limited number of monitoring locations (benchmarks or GNSS antennas). By exploiting the phase difference of the radar signals between or among a number (at least two) of satellite acquisitions over the same area, SAR-based techniques can provide movement information on millions of points scattered over a large region (10^4–10^5 km^2/scene), with a high spatial detail (image horizontal resolution is on the order of 25×25 m^2 or less depending on the acquisition mode), a sub-centimeter measurement accuracy, and a temporal sampling from a few days to one month (Ferretti et al. 2001). Therefore, levelling and GNSS have been less and less used over recent years to measure land subsidence. However, it must be emphasized that they remain of paramount importance in calibrating the SAR outcome (e.g., Tosi et al. 2016) and converting SAR differential displacements to absolute measurements (Wöppelmann et al. 2013).

Specific approaches have been developed over the recent years to increase the effectiveness of SAR-based applications to monitor land motion in coastal areas. Specifically, the low coherence in vegetated areas and the lack of radar reflectors in the most natural portions of lagoons and deltas has been overcome by establishing artificial reflectors (Strozzi et al. 2013). Moreover, various strategies have been proposed to obtain the best advantage from SAR sensors with different frequency bands by means of the integration of their outcomes (e.g., Cianflone et al. 2015; Tosi et al. 2016) (Fig. 4).

2.4 Land Cover and Land Use

Information on land cover/land use (LCLU) and their changes are of relevance for several coastal hazards. For instance, during coastal flooding (Sect. 4.1), water level attenuation due to hydrodynamic processes related to the land surface roughness, which depend on LCLU, can substantially reduce the areas, population and asset exposure to flooding (Vafeidis et al. 2019). LCLU products have been derived from satellite data for decades (Ban et al. 2015). For instance, the latest CORINE dataset is based on Sentinel-2 and Landsat-8 EO. Yet, mapping LCLU in the coastal zone with enough resolution (space and time) and relevant classification to better address coastal hazards and exposure is still needed. Characterizing the coastal geology (Fig. 5a, b) is also needed to assess the vulnerability

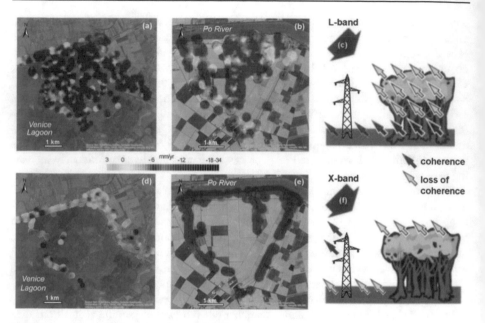

Fig. 4 Comparison of SAR outcomes on **a**, **d** the northernmost tip of Venice Lagoon and **b**, **e** a portion on the Po River delta, Italy. Positive values (in green) mean uplift while negative values (yellow to purple) mean land subsidence. The movements in **a**, **b** were obtained by SBAS on 16 L-band images acquired by ALOS-PALSAR between 2007 and 2010; those in **d**, **e** by PSI on 30 X-band Cosmo-SkyMed images from 2008 to 2011. The capability of L-band and X-band images to detect different radar targets are represented by the sketches in **c**, **f**: L-band acquisitions processed by SBAS provide information on fields and wetlands, X-band PSI on anthropogenic structures and infrastructures mainly located along roads and embankments. Notice how, on average, these latter subsided more in the Po Delta and less in the Venice Lagoon than the farmland and wetland surface, respectively. Subpanels **a**, **b**, **d**, **e** modified after Tosi et al. (2016)

of the shoreline. For instance, granite cliffs are less prone to erosion than sandy beaches. The detection of sandy beaches with EO is developing (e.g., Luijendijk et al. 2018). LCLU information is also relevant for eutrophication and water quality (Sect. 3.4) due to nutrient loading from, e.g., agricultural practices (Fig. 1).

3 Monitoring Relevant Variables for Marine Coastal Hazards

In addition to the above coastal zones characteristics, different metocean fields need to be monitored for a better assessment and monitoring of coastal hazards. The most important variables to monitor include sea surface temperature, sea level, ocean surface currents, near surface wind, ocean color, but also sea ice parameters, water turbidity, pH, nutrients and oxygen concentrations, not discussed in this article. These variables can directly drive coastal marine hazards, or their monitoring contributes to understand or estimate them. Table 1 exemplifies such relationships for a selection of marine coastal hazards discussed in Sect. 4. Various variables of the marine environment can be monitored with a global or quasi-global coverage and at synoptic scales thanks to different satellite-based instruments, such as optical imagers, spectroradiometers, infrared or microwave radiometers, altimeters, synthetic or real aperture radars, etc. (see Benveniste et al. in review). We briefly describe below EO of the most relevant environmental data for marine coastal hazards, which are also essential climate variables (ECV as defined by the Global Climate Observing System), with a focus over the coastal ocean.

3.1 Sea Level

Tide gauges provide the fundamental observations of local sea level changes at the coast (relative to the ground level, as they are grounded on land), with the first records dating back to the eighteenth century. Furthermore, tide gauges are the main source of high-frequency (< 1 h) sea level observations, which are essential for observing extreme water levels at the coast (e.g., Woodworth et al. 2017). However, tide gauge observations come with several limitations. The tide gauge coverage is rather sparse and inhomogeneous along the world's coastlines; some processes such as wave setup are not fully captured due to the location of tide gauges in wave-sheltered areas. GNSS stations tied to tide-gauges are required to provide vertical land motion over extended periods of time, but are often lacking (see Marcos et al. 2019).

Satellite altimetry, with its quasi-global coverage and rather high revisit time, complements tide gauge data. Sea surface height in the open ocean has been accurately monitored by satellite altimetry since late 1992 thanks to a suite of different altimetry missions (e.g., Cazenave et al. 2018). These space observations have largely improved our understanding of sea level variability, and have allowed a precise monitoring of sea level rise, with a global-mean rate of 3.35 ± 0.4 mm/yr and an acceleration of 0.12 ± 0.7 mm/yr^2 over 1993–2017 (WCRP Global sea level budget group 2018, Ablain et al. 2019). But despite the recognized invaluable contribution of satellite altimetry to better monitor sea level and ocean dynamics over most of the world's ocean, conventional satellite altimetry methods remain unable to monitor the sea level changes in the coastal zone. The poorer performances of satellite altimetry in the last kilometers (~ 20 km) off the coasts are due to land contamination in the radar footprint that modifies in a complex way the radar waveform, and to less accurate geophysical corrections (e.g., Birol et al. 2016; Cipollini et al. 2017; Benveniste et al. 2019). Yet, monitoring sea level close to the coast is much needed as sea level variability at the coast can differ from that of the open ocean (e.g., Vinogradov and Ponte 2011; Bingham and Hughes 2012; Calafat et al. 2012; Hughes et al. 2019), from weather event to multi-decadal timescales (Melet et al. 2018; Ponte et al. 2019), due to different dynamics and to the additional contribution of processes to these captured by satellite altimetry in the open ocean. Among them are tides, wind setup, surface atmospheric pressure effects, coastally trapped waves, wind-wave setup and swash, river runoffs or even more local processes such as seiches or meteotsunamis (Woodworth et al. 2019). Over the last years, coastal altimetry has substantially progressed thanks to specific waveform retracking (e.g., Passaro et al. 2014, 2018), improved geophysical corrections and data editing (Birol et al. 2016), or altimetry missions with smaller footprint (SARAL/AltiKa Ka-Band altimeter) or higher along-track resolution (CryoSat-2 and Sentinel-3A&B delay-Doppler Altimeters). Efforts are undertaken to combine specific waveform retracking, improved data editing and corrections to study sea level changes as close to the coast as possible (e.g., Marti et al. 2019).

In addition, altimetry faces sampling issues regarding high-frequency variations (e.g., associated to surges, tides, seiches, tsunamis) since revisit times are, at best, 10 days. Similarly, extreme events, which obviously matter in terms of coastal hazards, are highly localized in space and time while altimetry tracks are separated by tens of kilometers. As such, altimetry is not yet an alternative to tide gauges when it comes to monitoring high-frequency variations and sea level at the coast.

In practice, all mean and extreme sea level observations are limited in time and spatial resolutions: this includes satellite altimetry, tide gauges and other sea level monitoring approaches (e.g., SAR, Raucoules et al. 2018, GNSS reflectometry, Roussel et al. 2015). As no single approach can currently respond to the demand for information on mean and extreme sea levels, the synergic use of different observations and modeling approaches can be suggested as one way forward to progress in this area.

3.2 Sea Surface Temperature (SST)

Satellite-derived observations of SST are actually representative of different depths that depend on the passive sensor frequency. Infrared (IR) sensors measure the surface skin temperature $(O(10)\mu m)$, which can be affected by cool skin layer effects, especially at night, and warm layer effects in the daytime, in addition to potential diurnal warming. Microwave (MW) sensors measure the subskin temperature $(O(1)mm)$, below the thermal skin layer, and capture less the diurnal SST variability. MW sensors have a footprint of ~50–75 km, which limits the retrieval of SST data ~100 km inshore due to land contamination, but they present the advantage of being mostly insensitive to clouds. Infrared (IR) sensors on the other hand provide SST at a resolution of 1–4 km, but are strongly influenced by cloud emission and scattering. MW and IR SST retrievals are merged to provide global maps of SST. Retrieval of SST in the coastal zone also tends to be less accurate than in the open ocean due to a larger variability in atmospheric temperature, water vapor and aerosol concentrations, or due to potential alteration of the ocean surface emissivity related to contaminants, and to the lack of in situ data that are needed to validate SST retrievals. Diurnal warming of the sea surface has been relatively less studied in the coastal ocean than in the deep ocean. In the coastal zone, diurnal warming is influenced by several factors, such as sea breezes, bathymetric and tidal effects, in addition to the usual factors of solar insolation and surface winds. Diurnal warming can be of relevance for hazards related to ecosystems. For instance, coral bleaching events could be related to maximum daily temperatures. Improving EO derived SST quality in the coastal zone is a priority for SST developments in the next decade (see O'Carroll et al. 2019 for a review on SST EO).

3.3 Surface Wind-Waves

Significant wave height, the average height of the highest one-third of the waves, can be inferred from satellite altimetry (e.g., Ribal and Young 2019; Ardhuin ct al. 2019). Such observations suffer from similar limitations than those described for satellite altimetry sea level in the coastal zone (Sect. 3.1), and can similarly be improved with retracking, data filtering, and reduction of the radar footprint (Passaro et al. 2015; Quilfen et al. 2018; Dinardo et al. 2018). More complete information on the wave field can be obtained from directional wave spectrum derived from C-band SAR sensors (e.g., ERS-1, ERS-2, Envisat, Sentinel-1), enabling the monitoring of the wave directions and periods (e.g., Hasselmann et al. 2012), which are essential parameters in addition to significant wave height for wave-contributions to sea level at the coast. Yet, SAR wave-mode based wave period estimates are mostly reliable for long swells (with a wavelength longer than ~200 m, Ardhuin et al. 2019) due to blurring effects induced by wave orbital velocities. Another limitation in coastal zones comes from acquisition modes: for instance, although Sentinel-1

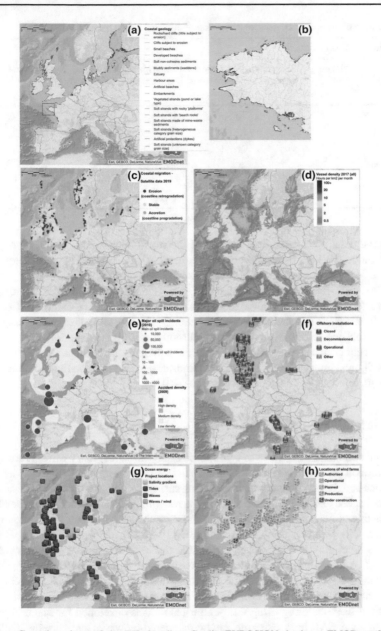

Fig. 5 **a** Coastal geology of the EU shoreline. Credit: EUROSION database, EMODnet. **b** Zoom over French Brittany. **c** Shoreline migration over 2007–2017 estimated from Sentinel-2, Landsat 5, 7 and 8 satellite data. Credit: EMODnet. **d** Shipping density in 1 km × 1 km cells for 2017 over all EU waters, based on AIS data and expressed as hours per square kilometer per month. Credit: Cogea, EMODnet. **e** Shading: Shipping accident density in the seas around the European Union for year 2009. Credit: EMSA, EMODnet. Brown circles and triangles: major oil spill accidents for years 2007–2008, units: tonnes. Credit: EMSA, EMODnet. **f** Offshore structures and facilities in a marine environment, usually for the production and transmission of electricity, oil, gas and other resources. Credit: OSPAR, EMODnet. **g** Marine energy project locations depending on the marine source of energy. Credit: EMODnet. **h** Location and status of offshore wind farms over EU water. Credit: EMODnet

Table 1 Relevance of metocean variables for a selection of different coastal hazards

EO observations	Coastal hazards					
	Flooding	Shoreline changes	Maritime security hazards	Maritime pollution	Degraded water quality	Ecosystem shifts
Sea level	x	x	x			x
SST				x		x
Wind-wave	x	x	x	x		
Surface currents		x	x	x		x
Surface winds	x	x	x	x		
Ocean color		x		x	x	x

acquires in the default wave-mode over most of the world open ocean, an interferometric wide-swath mode (IW) is operated over coastal zones. Algorithms are being developed to retrieve directional wave spectra from IW Sentinel-1 acquisitions.

The wave scatterometer SWIM on-board CFOSAT (launched in October 2018) recovers wave spectra for shorter waves, down to wavelengths of 70 m. Yet, C-band SAR or Ku-band RAR (such as SWIM) are still limited in enclosed basins and off eastern coasts, due to blurring associated with wave-induced orbital motions, short wave periods and heights, and wave direction close to the azimuthal direction (Alpers and Rufenach 1979; Dodet et al. 2019). Optical imagery (e.g., Sentinel-2, Landsat) can also provide the full directional wave spectrum in coastal areas (Kudryavtsev et al. 2017).

3.4 Near Surface Wind

The sea surface roughness is related to centimetric scale waves (capillary waves) formed by the wind stress on the ocean surface. As the sea surface reflectivity and emissivity depends on sea surface roughness, remote sensing of near surface winds exploits the relationships between wind speed and direction and sea surface roughness. Near surface wind speed and direction have both been retrieved and monitored over the global ocean by satellite radar scatterometers (e.g., Seawinds on-board QuikSCAT; SCAT onboard CFOSAT, HY2A, HY2B) (de Kloe et al. 2017). However, the spatial resolution of current scatterometers is of order 12–25 km, which is a limitation for coastal zone monitoring. Higher resolution wind retrieval, as close as 1–2 km off the coast, are possible thanks to SAR imagers. Yet, SAR only enables wind speed retrieval through a priori knowledge of wind direction and the inversion of backscattering using semi-empirical models developed for scatterometers. In the coastal zone, SAR backscattering can be hampered by waves and currents, bathymetry and topography (including capes and islands) and SST fronts. These phenomena can affect sea surface roughness (Zechetto et al. 2016) and can also generate small-scale features (< 25 km) in the near surface wind field (Chelton et al. 2004). Differential diurnal warming between land and ocean surfaces can also generate small-scale variability in the coastal zone wind field. Microwave radiometers and satellite altimetry can also provide long-term observations of wind speed, but do not provide wind direction and have limited performance in the coastal zone (Sects. 3.1, 3.3).

3.5 Ocean Surface Currents

Direct near-real time high-resolution 2D observations of surface currents can be provided by coastal high-frequency radars (HFR) (e.g., Ullman et al. 2006; Roarty et al. 2019). HFR have a measurement range of approximately 20–200 km offshore with a spatial resolution ranging from 300 m to 12 km, depending on their operating frequency. However, they cover comprehensive portions of the coastlines only for some countries, such as the USA. Geostrophic surface currents can be derived from satellite altimetry observations of sea level with a quasi-global coverage. Yet, altimeter-derived geostrophic currents might only be partly resolved in shelf seas, where the dominant spatial scale of ocean baroclinic mesoscale eddies, characterized by the first baroclinic Rossby deformation radius, is smaller than the effective resolution of altimeter products. Altimeter-derived geostrophic currents can then be complemented, and their resolution potentially increased, through synergetic use of information from surface winds to account for Ekman currents (e.g., Rio et al. 2014; Sudre et al. 2013; Dohan and Maximenko 2010; Bonjean and Lagerloef 2002) and from SST, by inverting the heat conservation equation, to partly account for other ageostrophic currents (e.g., Rio and Santoleri 2018). However, such techniques remain hampered by the limitations of satellite altimetry in the coastal zone (Sect. 3.1) and do not give access to the total surface currents. In particular, Stokes drift and tidal currents can account for a large fraction of the total currents in some areas (Fig. 6).

3.6 Ocean Color

Ocean color radiometry (OCR) results from the interaction of sunlight with the marine particles, which encompass pure water, phytoplankton through the common pigment (chlorophyll-a), the suspended particulate matter (SPM) and Colored Dissolved Organic Matter (CDOM), virus, bacteria (Antoine 1998; IOCCG 2000, Susik 2008). These marine particles can absorb and scatter more or less the sunlight in specific directions and at preferential wavelengths.

Remote sensing of OCR has revolutionized our vision of the distribution of the marine biomass. The first ocean color sensor, the Coastal Zone Coastal Scanner, was launched in 1978 and was a proof-of-concept. Since 1997, a new generation of remote sensors has been developed and put on satellite platforms leading to 20+-year records of satellite observations.

The primary parameter that can be estimated via remote sensing of ocean color is the so-called Remote Sensing Reflectance, R_{rs} (IOCCG 2000), which represents the quantity of light back-scattered by the ocean. The shape and magnitude of this spectra depend on the concentration and types of marine particles. From this parameter, a variety of bio-optical and biogeochemical parameters can be derived (IOCCG 2000, 2018) from empirical or semi-analytical algorithms (Matthews 2011; Odermatt et al. 2012; Blondeau-Patissier et al. 2014; IOCCG 2018; Werdell et al. 2018). In turn, these parameters provide very useful information for monitoring the water quality, especially over coastal zones: chlorophyll-a concentration and primary production, suspended particulate matter, particulate and dissolved organic carbon, particulate inorganic carbon, phytoplankton groups, transparency/turbidity (IOCCG 2008, 2018).

Historically, remote sensing of ocean color used sensor with high radiometric and spectral resolutions, medium-spatial resolution (300–1000 m) and high revisit times (typically, 3 days). But high-spatial-resolution sensors (10–100 m) with relevant radiometric and

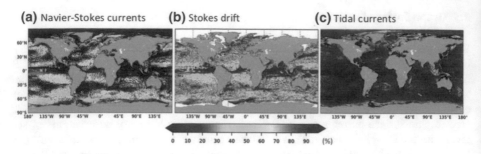

Fig. 6 Monthly averaged contribution in percentage to the magnitude of surface currents from **a** the ocean general circulation (stemming from Navier–Stokes equations), **b** wave-induced Stokes drift and **c** tidal currents for March 2018 in CMEMS product SMOC (product global-analysis-forecast-phy_001_024-hourly-merged-uv). Courtesy Stéphane Law Chune (Mercator Ocean International)

spectral resolution are now allowing water quality and pollution to be tracked over coastal areas (OLI/Landsat 8 and MSI/Sentinel-2 sensors). Efforts were undertaken these last years to adapt these sensors to aquatic science applications: water pixel extraction (Ngoc et al. 2019), atmospheric correction (Vanhellemont and Ruddick 2014, 2015, 2018), validation (Ody et al. 2016; Pahlevan et al. 2017a, b, 2018, 2019; Warren et al. 2019). Geostationary satellites (SEVERI, GOCI-I, Himawari-8) can also provide very useful information over coastal areas, even if some of them were not designed for ocean applications, as several images are taken per day with a current resolution ranging from 3 km (SEVIRI) to 500 m (GOCI-I) (Himawari-8: spatial resolution of 500 m to 1 km in the visible bands, 2 km in infrared bands) (Neukermans et al. 2009, 2012; Salama and Shen 2010; Ruddick et al. 2013; Choi et al. 2014; Doxaran et al. 2014; Vanhellemont et al. 2014; Huang et al. 2015; Kwiatkowska et al. 2016; Dorji and Fearns 2018). Issues still remain to have a complete system able to monitor coastal areas at high spatial and temporal resolutions (Mouw et al. 2015).

3.7 Integration of EO in Models and Forecasting

While satellite observations are an essential source of information to monitor metocean variables that are relevant to coastal hazards, they are often limited to the ocean surface, and do not provide a full spatial and temporal synoptic coverage of the ocean surface. Integrating satellite observations of the ocean surface and sparse in situ coastal observations into operational numerical global, regional or coastal models enables a synoptic 4D monitoring of the many ocean variables, including those non-observed variables, and improves the models' accuracy and prediction skills (e.g., Le Traon et al. 2015; Tonani et al. 2015). The more commonly assimilated satellite observations in ocean models are SST, sea level, Chlorophyll-a concentration, significant wave height (with progress regarding the assimilation of directional wave spectra), and sea-ice parameters. As for atmospheric models, used to force ocean models, they also assimilate EO data, notably scatterometer ocean vector winds, SST and significant wave height (e.g., ERA5 reanalysis, Hersbach et al. 2019).

Such integrated systems are the backbone of reanalyses (coherent historical representation of the ocean state over the past decades) and short-term forecasts (up to 10 days) of the physical and biogeochemical ocean states. The outputs can then be used for diverse applications related to coastal hazards. A focus on coastal sea level monitoring and forecasting with integrated systems is provided in Sect. 4.1.2.

In addition to constraining the predictions of the numerical models through data assimilation, EO are essential to guide the development and validation of numerical models of the ocean and to constrain the atmospheric conditions that are forcing ocean models.

4 Monitoring Marine Coastal Hazards

In this section, EO monitoring of main coastal hazards is discussed.

4.1 Coastal Erosion and Flooding

Being at the land-sea boundary, the coastal zone is prone to marine flooding due to the relative change of elevation between the sea level at the coast and the Earth's surface.

4.1.1 Total Water Level at the Coast

Relative sea level rise (RSLR, i.e., relative to the ground) and changes at the coast occur on a wide range of spatial and temporal scales, and are due to a complex combination of processes: processes causing sea level changes in the open ocean (e.g., mass addition from ice sheets, mountain glaciers, terrestrial water storage changes and the associated changes in gravitational and Earth rotation effects, steric effect and redistribution of heat, salt and mass by ocean circulations), additional processes causing sea level changes in the coastal ocean (e.g., tides, wave setup, storm surge, river discharges, wind driven upwelling) and vertical land motion in the coastal zone. Monitoring total water level changes at the coast for floods is therefore challenging, as it requires monitoring "offshore" sea level, as close as possible to the coast (Sect. 3.1), as well as ancillary fields that vary over a wide range of temporal and spatial scales: wind waves (Sect. 3.3), surface winds (Sect. 3,4), surface atmospheric pressure, river runoffs (Durand et al. 2019; Longuevergne et al. in review). Empirical formulae could be used to estimate wind-wave contributions to coastal sea level over large spatial scales and long time-scales (e.g., Melet et al. 2018) based on deep water wave characteristics (wave height, period and direction, Sect. 3.3) provided that nearshore bathymetry and beach profiles (Sects. 2.1, 2.2) are known, which is presently not the case over most of the coastal areas so that assumptions are needed (Dodet et al. 2019).

As the different processes contributing to total water level at the coast impact the coast differently, evaluating their relative importance is essential to assess the local coastline vulnerability to, e.g., flooding and erosion. However, interactions between these processes can be substantial (e.g., Idier et al. 2019). Monitoring the different relevant metocean variables to separately estimate the contribution from the main processes driving sea level changes at the coast is therefore not entirely satisfactory. Direct measurement of total water level at the coast, including the wind-wave setup and swash, is needed. Video-monitoring could offer possibilities in that regard but an adequate widespread observing system is still lacking.

Finally, in terms of flooding hazards, the sea level that matters is that relative to the land. Vertical land movements, and more specifically land subsidence, cause a relative rise of the mean sea level and have to be accounted for when evaluating coastal impacts (Ingebritsen and Galloway 2014). Coastal land subsidence, of both natural and anthropogenic origin, develops mainly at the local to regional scale. Despite the relatively small extent of the coast affected, land subsidence is becoming the dominant contributor to relative sea level rise and one of the main processes threatening low-lying deltaic zones worldwide, i.e., the coastal areas most prone to flooding (e.g., Shirzaei and Bürgmann 2018). Land subsidence in these landforms results from their continuously increasing overexploitation for urban and industrial settlements, crop production, land reclamation, and production of natural resources (e.g., hydrocarbons).

The measurements carried out over the last two decades by SAR interferometry show that land subsidence has caused in several deltas a loss of land elevation up to several tens of mm/yr, i.e., an amount much larger than the absolute sea level rise observed over the last decades. A few representative examples are provided in Fig. 7. The figure compares the sea level change observed by satellite altimetry from 1992 to 2014 with the vertical land motion in different deltas that are densely populated and characterized by a low elevation relative to the mean sea level.

SAR interferometry provides detailed maps of the vertical land motion (Sect. 2.3) and shows that the process is characterized by a significant heterogeneity that must be properly accounted for when past evolution of coastal zone is reconstructed (e.g., Corbau et al. 2019) and flooding scenarios in the future decades are developed (e.g., Shirzaei and Bürgmann 2018).

The subsurface fluid-pressure decline caused by pumping of groundwater is the main driver of land subsidence in coastal regions (e.g., Ng et al. 2012; Chaussard et al. 2013; Raucoules et al. 2013). However, natural processes, such as tectonics, glacial and sediment isostatic adjustment, natural compaction, and other anthropogenic activities like hydrocarbon production from deep reservoirs, land reclamation, marshlands drainage and conversion to farmlands, conversion of prime agricultural areas into residential and industrial, massive construction, reduction of sediment availability due to river damming and mining superpose to produce the observed displacements (e.g., Tosi et al. 2009; Ng et al. 2012). The simultaneous occurrence of processes characterized by different temporal and areal scale, together with the intrinsic heterogeneity of the hydro-geo-mechanical properties of the subsurface, explains the rationale of observed displacement variability.

Disentangling the various processes is of paramount importance. Understanding which are the main anthropogenic factors contributing to land subsidence allows defining, planning, and possibly implementing counter-measures to arrest or decrease it (e.g., Ingebritsen and Galloway 2014) and, consequently, reduce flooding hazard in coastal communities. This requires the use of specific numerical models (e.g., Zanello et al. 2011; Ye et al. 2016; Zoccarato et al. 2018) and the integration of remotely sensed information with in situ measurements provided, for example, by borehole extensometers, geotechnical lab tests on soil samples, piezometric records, geophysical logs, and sedimentation-erosion tables.

4.1.2 Monitoring and Forecasting Sea Level with Integrated Systems

As coastal sea level is both locally and remotely forced, dynamical downscaling of numerical ocean operational models is usually implemented to forecast sea level with a lead-time of a few days to a season (Wilandsky et al. 2017): coastal ocean forecasting systems (COFS, De Mey-Frémaux et al. 2019) are embedded in wider-area, coarser resolution regional operational forecasting systems (Sect. 3.7). COFS have to represent the coastal ocean continuum, from the open ocean, to shelf areas and estuaries or deltas. Coastal flooding primarily occurs during extreme sea level events, which are already more frequent (Menendez and Woodworth 2010) and are projected to become even more so in response to

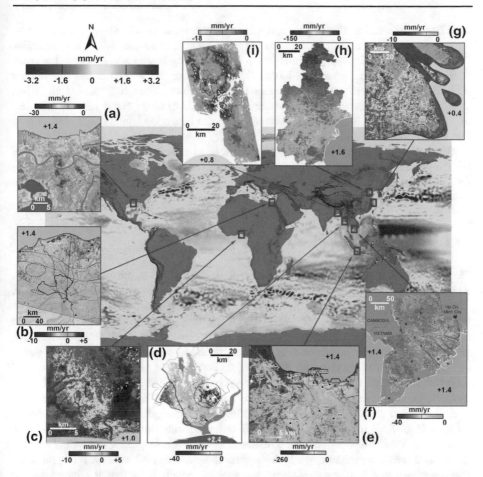

Fig. 7 Base map: total sea level change (in mm/yr) between 1992 and 2014, based on data collected from the TOPEX/Poseidon, Jason-1 and Jason-2 satellites. Blue regions are where sea level has gone down, and orange/red regions are where sea level has gone up (after NASA's Scientific Visualization Studio, https ://svs.gsfc.nasa.gov/4345). Insets **a–i**: vertical land motion (land subsidence is negative) detected by SAR interferometry in representative low-lying coastal regions: **a** New Orleans, USA (after Dixon et al. 2006); **b** Nile delta, Egypt (after Gebremichael et al. 2018); **c** Lagos, Nigeria (after Cian et al. 2019); **d** Yangon city, Myanmar (after van der Horst et al. 2018); **e** Jakarta, Indonesia (after Ng et al. 2012); **f** Mekong delta, Vietnam (after Erban et al. 2014); **g** Yangtze delta, China (after Yang et al. 2013); **h** Tianjin, China (after Zhang et al. 2016); and **i** Ganges–Bramaputra delta, Bangladesh (after Higgins et al. 2014). In each subpanel the sea level change in front of the coastline as quantified in the global map is provided for comparison

climate change (Vousdoukas et al. 2018). As extreme sea levels can be caused by the combination of different processes (Sect. 4.1.1), the regional and coastal forecasting systems should be as consistent as possible in terms of represented physical processes. The substantial interactions between the aforementioned processes (e.g., Idier et al. 2019), especially during extreme events, also advocates for coupled forecasting systems. These are under development for regional operational models (e.g., Staneva et al. 2016; Sotillo et al. 2019).

Satellite observations such as sea surface height for sea level (Sect. 3.1) and SST (Sect. 3.2) are routinely assimilated in most regional operational systems (Le Traon et al. 2019). Observing system evaluations (OSEs) have shown, in particular, the major contribution of satellite altimetry in constraining operational models and assimilating systems (e.g., Hamon et al. 2019). An accurate knowledge of the mean dynamic topography (MDT) is key for assimilating altimetric sea level in ocean forecasting systems (Le Traon et al. 2017). MDT can be inferred from the combination of altimetric and gravimetric (GOCE, GRACE missions) satellite data and in situ observations (e.g., Rio et al. 2014). Yet, MDT remains less accurate in the coastal ocean due to larger errors in altimetric data in coastal areas (Sect. 3.1) and to the scarcity of in situ data upon which scales smaller than 100 km are constrained in MDT estimates.

Given the complex combination and interactions of processes driving extreme sea levels, the short spatial scales and high-frequency nature of coastal zone evolution, forecasting total water level changes at the coast on spatial and temporal scales relevant for coastal zone management remains, however, challenging (Ponte et al. 2019).

4.1.3 From Total Water Level to Flooding

Flooding in low-lying areas can take place due to the following processes: (1) overflow, which is a flooding process during which extreme water levels exceed the heights of coastal natural or artificial defences; (2) overtopping, which occurs when extreme water levels do not exceed the height of coastal defences, and flooding occurs intermittently due to the effects of waves; (3) breaching of coastal defences or erosion of coastal dunes. Ultimately, water flows within the inundated area can accumulate in the lowest areas and the flooding can be enhanced by rain and groundwater. Today, these processes can be modeled accurately, provided that accurate bathymetry, topography (Sects. 2.1, 2.2), wind fields (Sect. 3.4) and land cover (Sect. 2.4) are available (USACE 1996; IGOS 2006; Bates et al. 2005; Le Roy et al. 2015; Ablain et al. 2016). Flooded areas can be monitored via satellite imagery (e.g., SAR data; Twele et al. 2016), and other relevant observations include for example the timing of the flood event, extreme water levels at gauges and along walls in the flooded area. Furthermore, while this article addresses coastal hazards, it can be noted that observations of damages after events are relevant as well, in particular because current flood vulnerability functions are highly uncertain and would deserve further calibration (e.g., Hallegatte et al. 2013; André et al. 2013; Hinkel et al. 2014).

Flooding hazards are leading to increased risks as the exposure of population and assets in the coastal zone are growing, sea level is rising in response to climate change, and high subsidence rates in many densely populated areas (e.g., Hallegatte et al. 2013; McGranahan et al. 2007; Neumann et al. 2015).

In addition to floods, relative sea level changes in the coastal ocean can have diverse adverse impacts such as salinization of aquifers, unsafe harbour operations, loss of coastal wetlands (e.g., salt marshes, mangroves), and degradation of ecosystems.

4.2 Shoreline Changes

While erosion and shoreline changes are often used interchangeably, they actually have different meanings (Le Cozannet et al. 2014). Erosion can be defined as a process involving morphological changes such as sediment transport or abrasion of rocks; shoreline changes commonly refer to the motion of a particular shoreline indicator, such as the base or the top of a dune or cliff, or the mean or high water lines on a beach

(Boak and Turner 2005). Hence, depending on the shoreline change proxy considered, coastal erosion may not necessarily result in shoreline changes.

Shoreline changes are typically monitored using a combination of in situ and remote sensing data, including historical charts and photographs, aerial photography, beach surveys (e.g., using GNSS or video imagery), and satellite images (Boak and Turner 2005). The manual and semi-automatic handling of such data within geographic information systems is extremely time consuming, so that automated shoreline detection algorithms using satellite images are receiving much attention. Where the shoreline evolves rapidly (i.e., more than 10 m between two acquisitions), these algorithms have demonstrated their capacity to quantify land area changes, such as in large deltas in South-East Asia (e.g., Shearman et al. 2013). However, most shorelines are currently evolving at rates not exceeding 1 m/year (Bird 1987). Reaching such accuracy is quite demanding, although promising results have been recently obtained by applying automated shoreline detection algorithms to the Google high-horizontal resolution imagery database (Luijendijk et al. 2018; Mentaschi et al. 2018; Fig. 5c).

Shoreline changes involve processes acting at multiple timescales ranging from multicentennial to extreme events (Ranasinghe 2016). The related hazards are generally seen as a threat to infrastructures and their stability and sometimes to coastal ecosystems. The associated prevention measures involve protecting the shorelines by means of hard or soft engineering (e.g., dikes, groins, sand nourishment), but may also require relocating assets such as buildings or transport infrastructures. There are two critical challenges associated with managing shoreline changes hazards today (Wong et al. 2014): first, shoreline retreat is now perceived as an increasing threat since it affects a number of urbanized beaches and cliffs; second, climate change and sea level rise will favor shoreline retreat over the twenty-first century and beyond, to an extent that remains largely unknown today.

Recent research in shoreline change monitoring has significantly improved the contribution of satellite remote sensing to managing shorelines and understanding the ongoing processes. First, recent global assessments of ongoing shoreline changes have confirmed the current paradigm that human intervention are the most obvious cause of observed shoreline changes worldwide (Mentaschi et al. 2018). Second, while sea level rise impacts on shoreline changes remain undetectable in most temperate and tropical areas (Cramer et al. 2014; Duvat 2019), early signs of transition toward shoreline changes are displayed in some coastal sites, which in some cases are suspected to be due to subsidence or to combined effects of sea level rise and waves (Albert et al. 2016; Garcin et al. 2016; Oppenheimer et al. 2019). This improved understanding of the current status of world's shorelines is based on a combination of field work and the analysis of aerial photographs and satellite images. Over the coming decades, breakthroughs in the area of shoreline change monitoring is expected to take place with the automatization of shoreline feature detection procedures. Furthermore, as accelerating sea level rise (Ablain et al. 2019; Oppenheimer et al. 2019) will favor coastal erosion, observations will be key to validate shoreline evolution models, which are still subject to large uncertainties (Toimil et al. 2020).

4.3 Maritime Security Hazards

Different human activities are conducted at sea, such as shipping, fishing, drilling, or production of marine energy (Fig. 5d–h).

Global trade is mostly conducted by shipping at sea, representing more than 80% of trade volume (UNCTAD 2017). Shipping lanes are becoming busier, and are dense in

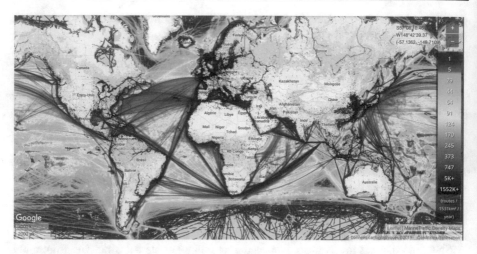

Fig. 8 2016–2017 density map of vessels. Map extracted from www.marinetraffic.com

coastal zones (Figs. 5d, 8). Industrial fishing occurs in more than 55% of the ocean area, with a spatial extent corresponding to more than four times that of agriculture (Kroodsma et al. 2018, Fig. 9).

Hazards related to shipping and fishing include accidental spills or operational discharges of oil and chemicals, dumping of waste (e.g., Liubartseva et al. 2018), and accidents (Fig. 1). Such hazards are rather concentrated in busy shipping lanes and ports (Figs. 8, 9), and are therefore affecting the coastal zone.

Accidents at sea can be induced by humans, but also by natural hazards such as large-amplitude or steep wave and sea-ice conditions. For instance, twenty-two super-carriers sunk due to collisions with rogue waves over the 1969–1994 period in the Pacific and Atlantic oceans, causing 525 fatalities (Kharif and Pelinovsky 2003). Accidents at sea are triggering search and rescue operations during which a search area has first to be determined, with the majority of rescue cases taking place less than 40 km from the shore (Breivik and Allen 2008).

The last known position of the vessel is an essential element to determine a search area. Different regulations require particular classes of vessels to be equipped with ship-borne transponders to transmit their identity and position at repeated intervals. Such tracking systems include the Automatic Identification System (AIS, IMO 2000), Long-Range Identification and Tracking (IMO 2006) and Vessel Monitoring System (FAO 1998). Although such tracking systems provide a surge of information (e.g., 22 billion AIS messages corresponding to more than 70,000 industrial fishing vessels where processed from 2012 to 2016 in Kroodsma et al. 2018), they do not provide a comprehensive detection of vessels (e.g., unidentified and non-cooperative ones). EO such as satellite-borne SAR data (e.g., Sentinel-1; Fig. 10; Greidanus et al. 2017; Santamaria et al. 2017) and optical satellite imagery have thus become essential to detect vessels (see a review by Kanjir et al. 2018). Yet, in the coastal zone, SAR-based ship detection is more prone to false alarms arising from azimuth ambiguities (re-occurrence of a target at a determined distance from the original target stemming from the pulse repetition frequency of the data sampling) due to the potential presence of strong reflectors on land and to low wind conditions in the coastal zone (e.g., Velotto et al. 2014; Fig. 10). Recently, the potential of ship detection by the Sentinel-5P Tropomi instrument (a high-resolution,

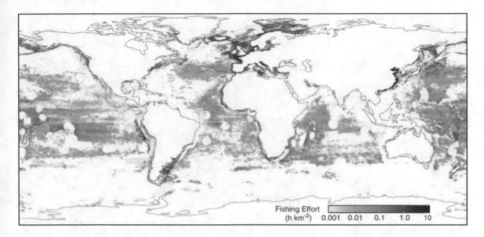

Fig. 9 Total fishing effort [hours fished per square kilometer (h km^{-2})] in 2016 by all vessels with AIS systems. Adapted from Kroodsma et al. (2018)

advanced multispectral imaging spectrometer), through their emission of nitrogen dioxide, has been highlighted. Yet, as the drift of an object lost at sea depends on its drift properties (notably the shape of the object), identifying the vessel type and size together with their position is also of importance. This is more challenging as it requires high-resolution satellite images to estimate different characteristics of the vessel. Monitoring fishing fleets' activity in particular could also support a sustainable use of the ocean by reducing overfishing and associated hazards of reduction of food production and biodiversity, and impairment of ecosystems functioning (UN 2015).

An object lost at sea is subject to drift forced by surface currents and drift predicted from leeway, the latter being defined in Breivik et al. (2013) as the motion of the object induced by wind (10 m reference height) and waves relative to the ambient current (between 0.3 and 1.0 m depth). Search and rescue operations therefore largely rely on Lagrangian pathways calculations, which depend on near-real time information on surface winds (Sect. 3.4), ocean surface currents (Sect. 3.5) and waves (Sect. 3.3) (van Sebille et al. 2017), with high spatiotemporal resolution. Indeed, the pathways are sensitive to currents on scales from meters to kilometers (e.g., D'Asaro et al. 2018). EO and/or their integration in operational models (Sect. 3.7) therefore provide essential information for the transport of objects lost at sea.

4.4 Maritime Pollution

4.4.1 Oil, HNS Spills, Plastics

Ships are transporting different substances such as oil and harmful and noxious substances (HNS) that, if spilled in the ocean, can create hazards to marine flora, fauna and/or human health, can interfere with other legitimate uses of the sea (tourism, fisheries, navigation) and can result in environmental damages. More than 2000 different HNS are regularly transported by sea, including acids, liquefied natural and petroleum gases, ammonia, benzene, palm and vegetable oils. Spills can occur from ships due to sinking, grounding or cracking, notably under bad weather conditions; from incidents on-board ships (e.g., fire, mechanical failure, human error); or from port installations. The Cedre database (http://

Surveys in Geophysics (2020) 41:1489–1534

Fig. 10 An example of ship detection with Sentinel-1, offshore Taragona's port (Spain). The green line shows the land/sea mask used by the detection algorithm. Echoes (bright spots in the image) on the ocean are mostly due to azimuth ambiguities from bright reflectors on land. Spots that were bright enough were detected by the detection algorithms (blue rectangles and green triangles). Red dots indicate targets that were automatically recognized as azimuth ambiguities. The only three real targets (green triangles) correspond to a buoy (at the image center) and to two ships (top right). From Greidanus et al. (2017)

wwz.cedre.fr) on spill incidents gives a median annual number of 29 spill incidents (with a volume > 10 m^3) over 2004–2013. For years 2013–2015, the total spill volume of 89,000 tons was distributed with 25% in offshore, 66% in inshore, 1% in estuaries and 8% in ports incidents. Spill incidents for HNS are less frequent than for oil. Despite increasing oil shipping over the last 50 years (e.g., from 1500 to more than 3000 million metric tons from 1970 to 2018, source: UNCTADStat), oil spill incidents have drastically decreased (with an average of nearly 80 spill incidents per year in the 1970s to 6 in the 2010s, source: ITOPF Oil Tanker Spill Statistics 2019). Once spilled, oil and HNS can evaporate, be dissolved, float or sink, depending on their properties.

Another source of pollution comes from marine debris. Plastics account for 80% of marine debris and cause environmental as well as health hazards (i.e., ingestion, harm to marine life, invasive marine bacteria or organisms that can disrupt ecosystems). Marine plastic sources are mostly land-based (i.e., runoff, sewer overflows, coastal tourism, industrial activities, paints, tyre dust; Jambeck et al. 2015) but marine-based sources (e.g., fishing and shipping litter, aquaculture) are not negligible. Only a small fraction of plastic litter floats close to the ocean surface. Most of plastic is fragmented into micro or nanoplastics through the action of solar radiation and marine conditions (wind, currents), sink in the ocean, is ingested by fishes or marine mammals, which represent a hazard for their health (e.g., Guerrini et al. 2019) or ends up on beaches (van Sebille et al. 2020).

As for search and rescue activities (Sect. 4.3), information on the sources of pollutants and marine debris, on the near surface wind, currents, and waves are crucial to perform Lagrangian calculations of their pathways (e.g., van Sebille et al. 2017; Maes et al. 2018). Additional properties for oil spill pathway modeling are needed, such as turbulent diffusivities or absorption by the coastal environment (De Dominicis et al. 2013).

EO can directly monitor oil spills thanks to the different emissivity, reflectance of water and oil (optical sensors), and to the reduction of the sea surface roughness induced by oil films (the oil dampens capillary waves generated by the winds on the ocean surface, that define sea surface roughness, Sect. 3.4), leading to changes in backscattering and brightness on SAR data under favorable metocean conditions (excluding low wind conditions). Visible, infrared, microwave and radar sensors such as spectroradiometers (e.g., Aqua-MODIS, Terra-MISR), spectrometers (ENVISAT-MERIS), multispectral radiometers (Sentinel-2) have thus been used to monitor oil spills, but SAR sensors have been primarily used (see a review on remote sensing of oil spill by Fingas and Brown 2018). An illustration of oil spill monitoring with SAR on-board Sentinel-1 is given on Fig. 11.

For more information on marine pollution and its monitoring from space, the reader is referred to Viatte et al. (2020).

4.4.2 Underwater Noise

The ocean is not quite a silent world. Water transmits sound efficiently and different natural sources of sound (breaking surface waves, rain, thunder, sediment movement, earthquakes or marine life, etc.) fill the ocean to create an ambient noise. Many marine species critically use sounds to navigate, communicate, find prey or reproduce. Yet, oceans are becoming noisier due to additional sources of noise induced by human activities such as shipping, use of sonar, pile driving, seismic surveys used to map the geological structure beneath the seabed for research or commercial purposes (e.g., prior to coastal construction, Fig. 5f), offshore marine energy development (wind farms, wave, tidal or current energy) (Fig. 5g–h), oil and gas exploration and operations, etc. Recreational shipping and industrial activities are particularly dense in coastal waters (Figs. 5, 8).

Anthropogenic underwater noise can induce adverse effects on marine life (Hildebrand 2005), although understanding and quantifying these impacts remains challenging (e.g., Popper and Hastings 2009; Slabbekoorn et al. 2010; Hawkins and Popper 2017). Effects depend on the noise intensity, duration and frequency, on marine animal characteristics (such as specie, sex, age) and range from temporary or permanent hearing impairment, displacement to quieter areas (which could lead animals to non-suitable ecosystems), acoustic masking (which could reduce ability to catch preys and communicate—including with offspring), increased stress levels, and can lead to the animal death. Underwater noise pollution has consequently been a growing concern and regulations are emerging at national (Merchant et al. 2016) and international levels (e.g., UN 2018; European Commission 2008—MSFD descriptor 11). As a result, noise mitigation and abatement recommendations and policies are also being created (e.g., Merchant 2019).

While EO cannot be used to directly assess underwater sound and related ecological impacts, they are beneficial for modeling and forecasting underwater noise propagation and

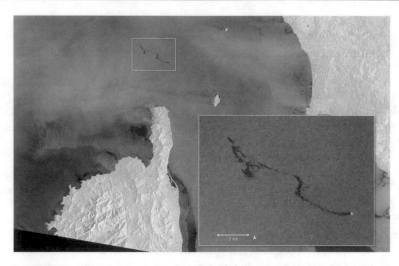

Fig. 11 Monitoring of an oil slick (dark area in the white rectangles) by Sentinel-1. The slick occurred after a collision between two merchant ships in the Mediterranean Sea on October 7, 2018. Credit: contains modified Copernicus Sentinel data (2018), processed by ESA, CC BY-SA 3.0 IGO

marine life exposure (e.g., Farcas et al. 2016; Roberts et al. 2016). Indeed, the propagation of sound is influenced by the 3D water density (i.e., temperature, salinity), which can be modeled and forecasted using operational ocean numerical models that assimilate satellite data (in particular satellite altimetry) (Sect. 3.7), and bathymetry, which can be derived by satellite (Sect. 2.1). Sources of noise can be inferred from maritime surveillance (AIS and SAR imagery, Sect. 4.3) and environmental data (e.g., for sea state, wind, Sects. 3.3 and 3.4). Maps of marine animal distribution data needed to estimate marine life exposure to noise can be inferred based on candidate covariates (Fig. 11). For instance, Roberts et al. 2016 used (i) physical oceanographic covariates such as SST, distance to SST fronts, wind speed, total and eddy kinetic energies, and distance to geostrophic eddies derived from sea surface height observations, and (ii) biological covariates such as chlorophyll concentration, primary production, and potential biomass and production of zooplankton and epipelagic micronekton. Most of these covariates were derived from satellite observations.

4.5 Water Quality

Although they represent only 7% of the total ocean surface, coastal and inland water zones produce up to 40% of the marine and freshwater biomasses inventoried today and 85% of the marine and freshwater resources exploited by humans. Moreover, inland waters provide key ecosystem services with direct linkages to human health (IOCCG 2018). Therefore, it is vital to study these waters in a systematic way and with a long-term perspective to characterize the variability of biogeochemical properties and to understand their impacts on the water quality. The EU Marine Strategy Framework Directive (MSFD) which is complementary to the European Water Framework Directive (WFD), concerning waters in estuarine transitional areas and very close to the coast, introduces the concept of "European waters" with the aim of reaching a good ecological state of the marine environment in its entirety, and the improvement of marine biodiversity conservation. Reaching this goal requires a global framework for coordinating state-members including a substantial effort in observations and surveys of waters to establish environmental baselines allowing us to understand ongoing and future developments in water quality.

The deterioration of surface water quality by contaminants, nutrients, excess heat, and other factors (Fig. 1) is arguably the greatest threat to future healthy water availability. Health studies show that water-related illnesses are a major global problem. Clearly suitable water quality is critical to sustaining life on our planet. The UN Millennium 2015 Goals highlight the importance of accessible freshwater and sanitation for human health. Clean water is critical not only to human health, but to overall ecosystem health. Monitoring and detecting change is critical for the protection of biodiversity in natural areas (e.g., Natura 2000[1]).

The monitoring of the water quality can be obtained by using ocean optics or ocean color radiometry (OCR), as described in Sect. 3.6 and recently reviewed in Chawla et al. (2020). Examples of the use of OCR to monitor water quality are provided in IOCCG (2018). Here we describe two examples over Vietnamese coastal waters (Loisel et al. 2014, 2017). These studies aimed at analyzing the spatiotemporal variability of the chlorophyll-a concentration and suspended particulate matter over Vietnamese coastal waters and Mekong delta respectively, using the MERIS archive between 2002 and 2012. The time series were decomposed into three terms (seasonal, trend, irregular term) and analyzed with regard to regional oceanographic and hydrologic conditions. The chlorophyll-a concentration showed a long-term monotonic trend from 2 to $> 5\%$ yr^{-1} in different coastal areas where aquaculture activities exhibited an increase in production weight (from 31 to 113%). For the Mekong delta, the suspended particulate matter concentration showed a long-term trend of about -5% yr^{-1} in the pro-delta area. This decreasing trend was linked to the decrease of the Mekong river sediment output during the high flow season. These studies show the interest of EO for surveying the changes in biogeochemical parameters that are proxies of water quality and to link their variabilities to physical forcings that are also obtained from EO. It highlights the interest of synergistic use of different EO observations to understand the variability of water quality in coastal zones.

4.6 Marine Ecosystems Shifts

Although not as evident as flooding or hurricanes, ecological shifts may be considered hazards, albeit of a subtle type. Ecological shifts—also known as regime shifts, critical or catastrophic transitions, phase shifts, abrupt shifts—are sudden, dramatic and persistent changes in the structure and function of an ecosystem. They signify a profound restructuring of the trophic web, often into a novel, or alternative state, difficult to reverse because of changes in the ecosystem internal feedback mechanisms. As a result, they often impact the ecosystem services, i.e., the benefits human populations obtain from ecosystems, and can disrupt economies and societies (Scheffer et al. 2001; Scheffer and Carpenter 2003; Scheffer 2009; Biggs et al. 2012; Conversi et al. 2015; Möllmann et al. 2015; Lauerburg et al. 2020; Swingedouw et al. in review).

Coastal marine habitats (beaches, salt-marshes and ponds, mangroves, estuaries, deltas and intertidal habitats, seagrass beds, kelp forests, coral reefs, salt-marshes and ponds, semi-enclosed seas) are degrading in many places (Hoegh-Guldberg et al. 2014).

---

[1] https://ec.europa.eu/environment/nature/natura2000/.

Ecological shifts have been found in all habitats where decadal time series exist (Moell-mann et al. 2011, 2015; Beaugrand et al. 2015; Gattuso et al. 2015). While the history of life reveals repeated planetary-scale tipping points, the risk of additional, possibly multiple, transitions is increasing (Hughes et al. 2013; Rocha et al. 2018; Steffen et al. 2018; Lenton et al. 2019). As combined human actions are now simultaneously increasing global temperatures and decreasing biodiversity, threatening 1 million species with extinction (https ://www.un.org/sustainabledevelopment/blog/2019/05/nature-decline-unprecedented-repor t/; IPBES 2019), we can reasonably expect more social-ecological regime shifts in the near future. We therefore need to use all available tools to monitor and forecast them. We discuss below how EO can help monitor and protect coastal ecosystems, and contribute to ecosystem shift early detection, by focusing on three examples: coral reefs, coastal marshes and pelagic ecosystems.

4.6.1 Coral Reefs

Coral reefs are some of the most biodiverse habitats of this planet, providing shelter to 25% of all fish species. However, in the last decades many coral reefs around the globe have shifted from the coral state to an alternate state of fleshy algae, from which recovery is rare (Hughes 1994; Bellwood et al. 2004). Bleaching occurs when waters are too warm and corals expel their symbiotic algae and turn white, and can be reversed if the warm period is short. Starting in the 1980s, mass bleaching phenomena have co-occurred around the globe, indicating the common warming stressor. Coral reefs are currently under multiple stress (ocean warming, acidification, local predators and diseases) in several locations worldwide, including the Great Barrier Reef (Bellwood et al. 2004, 2019; Normille 2016; Hughes et al. 2017). At the current rate of warming and associated predicted increases in marine heat waves (Oliver et al. 2019), it is under debate whether coral reefs will become extinct by the end of this century (Matz et al. 2018).

In the case of coral reefs, satellites have already contributed essential information for understanding some mechanisms, and they continue to monitor reef extension and to provide forecasts of the risk of bleaching (Nim et al. 2010; Kamenos and Hennige 2018). For example, analysis of the global satellite SST data (Sect. 3.2) has indicated the relative change in cool versus warm days: over 1985–2012 the warm-temperature periods have increased while the cool-temperature periods (which carry potential for reef recovery) have decreased (Heron et al. 2016). This information is essential for evaluating the risk of no-recovery. Global observation programs, such as the NOAA Coral Reef Watch, currently provide near-real-time satellite SST data and algorithms for monitoring thermal stress risk (Fig. 12).

4.6.2 Coastal Marshes

As human impacts grow and sea level rises, coastal marshes can become endangered. The Wadden Sea provides an emblematic example of this threat. This coastal strip located offshore The Netherlands, Germany and Denmark is the largest unbroken tidal flat area of the temperate world. It provides shelter and feeding grounds for numerous species and food provisions in this area are 10–20 times higher than in adjacent deeper waters (Reise et al. 2010). Because of its unique characteristics, it has been designed a UNESCO World Heritage Site (https://whc.unesco.org/en/list/1314/). Over time, human impacts (habitat destruction, overexploitation, pollution, sand extraction, climate change and associated sea level

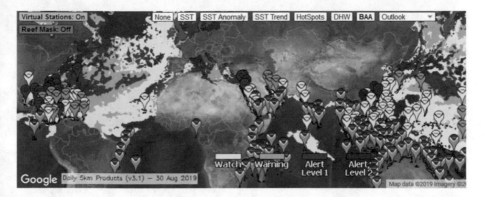

Fig. 12 Example of Bleaching Alert Area map—5 km Regional Virtual Stations, providing high resolution information on a global scale. From NOAA Coral Reef Watch, 2019, updated daily. NOAA Coral Reef Watch Version 3.1 Daily Global 5 km Satellite Coral Bleaching Heat Stress Alert Area Product overlaid on the Daily 5 km Satellite Regional Virtual Stations Map, Aug. 30, 2019. College Park, Maryland, USA: NOAA Coral Reef Watch. Data set accessed 30-Aug-2019 at https://coralreefwatch.noaa.gov/vs/map.php)

rise) have profoundly modified this ecosystem (Lotze et al. 2005; Eriksson et al. 2010). In particular, ecosystem regime shifts occurred in this sea in 1979 and 1988 (Weijerman et al. 2005; de Young et al. 2008; Conversi et al. 2010).

EO can provide essential large-scale (spatial) and continuous (temporal) information for monitoring and early detection of changes in coastal wetlands. For example, Copernicus Sentinel satellites images have been used to identify areas with mollusc abundance in the Wadden Sea (NEREUS/ESA/EC 2018). Copernicus Sentinel-1,2 and Sentinel-3 are currently providing data (on erosion and shoreline, and ocean color, respectively) for 3D assimilative models targeted to understanding shifting trends in this area (https://www.ecopotential-project.eu/).

4.6.3 Pelagic Ecosystems

Pelagic ecosystem shifts are more complex to investigate, as they do not pertain to fixed locations. While phytoplankton abundance has been monitored for decades via OCR using ocean color as a proxy for its abundance (Sect. 3.6), animal detection is more difficult—zooplankton is transparent and nekton is often well below the surface.

However, EO of proxies related to animal abundance can greatly extend the limits of in situ observations over the global ocean. For instance, global temperature data have been used in conjunction with the species thermal ecological niches to predict abrupt shifts (Beaugrand 2015; Beaugrand et al. 2015, 2019) and have shown that, although apparently rare, on a global scale ecological shifts can actually be widespread and ubiquitous (Fig. 13).

While essential, satellite data are currently generally underused by the biodiversity community. Three factors remain essential for a wider usage: continuity, affordability, and access of the data (Turner et al. 2015). For the detection and monitoring of tipping points, another factor is crucial: decadal duration. Indeed, ecological shifts are by definition a transition from one stable state, lasting several years, to another state, and at this scale an abrupt shift lasts a few years. As satellite data collection now spans decades for some

Fig. 13 Global abrupt shifts occurrences over the period 2010–2014, using species simulations based on METAL theory. Modified from Beaugrand et al. (2019). The color bar shows the percentage of individual species undergoing a significant shift in a given community: 50% means that half the simulated species exhibited a significant shift, according to the index of abruptness described in Beaugrand et al. (2019)

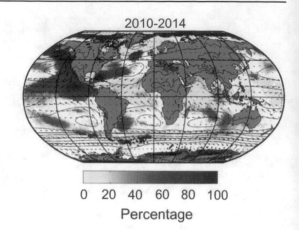

variables, we are confident that it will become increasingly used for ecological shifts detection and early warnings.

5 Conclusions and Perspectives

Satellite observations offer great potential for a long-term, synoptic, and rather high-frequency monitoring of the Earth's surface, thanks to a variety of sensors. The use of EO to monitor coastal metocean conditions, coastland hydro-geo-morphological setting, and hazards has significantly developed with the increasing number of satellites with radar sensors. Nowadays, Copernicus is the largest producer of freely and openly available EO data (Sentinels) in the world.

Recently, in the framework of OceanObs'19, numerous community white papers have provided a state of the art of EO of different metocean variables that are relevant for coastal marine hazards. In the present article, we focused on the coastal zone, and more specifically on EO monitoring of drivers of a set of coastal marine hazards, and of the corresponding hazards themselves. In spite of their obvious value, EO are challenged in the coastal zone for different reasons, including:

- Land contamination in the sensor footprint, preventing the retrieval of meaningful data very close to the coast. The size of the footprint determines the size of the land-contaminated area and depends on the sensor technology and frequency.
- Geophysical corrections tend to be less accurate in the coastal zone. For instance, there is a greater variability in atmospheric water vapor, temperature and aerosols in coastal areas than over the open ocean. Absorption and emissivity of these atmospheric components can alter the electromagnetic wave propagation and energy spectrum as received by the satellite-sensor.

In addition, the coastal ocean is a dynamic area, with high spatial and temporal variability, enhanced diurnal variability, and with other processes to be accounted for when compared to the open ocean. These features directly affect the different metocean variables that can be monitored by satellites (SST, SSH, wind, currents, waves, ocean color, etc.), and coastal hazards (e.g., pathways of floating debris and pollutants, dispersal of larvae, marine navigation safety, etc.). Coastal hazards can also be highly localized in space and time (e.g., marine floods due to cyclones are lasting O(1 day)). This leads to sampling issues, since catching localized events via along-track observations with a revisit time of several days is unlikely. Coastland geomorphological processes and features (e.g., land subsidence, topography) are generally characterized by a low temporal variability but a significant spatial heterogeneity.

These coastal ocean and hazard specificities guide requirements for future EO missions:

- Higher revisit frequencies (e.g., hourly to daily) to cover more systematically synoptic scales (e.g., using geostationary satellites),
- Higher spatial resolution to observe fine-scale coastal processes,
- Multi-band or multi-frame (e.g., stereo or tri-stereo over tens of seconds) observations to capture rapid evolutions and provide information on wave propagation, surface currents or drifting objects, harmful algal blooms.

A synergetic use of different observations can improve estimates of relevant variables for coastal hazards, as illustrated for surface currents through combined use of altimetric sea level, surface wind and SST observations. In addition, joint observations of coupled air–sea variables such as total surface currents, wind speed and direction, and directional wind-wave spectrum could bring further improvements in coastal ocean monitoring and further insights on various processes. In that respect, several proposals for satellite Doppler oceanography missions have been developed (e.g., SKIM, Ardhuin et al. 2018; WaCM, Rodriguez 2019; SEASTAR, Gommenginger 2019), each one targeting specific processes at different scales (e.g., Villas Bôas et al. 2019). Although these missions are not planned yet, they respond to actual requirements of coastal users.

However, acquiring EO data is not sufficient to monitor hazards in the dynamic coastal zones, which are characterized by shorter-scale variability. To provide relevant information for users, including coastal zone management, a complex added value chain starting from raw EO (satellite, in situ) has to be constructed. First, data have to be processed to provide level-3 or level-4 quality-controlled products. An across-variable, coherent and synoptic spatiotemporal monitoring of the coastal ocean can be reached when EO are assimilated jointly with in situ observations or other remotely sensed observations in operational, numerical models. Such integrated systems are the backbone of metocean condition forecasts in the open and coastal oceans which, by themselves or by backing early warning systems, are essential to reduce exposure to coastal hazards and related risks. Coastal forecasting skills have been shown to increase when near-real time EO are assimilated. They would further benefit from (i) joint observations of different variables, especially in the context of developing more integrated, more coupled operational monitoring and forecasting numerical systems, (ii) a better exploitation of EO by using more types of observations and more effective, potentially coupled data assimilation techniques, but also for model development and validation.

On top of this integrated system layer, additional layers can be added to go towards users. This includes the development of (i) relevant sectoral indicators, assessment and reporting activities, and (ii) service activities, such as user desk, user support and interaction with users to provide products that better address their needs. This is the rationale for the development of EU Copernicus Services, in addition to the Copernicus satellite component (Sentinels). Of particular importance for the monitoring of coastal metocean conditions and hazards are the marine and land services of Copernicus (aka CMEMS and CLMS, respectively). These two complementary services are developing their offer for coastal zone users. For instance, CLMS has launched the development of a thematic hotspot product for LCLU in the coastal zone, to capture landscape dynamics at high horizontal resolution over all EU coastal territory (landward distance of 10 km from the coastline, with 0.5 ha resolution, and more than 60 classes of LCLU that are adapted to the coastal zones). CMEMS is currently providing satellite level-3 and -4 data for the metocean variables addressed here, as well as global and regional reanalyses, nowcasts, and forecasts of the ocean physical (including waves) and biogeochemical states. CMEMS is continuously evolving with more coupled models, increased resolution, improved algorithms to better process satellite observations in the coastal zone, etc. CMEMS will keep evolving in the long-term to monitor and forecast the ocean at finer scales, to improve the monitoring of coastal zones, and to provide a more seamless and better forcing for very-high resolution coastal models (Le Traon et al. 2019). Finer-scale upper-ocean dynamics and surface currents monitoring and forecasting in particular is essential for key hazards such as maritime safety, maritime transport, search and rescue, drift modeling, riverine influence in the coastal environment, pollution monitoring and offshore operations. Yet, regional high-resolution modeling (e.g., (sub)kilometric scale) comes with challenges for the integration of high-resolution satellite (e.g., wide-swath altimetry) and in situ (e.g., HFR) observations. CMEMS and CLMS are also targeting a better dynamic characterization of the coastal zone (Sect. 2) and of river discharges monitoring and forecasting.

Artificial intelligence (AI) brings new perspectives in terms of processing of dense EO datasets in a drastically reduced time, but also in solving complex coastal environmental problems (Shulz et al. 2018) such as monitoring of worldwide coastal sea surface temperature and salinity at high-resolution (Medina-Lopez and Ureña-Fuentes 2019), classification of land use/cover (Aghighi et al. 2014), seagrass distribution in coastal waters (Perez et al. 2018), bathymetry inversion (Danilo and Melgani 2019; Benshila et al. 2020) and sediment transport and morphological evolution (Goldstein et al. 2019). All these methods are now applicable to spatial time series and can be transposed to most of EO satellite sensors when applied to coastal variables. In a more advanced scenario, these techniques can be applied to short-term forecasting such as for early warning systems, or even ENSO event predictions (Ham et al. 2019).

As some marine coastal hazards are becoming more frequent and widespread (e.g., coastal flooding, ecosystem degradation) and will be aggravated by climate change-induced ocean warming, acidification, deoxygenation, productivity decrease, sustained coastal zone monitoring programs, forecasting and early warning systems will be increasingly needed to reduce risks for the coastal zone environment and population and to guide coastal zone management.

Acknowledgements This article arose from the international workshop on "Natural and man-made hazards monitoring by the Earth Observation missions: current status and scientific gaps" held at the International Space Science Institute (ISSI), Bern, Switzerland, on April 15–18, 2019. The authors are grateful to Anny Cazenave and Teodolina Lopez for taking charge of the implementation of this special issue. The authors thank Thomas Folegot for helpful discussions, and Pierre-Yves Le Traon, Antonio Reppucci and two anonymous reviewers for comments leading to improvements in the manuscript.

References

Abileah R (2006) Mapping shallow water depth from satellite. In: Proceedings of the ASPRS annual conference, Reno, Nevada, pp 1–7

Ablain M, Becker M, Benveniste J, Cazenave A, Champollion N, Ciccarelli S, Jevrejeva S, Le Cozannet G, Leornardi N, Loisel H, et al (2016) Monitoring the evolution of coastal zones under various forcing factors using space-based observing systems, International Space Science Institute (ISSI) Forum reports. http://www.issibern.ch/forum/costzoneevo/wp-content/uploads/2016/11/ISSI-forum_Coastal_White-paper_18nov2016_Final.pdf. Accessed Aug 2019

Ablain M, Meyssignac B, Zawadzki L, Jugier R, Ribes A, Cazenave A, Picot N (2019) Uncertainty in satellite estimate of global mean sea level changes, trend and acceleration. Earth Syst Sci Data 11:1189–1202. https://doi.org/10.5194/essd-11-1189-2019

Aghighi H, Trinder J, Lim S, Tarabalka Y (2014) ISPRS annals of the photogrammetry, remote sensing and spatial information sciences. Gottingen 2(8):61–68. https://doi.org/10.5194/isprsannals-II-8-61-2014

Albert S, Leon JX, Grinham AR, Church JA, Gibbes BR, Woodroffe CD (2016) Interactions between sea-level rise and wave exposure on reef island dynamics in the Solomon Islands. Environ Res Lett 11(5):054011. https://doi.org/10.1088/1748-9326/11/5/054011

Almar R, Bergsma EWJ, Maisongrande P, de Almeida LPM (2019) Wave-derived coastal bathymetry from satellite video imagery: a showcase with pleiades persistent mode. Remote Sens Environ 231:111263. https://doi.org/10.1016/J.RSE.2019.111263

Almeida LP, Almar R, Bergsma EWJ, Berthier E, Baptista P, Garel E, Dada OA, Alves B (2019) Deriving high spatial-resolution coastal topography from sub-meter satellite stereo imagery. Remote Sens 11(5):590. https://doi.org/10.3390/rs11050590

Alpers W, Hennings I (1984) A theory of the imaging mechanism of underwater bottom topography by real and synthetic aperture radar. J Geophys Res 89(C6):10529–10546

Alpers WR, Rufenach CL (1979) The effect of orbital motions on synthetic aperture radar imaging of ocean waves. IEEE Trans Antennas Propag 27:685–690

André C, Monfort D, Bouzit M, Vinchon C (2013) Contribution of insurance data to cost assessment of coastal flood damage to residential buildings: insights gained from Johanna (2008) and Xynthia (2010) storm events. Nat Hazards Earth Syst Sci 13(8):2003–2012. https://doi.org/10.5194/nhess-13-2003-2013

Antoine D (1998) Apports de la télédétection spatiale à la "couleur de l'océan" à l'océanographie. Océanis 24:81–150

Ardhuin F, Aksenov Y, Benetazzo A, Bertino L, Brandt P, Caubet E et al (2018) Measuring currents, ice drift, and waves from space: the sea surface kinematics multiscale monitoring (SKIM) concept. Ocean Sci 14:337–354. https://doi.org/10.5194/os-2017-65

Ardhuin F, Stopa JE, Chapron B, Collard F, Husson R, Jensen RE, Johannessen J, Mouche A, Passaro M, Quartly GD, Swail V, Young I (2019) Observing sea states. Front Mar Sci 6:124. https://doi.org/10.3389/fmars.2019.00124

Ban Y, Gong P, Giri C (2015) Global land cover mapping using Earth observation satellite data: recent progresses and challenges. ISPRS J Photogramm Remote Sens 103:1–6. https://doi.org/10.1016/j.isprs.2015.01.001

Bates PD, Dawson RJ, Hall JW, Horritt MS, Nicholls RJ, Wicks J, Hassan MAAM (2005) Simplified two-dimensional numerical modelling of coastal flooding and example applications. Coast Eng 52(9):793–810. https://doi.org/10.1016/j.coastaleng.2005.06.001

Beaugrand G (2015) Theoretical basis for predicting climate-induced abrupt shifts in the oceans. Philos Trans R Soc Lond B Biol Sci 370:20130264. https://doi.org/10.1098/rstb.2013.0264

Beaugrand G, Conversi A, Chiba S, Edwards M, Fonda-Umani S, Greene C, Mantua N, Otto SA, Reid PC, Stachura MM, Stemmann L, Sugisaki H (2015) Synchronous marine pelagic regime shifts in the Northern Hemisphere. Philos Trans R Soc B 370:20130272. https://doi.org/10.1038/s41558-019-0420-1

Beaugrand G, Conversi A, Atkinson A, Cloern J, Chiba S, Fonda-Umani S, Kirby RR, Greene CH, Goberville E, Otto SA (2019) Prediction of unprecedented biological shifts in the global ocean. Nat Clim Change 9:237. https://doi.org/10.1038/s41558-019-0420-1

Bellwood DR, Hughes TP, Folke C, Nyström M (2004) Confronting the coral reef crisis. Nature 429:827. https://doi.org/10.1038/nature02691

Bellwood DR, Pratchett MS, Morrison TH, Gurney GG, Hughes TP, Álvarez-Romero JG, Day JC, Grantham R, Grech A, Hoey AS (2019) Coral reef conservation in the Anthropocene: confronting spatial mismatches and prioritizing functions. Biol Cons 236:604–615. https://doi.org/10.1016/j.biocon.2019.05.056

Benshila R et al. (2020) A deep learning approach for estimation of the nearshore bathymetry. In: Malvárez G, Navas, F. (eds) Proceedings from the international coastal symposium (ICS) 2020 (Seville, Espagne). J Coast Res, 11–15

Benveniste J, Mandea M, Melet A, Ferrier P (in review) Earth observations for coastal hazards monitoring and international services. Surv Geophys

Benveniste J, Cazenave A, Vignudelli S, Fenoglio-Marc L, Shah R, Almar R, Andersen O, Birol F, Bonnefond P, Bouffard J, Calafat F, Cardellach E, Cipollini P, Le Cozannet G, Dufau C, Fernandes MJ, Frappart F, Garrison J, Gommenginger C, Han G, Høyer JL, Kourafalou V, Leuliette E, Li Z, Loisel H, Madsen KS, Marcos M, Melet A, Meyssignac B, Pascual A, Passaro M, Ribó S, Scharroo R, Song YT, Speich S, Wilkin J, Woodworth P, Wöppelmann G (2019) Requirements for a coastal hazards observing system. Front Mar Sci 6:348. https://doi.org/10.3389/fmars.2019.00348

Berardino P, Fornaro G, Lanari R, Sansosti E (2002) A new algorithm for surface deformation monitoring based on small baseline differential SAR interferograms. IEEE Trans Geosci Remote Sens 40(11):2375–2383. https://doi.org/10.1109/TGRS.2002.803792

Bergmann M, Durand F, Krien Y, Jamal Uddin Khan M, Ishaque M, Testut L, Calmant S, Maisongrande P, Saiful Islam AKM, Papa F, Ouillon S (2018) Topography of the intertidal zone along the shoreline of Chittagong (Bangladesh) using PROBA-V imagery. Int J Remote Sens. https://doi.org/10.1080/01431161.2018.1504341

Bergsma E, Almar R (2020) Coastal coverage of ESA's Sentinel 2 mission. Adv Space Res. https://doi.org/10.1016/j.asr.2020.03.001 (in press)

Bergsma EWJ, Almar R, Maisongrande P (2019) Radon-augmented Sentinelii satellite imagery to derive wave-patterns and regional bathymetry. Remote Sens 11(16):1918. https://doi.org/10.3390/rs11161918

Biggs R, Blenckner T, Folke C, Gordon L, Norström A, Nyström M, Peterson G (2012) Regime shifts. In: Hastings A, Gross L (eds) Encyclopedia of theoretical ecology. University of California Press, Berkeley

Bindoff N et al (2019) Changing ocean, marine ecosystems, and dependent communities. In: IPCC special report on the ocean and cryosphere in a changing climate

Bingham RJ, Hughes CW (2012) Local diagnostics to estimate density-induced sea level variations over topography and along coastlines. J Geophys Res 117(C1):C01013

Bird ECF (1987) The modern prevalence of beach erosion. Mar Pollut Bull 18(4):151–157. https://doi.org/10.1016/0025-326X(87)90238-4

Birol F, Fuller N, Lyard F, Cancet M, Niño F, Delebecque C, Fleury S, Toublanc F, Melet A, Saraceno M, Leger F (2016) Coastal applications from nadir altimetry: example of the X-TRACK regional products. Adv Space Res. https://doi.org/10.1016/j.asr.2016.11.005

Blondeau-Patissier D, Gower JFR, Dekker AG, Phinn SR, Brando VE (2014) A review of ocean color remote sensing methods and statistical techniques for the detection, mapping and analysis of phytoplankton blooms in coastal and open oceans. Prog Oceanogr 123:124–144

Boak EH, Turner IL (2005) Shoreline definition and detection: a review. J Coast Res 21:688–703. https://doi.org/10.2112/03-0071.1

Bonjean F, Lagerloef GS (2002) Diagnostic model and analysis of the surface currents in the tropical Pacific Ocean. J Phys Oceanogr 32:2938–2954

Breivik Ø, Allen A (2008) An operational search and rescue model for the Norwegian Sea and the North Sea. J Mar Syst 69(1–2):99–113. https://doi.org/10.1016/j.jmarsys.2007.02.010

Breivik Ø, Allen AA, Maisondieu C, Olagnon M (2013) Advances in search and rescue at sea. Ocean Dyn 63:83–88. https://doi.org/10.1007/s10236-012-0581-1

Burke L, Kura Y, Kasem K, Revenga C, Spalding M, McAllister D (2001) Coastal ecosystems. World Resources Institute, Washington

Caballero I, Stumpf RP (2019) Retrieval of nearshore bathymetry from Sentinel-2A and 2B satellites in South Florida coastal waters. Estuar Coast Shelf Sci 226:106277. https://doi.org/10.1016/j.ecss.2019.106277

Calafat FM, Chambers DP, Tsimplis MN (2012) Mechanisms of decadal sea level variability in the eastern North Atlantic and the Mediterranean Sea. J Geophys Res 117:C09022

Cazenave A, Le Cozannet G, Benveniste J, Woodworth PL, Champollion N (2017) Monitoring coastal zone changes from space. Eos. https://doi.org/10.1029/2017EO085581

Cazenave A, Palanisamy H, Ablain M (2018) Contemporary sea level changes from satellite altimetry: what have we learned? What are the new challenges? Adv Space Res 62:1639–1653

Chaussard E, Amelung F, Abidin H, Hong S-H (2013) Sinking cities in Indonesia: ALOS PALSAR detects rapid subsidence due to groundwater and gas extraction. Remote Sens Environ 128:150–161. https://doi.org/10.1016/j.rse.2012.10.015

Chawla I, Karthikeyan L, Mishra AK (2020) A review of remote sensing applications for water security: quantity, quality and extremes. J Hydrol 585:124826

Chelton DB, Schlax MG, Freilich MH, Milliff RF (2004) Satellite measurements reveal persistent small-scale features in ocean winds. Science 303:978–983. https://doi.org/10.1126/science.1091901

Choi C, Kim DJ (2018) Optimum baseline of a single-pass In-SAR system to generate the best DEM in tidal flats. IEEE J Sel Top Appl Earth Obs Remote Sens 11(3):919–929. https://doi.org/10.1109/JSTAR S.2018.2795107

Choi J-K, Park Y-J, Lee BR, Eom J, Moon J-E, Ryu J-H (2014) Application of the Geostationary Ocean Color Imager (GOCI) to mapping the temporal dynamics of coastal water turbidity. Remote Sens Environ 146:24–35. https://doi.org/10.1016/j.rse.2013.05.032

Cian F, Delgado Blasco JM, Carrera L (2019) Sentinel-1 for monitoring land subsidence of coastal cities in Africa using PSInSAR: a methodology based on the integration of SNAP and StaMPS. Geosciences 9:124. https://doi.org/10.3390/geosciences9030124

Cianflone G, Tolomei C, Brunori CA, Dominici R (2015) InSAR time series analysis of natural and anthropogenic coastal plain subsidence: the case of Sibari (Southern Italy). Remote Sens 7:16004–16023. https://doi.org/10.3390/rs71215812

Cicin-Sain B, Bernal P, Vandeweerd V, Belfiore S, Goldstein K (2002) A guide to oceans, coasts, and islands at the world summit on sustainable development. Center for the Study of Marine Policy, Newark, Delaware, August 2002. http://hdl.handle.net/1834/301. Accessed May 2019

Cipollini P, Calafat FM, Jevrejeva S, Melet A, Prandi P (2017) Monitoring sea level in the coastal zone with satellite altimetry and tide gauges. Surv Geophys 38:33–57. https://doi.org/10.1007/s1071 2-016-9392-0

Collin A, Hench JL, Pastol Y, Planes S, Thiault T, Schmitt RJ, Holbrook SJ, Davies N, Troyer M (2018) High resolution topobathymetry using a Pleiades-1 triplet: moorea island in 3D. Remote Sens Environ 208:109–119. https://doi.org/10.1016/j.rse.2018.02.015

Commission European (2008) Directive 2008/56/EC of the European Parliament and of the Council of 17 June 2008, establishing a framework for community action in the field of marine environmental policy (Marine Strategy Framework Directive). Off J Eur Union L 164:19–40

Conversi A, Umani SF, Peluso T, Molinero JC, Santojanni A, Edwards M (2010) The Mediterranean Sea regime shift at the end of the 1980s, and intriguing parallelisms with other European basins. PLoS ONE 5(5):e10633. https://doi.org/10.1371/journal.pone.0010633

Conversi A, Dakos V, Gardmark A, Ling S, Folke C, Mumby PJ, Greene C, Edwards M, Blenckner T, Casini M, Pershing A, Moellmann C (2015) A holistic view of marine regime shifts. Philos Trans R Soc B Biol Sci 370:8. https://doi.org/10.1098/rstb.2013.0279

Corbau C, Simeoni U, Zoccarato C, Mantovani G, Teatini P (2019) Coupling land use evolution and subsidence in the Po Delta, Italy: revising the past occurrence and prospecting the future management challenges. Sci Total Environ 654:1196–1208. https://doi.org/10.1016/j.scitotenv.2018.11.104

Cramer W, Yohe GW, Auffhammer M, Huggel C, Molau U, da Silva Dias MAF, Solow A, Stone DA, Tibig L (2014) Detection and attribution of observed impacts. In: Field CB, Barros VR, Dokken DJ, Mach KJ, Mastrandrea MD, Bilir TE, Chatterjee M, Ebi KL, Estrada YO, Genova RC, Girma B, Kissel ES, Levy AN, MacCracken S, Mastrandrea PR, White LL (eds) Climate change 2014: impacts, adaptation, and vulnerability. Part A: global and sectoral aspects. Contribution of Working Group II to the Fifth Assessment Report of the Intergovernmental Panel on Climate Change. Cambridge University Press, Cambridge, pp 979–1037

Crossland C, Baird D, Ducrotoy J-P, Lindeboom H, Buddemeier R, Dennison W et al (2005) The coastal zone—a domain of global interactions. In: Crossland C, Kremer H, Lindeboom H, Marshall Crossland J, Tissier MA (eds) Coastal fluxes in the Anthropocene. Springer, Berlin, pp 1–37. https://doi. org/10.1007/3-540-27851-6

D'Asaro E, Shcherbina AY, Klymak JM, Molemaker J, Novelli G, Guigand CM et al (2018) Ocean convergence and the dispersion of flotsam. Proc Natl Acad Sci USA 115:201718453. https://doi. org/10.1073/pnas.1718453115

Da Lio C, Teatini P, Strozzi T, Tosi L (2018) Understanding land subsidence in salt marshes of the Venice Lagoon from SAR Interferometry and ground-based investigations. Remote Sens Environ 205:56–70. https://doi.org/10.1016/j.rse.2017.11.016

Danilo C, Binet R (2013) Bathymetry estimation from wave motion with optical imagery: influence of acquisition parameters. IEEE Trans Geosci Remote Sens. https://doi.org/10.1109/OCEANS-Berge n.2013.6608068

Danilo C, Melgani F (2019) High-coverage satellite-based coastal bathymetry through a fusion of physical and learning methods. Remote Sens 11:376

De Dominicis M, Pinardi N, Zodiatis G, Archetti R (2013) MEDSLIK-II, a Lagrangian marine surface oil spill model for short-term forecasting—part 2: numerical simulations and validations. Geosci Model Dev 6:1871–1888. https://doi.org/10.5194/gmd-6-1871-2013

de Kloe J, Stoffelen A, Verhoef A (2017) Improved use of scatterometer measurements by using stress-equivalent reference winds. IEEE J Sel Top Appl Earth Observ Remote Sens 10:2340–2347. https ://doi.org/10.1109/JSTARS.2017.2685242

De Mey-Frémaux P, Ayoub N, Barth A, Brewin R, Charria G, Campuzano F, Ciavatta S, Cirano M, Edwards CA, Federico I, Gao S, Garcia Hermosa I, Garcia Sotillo M, Hewitt H, Hole LR, Holt J, King R, Kourafalou V, Lu Y, Mourre B, Pascual A, Staneva J, Stanev EV, Wang H, Zhu X (2019) Model-observations synergy in the coastal ocean. Front Mar Sci 6:436. https://doi.org/10.3389/ fmars.2019.00436

de Young B, Barange M, Beaugrand G, Harris R, Perry RI, Scheffer M, Werner F (2008) Regime shifts in marine ecosystems: detection, prediction and management. Trends Ecol Evol 23:402–409. https ://doi.org/10.1016/j.tree.2008.03.008

Dinardo S, Fenoglio-Marc L, Buchhaupt C, Becker M, Scharroo R, Fernandes MJ, Benveniste J (2018) Coastal SAR and PLRM altimetry in German Bight and West Baltic Sea. Adv Space Res 62(6):1371–1404. https://doi.org/10.1016/j.asr.2017.12.018

Dixon TH, Amelung F, Ferretti A, Novali F, Rocca F, Dokka R, Sella G, Kim S-W, Wdowinski S, Whitman D (2006) Subsidence and flooding in New Orleans. Nature 441:587–588. https://doi. org/10.1038/441587a

Dodet G, Melet A, Ardhuin F, Bertin X, Idier D, Almar R (2019) The contribution of wind-generated waves to coastal sea-level changes. Surv Geophys 40(6):1563–1601. https://doi.org/10.1007/s1071 2-019-09557-5

Dohan K, Maximenko N (2010) Monitoring ocean currents with satellite sensors. Oceanography 23:94–103. https://doi.org/10.5670/oceanog.2010.08

Dorji P, Fearns P (2018) Atmospheric correction of geostationary Himawari-8 satellite data for Total Suspended Sediment mapping: a case study in the Coastal Waters of Western Australia. ISPRS J Photogramm Remote Sens 144:81–93. https://doi.org/10.1016/j.isprsjprs.2018.06.019

Doxaran D, Lamquin N, Park Y-J, Mazeran C, Ryu J-H, Wang M, Poteau A (2014) Retrieval of the seawater reflectance for suspended solids monitoring in the East China Sea using MODIS, MERIS and GOCI satellite data. Remote Sens Environ 146:36–48. https://doi.org/10.1016/j. rse.2013.06.020

Durand F, Piecuch CG, Becker M, Papa F, Raju SV, Khan JU, Ponte RM (2019) Impact of continental freshwater runoff on coastal sea level. Surv Geophys 40:1437. https://doi.org/10.1007/s1071 2-019-09536-w

Duvat VK (2019) A global assessment of atoll island planform changes over the past decades. Wiley Interdiscip Rev Clim Change 10(1):e557. https://doi.org/10.1002/wcc.557

Erban LE, Gorelick SM, Zebker HA (2014) Groundwater extraction, land subsidence, and sea-level rise in the Mekong Delta, Vietnam. Environ Res Lett 9:084010. https://doi.org/10.1088/1748-9326/9/8/084010

Eriksson BK, van der Heide T, van de Koppel J, Piersma T, van der Veer HW, Olff H (2010) Major changes in the ecology of the Wadden Sea: human impacts, ecosystem engineering and sediment dynamics. Ecosystems 13:752–764. https://doi.org/10.1007/s10021-010-9352-3

Ettritch G, Hardy A, Bojang L, Cross D, Bunting P, Brewer P (2018) Enhancing digital elevation models for hydraulic modelling using flood frequency detection. Remote Sens Environ 217:506–522. https ://doi.org/10.1016/j.rse.2018.08.029

Evagorou EG, Mettas C, Agapiou A, Themistocleous K, Hadjimitsis DG (2019) Bathymetric maps from multi-temporal analysis of Sentinel-2 data: the case study of Limassol. Cyprus. Adv. Geosci. 45:397–407

FAO (1998) FAO Technical guidelines for responsible fisheries—fishing operations—1 Suppl. 1-1. In: Vessel monitoring systems. Food and Agriculture of the United Nations, M-41, ISBN 92-5-104179-2, Rome

Farcas A, Thompson PM, Merchant ND (2016) Underwater noise modelling for environmental impact assessment. Environ Impact Assess Rev 57:114–122. https://doi.org/10.1016/j.eiar.2015.11.012

Farr TG et al (2007) The Shuttle radar topography mission. Rev Geophys 45:RG2004. https://doi.org/10.1029/2005RG000183

Ferretti A, Prati C, Rocca F (2001) Permanent scatterers in SAR interferometry. IEEE Trans Geosci Remote Sens 39(1):8–20. https://doi.org/10.1109/36.898661

Ferretti A, Fumagalli A, Novali F, Prati C, Rocca F, Rucci A (2011) A new algorithm for processing interferometric data-stacks: Squee-SAR™. IEEE Trans Geoscid Remote Sens 49(9):3460–3470

Fingas M, Brown CE (2018) A review of oil spill remote sensing. Sensors 18:91. https://doi.org/10.3390/s18010091

Gabriel AK, Goldstein RM, Zebker HA (1989) Mapping small elevation changes over large areas: differential radar interferometry. J Geophys Res 94(B7):9183–9191. https://doi.org/10.1029/JB094iB07p09183

Gao J (2009) Bathymetric mapping by means of remote sensing: methods, accuracy and limitations. Prog Phys Geogr 33:103–116. https://doi.org/10.1177/0309133309105657

Garcin M, Vendé-Leclerc M, Maurizot P, Le Cozannet G, Robineau B, Nicolae-Lerma A (2016) Lagoon islets as indicators of recent environmental changes in the South Pacific—The New Caledonian example. Cont Shelf Res 122:120–140. https://doi.org/10.1016/j.csr.2016.03.025

Gattuso JP, Magnan A, Billé R, Cheung WW, Howes EL, Joos F, Allemand D, Bopp L, Cooley SR, Eakin CM, Hoegh-Guldberg O (2015) Contrasting futures for ocean and society from different anthropogenic CO_2 emissions scenarios. Science 349(6243):4722. https://doi.org/10.1126/science.aac4722

Gebremichael E, Sultan M, Becker R, El Bastawesy M, Cherif O, Emil M (2018) Assessing land deformation and sea encroachment in the Nile Delta: a radar interferometric and inundation modeling approach. J Geophys Res Solid Earth 123:3208–3224. https://doi.org/10.1002/2017JB015084

Goldstein EB, Coco G, Plant NG (2019) A review of machine learning applications to coastal sediment transport and morphodynamics. Earth Sci Rev 194:97–108. https://doi.org/10.1016/j.earscirev.2019.04

Gommenginger C (2019) SEASTAR: a mission to study ocean submesoscale dynamics and small-scale atmosphere-ocean processes in coastal, shelf and polar seas. Front Mar Sci. https://doi.org/10.13140/RG.2.2.20415

Greidanus H, Alvarez M, Santamaria C, Thoorens FX, Kourti N, Argentieri P (2017) The SUMO ship detector algorithm for satellite radar images. Remote Sens 9;246. https://doi.org/10.3390/rs9030246

Guerrini F, Mari L, Casagrandi R (2019) Modeling plastics exposure for the marine biota: risk maps for fin whales in the Pelagos Sanctuary (North-Western Mediterranean). Frontiers in Marine Science. https://doi.org/10.3389/fmars.2019.00299

Hallegatte S, Green C, Nicholls RJ, Corfee-Morlot J (2013) Future flood losses in major coastal cities. Nat Clim Change 3(9):802. https://doi.org/10.1038/nclimate1979

Ham Y-G, Kim J-H, Luo J-J (2019) Deep learning for multi-year ENSO forecasts. Nature 573:568–572

Hamon M, Greiner E, Remy E, Le Traon PY (2019) Impact of multiple altimeter data and mean dynamic topography in a global analysis and forecasting system. J Atmos Ocean Technol. https://doi.org/10.1175/JTECH-D-18-0236.1

Hasselmann K, Chapron B, Aouf L, Ardhuin F, Collard F, Engen G et al (2012) The ERS SAR wave mode: a breakthrough in global ocean wave observations. In: ERS Missions: 20 years of observing earth. European Space Agency, Noordwijk, pp 165–198

Hawkins AD, Popper AN (2017) A sound approach to assessing the impact of underwater noise on marine fishes and invertebrates. ICES J Mar Sci 74(3):635–651. https://doi.org/10.1093/icesjms/fsw205

Heron SF, Maynard JA, van Hooidonk R, Eakin CM (2016) Warming trends and bleaching stress of the world's coral reefs 1985–2012. Sci Rep 6:38402. https://doi.org/10.1038/srep38402

Hersbach H, Bell B, Berrisford P, Horányi A, Muñoz Sabater J, Nicolas J, Radu R, Schepers D, Simmons A, Soci C, Dee D (2019) Global reanalysis: Goodbye ERA-interim, hello ERA5. ECMWF Newsletter No. 159:17-24. http://dx.doi.org/10.21957/vf291hehd7

Heygster G, Dannenberg J, Notholt J (2010) Topographic mapping of the german tidal flats analyzing SAR images with the waterline method. IEEE Trans Geosci Remote Sens 48(3):1019–1030

Higgins SA, Overeem I, Steckler MS, Syvitski JPM, Seeber L, Akhter SH (2014) InSAR measurements of compaction and subsidence in the Ganges–Brahmaputra Delta, Bangladesh. J Geophys Res Earth Surf 119:1768–1781. https://doi.org/10.1002/2014JF003117

Hildebrand JA (2005) Impacts of anthropogenic sound. In: Reynolds JE III, Perrin WF, Reeves RR, Montgomery S, Ragen TJ (eds) Marine mammal research: conservation beyond crisis. The John Hopkins University Press, Baltimore, pp 101–124

Hinkel J, Lincke D, Vafeidis AT, Perrette M, Nicholls RJ, Tol RS, Marzeion B, Fettweis X, Ionescu C, Levermann A (2014) Coastal flood damage and adaptation costs under 21st century sea-level rise. Proc Natl Acad Sci 111(9):3292–3297. https://doi.org/10.1073/pnas.1222469111

Hoegh-Guldberg O, Cai R, Poloczanska ES, Brewer PG, Sundby S, Hilmi K, Fabry VJ, Jung S (2014) The Ocean. In: Barros VR, Field CB, Dokken DJ, Mastrandrea MD, Mach KJ, Bilir TE, Chatterjee M, Ebi KL, Estrada YO, Genova RC, Girma B, Kissel ES, Levy AN, MacCracken S, Mastrandrea PR, White LL (eds) Climate change 2014: impacts, adaptation, and vulnerability. Part B: regional aspects. Contribution of Working Group II to the Fifth Assessment Report of the Intergovernmental Panel on Climate Change. Cambridge University Press, Cambridge, pp 1655–1731

Holling CS (1973) Resilience and stability of ecological systems. Annu Rev Ecol Syst 4:1–23. https://doi.org/10.1146/annurev.es.04.110173.000245

Holman RA, Lalejini D, Holland KT (2016) A parametric model for barred equilibrium beach profiles: two-dimensional implementation. Coast Eng 117:166–175. https://doi.org/10.1016/j.coastaleng.2016.07.010

Huang C, Yang H, Zhu AX, Zhang M, Lü H, Huang T et al (2015) Evaluation of the Geostationary Ocean Color Imager (GOCI) to monitor the dynamic characteristics of suspension sediment in Taihu Lake. Int J Remote Sens 36:3859–3874. https://doi.org/10.1080/01431161.2015.1070323

Hughes TP (1994) Catastrophes, phase shifts, and large-scale degradation of a Caribbean coral reef. Science 265:1547–1551

Hughes TP, Carpenter S, Rockström J, Scheffer M, Walker B (2013) Multiscale regime shifts and planetary boundaries. Trends Ecol Evol 28:389–395. https://doi.org/10.1016/j.tree.2013.05.019

Hughes TP, Kerry JT, Álvarez-Noriega M, Álvarez-Romero JG, Anderson KD, Baird AH, Babcock RC, Beger M, Bellwood DR, Berkelmans R, Bridge TC, Butler IR, Byrne M, Cantin NE, Comeau S, Connolly SR, Cumming GS, Dalton SJ, Diaz-Pulido G, Eakin CM, Figueira WF, Gilmour JP, Harrison HB, Heron SF, Hoey AS, Hobbs J-PA, Hoogenboom MO, Kennedy EV, Kuo C-y, Lough JM, Lowe RJ, Liu G, McCulloch MT, Malcolm HA, McWilliam MJ, Pandolfi JM, Pears RJ, Pratchett MS, Schoepf V, Simpson T, Skirving WJ, Sommer B, Torda G, Wachenfeld DR, Willis BL, Wilson SK (2017) Global warming and recurrent mass bleaching of corals. Nature 543:373–377. https://doi.org/10.1038/nature21707

Hughes CW, Fukumori I, Griffies SM, Huthnance JM, Minobe S, Spence P, Thompson KR, Wise A (2019) Sea level and the role of coastal trapped waves in mediating the influence of the open ocean on the coast. Surv Geophys 40:1467–1492

Hugo G (2011) Future demographic change and its interactions with migration and climate change. Glob Environ Change 21S:S21–S33. https://doi.org/10.1016/j.gloenvcha.2011.09.008

Idier D, Bertin X, Thompson P, Pickering MD (2019) Interactions between mean sea level, tide, surge, waves and flooding: mechanisms and contributions to sea level variations at the coast. Surv Geophys. https://doi.org/10.1007/s10712-019-09549-5

IGOS (2006) A coastal theme for the IGOS partnership—for the monitoring of our environment from space and from earth. Paris, UNESCO 2006. 60 pp. (IOC Information document No. 1220)

IMO (2000) SOLAS Convention 1974, Chapter V, Regulation 19 (Amended 6 Dec 2000)

IMO (2006) SOLAS Convention 1974, Chapter V, Regulation 19-1 (Adopted 19 May 2006)

Ingebritsen SE, Galloway DL (2014) Coastal subsidence and relative sea level rise. Environ Res Lett 9:091002. https://doi.org/10.1016/j.gloenvcha.2011.09.008

IOCCG (2000) Remote sensing of ocean colour in coastal, and other optically-complex, waters. In: Sathyendranath S (ed) Reports of the International Ocean-Colour Coordinating Group, No. 3, IOCCG, Dartmouth, Canada

IOCCG (2008) Why ocean colour? The societal benefits of ocean-colour technology. In: Platt T, Hoepffner N, Stuart V, Brown C (eds) Reports of the International Ocean-Colour Coordinating Group, No. 7, IOCCG, Dartmouth, Canada

IOCCG (2018) Earth observations in support of global water quality monitoring. In: Greb S, Dekker A, Binding C (eds) IOCCG Report Series, No. 17, International Ocean Colour Coordinating Group, Dartmouth, Canada

IPBES (2019) Global assessment report on biodiversity and ecosystem services of the Intergovernmental Science-Policy Platform on Biodiversity and Ecosystem Services. In: Brondizio ES, Settele J, Díaz S, Ngo HT (eds) IPBES secretariat, Bonn, Germany

Jambeck JR, Geyer R, Wilcox C, Siegler TR, Perryman M, Andrady A, Narayan R, Law KL (2015) Plastic waste inputs from land into the ocean. Science 347(6223):768–771

Jawak SD, Vadlamani SS, Luis AJ (2015) A synoptic review on deriving bathymetry information using remote sensing technologies: models, methods and comparisons. Adv Remote Sens 4:147–162. https://doi.org/10.4236/ars.2015.42013

Kamenos NA, Hennige SJ (2018) Reconstructing four centuries of temperature-induced coral bleaching on the Great Barrier Reef. Front Mar Sci 5:283. https://doi.org/10.3389/fmars.2018.00283

Kanjir U, Greidanus H, Oštir C (2018) Vessel detection and classification from spaceborne optical images: a literature survey. Remote Sens Environ 207:1–26

Karunarathna H, Horrillo-Caraballo J, Kuriyama Y, Mase H, Ranasinghe R, Reeve DE (2016) Linkages between sediment composition, wave climate and beach profile variability at multiple timescales. Mar Geol 381:194–208

Khan MJU, Ansary MN, Durand F, Testut L, Ishaque M, Calmant S, Krien Y, Islam AS, Papa F (2019) High-resolution intertidal topography from Sentinel-2 multi-spectral imagery: synergy between remote sensing and numerical modeling. Remote Sens 11:2888

Kharif C, Pelinovsky E (2003) Physical mechanisms of the rogue wave phenomenon. Eur J Mech B/Fluids 22:603–634

Kroodsma DA, Mayorga J, Hochberg T, Miller NA, Boerder K, Ferretti F, Wilson A, Bergman B, White TD, Block BA, Woods P, Sullivan B, Costello C, Worm B (2018) Tracking the global footprint of fisheries. Science 359:904–908

Kudryavtsev V, Yurovskaya M, Chapron B, Collard F, Donlon C (2017) Sun glitter imagery of ocean surface waves. Part 1: directional spectrum retrieval and validation. J Geophys Res (Oceans) 122(2):1369–1383

Kulp SA, Strauss BH (2018) CoastalDEM: a global coastal digital elevation model improved from SRTM using a neural network. Remote Sens Environ 206:231–239. https://doi.org/10.1016/j.rse.2017.12.026

Kwiatkowska EJ, Ruddick K, Ramon D, Vanhellemont Q, Brockmann C, Lebreton C, Bonekamp HG (2016) Ocean colour opportunities from Meteosat Second and Third Generation geostationary platforms. Ocean Sci 12:703–713. https://doi.org/10.5194/os-12-703-2016

Lauerburg RAM, Diekmann R, Blanz B, Gee K, Held H, Kannen A, Möllmann C, Probst WN, Rambo H, Cormier R, Stelzenmüller V (2020) Socio-ecological vulnerability to tipping points: a review of empirical approaches and their use for marine management. Sci Total Environ 705:135838

Le Cozannet G, Garcin M, Yates M, Idier D, Meyssignac B (2014) Approaches to evaluate the recent impacts of sea-level rise on shoreline changes. Earth Sci Rev 138:47–60. https://doi.org/10.1016/j.earscirev.2014.08.005

Le Cozannet G, Kervyn M, Russo S, Ifejika-Speranza C, Ferrier P, Foumelis M, Modaressi H (2020) Space-based earth observations for disaster risk management. Surv Geophys. https://doi.org/10.1007/s10712-020-09586-5

Le Roy S, Pedreros R, André C, Paris F, Lecacheux S, Marche F, Vinchon C (2015) Coastal flooding of urban areas by overtopping: dynamic modelling application to the Johanna storm (2008) in Gâvres (France). Nat Hazards Earth Syst Sci 15(11):2497–2510. https://doi.org/10.5194/nhess-15-2497-2015

Le Traon PY (2018) Satellites and operational oceanography. In: Chassignet E, Pascual A, Tintoré J, Verron J (eds) New frontiers in operational oceanography. GODAE OceanView, Met Office, pp 161–190. https://doi.org/10.17125/gov2018.ch07

Le Traon PY, Antoine D, Bentamy A, Bonekamp H, Breivik LA, Chapron B et al (2015) Use of satellite observations for operational oceanography: recent achievements and future prospects. J Oper Oceanogr 8:s12–s27

Le Traon PY, Dibarboure G, Jacobs G, Martin M, Remy E, Schiller A (2017) Use of satellite altimetry for operational oceanography. In: Stammer D, Cazenave A (eds) Satellite altimetry over oceans and land surfaces. CRC Press, Boca Raton

Le Traon PY et al (2019) From observation to information and users: the copernicus marine service perspective. Front Mar Sci 6:234. https://doi.org/10.3389/fmars.2019.00234

Lee SK, Ryu JH (2017) High-accuracy tidal flat digital elevation model construction using TanDEM-X science phase data. IEEE J Sel Top Appl Earth Obs Remote Sens 10(6):2713–2724. https://doi.org/10.1109/JSTARS.2017.2656629

Lee ZP, Carder KL, Mobley CD, Steward RG, Patch JS (1998) Hyperspectral remote sensing for shallow waters: deriving bottom depths and water properties by optimization. Appl Opt 38(18):3831–3843. https://doi.org/10.1364/AO.38.003831

Lenton TM, Rockström J, Gaffney O, Rahmstorf S, Richardson K, Steffen W, Schellnhuber HJ (2019) Climate tipping points—too risky to bet against. Nature 575:592–595. https://doi.org/10.1038/d41586-019-03595-0

Li Z, Heygster G, Notholt J (2014) Intertidal topographic maps and morphological changes in the German Wadden Sea between 1996–1999 and 2006–2009 from the waterline method and SAR images. IEEE J Sel Top Appl Earth Obs Remote Sens 7(8):3210–3224

Liubartseva S, Coppini G, Lecci R, Clementi E (2018) Tracking plastics in the Mediterranean: 2D Lagrangian model. Mar Pollut Bull 129:151–162

Loisel H, Mangin A, Vantrepotte V, Dessailly D, Dinh DN, Garnesson P, Ouillon S, Lefebvre JP, Mériaux X, Phan TM (2014) Variability of suspended particulate matter concentration in coastal waters under the Mekong's influence from ocean color (MERIS) remote sensing over the last decade. Remote Sens Environ 150:218–230. https://doi.org/10.1016/j.rse.2014.05.006

Loisel H, Vantrepotte V, Ouillon S, Ngoc DD, Herrmann M, Tran V et al (2017) Assessment and analysis of the chlorophyll-a concentration variability over the Vietnamese coastal waters from the MERIS ocean color sensor (2002–2012). Remote Sens Environ 190:217–232. https://doi.org/10.1016/j.rse.2016.12.016

Longuevergne et al. (in review)

Lotze HK, Reise K, Worm B, van Beusekom J, Busch M, Ehlers A, Heinrich D, Hoffmann RC, Holm P, Jensen C, Knottnerus OS, Langhanki N, Prummel W, Vollmer M, Wolff WJ (2005) Human transformations of the Wadden Sea ecosystem through time: a synthesis. Helgol Mar Res 59:84–95. https://doi.org/10.1007/s10152-004-0209-z

Luijendijk A, Hagenaars G, Ranasinghe R, Baart F, Donchyts G, Aarninkhof S (2018) The state of the world's beaches. Sci Rep 8(1):6641. https://doi.org/10.1038/s41598-018-24630-6

Lyzenga DR (1978) Passive remote sensing techniques for mapping water depth and bottom features. Appl Opt 17(3):379–383. https://doi.org/10.1364/AO.17.000379

Maes C, Grima N, Blanke B, Martinez E, Paviet-Salomon T, Huck T (2018) A surface "superconvergence" pathway connecting the South Indian Ocean to the subtropical South Pacific gyre. Geophys Res Lett 45:1915–1922. https://doi.org/10.1002/2017GL076366

Marcos M, Wöppelmann G, Matthews A, Ponte RM, Birol F, Ardhuin F, Coco G, Santamaria-Gomez A, Ballu V, Testut L, Chambers D, Stopa JE (2019) Coastal sea level and related fields from existing observing systems. Surv Geophys 40(6):1293–1317

Marti F, Cazenave A, Birol F, Passaro M, Leger F, Nino F, Almar R, Benveniste J, Legeais JF (2019) Altimetry-based sea level trends along the coasts of western Africa. Adv Space Res. https://doi.org/10.1016/j.asr.2019.05.033 (in press)

Martínez ML, Intralawan A, Vázquez G, Pérez-Maqueo O, Sutton P, Landgrave R (2007) The coasts of our world: ecological, economic and social importance. Ecol Econ 63:254–272. https://doi.org/10.1016/j.ecolecon.2006.10.022

Mason DC, Davenport IJ, Robinson GJ, Flather RA, Mccartney BS (1995) Construction of an inter-tidal digital elevation model by the "water-line" method. Geophys Res Lett 22(23):3187–3190

Matthews MW (2011) A current review of empirical procedures of remote sensing in inland and near-coastal transitional waters. Int J Remote Sens 32:6855–6899. https://doi.org/10.1080/01431161.2010.512947

Matz MV, Treml EA, Aglyamova GV, Bay LK (2018) Potential and limits for rapid genetic adaptation to warming in a Great Barrier Reef coral. PLoS Genet 14:e1007220. https://doi.org/10.1371/journal.pgen.1007220

Mayer L, Jakobsson M, Allen G, Dorschel B, Falconer R, Ferrini V et al (2018) The Nippon foundation—GEBCO seabed 2030 project: the quest to see the world's oceans completely mapped by 2030. Geosciences 8:63. https://doi.org/10.3390/geosciences8020063

McCarthy BL (2010) Coastal bathymetry using satellite observation in support of intelligence preparation of the environment, Calhoun thesis: The NPS Institutional Archive DSpace Repository

McGranahan G, Balk D, Anderson B (2007) The rising tide: assessing the risks of climate change and human settlements in low elevation coastal zones. Environ Urban 19:17–37. https://doi.org/10.1177/0956247807076960

Medina-Lopez E, Ureña-Fuentes L (2019) High-resolution sea surface temperature and salinity in coastal areas worldwide from raw satellite data. Remote Sens 11(19):2191. https://doi.org/10.3390/rs11192191

Melet A, Meyssignac B, Almar R, Le Cozannet G (2018) Under-estimated wave contribution to coastal sea-level rise. Nat Clim Change 8(3):234. https://doi.org/10.1038/s41558-018-0088-y

Menendez M, Woodworth P (2010) Changes in extreme high water levels based on a quasi-global tide-gauge data set. J Geophys Res 115:C10011. https://doi.org/10.1029/2009JC005997

Mentaschi L, Vousdoukas MI, Pekel JF, Voukouvalas E, Feyen L (2018) Global long-term observations of coastal erosion and accretion. Sci Rep 8(1):12876. https://doi.org/10.1038/s41598-018-30904-w

Merchant ND (2019) Underwater noise abatement: economic factors and policy options. Environ Sci Policy 92:116–123. https://doi.org/10.1016/j.envsci.2018.11.014

Merchant ND, Brookes KL, Faulkner RC, Bicknell AWJ, Godley BJ, Witt MJ (2016) Underwater noise levels in UK waters. Sci Rep 6:36942. https://doi.org/10.1038/srep36942

Moellmann C, Conversi A, Edwards M (2011) Comparative analysis of European wide marine ecosystem shifts: a large-scale approach for developing the basis for ecosystem-based management. Biol Lett 7:484–486. https://doi.org/10.1098/rsbl.2010.1213

Möllmann C, Folke C, Edwards M, Conversi A (2015) Marine regime shifts around the globe: theory, drivers and impacts. Philos Trans R Soc Lond B Biol Sci 370(1659):20130260. https://doi.org/10.1098/rstb.2013.0260

Monteys X, Harris P, Caloca S, Cahalane C (2015) Spatial prediction of coastal bathymetry based on multispectral satellite imagery and multibeam data. Remote Sens 7(10):13782–13806. https://doi.org/10.3390/rs71013782

Mouw CB, Greb S, Aurin D, DiGiacomo PM, Lee Z, Twardowski M, Binding C, Hu C, Ma R, Moore T, Moses W, Craig SE (2015) Aquatic color radiometry remote sensing of coastal and inland waters: challenges and recommendations for future satellite missions. Remote Sens Environ 160:15–30. https://doi.org/10.1016/j.rse.2015.02.001

Nayak PK, Armitage D (2018) Social-ecological regime shifts (SERS) in coastal systems. Ocean Coast Manag 161:84–95. https://doi.org/10.1016/j.ocecoaman.2018.04.020

NEREUS/ESA/EC (2018) The ever growing use of copernicus across Europe's regions. https://www.copernicus.eu/sites/default/files/2018-10/copernicus4regions.pdf. Accessed Sept 2019

Neukermans G, Ruddick K, Bernard E, Ramon D, Nechad B, Deschamps P-Y (2009) Mapping total suspended matter from geostationary satellites: a feasibility study with SEVIRI in the Southern North Sea. Opt Express 17:14029–14052. https://doi.org/10.1364/OE.17.014029

Neukermans G, Ruddick KG, Greenwood N (2012) Diurnal variability of turbidity and light attenuation in the southern North Sea from the SEVIRI geostationary sensor. Remote Sens Environ 124:564–580. https://doi.org/10.1016/j.rse.2012.06.003

Neumann B, Vafeidis AT, Zimmermann J, Nicholls RJ (2015) Future coastal population growth and exposure to sea-level rise and coastal flooding—a global assessment. PLoS ONE 10(3):e0118571. https://doi.org/10.1371/journal.pone.0118571

Ng AH-M, Ge L, Li X, Abidin HZ, Andreas H, Zhang K (2012) Mapping land subsidence in Jakarta, Indonesia using persistent scatterer interferometry (PSI) technique with ALOS PALSAR. Int J Appl Earth Obs Geoinf 18:232–242. https://doi.org/10.1016/j.jag.2012.01.018

Ngoc DD, Loisel H, Jamet C, Vantrepotte V, Duforêt-Gaurier L, Minh CD, Mangin A (2019) Coastal and inland water pixels extraction algorithm (WiPE) from spectral shape analysis and HSV transformation applied to Landsat 8 OLI and Sentinel-2 MSI. Remote Sens Environ 223:208–228

Nim CJ, Skirving WJ, Eakin CM (2010) Satellite monitoring of reef vulnerability in a changing climate. NOAA Technical Report CRCP 1. NOAA Coral Reef Conservation Program. Silver Spring, MD. http://doi.org/10.1126/science.aaf9933

Normille N (2016) Survey confirms worst-ever coral bleaching at Great Barrier Reef. Science. https://doi.org/10.1126/science.aaf9933

O'Carroll AG, Armstrong EM, Beggs HM, Bouali M, Casey KS, Corlett GK, Dash P, Donlon CJ, Gentemann CL, Høyer JL, Ignatov A, Kabobah K, Kachi M, Kurihara Y, Karagali I, Maturi E, Merchant CJ, Marullo S, Minnett PJ, Pennybacker M, Ramakrishnan B, Ramsankaran R, Santoleri R, Sunder S, Saux Picart S, Vázquez-Cuervo J, Wimmer W (2019) Observational needs of sea surface temperature. Front Mar Sci 6:420. https://doi.org/10.3389/fmars.2019.00420

Odermatt D, Gitelson A, Brando VE, Schaepman M (2012) Review of constituent retrieval in optically deep and complex waters from satellite imagery. Remote Sens Environ 118:116–126. https://doi.org/10.1016/j.rse.2011.11.013

Ody A, Doxaran D, Vanhellemont Q, Nechad B, Novoa S, Many G, Bourrin F, Verney R, Pairaud I, Gentili B (2016) Potential of high spatial and temporal ocean color satellite data to study the dynamics of suspended particles in a micro-tidal river plume. Remote Sens 8(3):245

OECD (2016) The Ocean Economy in 2030. OECD Publishing, Paris. https://doi.org/10.1787/9789264251724-en

Oliver EC, Burrows MT, Donat MG, Sen Gupta A, Alexander LV, Perkins-Kirkpatrick S, Benthuysen JA, Hobday AJ, Holbrook NJ, Moore PJ, Thomsen MS, Wernberg T, Smale DA (2019) Projected marine heatwaves in the 21st century and potential for ecological impact. Front Mar Sci 6:734

Oppenheimer M, Glacovic B, Hinkel J, Van De Wal R, Magnan A, Abd-Elgawad A, Cai R, Cifuentes-Jara M, Deconto R, Ghosh T, Hay J, Isla F, Marzeion B, Meyssignac M, Sebesvari Z (2019) Sea-level rise and Implications for low lying islands, coasts and communities. Special report on the ocean and cryosphere in a changing climate (SROCC). https://report.ipcc.ch/srocc/pdf/SROCC_FinalDraft_Chapter4.pdf. **(in press)**

Pahlevan N, Sarkar S, Franz BA, Balasubramanian SV, He J (2017a) Sentinel-2 MultiSpectral Instrument (MSI) data processing for aquatic science applications: demonstrations and validations. Remote Sens Environ 201:47–56. https://doi.org/10.1016/j.rse.2017.08.033

Pahlevan N, Schott JR, Franz BA, Zibordi G, Markham B, Bailey S, Schaff CB, Ondrusek M, Greb S, Strait CM (2017b) Landsat 8 remote sensing reflectance (Rrs) products: evaluations, intercomparisons and enhancements. Remote Sens Environ 190:289–301. https://doi.org/10.1016/j.rse.2016.12.030

Pahlevan N, Balasubramanian SV, Sarkar S, Franz BA (2018) Toward long-term aquatic science products from heritage Landsat missions. Remote Sens 10:23. https://doi.org/10.3390/rs10091337

Pahlevan N, Chittimalli SK, Balasubramanian SV (2019) Sentinel-2/Landsat-8 product consistency and implications for monitoring aquatic systems. Remote Sens Environ 220:19–29. https://doi.org/10.1016/j.rse.2018.10.027

Passaro M, Cipollini P, Vignudelli S, Quartly G, Snaith H (2014) ALES: a multi-mission subwaveform retracker for coastal and open ocean altimetry. Remote Sens Environ 145:173–189. https://doi.org/10.1016/j.rse.2014.02.008

Passaro M, Fenoglio-Marc L, Cipollini P (2015) Validation of significant wave height from improved satellite altimetry in the German bight. IEEE Trans Geosci Remote Sens 53(4):2146–2156. https://doi.org/10.1109/TGRS.2014.2356331

Passaro M, Zulfikar Adlan N, Quartly GD (2018) Improving the precision of sea level data from satellite altimetry with high-frequency and regional sea state bias corrections. Remote Sens Environ 218:245–254. https://doi.org/10.1016/j.rse.2018.09.007

Perez D, Islam K, Hill V, Zimmerman R, Schaeffer B, Li J (2018) Deepcoast: quantifying seagrass distribution in coastal water through deep capsule networks. In: Chinese conference on pattern recognition and computer vision (PRCV). Springer, pp 404–416

Pleskachevsky A, Lehner S, Heege T, Mott C (2011) Synergy and fusion of optical and synthetic aperture radar satellite data for underwater topography estimation in coastal areas. Ocean Dyn 61(12):2099–2120. https://doi.org/10.1007/s10236-011-0460-1

Ponte R, Carson M, Cirano M, Domingues C, Jevrejeva S, Marcos M, Mitchum G, van de Wal R, Woodworth P et al (2019) Towards comprehensive observing and modeling systems for monitoring and predicting regional to coastal sea level. Front Mar Sci. https://doi.org/10.3389/fmars.2019.00437

Popper AN, Hastings MC (2009) Effects of pile driving and other anthropogenic sounds on fish: part 1-critical literature review. J Fish Biol 75:455–489. https://doi.org/10.1111/j.1095-8649.2009.02319.x

Poupardin A, Idier D, De Michele M, Raucoules D (2015) Water depth inversion from a single SPOT-5 dataset. IEEE Trans Geosci Remote Sens 54(4):2329–2342. https://doi.org/10.1109/TGRS.2015.2499379

Poupardin A, Idier D, de Michele M, Raucoules D (2016) Water depth inversion from a single SPOT-5 dataset. IEEE Trans Geosci Remote Sens. https://doi.org/10.1109/TGRS.2015.2499379

Quilfen Y, Yurovskaya M, Chapron B, Ardhuin F (2018) Storm waves focusing and steepening in the Agulhas current: satellite observations and modeling. Remote Sens Environ 216:561–571

Ranasinghe R (2016) Assessing climate change impacts on open sandy coasts: a review. Earth-Sci Rev 160:320–332. https://doi.org/10.1016/j.earscirev.2016.07.011

Raucoules D, Le Cozannet G, Wöppelmann G, De Michele M, Gravelle M, Daag A, Marcos M (2013) High nonlinear urban ground motion in Manila (Philippines) from 1993 to 2010 observed by DInSAR: implications for sea-level measurement. Remote Sens Environ 139:386–397. https://doi.org/10.1016/j.rse.2013.08.021

Raucoules D, Le Cozannet G, de Michele M, Capo S (2018) Observing water-level variations from space-borne high-resolution Synthetic Aperture Radar (SAR) image correlation. Geocarto Int 33(9):977–987

Reise K, Baptist M, Burbridge P, Dankers N, Fischer L, Flemming B, Oost AP, Smit C (2010) The Wadden Sea-a universally outstanding tidal wetland. The Wadden Sea 2010 Common Wadden Sea Secretariat (CWSS); Trilateral Monitoring and Assessment Group: Wilhelmshaven (Wadden Sea Ecosystem; 29/editors, Harald Marencic and Jaap de Vlas), Book 7

Ribal A, Young IR (2019) 33 years of globally calibrated wave height and wind speed data based on altimeter observations. Sci Data 6:77. https://doi.org/10.1038/s41597-019-0083-9

Rio MH, Santoleri R (2018) Improved global surface currents from the merging of altimetry and sea surface temperature data. Remote Sens Environ 216:770–785. https://doi.org/10.1016/j.rse.2018.06.003

Rio MH, Mulet S, Picot N (2014) Beyond GOCE for the ocean circulation estimate: synergistic use of altimetry, gravimetry, and in situ data provides new insight into geostrophic and Ekman currents. Geophys Res Lett 41:8918–8925. https://doi.org/10.1002/2014GL061773

Roarty H, Cook T, Hazard L, George D, Harlan J, Cosoli S, Wyatt L, Alvarez Fanjul E, Terrill E, Otero M, Largier J, Glenn S, Ebuchi N, Whitehouse B, Bartlett K, Mader J, Rubio A, Corgnati L, Mantovani C, Griffa A, Reyes E, Lorente P, Flores-Vidal X, Saavedra-Matta KJ, Rogowski P, Prukpitikul

S, Lee S-H, Lai J-W, Guerin C-A, Sanchez J, Hansen B, Grilli S (2019) The global high frequency radar network. Front Mar Sci 6:164. https://doi.org/10.3389/fmars.2019.00164

Roberts J, Best BD, Mannocci L, Fujioka E, Halpin PN, Palka DL, Garrison LP, Mullin KD, Cole TVN, Khan CB, McLellan WA, Pabst DA, Lockhart GG (2016) Habitat-based cetacean density models for the U.S. Atlantic and Gulf of Mexico. Sci Rep 6:22615. https://doi.org/10.1038/srep22615

Rocha JC, Peterson G, Bodin Ö, Levin S (2018) Cascading regime shifts within and across scales. Science 362:1379

Rodriguez E (2019) The winds and currents mission concept. Front. Mar. Sci. https://doi.org/10.3389/fmars.2019.00438

Roussel N, Ramillien G, Frappart F, Darrozes J, Gay A, Biancale R, Striebig N, Hanquiez V, Bertin X, Allain D (2015) Sea level monitoring and sea state estimate using a single geodetic receiver. Remote Sens Environ 171:261–277

Ruddick K, Neukermans G, Vanhellemont Q, Jolivet D (2013) Challenges and opportunities for geostationary ocean colour remote sensing of regional seas: a review of recent results. Remote Sens Environ 146:63–76. https://doi.org/10.1016/j.rse.2013.07.039

Sagawa T, Yamashita Y, Okumura T, Yamanokuchi T (2019) Satellite derived bathymetry using machine learning and multi-temporal satellite images. Remote Sensing 11:1155

Salama MS, Shen F (2010) Simultaneous atmospheric correction and quantification of suspended particulate matters from orbital and geostationary earth observation sensors. Estuar Coast Shelf Sci 86:499–511. https://doi.org/10.1016/j.ecss.2009.10.001

Salameh E, Frappart F, Almar R, Baptista P, Heygster G, Lubac B, Raucoules D, Almeida LP, Bergsma EWJ, Capo S, De Michele M, Idier D, Li Z, Marieu V, Poupardin A, Silva PA, Turki I, Laignel B (2019) Monitoring beach topography and nearshore bathymetry using spaceborne remote sensing: a review. Remote Sens 11(18):2212. https://doi.org/10.3390/rs11192212

Santamaria C, Alvarez M, Greidanus H, Syrris V, Soille P, Argentieri P (2017) Mass processing of Sentinel-1 images for maritime surveillance. Remote Sens 9(7):678. https://doi.org/10.3390/rs9070678

Scheffer M (2009) Critical transitions in nature and society. Princeton University Press, Princeton

Scheffer M, Carpenter SR (2003) Catastrophic regime shifts in ecosystems: linking theory to observation. Trends Ecol Evol 18:648–656. https://doi.org/10.1016/j.tree.2003.09.002

Scheffer M, Carpenter S, Foley JA, Folke C, Walker B (2001) Catastrophic shifts in ecosystems. Nature 413:591–596. https://doi.org/10.1038/35098000

Shearman P, Bryan J, Walsh JP (2013) Trends in deltaic change over three decades in the Asia-Pacific region. J Coastal Res 29(5):1169–1183. https://doi.org/10.2112/JCOASTRES-D-12-00120.1

Shirzaei M, Bürgmann R (2018) Global climate change and local land subsidence exacerbate inundation risk to the San Francisco Bay Area. Science Advances 4(3):eaap923. https://doi.org/10.1126/sciadv.aap9234

Shulz K, Hansch R, Sorgel U (2018) Machine learning methods for remote sensing applications: an overview. In: Earth resources and environmental remote sensing/GIS applications IX, vol 10790. International Society for Optics and Photonics, p 1079002

Slabbekoorn H, Bouton N, van Opzeeland I, Coers A, ten Carte C, Popper AN (2010) A noisy spring: the impact of globally rising underwater sound levels on fish. Trends Ecol Evol 25:419–427. https://doi.org/10.1016/j.tree.2010.04.005

Small C, Nicholls RJ (2003) A global analysis of human settlement in coastal zones. J Coast Res 19(3):584–599. https://doi.org/10.2307/4299200

Smith WHF, Sandwell DT (1997) Global seafloor topography from satellite altimetry and ship depth soundings. Science 277:1957–1962

Sosik HM (2008) Characterizing seawater constituents from optical properties. In: Babin M, Roesler CS, Cullen JJ (eds) Real-time coastal observing systems for ecosystem dynamics and harmful algal blooms. UNESCO, Paris, pp 281–329. http://hdl.handle.net/11329/305

Sotillo MG, Cerralbo P, Lorente P, Grifoll M, EspinoM Sanchez-Arcilla A, Álvarez- Fanjul E (2019) Coastal ocean forecasting in Spanish ports: the SAMOA operational service. J Oper Oceanogr. https://doi.org/10.1080/1755876X.2019.1606765

Staneva J, Wahle K, Koch W, Behrens A, Fenoglio-Marc L, Stanev EV (2016) Coastal flooding: impact of waves on storm surge during extremes—a case study for the German Bight. Nat Hazards Earth Syst Sci 16:2373–2389

Steffen W, Rockström J, Richardson K, Lenton TM, Folke C, Liverman D, Summerhayes CP, Barnosky AD, Cornell SE, Crucifix M, Donges JF, Fetzer I, Lade SJ, Scheffer M, Winkelmann R, Schellnhuber HJ (2018) Trajectories of the earth system in the anthropocene. Proc Natl Acad Sci 115:8252

Strozzi T, Teatini P, Tosi L, Wegmüller U, Werner C (2013) Land subsidence of natural transitional environments by satellite radar interferometry on artificial reflectors. J Geophys Res Earth Surf 118:1177–1191. https://doi.org/10.1002/jgrf.20082

Sudre J, Maes C, Garçon V (2013) On the global estimates of geostrophic and Ekman surface currents. Limnol Oceanogr Fluids Environ 3:1–20. https://doi.org/10.1215/21573689-2071927

Swingedouw D, Ifejika-Speranza C, Bartsch A, Durand G, Jamet C, Beaugrand G, Conversi A (in review) Early warning from space for a few key tipping points in physical and biological systems, and in societies. Surv Geophys

Tadono T, Nagai H, Ishida H, Oda F, Naito S, Minakawa K, Iwamoto H (2016) Generation of the 30-m mesh global digital surface model by ALOS PRISM. Int Arch Photogramm Remote Sens Spat Inf Sci XLI-B4:157–162. https://doi.org/10.5194/isprsarchives-XLI-B4-157-2016

Tateishi R, Akutsu A (1992) Relative DEM production from SPOT data without GCP. Int J Remote Sens 13(14):2517–2530. https://doi.org/10.1080/01431169208904061

Teatini P, Tosi L, Strozzi T, Carbognin L, Wegmüller U, Rizzetto F (2005) Mapping regional land displacements in the Venice coastland by an integrated monitoring system. Remote Sens Environ 98(4):403–413. https://doi.org/10.1016/j.rse.2005.08.002

Toimil A, Camus P, Losada IJ, Le Cozannet G, Nicholls RJ, Idier D, Maspataud A (2020) Climate change driven coastal erosion modelling in temperate sandy beaches: methods and uncertainty treatment. Earth Sci Rev 202:103110. https://doi.org/10.1016/j.earscirev.2020.103110

Tonani M, Balmaseda M, Bertino L, Blockley E, Brassington G, Davidson F et al (2015) Status and future of global and regional ocean prediction systems. J Oper Oceanogr 8:s201–s220. https://doi.org/10.1080/1755876X.2015.1049892

Tosi L, Teatini P, Carbognin L, Brancolini G (2009) Using high resolution data to reveal depth-dependent mechanisms that drive land subsidence: the Venice coast, Italy. Tectonophysics 474:271–284. https://doi.org/10.1016/j.tecto.2009.02.026

Tosi L, Da Lio C, Strozzi T, Teatini P (2016) Combining L- and X-band SAR interferometry to assess ground displacements in heterogeneous coastal environments: the Po River Delta and Venice Lagoon, Italy. Remote Sens 8:308. https://doi.org/10.3390/rs8040308

Tozer B, Sandwell DT, Smith WHF, Olson C, Beale JR, Wessel P (2019) Global bathymetry and topography at 15 arc sec: SRTM15+. Earth Space Sci. https://doi.org/10.1029/2019EA000658

Turner W, Rondinini C, Pettorelli N, Mora B, Leidner AK, Szantoi Z, Buchanan G, Dech S, Dwyer J, Herold M, Koh LP, Leimgruber P, Taubenboeck H, Wegmann M, Wikelski M, Woodcock C (2015) Free and open-access satellite data are key to biodiversity conservation. Biol Conserv 182:173–176. https://doi.org/10.1016/j.biocon.2014.11.048

Twele A, Cao W, Plank S, Martinis S (2016) Sentinel-1-based flood mapping: a fully automated processing chain. Int J Remote Sens 37(13):2990–3004. https://doi.org/10.1080/01431161.2016.1192304

Ullman DS, O'Donnell J, Kohut J, Fake T, Arthur Allen A (2006) Trajectory prediction using HF radar surface currents: Monte Carlo simulations of prediction uncertainties. J Geophys Res 111:C12005. https://doi.org/10.1029/2006JC003715

UNCTAD (2017) Review of maritime transport 2017. United Nations conference on trade and development. United Nations, Geneva, Switzerland. http://unctad.org/en/PublicationsLibrary/rmt2017_en.pdf

United Nations (UN) (2015) Transforming our world: the 2030 agenda for sustainable development; Resolution adopted by the General Assembly. UN, Rome, Italy

United Nations (UN) (2018) Nineteenth meeting of the united nations open-ended informal consultative process on oceans and the law of the sea: anthropogenic underwater noise. 18–22 June 2018, New York

USACE (1996) Risk-based analysis for flood damage reduction studies, Report EM1110-2-1619. United States Army Corps of Engineers, Washington

Vafeidis AT, Schuerch M, Wolff C, Spencer T, Merkens JL, Hinkel J, Lincke D, Brown S, Nicholls RJ (2019) Water-level attenuation in global-scale assessments of exposure to coastal flooding: a sensitivity analysis. Nat Hazards Earth Syst Sci 19:973–984. https://doi.org/10.5194/nhess-19-973-2019

van der Horst T, Rutten MM, van de Giesen NC, Hanssen RF (2018) Monitoring land subsidence in Yangon, Myanmar using Sentinel-1 persistent scatterer interferometry and assessment of driving mechanisms. Remote Sens Environ 217:101–110. https://doi.org/10.1016/j.rse.2018.08.004

van Sebille E, Griffies SM, Abernathey R, Adams TP, Berloff P, Biastoch A et al (2017) Lagrangian ocean analysis: fundamentals and practices. Ocean Model 121:49–75. https://doi.org/10.1016/j.ocemod.2017.11.008

van Sebille E et al (2020) The physical oceanography of the transport of floating marine debris. Environ Res Lett 15:023003

Vanhellemont Q, Ruddick K (2014) Turbid wakes associated with offshore wind turbines observed with Landsat 8. Remote Sens Environ 145:105–115. https://doi.org/10.1016/j.rse.2014.01.009

Vanhellemont Q, Ruddick K (2015) Advantages of high quality SWIR bands for ocean colour processing: EXAMPLES from Landsat-8. Remote Sens Environ 161:89–106. https://doi.org/10.1016/j.rse.2015.02.007

Vanhellemont Q, Ruddick K (2018) Atmospheric correction of metre-scale optical satellite data for inland and coastal water applications. Remote Sens Environ 216:586–597. https://doi.org/10.1016/j.rse.2018.07.015

Vanhellemont Q, Neukermans G, Ruddick K (2014) Synergy between polar-orbiting and geostationary sensors: remote sensing of the ocean at high spatial and high temporal resolution. Remote Sens Environ 146:49–62. https://doi.org/10.1016/j.rse.2013.03.035

Velotto D, Soccorsi M, Lehner S (2014) Azimuth ambiguities removal for ship detection using full polarimetric X-band SAR data. IEEE Trans Geosci Remote Sens 52(1):76–88

Viatte C, Clerbaux C, Maes C, Daniel P, Garello R, Safieddine S, Ardhuin F (2020) Air and sea pollution seen from space. Surv Geophys. https://doi.org/10.1007/s10712-020-09599-0

Villas Bôas AB, Ardhuin F, Ayet A, Bourassa MA, Brandt P, Chapron B, Cornuelle BD, Farrar JT, Fewings MR, Fox-Kemper B, Gille ST, Gommenginger C, Heimbach P, Hell MC, Li Q, Mazloff MR, Merrifield ST, Mouche A, Rio MH, Rodriguez E, Shutler JD, Subramanian AC, Terrill EJ, Tsamados M, Ubelmann C, van Sebille E (2019) Integrated observations of global surface winds, currents, and waves: requirements and challenges for the next decade. Front Mar Sci 6:425. https://doi.org/10.3389/fmars.2019.00425

Vinogradov SV, Ponte RM (2011) Low-frequency variability in coastal sea level from tide gauges and altimetry. J Geophys Res 116:C07006. https://doi.org/10.1029/2011JC007034

Vousdoukas MI, Mentaschi L, Voukouvalas E, Verlaan M, Jevrejeva S, Jackson LP, Feyen L (2018) Global probabilistic projections of extreme sea levels show intensification of coastal flood hazard. Nat Commun 9:2360. https://doi.org/10.1038/s41467-018-04692-w

Warren MA, Simis SGH, Martinez-Vicente V, Poser K, Bresciani M, Alikas K, Spyrakos E, Giardino C, Ansper A (2019) Assessment of atmospheric correction algorithms for the Sentinel-2A MultiSpectral Imager over coastal and inland waters. Remote Sens Environ 225:267–289. https://doi.org/10.1016/j.rse.2019.03.018

WCRP Global Sea Level Budget Group (2018) Global sea-level budget 1993–present. Earth Syst Sci Data 10:1551–1590

Wegmüller U, Werner C, Strozzi T, Wiesmann A (2004) Multi-temporal interferometric point target analysis. Ser Remote Sens Anal Multi-Tempor Remote Sens Images. https://doi.org/10.1142/97898 12702630_0015

Wegmüller U, Santoro M, Werner C, Strozzi T, Wiesmann A, Lengert W (2009) DEM generation using ERS–ENVISAT interferometry. J Appl Geophys 69:51–58. https://doi.org/10.1016/j.jappgeo.2009.04.002

Weijerman M, Lindeboom H, Zuur AF (2005) Regime shifts in marine ecosystems of the North Sea and Wadden Sea. Mar Ecol Prog Ser 298:21–39. https://doi.org/10.3354/meps298021

Wells S, Ravilious C, Corcoran E (2006) In the front line: shoreline protection and other ecosystem services from Mangroves and Coral Reefs. vol 36. UNEP-WCMC

Werdell PJ, McKinna LI, Boss E, Ackleson SG, Craig SE, Gregg WW, Stramski D et al (2018) An overview of approaches and challenges for retrieving marine inherent optical properties from ocean color remote sensing. Prog Oceanogr 160:186–212

Wiehle S, Pleskachevsky A (2018) Bathymetry derived from Sentinel-1 Synthetic Aperture Radar. EUSAR 2018, https://elib.dlr.de/118174/. Accessed Sept 2019

Wilandsky MJ, Marra JJ, Chowdhury MR, Stephens SA, Miles ER, Fauchereau N, Spillman CM, Smith G, Beard G, Wells J (2017) Multimodel ensemble sea level forecasts for tropical pacific islands. J Appl Meteorol Climatol 56:849–862

Wölfl AC, Snaith H, Amirebrahimi S, Devey CW, Dorschel B, Ferrini V, Huvenne VAI, Jakobsson M, Jencks J, Johnston G, Lamarche G, Mayer L, Millar D, Pedersen TH, Picard K, Reitz A, Schmitt T, Visbeck M, Weatherall P, Wigley R (2019) Seafloor mapping—the challenge of a truly global ocean bathymetry. Front Mar Sci 6:283. https://doi.org/10.3389/fmars.2019.00283

Wong PP, Losada IJ, Gattuso J-P, Hinkel J, Khattabi A, McInnes KL, Saito Y, Sallenger A (2014) Coastal systems and low-lying areas. In: Field CB, Barros VR, Dokken DJ, Mach KJ, Mastrandrea MD, Bilir TE, Chatterjee M, Ebi KL, Estrada YO, Genova RC, Girma B, Kissel ES, Levy AN, MacCracken S, Mastrandrea PR, White LL (ed) Climate change 2014: impacts, adaptation, and vulnerability. Part A: global and sectoral aspects. Contribution of Working Group II to the Fifth

Assessment Report of the Intergovernmental Panel on Climate Change. Cambridge University Press, Cambridge, pp 361-409

Woodworth P, Hunter JR, Marcos M, Caldwell P, Menéndez M, Haigh I (2017) Towards a global higher-frequency sea level dataset. Geosci Data J 3:50–59

Woodworth P, Melet A, Marcos M, Ray RD, Wöppelmann G, Sasaki YN, Cirano M, Hibbert A, Huthnance JM, Montserrat S, Merrifield MA (2019) Forcing factors affecting sea level changes at the coast. Surv Geophys 40(6):351–1397. https://doi.org/10.1007/s10712-019-09531-1

Wöppelmann G, Le Cozannet G, De Michele M, Raucoules D, Cazenave A, Garcin M, Hanson S, Marcos M, Santamaría-Gómez A (2013) Is land subsidence increasing the exposure to sea level rise in Alexandria, Egypt? Geophys Res Lett 40(12):2953–2957. https://doi.org/10.1002/grl.50568

Yang Y, Pepe A, Manzo M, Bonano M, Liang DN, Lanari R (2013) A simple solution to mitigate noise effects in time-redundant sequences of small baseline multi-look DInSAR interferograms. Remote Sens Lett 4(6):609–618

Ye S, Luo Y, Wu J, Yan X, Wang H, Jiao X, Teatini P (2016) Three-dimensional modeling of land subsidence in Shanghai, China. Hydrogeol J 24(3):695–709. https://doi.org/10.1007/s10040-016-1382-2

Zanello F, Teatini P, Putti M, Gambolati G (2011) Long term peatland subsidence: experimental study and modeling scenarios in the Venice coastland. J Geophys Res Earth Surface 116:F04002. https://doi.org/10.1029/2011JF002010

Zechetto S, De Biasio F, Della Valle A, Cucco A, Quattrocchi G, Cadau E (2016) Wind fields from COSMO-SKYMED and Radarsat-2 SAR in coastal areas. In: IEEE international geoscience and remote sensing symposium (IGARSS), IEEE, pp 1535–1538. https://doi.org/10.1109/IGARSS.2015.7326073

Zhang Y, Wang C, Wu J, Qi J, Salas WA (2009) Mapping paddy rice with multitemporal ALOS/PALSAR imagery in southeast China. Int J Remote Sens 30(23):6301–6315. https://doi.org/10.1080/01431160902842391

Zhang Y, Wu H, Kang Y, Zhu C (2016) Ground subsidence in the Beijing–Tianjin–Hebei region from 1992 to 2014 revealed by multiple SAR stacks. Remote Sens 8(8):675. https://doi.org/10.3390/rs8080675

Zoccarato C, Minderhoud PSJ, Teatini P (2018) The role of sedimentation and natural compaction in a prograding delta: insights from the mega Mekong delta, Vietnam. Sci Rep 8:11437. https://doi.org/10.1038/s41598-018-29734-7

Publisher's Note Springer Nature remains neutral with regard to jurisdictional claims in published maps and institutional affiliations.

Affiliations

A. Melet[1] · P. Teatini[2] · G. Le Cozannet[3] · C. Jamet[4] · A. Conversi[5] · J. Benveniste[6] · R. Almar[7]

[1] Mercator Ocean International, 8 rue Hermès, 31520 Ramonville-Saint-Agne, France

[2] Department of Civil, Environmental and Architectural Engineering, University of Padova, Via Marzolo 9, 3531 Padua, Italy

[3] BRGM, 45000 Orléans, France

[4] Univ. Littoral Cote d'Opale, Univ. Lille, CNRS, UMR 8187, LOG, Laboratoire d'Océanologie et de Géosciences, 62930 Wimereux, France

[5] National Research Council of Italy, CNR-ISMAR-Lerici, Forte Santa Teresa, Loc. Pozzuolo, 19032 Lerici, SP, Italy

[6] European Space Agency (ESA-ESRIN), Largo Galileo Galilei, 1, 00044 Frascati, Italy

[7] LEGOS, Université de Toulouse, CNES, CNRS, IRD, UPS, 31400 Toulouse, France

Surveys in Geophysics (2020) 41:1535–1581
https://doi.org/10.1007/s10712-020-09616-2

Contributions of Space Missions to Better Tsunami Science: Observations, Models and Warnings

H. Hébert[1] · G. Occhipinti[2] · F. Schindelé[1] · A. Gailler[1] · B. Pinel-Puysségur[1] ·
H. K. Gupta[3] · L. Rolland[4] · P. Lognonné[2] · F. Lavigne[5] · E. Meilianda[6] ·
S. Chapkanski[5] · F. Crespon[7] · A. Paris[1] · P. Heinrich[1] · A. Monnier[1] · A. Jamelot[8] ·
D. Reymond[8]

Received: 13 December 2019 / Accepted: 22 September 2020 / Published online: 9 October 2020
© Springer Nature B.V. 2020

Abstract

Most tsunamis occur after large submarine earthquakes, particularly in the Pacific Ocean. However, following the 2004 tsunami in the Indian Ocean, tsunami hazard awareness was significantly raised at the global scale, and warning systems were developed in many other regions, where large tsunamis are rarer but can also produce large catastrophes. Here we first review the basic physics of a tsunami, from its triggering to its coastal impact, and we offer a review of the geophysical and sea-level data that can describe the various processes operating during a tsunami. Global Navigation Satellite System (GNSS) data have a key role in better describing the ground deformation following a tsunamigenic earthquake close to the coast. The GNSS observations complement seismological data to constrain the rupture model rapidly and robustly. Interferometric Synthetic Aperture Radar (SAR) also contributes to this field, as well as optical imagery, relevant to monitoring elevation changes following subaerial landslides. The observation of the sea-level variations, in the near field and during the propagation across the ocean, can also increasingly benefit from GNSS data (from GNSS buoys) and from robust satellite communication: pressure gauges anchored on the seafloor in the deep ocean contribute to warning systems only by data continuously transmitted through satellites. The sounding of ionospheric Total Electron Content (TEC) variations through GNSS, altimetry, or a ground-based airglow camera, is a promising way to record tsunami initiation and propagation indirectly. Finally, GNSS, optical and SAR imagery are essential to map and quantify the damage following tsunami flooding. Satellite data are expected to contribute more to operational systems in the future provided they are reliably available and analysed in real time.

Keywords Tsunami · Early Warning · GNSS · Altimetry · SAR · Ionosphere

✉ H. Hébert
   helene.hebert@cea.fr

Extended author information available on the last page of the article

1 Introduction to Current Challenges in Tsunami Science and Warnings

The twenty-first century started with giant catastrophic tsunamis, among them the most massive natural disasters observed in 50 years. Following subduction earthquakes of extreme magnitudes, in 2004 off Sumatra (Indonesia, Mw = 9.1), and in 2011 off Tohoku (Japan, Mw = 9.1), tsunami waves were triggered and propagated across the global ocean, after having caused enormous destruction on the local coastal areas, nearshore, and as far as several kilometres inland in some locations. These events were the starting point of a general rise of awareness of this geophysical hazard, which, although known about in a limited scientific community in the 20th century, had very rarely occurred with that level of force since the 1960s.

Tsunamis, which find their origin in geological processes of the Earth, were relatively well understood and characterized in the scientific community before 2004, especially around the Pacific Ocean, where they occur most frequently. Nevertheless, the 2004 and 2011 events surprised most of the tsunami scientific community, essentially because they took place in underestimated, hazardous areas. Indeed, from a physical point of view, any subduction earthquake with a large enough magnitude, shallow depth and significant thrusting component in its focal mechanism, is expected to trigger a tsunami, and available tsunami catalogues underline the tsunami source areas as correlated to subductions (Fig. 1). However, the fact that such extreme earthquakes occurred off Sumatra and of Tohoku was underrated in the beginning of the 21st century, partly because of very long return periods for such events (above 500 years). The subsequent consequences were due to the lack of or the failure of all or part of the basic components of tsunami warning and mitigation, despite being well known at that time (Bernard 2001), i.e. a proper estimation of the geophysical hazard level, an efficient warning system, and a high level of coastal preparedness: in 2004, the lack of regional and national warning systems, and, in 2011, the underestimation of the earthquake magnitude and of the protection design at the Fukushima Daiichi nuclear power plant, in an otherwise well-prepared country.

Thus, in 2004 and 2011, the human, industrial and natural coastal vulnerabilities were particularly and tragically highlighted. Yet tsunami science can benefit from favourable conditions which allow the prediction, as soon as the origin has been detected, of propagation times that give sufficient time for preventative measures. This is essentially the basic principle of the tsunami warning, which can be operated mainly for earthquake-related tsunamis. It is worth noting that, by contrast, landslide-related tsunamis cannot often take advantage of the source detection. While regional seismic networks can detect major mass failures, smaller events with significant tsunamigenic potential are challenging to characterize unless local dense geophysical networks are available.

The first regional system was established in 1965 in the Pacific Ocean in the general framework and governance of the UNESCO Intergovernmental Oceanographic Commission (IOC 1965). This organization has strengthened cooperation between systems deployed since the 1940s in Japan and the USA, in the 1950s in Russia, and in the 1960s in French Polynesia (e.g., Kong et al. 2015). In December 2004, seismologists on duty in the Pacific Tsunami Warning Center (PTWC) tried to quickly share information with institutions or authorities in the Indian Ocean. Still, no operational system was in place to address tsunami warnings, the tsunami hazard being underrated at that time. Soon afterwards, the international community had fostered the setting up of Tsunami Warning Systems (hereafter TWS) in every oceanic basin exposed to such a hazard (IOC 2005a,

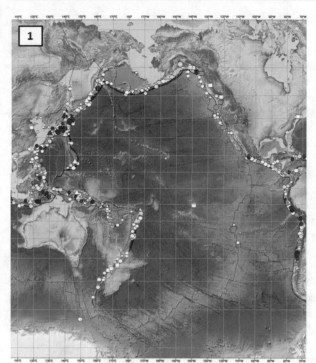

Pacific Ocean

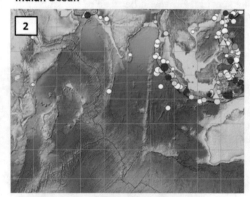

Indian Ocean

North East Atlantic and Mediterranean (NEAM)

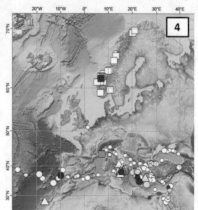

Carribean

Fig. 1 Map of the major tsunami sources from 1610 B.C. to A.D. 2017, triggered by earthquakes (circles), volcanic eruptions (triangles), landslides (squares), and other causes (NCEI 2017), displayed for the four regions monitored by Tsunami Warning Systems: (1) Pacific Ocean, (2) Indian Ocean, (3) Caribbean and (4) North-East Atlantic and Mediterranean (NEAM) region. Most of the tsunamis are triggered by earthquakes (circles)

b, c). Based on national initiatives and funding, warning systems have been developed in numerous countries and oceanic basins since then.

The different systems are building their capacities and managing the operation procedures through four Intergovernmental Coordination Groups (ICGs). Each of them is responsible for a basin-wide policy (Pacific Ocean, Indian Ocean, Caribbean, North-East Atlantic and Mediterranean Sea) which is built with global interoperability (Fig. 2). For such natural transoceanic hazards, interoperability and data exchanges are naturally essential, and much progress has been made since the 1960s when the Pacific system was in its early stages and when communication means were mostly by fax, phone, and email since the 1990s.

This specific domain of global telecommunication is a prominent topic where satellite usage drastically improved the efficiency of the real-time systems. This, of course, does not constitute the only segment where such data are nowadays vital. In the following examples, we aim to illustrate how space missions significantly contribute to the improvement in tsunami science, especially under the three main headings mentioned above (hazard, warning, mitigation and preparedness):

- Optical imagery may have been the first historical contribution to tsunami science, naturally helping to draw post-disaster surveys and to establish mitigation plans and risk maps. It is often complemented by SAR (Synthetic Aperture Radar) imagery, which also helps to characterize the coseismic deformation, as well as the mapping of coastal damage;
- Altimetry has dramatically contributed to oceanography and oceanic geodynamics since the 1980s. As such a technique maps the oceanic surface, it may be expected to significantly contribute to tsunami remote sensing. However, because of the small number of satellites and the small amplitude of the majority of tsunamis in the deep ocean, it has very rarely proved successful on this issue;

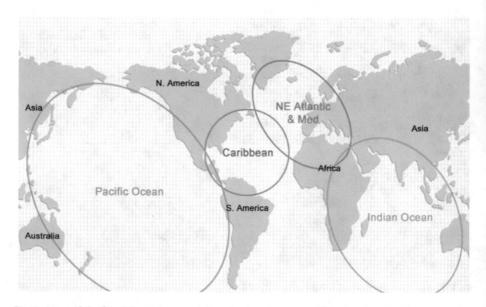

Fig. 2 Areas of the four Intergovernmental Coordination Groups established under UNESCO/IOC framework, following the 2004 tsunami (version from UNESCO/IOC 2008)

- GNSS data (Global Navigation Satellite System, such as Global Positioning System or GPS, Galileo, Glonass, Beidou) are probably among the most important contributors to tsunami science, helping to better map the post-disaster damages and the vulnerable areas with the objective of improving vulnerability assessment and prevention plans, providing more accurate bathymetric data with improved positioning of ship soundings, and, in the event of a tsunami, promptly contributing to a subtle coseismic source imagery and to nearshore measures of sea-level variations. Recently GNSS-reflectometry (GNSS-R) can estimate multiple water height measurements in the case of an optimal GNSS network;
- More recently, the estimation of the variations of the ionospheric TEC (Total Electron Content) (from GNSS, altimetry, airglow camera), due to the coupling of tsunami waves upwards in the atmosphere and ionosphere, has offered convincing pictures of tsunami propagation and may contribute to a promising real-time perspective;
- Hyperspectral data provide a promising way of mapping shallow water depths for numerical modelling dedicated to prevention plans.

Not all these various aspects can be efficient in real-time, depending on how long the data processing and communication means can deliver early information, and the extent to which they rely on ground receivers which are more or less sparsely located. However, they all should play a growing role in the future, not only to improve scientific knowledge but also to contribute to warning and emergency services.

In this paper, we first introduce basic tsunami physics and modelling methods (Sect. 2), then we show how space data increasingly contribute to the tsunami source characterization (Sect. 3), to tsunami height measurements (Sect. 4), and to post event surveys (Sect. 5). Finally we show which selected space techniques can play a growing role in tsunami warning systems (Sect. 6), once they have been validated and made available in real time. Finally we synthesize in the conclusions the maturity level of the different techniques with respect to the various goals addressed, whether scientific, modelling or warning.

2 Basic Physics of Tsunamis

2.1 How a Tsunami is Triggered and Propagates

A tsunami consists of gravity waves propagating across the ocean following a geological event affecting the ocean floor or coastal seafloor, such as an earthquake, a landslide, or a volcanic eruption, all of which can transmit potential energy to the oceanic layer (Fig. 3). The periods of the triggered tsunami waves range from a few tens of seconds (for landslides) to a few tens of minutes (large earthquakes). Their oceanic propagation obeys the gravity wave dispersion equation. The tsunami wavelengths are typically several tens of kilometres that are generated by great earthquakes. This is thus much larger than the water depth (from a few hundred to thousands of metres). In that case, the shallow water assumption leads to a speed $c = \sqrt{gh}$ where g is the acceleration due to gravity and h the water depth.

Propagation of tsunami waves across the oceans takes several hours to more than a day, with velocities reaching 500–800 km/h in the far deep sea or ocean. Tsunami frequency dispersion, especially for smaller wavelengths, corresponds mostly to small landslide sources. Shorter wavelength waves travel slower, and sometimes arrive hours

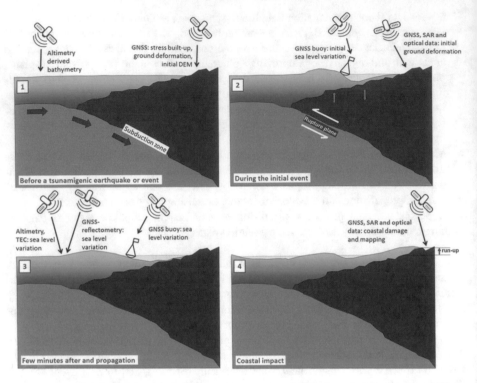

Fig. 3 Different stages of a tsunami following a subduction earthquake, from the stress build-up (1), the coseismic triggering (2), the first minutes after and the subsequent propagation (3), and the coastal impact (4). Each step can be monitored through various space data

after the first arrival at very far distances. The tsunamis with shortest wavelengths, such as those due to landslides, are essentially confined to short distances from the source due to rapid attenuation (e.g., Labbé et al. 2012; Paris et al. 2019).

Tsunamis also trigger perturbations upwards into the atmosphere and subsequently into the ionosphere (Fig. 4). Indeed, tsunami long wavelengths favour a vertical coupling and then propagation of the induced gravity waves in the atmosphere up to the ionosphere, where they in turn induce variations in the TEC (Total Electron Content). The large decrease of the air density with the altitude causes a significantly stronger tsunami-coupled atmospheric wave. Amplified wavefield travels in the ionosphere after coupling between the neutral atmosphere and the ionospheric plasma (e.g., Artru et al. 2005; Occhipinti et al. 2006). The ionospheric record of this upward propagation is made possible by the amplification of the electron density perturbation in the ionospheric plasma at the F-region (at about 200-300 km height). The efficiency of this last coupling strongly depends on the geomagnetic field inclination (Occhipinti et al. 2008): the tsunamigenic perturbation in the F-region plasma is more easily observed at equatorial and mid-latitudes than at the high latitudes. However, where efficient, this underlines how global a tsunami can be, throughout the surface of the oceans as well as in the various atmospheric layers above.

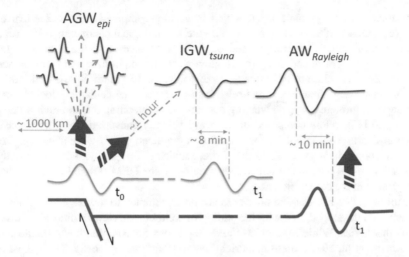

Fig. 4 Atmospheric/ionospheric waves generated by earthquakes and tsunamis: the vertical displacement at the epicentral area or at a teleseismic distance (induced by tsunamis and Rayleigh waves) produces, in the near field, an acoustic-gravity wave coupled with the uplift at the source (AGWepi); in the far-field, an internal gravity wave coupled with the tsunami (IGWtsuna), and a pure acoustic wave coupled with the Rayleigh wave (AWRayleigh). During the upward propagation, the AGWepi, IGWtsuna and AWRayleigh are strongly amplified by the effect of the exponential decrease in the air density with increasing height (from Occhipinti 2015)

2.2 Spatial and Time Scales of a Tsunami

This basic review of a tsunami's life indicates that various temporal and spatial scales are at play during the event. While the original source takes a few minutes to couple with the oceanic layer, and implying dimensions from several hundreds of metres to more than a thousand kilometres at most, the subsequent propagation involves hours of sea-level variations over distances exceeding several thousands of kilometres (in some cases related to large earthquakes) while tsunamis initiated from landslides propagate only a shorter distance.

While the tsunami offshore propagation is characterized by wave amplitudes smaller than one metre in most cases, the nonlinear amplification close to the coasts leads to very large wave amplitudes, damaging the coastlines and sometimes reaching as far as several kilometres inland. The powerful flow of water flooding, arriving on the coast with velocities from 20 to 40 km/h, is responsible for the serious destruction of vulnerable buildings and facilities. It is able to cause casualties as soon as the water height is greater than about 50 cm, and also to reach several kilometres inland for the largest tsunamis.

2.3 Modelling Approach

Tsunami hazards can be efficiently studied through numerical methods that have considerably improved since the first developments in the 1980s. The tsunami wavelengths are large compared to the water depth (several hundreds of metres) in most of the different stages of the tsunami's life. The initiation is begun by an instantaneous coupling of the ground deformation with the oceanic layer, which is usually computed with a classical elastic model (e.g., Okada 1985) taking into account the seismological parameters of the rupture.

Numerical methods, generally based on finite differences or finite element schemes, have mostly been developed using the shallow water assumption, which can also account for nonlinear terms close to the shore (e.g., Synolakis and Bernard 2006). Under the shallow water approach, there is a direct relationship between the tsunami velocity c and the water depth h, through $c = \sqrt{gh}$ where g denotes the acceleration due to gravity. The propagation grid is therefore composed of the wave velocity converted from the water depth at each grid cell. To account for the nearshore amplification and wave shoaling, the grid resolution should increase close to the shore, requiring finer bathymetry. Eventually, the computation of the possible run-up values requires a fine DEM (Digital Elevation Model) of the ground in the case of flooding.

As for landslide sources, the processes are more complex to model, and various methods have been developed, first with the sliding of solid blocks, or with fluid-like flows with various rheologies. While 3D Navier–Stokes equations for complex flows should be used (e.g., Abadie et al. 2012), these numerical methods are usually costly. Thus, most of the methods use 2D depth-integrated approaches, when it is relevant (e.g., Heinrich et al. 2001; Løvholt et al. 2015), with constant viscosity, or depending on the flow dynamics such as shear stress evolution (Assier-Rzadkiewicz et al. 1997); internal friction laws for granular flow can also be applied (Mangeney et al. 2000).

The coupling with water waves requires addressing frequency dispersion induced by the short wavelengths of the triggered tsunami waves, as observed in various experiments (Fritz et al. 2004, Mohammed and Fritz 2012), and handled by solving weakly nonlinear Boussinesq equations (e.g., Glimsdal et al. 2013). Numerical methods addressing both the landslide and the wave propagation are efficient for case studies, either for past or recent events (Poupardin et al. 2017) and/or for hazard scenarios (Paris et al. 2019).

While numerical modelling has been crucial in better estimating tsunami hazards, very detailed bathymetric and topographic data are required to accurately simulate coastal amplification. In addition, the parameters input into the models also need fine characterization of the geophysical source. In the following section, we shall show how space data can contribute to all these various aspects.

3 Contributions of Space Data to Improved Tsunami Source Assessment

Tsunami hazard prevention partly relies on numerical modelling which has been greatly improved in recent years, not only thanks to high-performance computing, but also to more accurate bathymetric and topographic data, some of which is directly derived from satellite data. However, one should not forget that numerical techniques should be controlled in their various steps, using robust data, validated schemes and keeping in mind their uncertainties. One point is to rely on validated benchmarked problems (and their progress is constantly noteworthy, e.g., Synolakis and Bernard 2006; Lynett et al. 2017); another point is undoubtedly to enter the correct data and parameters into the models. In addition, a tsunami results from the coupling between a specific submarine ground deformation, and the overlying oceanic layer. Thus, two distinct layers (crust and ocean) need to be scrutinized in order to gather data describing the physical processes involved. Space missions can help to address this challenge—to better understand the initiation, the propagation, and the flooding related to a given tsunami. In this section, we introduce the main techniques used nowadays to characterize tsunami source, independently from their possible operational use addressed in Sect. 6.

Hazard studies first need a relevant description of tsunami sources, based on catalogues and/or syntheses of recent well-documented events. The fact that hazard studies may be conducted under deterministic or probabilistic approaches is not discussed in this paper, since both need the same kind of data anyway. Then a fine description and measurement of tsunami heights area is needed to better describe the propagation pattern and validate the models. Finally, identifying hazardous coastal areas requires the establishment of observational databases in the aftermath of tsunami flooding, or based on paleotsunami studies.

3.1 Study of the Seismic Cycle and Plate Coupling to Design Possible Tsunami Sources

Earthquakes are not predictable at the present time. However, a number of techniques based on satellite geodesy (GNSS, SAR) allow the monitoring of specific fault areas and seismogenic zones in order to better understand processes possibly linked to the trigger of seismic ruptures. Significant advances have been made since the 1990s in various seismically active areas (see Elliott et al. 2016; Gupta and Gahalaut 2015). Studies from satellite geodesy contribute to a better understanding of the different stages of the seismic cycle and offer a way to monitor the deformation rates, including time-dependent phenomena, such as slow slip events or transients that are observed in subduction zones and that are probably related to seismogenesis.

For submarine seismic areas, satellite geodesy naturally only gives a partial subaerial view, but is possibly also relevant for detailed coastal processes (e.g., Polcari et al. 2018). However, active submarine coseismic areas can also be monitored through a growing number of sensors, including ocean bottom seismic arrays, submarine cables (Howe et al. 2019), acoustic ranging submarine geodesy and promising optical fibre (Ajo-Franklin et al. 2019; Sladen et al. 2019), defining the so-called seafloor geodesy. Though primarily based on submarine networks, most of the methods use kinematic GNSS satellites to be tied to the global reference frame, and they provide promising insights into the estimation of accurate seafloor deformation, including some possible signals close in space and time to the largest tsunamigenic ruptures, such as in 2011 in Tohoku (Bürgmann and Chadwell 2014).

3.2 Characterization and Extent of the Rupture for an Earthquake

Following an earthquake, seismic networks at regional, national and global scales rapidly define the location, depth and magnitude of the event. However, the magnitude, location and depth of the earthquakes are not sufficient to characterize the tsunami hazard. This must be complemented with information on the rupture extent (whenever it was potentially submarine) and pattern (whenever it involved a dip slip component to efficiently couple with the oceanic layer). Seismology is able to provide such information, although the ambiguity on the double couple resolved through seismic tensor inversion does not define the accurate fault mechanism, unless local tectonics is well known, as in typical subductions, a task that is sometimes difficult under the sea.

The elastic deformation and the crustal rupture due to an earthquake imply significant ground deformation that can be measured by satellite data. The recent progress made in the GNSS data processing offers the more plausible and precise technique allowing for a quick computation of a ground deformation map/picture and thus the ability to extrapolate the possible extent of the rupture beneath the sea, depending on the rupture duration and GNSS network location and density close to the rupture zone. Such data contributes to joint inversions providing a finer pattern for the rupture and contributing to magnitude estimates (see, for example, for recent tsunamigenic earthquakes, Ozawa et al. 2011; Wang et al. 2013; Grandin et al. 2016) in addition to giving interpretation of the successive ruptures within the seismic cycle of a subducting plate (Moreno et al. 2010). GNSS data also bring constraints on the post-seismic sequence following tsunamigenic ruptures (Raharja et al. 2016), even in the specific case of slow earthquakes that produce tsunamis much larger than the magnitude level indicates, as in the case of the Java 2006 earthquake and tsunami (Hébert et al. 2012).

Interferometric SAR imagery analysis also gives valuable information for the coastal deformation induced by an earthquake, though in general offering only a truncated view since the technique does not reveal the seafloor deformation. In specific cases, such as in narrow bays surrounded by subaerial ground, this applies admirably, as in the recent Palu, 2018, earthquake and tsunami (Jamelot et al. 2019). As GNSS data, such ground deformation data can thus be inverted in terms of source pattern characterization.

However, recent major tsunamigenic earthquakes could not all be easily monitored through GNSS or InSAR data, being located far from coastal ground places where these techniques are able to provide measurements (Ammon et al. 2008; Lay et al. 2010), and seismological methods remain unavoidable and quite unique in these cases too. Only seafloor geodesy could complement ground-based GNSS data there.

Finally, the recently developed ionospheric seismology has been supported by the growing number of GNSS networks allowing for imaging the TEC variations induced by the seismic shaking. In the case of submarine tsunamigenic earthquakes, this technique brings an original view of the various processes occurring in the near field, during the minutes when the tsunami is initiated. This concept will be developed in Sect. 4.

3.3 Identification of Tsunamigenic Landslides

Landslide-prone sections in coastal places can also be monitored, using not only dense seismic arrays and tilt meters but also satellite methods as well. Geodetic techniques, including space borne and ground-based SAR data, are currently common methods of monitoring active volcanoes, as well as for major landslides threatening valleys (e.g., Gaffet et al. 2010; Carlà et al. 2019b). The data provides information on the ongoing ground deformation, which can be compared to mechanical models, and it also possibly contributes to operational warning when case values exceeding relevant thresholds are detected, such as on Stromboli (see Sect. 6).

In addition to GNSS and SAR methods useful to estimate ground deformation, the computation of differential DEM through ground space image matching (e.g., Guérin et al. 2014) also estimates change detection and helps to monitor landslide and slope deformation prior to collapses that can be tsunamigenic when they are at the coast. From that perspective, the recent study of the 2017 Greenland collapse and tsunami allowed the identification of a suspected possible rock volume (Fig. 5) that could be destabilized in the vicinity of the previous tsunamigenic failure in 2017, which produced flooding in a village approximately 30 km away from the source (Paris et al. 2019). Such a volume was consistent with estimates made from the analysis of seismic waveforms (Bessette-Kirton et al. 2017; Chao et al. 2018).

Fig. 5 Following the 2017 landslide and tsunami in western Greenland, difference between the "before" (Spot6 2013) and "after" (Pleiades 2017) digital surface models (DSM), superimposed on the "after" DSM, computed from the Pleiades tri-stereoscopic images from 2017. The variations of the difference elevation values (in m) are represented with the colour bar. The red ellipse indicates a potentially unstable area exposed to possible future failure (Figure adapted from Paris et al. 2019)

However, many tsunamigenic landslides have their origins solely on the seafloor, for which context very few detailed maps are generally available. In some cases, seismic or hydroacoustic networks operated in the near field are able to record seismic signals due to the destabilization and sliding of the volume on a rough interface (e.g., Caplan-Auerbach et al. 2001). But generally high-resolution bathymetric data available prior to and after the event naturally help to better characterize the landslide shape and volume, as in the recent case of the 2018 Krakatau submarine collapse (Paris et al. 2020), complemented by satellite optical or SAR imagery for the subaerial portion (Ye et al. 2020). As proper data are generally only available for intermediate or shallow depths, altimetry derived bathymetry is not of much use, and only high-resolution sounding data from ships can provide the necessary information.

3.4 Bathymetric and Topographic Data Requirements for Numerical Modelling

Initial attempts to model tsunami propagation relied on ray theory derived from geometrical optics and accounting for diffraction when strong velocity gradients occurred. For that purpose, the first global bathymetric datasets, derived from marine cruises and thus only roughly accounting for seafloor details (Smith 1993), were used in the first numerical methods which proved very useful to compute tsunami arrival times and to examine the first order amplifying bathymetric patterns (Satake 1988; Mader and Curtis 1991). The development of global bathymetry derived from satellite altimetry (Smith and Sandwell

1997) at that time provided strikingly more detailed maps from the ocean seafloor, and thus further enabled a more ocean-wide and accurate coverage, useful for tsunami numerical models (Titov and González 1997; Hébert et al. 2001).

Nonlinear effects leading to increased tsunami impacts along the coast require finer bathymetry close to the shore to constrain numerical models, whereas altimetry derived datasets are out of their validity domain. Such shallow bathymetry can be evaluated through the inversion of ocean wave refraction by innovative methods (Poupardin et al. 2014; Li et al. 2016). Although direct local measurements using airborne LIDAR are now contributing to very high-resolution datasets (for example the Litto3D® project in France), methods based on satellite hyperspectral data also offer an interesting perspective (Ma et al. 2014).

Merging fine bathymetry with fine low topography is essential to make tsunami modelling robust in coastal areas. Generally, tsunami models provide accurate estimations only when using grid cell sizes typically consisting of a few metres. Recent advances on the acquisition of coastal datasets, with a large numbers of methods (see, for example, Salameh et al. 2019, for a review) indicate that such models should be more and more robust in the areas at risk, and much more useful to tsunami hazard studies, not only for better preparedness, but also from the perspective of computing real-time forecasting (see Sect. 6).

4 Measuring Tsunami Wave Heights and Periods

Once the tsunami has been generated, it is able to propagate for hours, and to produce damaging impact thousands of kilometres away from its source. Transoceanic propagation is currently essentially monitored through conventional tide gauges in harbours, and, in the deep ocean, through deep ocean pressure gauges (such as the Deep Ocean Assessment and Reporting of Tsunamis, DART, Milburn et al. 1996). These data, complemented by tide gauges deployed in harbours, are the key sensors providing the physical features of the tsunami wave trains.

4.1 Importance of GNSS Data for Sea-Level Measurement Techniques

The network of usual tide gauges deployed in harbour areas was considerably developed on a global scale after the catastrophic 2004 tsunami (Angove et al. 2019), and sensors were also improved. Indeed, most of the coastal sea-level gauges now consist of radars which avoid floating devices and management of probing tubes in the sea, the way most tsunamis were measured in the past (e.g., Hébert et al. 2007, in La Réunion island).

In addition, at some distance from the coast, where measurements can be conducted on buoys and anchored devices, new gauges have been developed based on GNSS data. For example, Japan developed the GNSS buoy system during the late 1990s (Kato et al. 2010; Terada 2015) (Fig. 6 top). This tsunami monitoring system uses Real-Time Kinematic GPS technology to position a GNSS receiver mounted on top of a buoy, floating at the sea's surface, in relation to a land-based GNSS receiver, thus taking advantage of differential processing. After various successful experimental systems, in 2008, the Nationwide Ocean Wave information network for Ports and HArbourS (NOWPHAS 2020, Japan) was originally established with 10 GNSS buoys (extended to 18 sensors as of 2019), mostly along the Sanriku coast, an area identified as being prone to frequent and large tsunamis.

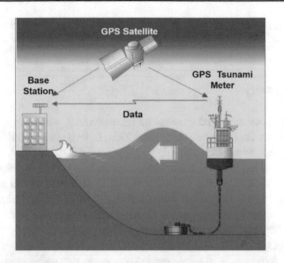

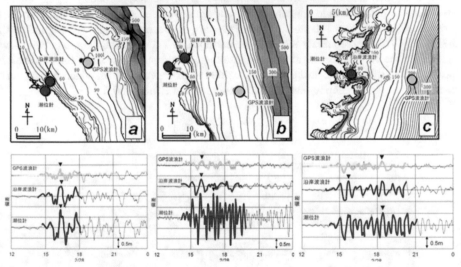

Fig. 6 Top: device on GNSS buoys deployed nearshore (figure taken from Kato et al. 2018). Such a buoy off Iwate prefecture was able to detect an unexpected 7 m high wave, 24 min after the March 2011 Tohoku earthquake, allowing for an updated warning level 4 min later (Ozaki 2012). Bottom: Subsets a, b and c show the deployment of such GNSS buoys (yellow dots) to complement conventional tide gauges located at the entrance of the bay (blue dots) and the bottom of the bay (red dots), for three different bays in north Japan (from Kawai et al. 2012). Different amplifications are evidenced for the tsunami coming from the Chile 2010 earthquake, about 22 h after the triggering, and showing different kinds of site effect amplification depending on the bay

These sensors also contribute to refined studies on the coastal amplification response to a tsunami arrival. For instance, the tsunami generated by the February 2010 Chilean earthquake was recorded on the Japanese coast after a 22-h propagation time, on tide gauges and GNSS buoys (Kawai et al. 2012) (Fig. 6 bottom). The data record illustrates how the tsunami amplification takes place on three different bays, and a quantification of the amplification factor (that also probably depends on the frequency) is enabled: each bay exhibits a specific site effect, related to its dimensions and depth. A better identification of these various coastal amplification patterns obviously plays a key role for the improvement in hazard estimation.

4.2 Satellite Altimetry

Due to their global coverage and continuous measurement, altimetry satellites have been considered to detect tsunami waves since the 1990s. Indeed such techniques aimed at studying the oceanic surface could intuitively produce useful data to monitor tsunami heights. However, the very small tsunami amplitudes expected in the deep ocean made the investigation very uncertain, since the altimetry signal is much hidden in mesoscale oceanic processes. The latter, however, usually implies spatial scales larger than those of tsunami waves, which can therefore be extracted as short wavelengths. A tentative detection was proposed during the 1992 Nicaragua tsunami, which was supported by tsunami numerical modelling (Okal et al. 1999).

In 2004, the detection was very impressive since the raw altimetry signal clearly exceeded the mean oceanic ambient signal and thus provided a crest-to-trough displacement amplitude of 80 to 100 cm on Topex/Poseidon and Jason platforms (Occhipinti et al. 2006) (Fig. 7), observed about 2 h after the earthquake. These outstanding data were also unique for the further investigation of seismological interpretation, since their inversion gave a picture of the slip distribution for the 2004 Sumatra earthquake, which proved very consistent with other geophysical inversions (Sladen and Hébert 2008).

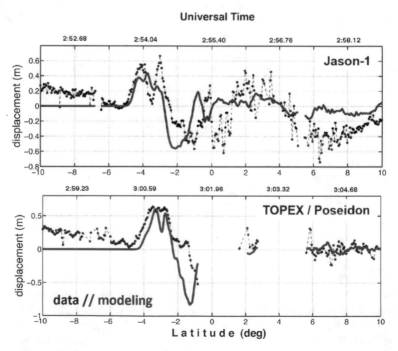

Fig. 7 Altimetry signatures of the 2004 Sumatra tsunami, with measurements of the ocean surface (black) for the Jason-1 (top) and Topex/Poseidon (bottom). The synthetic ocean displacements are shown in red (Figure from Occhipinti et al. 2006)

This observation still remains exceptional, and the 2011 tsunami could not be recorded at such an early stage of the propagation, not necessarily because of smaller amplitudes, but due to satellite tracks not properly crossing the tsunami front waves in their temporal distribution. The available maximal amplitudes, from 20 to 60 cm, were available on Envisat, Jason 1 and Jason 2, from 5:20 to 8:25 h after the earthquake (Song et al. 2012), and were not as useful as in 2004 to retrieve any information on the source processes.

It should also be kept in mind that the altimetry profiles give a picture of the tsunami propagation along tracks, crossing the area by chance. Even though almost static compared to the tsunami propagation time scale, they do not provide full insight into the accurate description of the waveforms. The full temporal description of amplitudes and phases are usually more efficiently recovered through measurements using sea-level sensors, though only located at specific fixed locations.

4.3 Atmospheric Gravity Waves

Tsunamis can trigger acoustic and gravity waves that propagate upward into the atmosphere. Infrasound waves were detected following the 2004 tsunami, thanks to infrasound measurements made by the International Monitoring System (Comprehensive Nuclear-Test-Ban Treaty) (Blanc et al. 2018; Marty 2019). They were associated with the possible coastal impact in the Bay of Bengal, during the shoaling phenomenon (Le Pichon et al. 2005).

Gravity waves of lower frequency are also observed at ground level thanks to infrasound networks (e.g., Hupe et al. 2019). In space, those induced by tsunami waves triggered by the 2011 Tohoku earthquakes were reportedly detected thanks to GOCE (Gravity Field and Steady-State Ocean Circulation Explorer) accelerometer data (Garcia et al. 2014). High-frequency perturbations of the associated parameters were observed during three crossings of the tsunami waves by the GOCE trajectory. These rare signals could be explained consistently with a gravity wave model and also provided the propagation azimuth of the gravity waves. Other missions dedicated to the measurement of the variations of Earth's gravity field such as GRACE (Gravity Recovery And Climate Experiment) could also theoretically produce signals related to the coupling of tsunami waves upwards in the atmosphere.

4.4 Total Electron Content in the Ionosphere from GNSS and Altimetry

An outstanding capacity for measuring tsunamis coupled in the atmosphere now relies on GNSS data. Permanent GNSS networks dedicated to tectonic observations and real-time precise positioning are able to measure ionospheric Total Electron Content (TEC) variations (Mannucci et al. 1998; Lognonné et al. 2006) induced by a tsunami (Fig. 4). The TEC represents the ionospheric electron density integrated along the ray-path between ground-based or space-borne receivers and satellites. Precise GNSS positioning makes use of two radio frequencies (1.2 and 1.5 GHz) to correct for the dispersive delay induced by the ionospheric plasma on radio signals, thus providing TEC measurement. As the density of the ionospheric plasma is strongly peaked, the TEC measurements are usually considered to be located at the altitude of maximum electron density, that is, at around 300 km altitude.

Ionospheric seismology (Lognonné 2009; Jin et al. 2015; Occhipinti 2015; Jin 2019; Astafyeva 2019), in essence the branch of seismology studying the effect of earthquakes and tsunamis in the atmosphere and detectable in the ionosphere, took great advantage of the measurement of the TEC by GNSS. A first detection of tsunami propagation, and subsequent coastal impact when reaching Japan about 22 h after the generation, was achieved after the tsunami following the 2001 Peru earthquake (Mw = 8.1), thanks to the dense Japanese GPS network GEONET (Artru et al. 2005).

Today dense GNSS arrays in Japan (GEONET network), USA (PBO network) or New Zealand (GEONET NZ network) allow the imaging of the ionosphere finely and widely (Fig. 8). These measurements have been completed by satellite to satellite occultations, such as using the COSMIC system, which have been able also to detect tsunami ionospheric signals, but with very different sounding geometries (Coïsson et al. 2015, Yan et al. 2018).

In the near field, the detection of acoustic-gravity waves coupled with the uplift at the source (AGW_{epi}) by GNSS-TEC permits the imaging of the source extent about 8 min after the rupture (Astafyeva et al. 2011, 2013, 2019) (Fig. 4). It visualizes the source propagation pattern of the AGWepi, and also in the far field the Rayleigh waves ($AGW_{Rayleigh}$) and the tsunami (IGW_{tsuna}) signatures in the ionosphere (Artru et al. 2005; Occhipinti et al. 2013), including the oceanic region overlooking the rupture not covered by any other sensors, or rarely by GNSS buoys giving local measurements only.

In the same way as GNSS, altimetry data also make it possible to estimate the total electron content. Indeed, the TEC is required to remove the ionospheric effects from the altimetric data (Bilitza and Rawer 1996) for reliable measurements.

In the case of the 2004 Sumatra tsunami, the Topex/Poseidon and Jason-1 satellites showed the tsunami signature at the same time on the sea surface (Fig. 7) and in the ionosphere (Fig. 9). Using a three-dimensional numerical modelling (based on the linearized momentum and continuity equations, for irrotational, inviscid and incompressible flow), Occhipinti et al. (2006) created a pseudo-spectral propagator to transfer the oceanic displacement (from a tsunami model) to the atmosphere/ionosphere to compute the IGWtsuna generated by the Sumatra tsunami. This quantitative approach reproduced the TEC observed by Topex/Poseidon and Jason-1 over the Indian Ocean on 26 December 2004, thereby validating the detectability of tsunamis by ionospheric sounding.

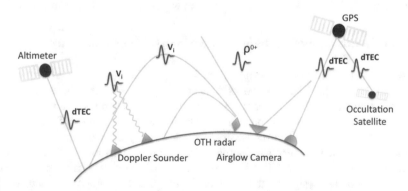

Fig. 8 The interaction of the AGW_{epi}, IGW_{tsuna} and $AW_{Rayleigh}$ with the ionosphere produces strong variations in the plasma velocity and plasma density observable by ionospheric sounding techniques. Nominally, the altimeters and GNSS measuring the total electron content (TEC) (expressed in TEC units (TECU); 1 $TECU = 10^{16}$ e $-/m^2$); the Doppler sounders and Over-The-Horizon (OTH) radars measuring the vertical displacement of the reflection layer in the ionosphere; and the airglow camera at 630 nm measuring a photon produced by the recombination of the oxygen, at 250 km of altitude, and directly related to the O^+ density. (Figure from Occhipinti 2015)

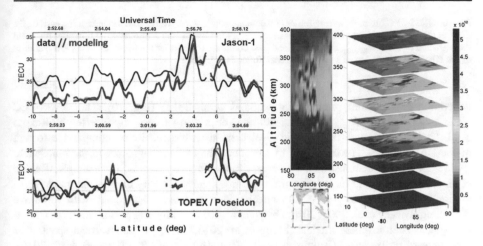

Fig. 9 Sumatra earthquake, 2004: the modelled and observed TEC are shown for Jason-1 and Topex/Poseidon satellites: data (black), synthetic TEC without production-recombination-diffusion effects (blue), with production-recombination (red), and production-recombination-diffusion (green). Right: electron density perturbation produced in the ionosphere by the tsunami gravity wave to model the TEC perturbation (vertical integration of the electron density along the line of sight) (From Occhipinti et al. 2006)

The results were later reproduced by Mai and Kiang (2009), and other theoretical works followed Occhipinti et al. (2006) to calculate the effects of dissipation, viscosity and thermal conduction on the IGW_{tsuna} (Hickey et al. 2009). More recently, electromagnetic field perturbations were accounted for in addition to viscosity and compressibility (Kherani et al. 2012); when complemented with a tsunami normal mode model (Coïsson et al. 2015), it provides a full direct coupling method from the ocean up to the TEC signal which proved successful in modelling (Rakoto et al. 2017). Although this kind of model uses a flat bathymetry, the agreement between synthetic and observed waveforms (DART and GNSS-TEC) was very satisfactory for several recent tsunamis (Kuril 2006; Haida Gwai 2012; Tohoku 2011), at least for the first arrivals.

Rolland et al. (2010) generalized the ionospheric detection of IGW_{tsuna} for moderate tsunami events (Kuril, 2006, $M = 8.3$; Samoa, 2009, $M = 8.1$; Chile, 2010, $M = 8.8$), analysing the GNSS-TEC observed by 50 receivers in Hawaii during the far-field tsunami propagation. They were able to reduce the empirical detection threshold to 2 cm, a figure similar to the height of the Samoa tsunami offshore Hawaii. Comparison between the GNSS-TEC observations and the oceanic DART data showed similarities in the waveform as well as in the spectral signature of the ionospheric and oceanic data. This result showed again that the ionosphere is a sensitive medium for detecting tsunami propagation. The study was confirmed by Galvan et al. (2011), while Grawe and Makela (2015) later highlighted the combined effects of the geomagnetic field and GNSS-TEC observation geometry in the cases of the 2011 Tohoku-Oki tsunami and the 2012 Haida Gwaii tsunami (induced by an $M = 7.7$ earthquake). Similar results using the Hawaiian GNSS network have been obtained by Savastano et al. (2017) in a real-time scenario, basically using only the real-time information available from GPS receivers and satellites. The signature of the Tohoku tsunami (IGW_{tsuna}) in the TEC has been also detected using COSMIC radio-occultation (Coïsson et al. 2015), showing the possibility of additionally enlarging the coverage.

The catastrophic tsunamigenic earthquake in Tohoku, Japan, in 2011, reaffirmed the potential advantage of the GNSS-TEC ionospheric sounding to visualize tsunami

Fig. 10 GNSS-TEC observed during the 2011 Tohoku event by GEONET. From top left: at the earthquake ▶ time; at the first arrival in the ionosphere of the AGW_{epi} after 8 min, and 1 min later; 27 min after the event the signature of the $AW_{Rayleigh}$ is observed and, last column, 56 min after the event, there is clear evidence of the IGW_{tsuna}. Adapted from a video related to Rolland et al. (2011)

generation and tsunami propagation (Rolland et al. 2011) (Fig. 10). The GNSS-TEC observation of the IGW_{tsuna} in the far-field has been also successfully inverted by Rakoto et al. (2018) to estimate the oceanic displacement with a resolution comparable to that of the DART buoys. The capability to reproduce the tsunami signature in the ionosphere (IGW_{tsuna}) is now demonstrated by several modelling approaches: for example, Rakoto et al. (2017) used a 1D model, integrating atmosphere variability and normal mode summation (Lognonné et al. 1998). Other techniques have been developed, with full wave modelling (Hickey et al. 2009, 2010a, b; Yu et al. 2017), with wave perturbations of Global Ionosphere-Thermosphere models (Meng et al. 2015, 2018) or with vertical spectral or time upward propagation (Broutman et al. 2014; Huba et al. 2015; Vadas et al. 2015), among others.

The estimation of the oceanic displacement by GNSS-TEC observed during the Tohoku earthquake has also been generalized to several events (Rakoto et al. 2018), who demonstrated the capability of such modelling to estimate ocean height estimation to ~ 10% of a posteriori errors. However, these approaches have to be generalized for all kinds of tsunami heights and wavelengths.

4.5 Airglow Signal

Although not based on satellite platforms so far, additional works theoretically explored the detection capability of IGW_{tsuna} by a ground-based airglow camera (Hickey et al. 2010b) and over-the-horizon radar (Coïsson et al. 2011). During the Tohoku tsunami, in addition to TEC-GNSS measurements, the 630 nm airglow camera located in Hawaii, observing the O^+ night density variation, showed, for the first time, the internal gravity waves forced by the tsunami (IGW_{tsuna}) in a zone of 180×180 km^2 around the island (Fig. 11). This method showed the observation of the tsunami ionospheric signature with an unprecedented high spatial density, visualizing the tsunami propagation across a wide area, rather than using local buoy measurements (Makela et al. 2011; Occhipinti et al. 2011).

The tsunami signature was successfully identified thanks to the matching of the waveform reproduced by numerical modelling (Occhipinti et al. 2011), a fact that also highlights the role of the bathymetry: approaching the Hawaiian archipelago, the tsunami propagation is slowed down (due to the reduction of the sea depth), but the IGW_{tsuna}, already formed and propagating in the atmosphere/ionosphere conserves its speed and arrives before the tsunami wave-front, slowing down near the shore (white-dotted-lines, Fig. 11). The Tohoku tsunami has maximum energy at two main periods of 14 min and 26 min, with the longer period wave propagating faster (Occhipinti et al. 2011).

A visual 2D cross-correlation between the observation and modelling highlights a shift of around 2° in both directions, potentially explained by the effect of the wind (which is not included in the modelling), as anticipated by Occhipinti et al. (2006). The airglow camera in Hawaii was unfortunately not calibrated, consequently, the comparison between data and synthetic results is only qualitative (based on the shape of waveform and arrival time) and does not allow the explicit connection between the observed number of photons and the amplitude of the tsunami. The tsunami signature

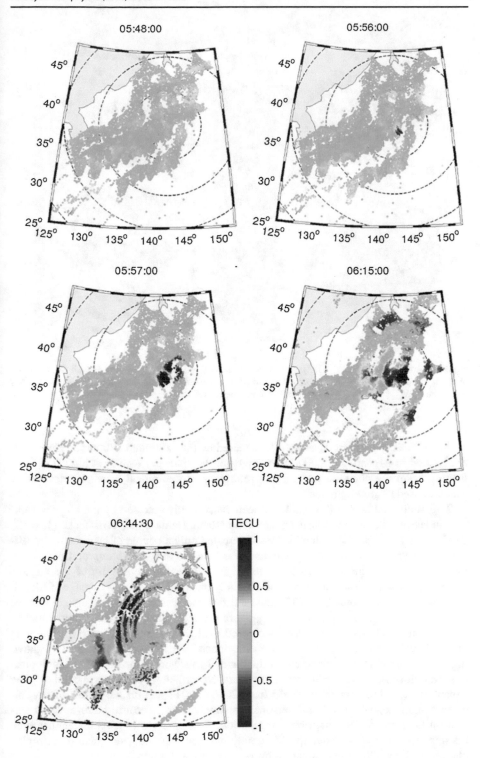

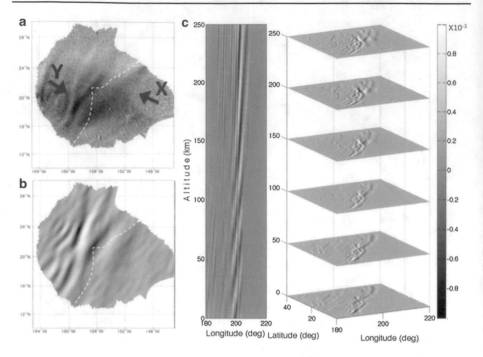

Fig. 11 Tohoku tsunami (2011) IGW_{tsuna} observed by the 630 nm airglow camera in Hawaii (**a**), modelled at the layer of observation (250 km) (**b**), and full 3D (**c**). Dotted-white-line shows tsunami wave-front, Y and X the IGW_{tsuna} at periods of 14 min and 26 min, respectively (red arrows highlight the position, not the direction) (Figure from Occhipinti et al. 2011)

in the O^+ density was also successfully detected by the calibrated 630 nm airglow camera located in Argentina, 18,000 km away from the epicenter (Smith et al. 2015), confirming again the potential interest and reliability of the optical tsunami detection by ground-based airglow cameras.

Two additional tsunami events have been successfully detected by the airglow cameras in Hawaii: the moderate Queen Charlotte Island, Haida Gwai, earthquake ($M = 7.7$, 28 October 2012, British Colombia) which created only a couple of centimetres of tsunami wave amplitude and no damage, and was detected despite the unpropitious observation conditions (quasi-full moon whose light reduces the sensitivity of the airglow camera) (Grawe and Makela 2015); and secondly the tsunamigenic earthquake in Illapel, Chile ($M = 8.3$, 16 September 2015, few fatalities and several areas of damage) (Grawe and Makela 2017). In any case, $AW_{Rayleigh}$ are routinely detected by Doppler sounders (Artru et al. 2004) and OTH-radar (Occhipinti et al. 2010) for events with a magnitude larger than 6.5. Occhipinti et al. (2018) introduced the ionospheric magnitude showing that ionospheric monitoring is a valuable and trustworthy method of seismological observation sensitive to the source parameters. Nevertheless, the experimental airglow camera deployed in Pamatai in Tahiti from 2016 to 2017 was unable to record any relevant tsunami signal (Reymond, personal communication), because there were no significant tsunamis during that observing period. It also serves as a reminder that such instruments detect only under specific conditions that must be guaranteed (dark nights and obviously without contamination by, e.g., nearby forest trees).

The first satellite-based observations of tsunami-induced airglows (Yang et al. 2017) therefore create the potential for future space-based observation to be able to obtain better continuity of the ionospheric monitoring from above the Earth.

4.6 Global Navigation Satellite System Reflectometry (GNSS-R)

Finally, GNSS-R is an innovative way of using GNSS signals for remote sensing. It uses ocean reflected GNSS signals for sea surface altimetry. But, in contrast to conventional satellite radar altimetry, multiple height measurements within a wide field of view can be made simultaneously.

Two GNSS-R altimetry approaches, the code altimetry and the carrier phase altimetry, can be distinguished from each other. In the code altimetry or PAssive Reflectometry and Interferometry System (PARIS) concept proposed by Martin-Neira (1993), the time delays between the direct and reflected signals are measured, which can be translated into the absolute height of the reflecting surface at the specular point. In ground-based GNSS reflection experiments above an artificial pond, height accuracies of 1 cm can be obtained (Martin-Neira et al. 2002). In a series of airplane and balloon experiments, this technique was successfully applied in airborne campaigns (Rius et al. 2002; Ruffini et al. 2004). In these experiments, sea-level heights with accuracies of up to 5 cm, as well as the relation between C/A-code correlation function and significant wave heights, were determined using dedicated delay mapping GPS receivers.

It was then proposed (Stosius et al. 2010) that strong tsunamis induced by earthquakes of magnitude $M > 8.5$ could be detected with certainty from any orbit altitude within 15–25 min by a 48/8 or 81/9 Walker constellation (Walker 1984), given the idea that tsunami waves of 20 cm or more can be detected by space-borne GNSS-R.

If the three GNSS networks (GPS, GLONASS and Galileo) are combined, a better detection performance can be expected for all scenarios considered (Stosius et al. 2011). In this case, an 18- satellite constellation would have detected the 2004 Sumatra tsunami within 17 min with certainty, while it takes 53 min if only GPS is considered, showing the growing interest in utilizing multiple GNSS systems. With a dedicated Low Earth Orbit (LEO) constellation of satellites equipped with GNSS-R receivers, densely spaced sea surface height measurements could be established to detect tsunamis in the future. Recovering accurate tsunami heights from such a technique remains, however, uncertain.

4.7 Conclusions on the Various Techniques Available to Measure Tsunamis

In the earlier sections, we have shown how the usual ways of characterizing possible tsunami sources and propagation were significantly improved thanks to major improvements appearing from the satellite data. Previously based on a limited number of tide gauges deployed in harbours, the tsunami waveforms are now also made available during the propagation, not only from pressure gauges deployed since the 1990s on the seafloor (DART buoys), but also on GNSS buoys anchored closer to the shore. Space missions, not based on ground sensors, such as altimetry, have provided limited insight into tsunami propagation so far. By contrast, GNSS data relying on ground sensors brought a striking improvement to indirectly accessed tsunami characteristics obtained through TEC measurements.

5 Coastal Effects of a Tsunami and Improved Risk Prevention

Measurements of sea-level variations that offer a quantitative view of the tsunami wave-forms at several coastal locations are very useful to tsunami scientists, as seismograms are essential to seismologists. But civil protection and emergency services rapidly need, through interpreted data, an assessment of the areas that have been affected by the tsunami waves, to quantify the structural and human damage over the flooded area. Satellite imagery thus gives additional constraints on the flooding extent, the damage intensity and the coastal vulnerability (Borrero et al. 2006; Chen et al. 2006). International cooperation is efficient in case of catastrophic disasters, in order to open data useful for the damage detection and to help the crisis management. This effort is effective since 2000 by the International Charter Space and Major Disasters composed of space agencies and space system operators all over the world (Bessis et al. 2004).

5.1 Damage Assessment Through Optical Imagery

One example of the contribution of satellite imagery to post-tsunami studies was crucial in the aftermath of the 2004 disaster, particularly in Indonesia's Aceh Province. Indeed, at the time, this province was one of the most isolated in Southeast Asia because it was in the grip of a civil war between the Indonesian Army and separatist forces. As the province was closed to foreigners and journalists, the damage caused by the tsunami was initially assessed by satellite imagery, in addition to an unprecedented amount of video footage and pictures.

Several images were quickly posted on the internet, including a 60-cm resolution Quickbird image produced by Digital Globe on 28 December 2004, 2 days after the tsunami. This was compared with the previous one dated 23 June 2004 (Fig. 12, a and b). In the weeks and months following the disaster, field studies organized as part of the various field trips coordinated by UNESCO (International Tsunami Survey Team-ITST) or specific research programs such as TSUNARISK (Lavigne et al. 2009) gave the opportunity to refine the first observations from aerial imagery and to provide more precise data of altimetry, wave height and run-up. Figure 12a, b shows the damage in the harbour area of Ulee Lheue, northwest of Banda Aceh, resulting from the combined action of subsidence caused by the coseismic vertical deformation, and by erosion caused by the tsunami. In this area, the subsidence reached an amplitude of about 1–1.5 m, which partly explains the retreat of the coastline and the flooding of the fish ponds (tambak) which had gradually replaced the mangrove.

The comparative study of these satellite images also offers researchers to study the damage to buildings. In a 7 km² area between the Ulee Lheue harbour and downtown Banda Aceh, Leone et al. (2010) conducted a systematic diagnosis of the state of the buildings approximately 1 year after the tsunami. Their method of damage spatial analysis was based on a combination of field surveys, photo interpretations and GIS. They built a new "macro-tsunamic" intensity scale based on a typology of building types and damage done.

Beyond the short-term impacts of tsunamis, satellite imagery also gives us the ability to study longer-term resilience, whether environmental or societal (i.e. reconstruction dynamics). Thus, 12 years after the 2004 tsunami, a shoreline mapping was reconstructed from an IKONOS image of 10 January 2017 at 1.5 m resolution and compared to previous ones. Figure 12c shows that the barrier island north of the port of Ulee Lheue has been almost abandoned.

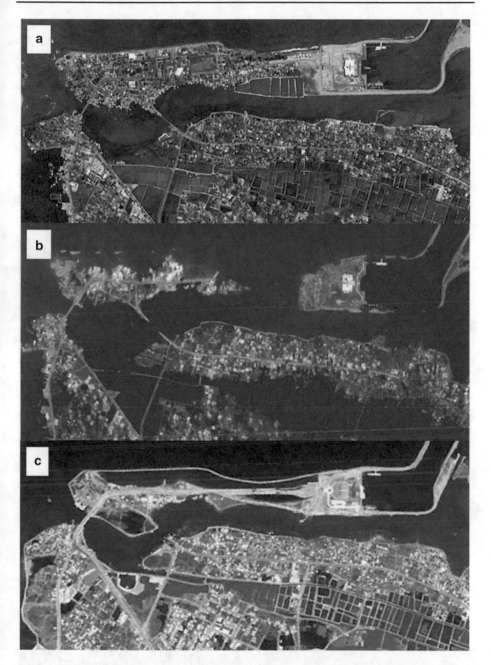

Fig. 12 Landscape modifications at Ulee Lheue harbour region, north western Banda Aceh. **a** Six months before the tsunami impact (Image Digital Globe on 23 June 2004, 60 cm resolution). **b** Two days after the tsunami impact (Image Digital Globe 60 cm resolution). **c** IKONOS image on 10 January 2017 (resolution: 1.5 m)

However, the area behind it has been completely rebuilt. Further west of the port, in the Lambadeuk district, large areas submerged by the tsunami have remained flooded to this day (Fig. 13a). In contrast, in the Kuala Gigieng area east of Banda Aceh (Fig. 13b), the shoreline has gradually returned to its original location, as subsidence may have been lower (around 60 cm), but mainly due to a significant volume of sediment deposited by the Aceh River.

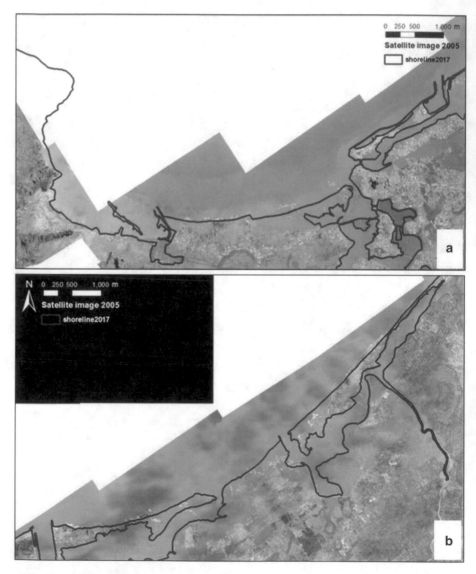

Fig. 13 Shoreline changes at the Banda Aceh coast between 2005 and 2017 (Meilianda et al. 2019). **a** Lambadeuk area, west of Banda Aceh, was more affected by the subsidence and has definitely lost several hectares of land. **b** Kuala Gigieng area, east of Banda Aceh, has now two relatively permanently open inlets, partly due to post-tsunami coastal aggradation

Satellite imagery is also of great help when mapping is required in remote areas where in situ measurements are not easily available. For instance, after the 2006 Java earthquake and tsunami (a so-called tsunami earthquake, provoking a tsunami much larger than the magnitude level could indicate, generally with a very slow seismic rupture, as defined in Kanamori 1972), the sector with suspected maximum run up values, reaching locally up to 20 m, was expected in an area with restricted access. Thanks to satellite imagery, the horizontal mapping of the inland flooding could be defined (Hébert et al. 2012). Run-up values which can then be estimated from the vertical heights only are available, which was in that case possible via a stereographic approach such as the one used to define DEMs.

5.2 Damage Assessment Through SAR Imagery

SAR imagery is obtained by the measurement of the backscattering of an electromagnetic wave emitted by a satellite. One of the main advantages of SAR compared to optical imagery is its all-weather capacity, i.e., images of the ground surface can be acquired even in cloudy conditions. SAR imagery is widely used for tsunami-related damage detection, using different approaches. For example, simple coloured RGB (red–green–blue) compositions can be produced with images acquired, respectively, before and after the event, in order to obtain a quick assessment of flooded or damaged areas. Such maps do not indicate any automatic classification of damaged or flooded pixels; nonetheless, they are very useful for civil protection purposes as their interpretation is simple.

More complex techniques permit the automatic detection of flooded (Iyyappan et al. 2018) and damaged areas, or even the automatic classification of the level of damage due to a tsunami. Some authors use SVM (Support Vector Machine) classification schemes (Endo et al. 2018), sometimes applied to polarimetric SAR data which gives valuable information on urban areas (Ji et al. 2018). Indeed, as polarimetry brings information on different backscattering mechanisms, it is possible to characterize the condition of buildings. Recently, some studies have been aimed at rapid damage assessment using Deep Neural Networks on SAR images (Bai et al. 2018) or data fusion between optical and SAR data (Adriano et al. 2019). These assessments could then be used for civil protection.

Below we show a simple example of a coloured composition on the 2011 Tohoku great tsunami. The German satellite TerraSAR-X, which operates in X-band (about 9 GHz), acquired images over the Sendai region before and after the tsunami. Figure 14 shows the RGB composition of two images acquired on 20 October 2010 and 12 March 2011. The first image is in the blue and green channels (or equivalently a cyan channel). The second image is in the red channel. The areas where no change occurred between the two dates appear in grey levels, from black (for low backscatter pixels) to white (for high backscatter pixels). The pixels where backscattering is higher (respectively lower) on the first date than on the second one, appear in cyan (respectively in red), thus highlighting a change.

Land areas appearing in cyan are flooded areas. Indeed, as can be seen on the upper inset, the backscattering of the radar wave on water is very low: since it is lower than the backscattering of land, flooded areas appear in cyan. These areas can very easily be detected from this map: it can be observed that very large areas have been flooded, in particular between the shoreline and Sendai city. On the sea, the cyan structures on the upper inset correspond to shellfish farming that was destroyed during the tsunami. Red pixels correspond to debris with high backscattering, coming from shellfish farming (along the shoreline on the upper inset) or other debris (as on the airport area).

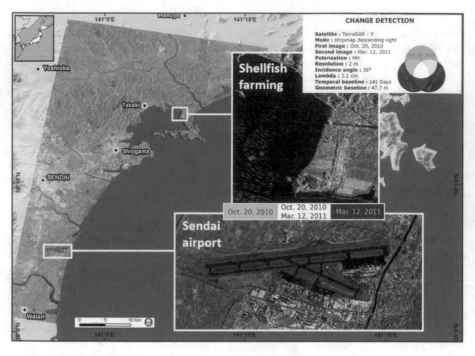

Fig. 14 Change detection obtained from TerraSAR-X images acquired on 20 October 2010 and 12 March 2011, showing the impact of the 2011 Tohoku tsunami in the area of the Sendai plain. The upper inset shows destroyed shellfish farms while the lower inset shows debris on the Sendai airport

Such SAR analyses are usually carried out after the events, to estimate damage in addition to ground deformation. The need for an available image prior to the event, and for real-time processing, makes this technique still poorly available in real time.

5.3 Towards Improved Risk Prevention and Resilience Capacity

Tsunami data acquired following disasters helps local risk prevention plans to be established for the benefit of the coastal communities threatened by such events. Space data, and especially GNSS positioning, have continuously improved the technologies used to draw maps with the required accuracy and resolution. Combined with hazard maps deduced from appropriate numerical modelling, the tsunami risk maps now integrate many products and types of information derived from satellite data. The same applies for the design of evacuation routes and assembly areas. Such crucial maps can also be continuously updated through the use of optical and radar imagery that monitors major changes in land use.

As satellite data play a growing role in understanding the processes and helping to create better preparedness for tsunami damage, it is easy to understand that it now deeply integrates the various stages involved in a tsunami warning system.

6 Which Space Data are Required for Tsunami Warning Systems (TWSs)?

In this section, we propose to investigate which satellite mission could be useful in the frame of tsunami real-time warning. To this end, data from such space missions should not only be available in real time and over all the areas exposed to tsunamis, but their processing or analysis should also be made available to decision makers very rapidly. Consequently among the techniques exposed previously, it is highly improbable that InSAR, sea-level measurement by altimetry or GNSS-R techniques may be useful in the first minutes. Conversely, tsunami scientists are more and more interested in data derived from GNSS techniques, whatever for sea-level measurements on buoys, or TEC measurements, or for the quick description of surface ground deformation that allows for source inversion. These data will play a growing role in tsunami warning, which however still essentially relies on seismology and satellite robust communication.

6.1 Principle of TWSs and the Situation After 2004

The essence of a tsunami warning system (TWS) is the prompt detection of the geophysical event prior to possibly triggered tsunami waves, combined with the measurement and monitoring of sea wave heights throughout the oceanic basin where they would propagate. Due to the time scales of a tsunami, such monitoring can last for hours; however, the role of the TWS is crucial in the first few minutes following the initiation, in order to warn the authorities.

Early detection is more effective for an earthquake than for a landslide, since seismic networks arc now very efficient at providing a quantitative estimation of the main parameters of the mainshock. For a landslide, however, it still remains a challenge to have the equivalent characterization, unless very dense networks are available near every landslide-prone area.

Since the tsunami waves propagate more slowly than the seismic waves, the tsunami can then be monitored along its propagation in the minutes and hours following the earthquake, provided data can be acquired and transmitted. Originally relying on tide gauges deployed in harbours and locally transmitted through radio and progressively upward to satellites, the monitoring confirmed the occurrence of a tsunami and the ability to keep on updating warning status (while tsunami amplitude remained greater than 1 m). With the development of global seismic networks, a system such as the one designed in the Pacific Ocean back in the 1960s would have also been able to monitor the global seismicity and to give assistance for oceanic basins without an operational warning system. However, in the early 2000s, no global international architecture was established, and national initiatives did not exist to consider setting up tsunami warning systems. Thus, in 2004, there were no international procedures or contact points to receive and interpret the messages that the PTWC in Hawaii could send within the Indian Ocean.

Following the 2004 disaster, the Indian Tsunami Early Warning System (ITEWS), the set-up of which was started in 2005 (IOC 2005a), was completed by September 2007 (Nayak and Srinivas Kumar 2008a, b; Uma Devi et al. 2016). The ITEWS monitors seismically active regions of the Sumatra-Java subduction (4000 km long) and of the Makran coast (500 km) in the Arabian Sea (Gupta 2005, 2008). In addition to the essential

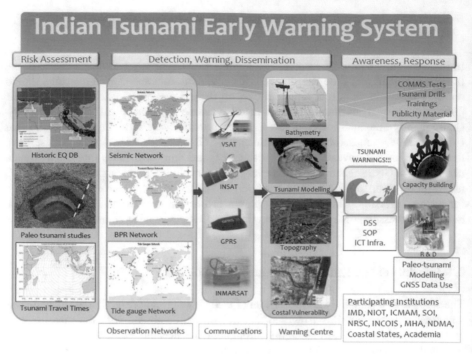

Fig. 15 Various components of the Indian Tsunami Early Warning System (ITEWS). Three approaches contribute to an efficient tsunami warning, a view shared among all TWS: the risk assessment, the operational detection and dissemination, and the high level of awareness and response capacity. Most of the components partly rely on space data or satellite means (courtesy, Indian National Center for Ocean Information and Services (INCOIS), Hyderabad, India)

components of a typical TWS mentioned above (Fig. 15), the ITEWS also includes ocean Bottom Pressure Recorders (BPRs) that encompass the two sources of tsunamigenic earthquakes.

The whole ITEWS makes use of Earth observations in multiple ways and illustrates well the three main pillars of an efficient system: hazard assessment, efficient warning procedures, and preparedness (Fig. 15). ITEWS has been working efficiently since 2007, complying with the IOC requirement of issuing any advisory within 10–15 min of time after an earthquake, in 8 min on average. False alarms are avoided, reminding us that before the establishment of the ITEWS, the 28 March 2005 Nias Mw = 8.7 earthquake occurred in the vicinity of the 26 December 2004 Sumatra earthquake, and that a tsunami warning was issued by the Pacific Tsunami Warning Center (PTWC) and by the JMA (Japan Meteorological Agency). Consequently, millions of citizens living on the east coast of India were evacuated away from the shoreline, while this large earthquake did not trigger any large tsunami due to a particular low coupling with the ocean (Kerr 2005).

In the Caribbean and in the North-East Atlantic and Mediterranean (NEAM) regions, seismic networks and sea-level stations have been greatly developed since 2005, with a growing number of real-time data exchanges and capacity building of the national systems.

6.2 Seismological Challenge

Based on seismic detection, the early warning system has always been essentially seismological, and this remains the same up to the present time. In the first minutes after an earthquake, magnitude, depth and location are quickly available depending on the networks, but it is difficult to estimate the direction, duration, size (length and width) and exact location of the rupture (relative to the hypocenter). Yet these parameters are major influences of the tsunami triggering, and generally the complete seismological solutions start being stable after about 30 min.

In the first few minutes after the 26 December 2004 earthquake, its magnitude was estimated to be 8.5. This was an underestimate, due to the low frequency content of the seismic waves that was not properly taken into account in the numerical methods routinely used (Lay et al. 2005). Indeed such a magnitude had never been reached since the rise of modern seismic networks and since the last equivalent earthquakes in the 1960s in the Pacific Ocean. Again in 2011, in a well-prepared operational context, the first estimates routinely used in operation by JMA were able to produce a magnitude Mj of 7.9 only (based on broad band saturated seismic data), which was revised to 8.8 about 50 min later (JMA 2013).

Nevertheless it is necessary for some recent advances to be stressed, especially those which helped to reduce delays before obtaining focal mechanisms. For example, the low-frequency W seismic phase (Kanamori and Rivera 2008) helps to quickly provide the first estimations of the focal mechanism, which is a key information on the rupture pattern. A finer examination of the first minutes of the seismic signal influenced by the gravity changes also offers a rapid indication of the possible large magnitude in case of mega earthquakes (Vallée et al. 2017) prone to trigger tsunamis. Most of the TWS have adopted such rapid W phase techniques (see Duputel et al. 2011 for the Pacific Ocean; Roch et al. 2016 for NEAM region) that prove efficient in the near field. Warning systems watching more distant shorelines still follow methods adapted from long period surface wave inversions and are numerically made more efficient (Clément and Reymond 2015, for French Polynesia); although they are less rapid, they produce a robust and accurate focal mechanism.

In this early part of tsunami warning, it is essential to share, via satellite communication, seismic data, information and naturally to disseminate the first messages. Later on, when the tsunami propagates across the ocean, sea-level data again must be transmitted on the most robust basis.

6.3 Robust Communication for Scientific Data and Message Dissemination

The Global Telecommunication System (GTS) satellites (WMO 2018) are used by the tsunami warning systems for two purposes. First, the transmission of the tide gauge data measuring the tsunami waves on coastlines are sent through GTS. The sampling time is generally 1 or 2 min and the latency varies from 6 min to 1 h. Secondly, the Tsunami Service Providers (TSP) send the alert or threat messages, via GTS to the services and tsunami warning centres connected to GTS. The time delay is generally very short, less than 1–3 min. One of the recent requests is the possibility to add files to GTS messages, and this option would enhance the capacities of the TWS.

The GTS transmission is essential, especially when it may be the only available telecommunication in remote places where the local communication system is not robust enough.

On the other hand, places that could be impacted by earthquakes and where ground communication could be damaged, including due to the lack of an electrical power supply, need satellite transmission for alert messages and information on sea-level data.

6.4 Increasing Use of Dense GNSS Networks to Characterize the Tsunami Source

As previously mentioned, GNSS data are now almost unavoidable to characterize earthquake and measure ground deformation where dense enough networks are available, and then also to invert for the fault source pattern. Using such data under operational constraints remains generally challenging. Japan is one of the few examples where dense GNSS networks can rapidly produce insights into the coseismic ruptures. Chile (Barrientos et al. 2018) and New Zealand networks are also being improved for that purpose.

In Japan, the March 2011 earthquake and tsunami in Tohoku, Japan, have shown how relevant dense GNSS networks can be, to compute, within minutes, the extent and amplitude of the ground deformation (Ozawa et al. 2011; Song et al. 2012). GNSS networks complement seismological analysis in capturing the fault rupture extension and location, and its slip amount (Yamamoto et al. 2016; Ohta 2012; Ohta et al. 2018). More recently, Melgar and Hayes (2019) highlighted the capacities of GNSS data in characterizing large earthquakes before the end of the rupture process.

However, considering areas with rare seismicity, such as the Ligurian Sea or the North African margin, it is questionable whether existing networks could contribute. The potential detectability through GNSS sensors can be investigated to define elements to optimize a possible GNSS network geometry. Potential tsunami scenarios related are compiled here, based on the operational databases in place of the French TWS part of the NEAMTWS (Gailler et al. 2013).

The tsunami pre-computed scenarios, under a classical approach followed in a TWS (Titov et al. 2005), rely on state-of-the-art regional seismotectonics, consisting of unit faults 25 kilometres long and 20 kilometres wide (Fig. 16) that can be combined to achieve magnitudes beyond 6.9 (Gailler et al. 2013). A threshold of 10 cm vertically is considered to be the minimum value able to be detected in real time by a dense GNSS network; recent improvements in the technology probably will allow a lower value, though the vertical component remains the most difficult to resolve.

In the Ligurian Basin, faults localized on-land are mainly normal, with a few thrust faults into the French part. In the Tyrrhenian Sea, there is also a major region of normal faults and in the Ligurian Sea, faults are mainly thrust and strike-slip (Fig. 16, top). Some 20 events are identified for which the induced deformation (positive or negative) is higher than 10 cm. Along the coast, sectors capable of detecting events leading to more than 10 cm of deformation can be identified (areas of La Spezia, Aleria, Piombino in Italy and at the French Italian border, between Cannes and Imperia).

In Algeria, most of the faults localized on-land are strike-slip, with a few thrust faults on the West part of Jijel area (e.g., Kherroubi et al. 2009). Offshore, all the faults are thrust faults (Fig. 16, bottom). There is a maximum number of 21 events for which the induced deformation (positive or negative) is higher than 10 cm, and again these areas can be mapped, between Annaba and Chetaïbi, and for the area near Collo (Fig. 16).

Note also that including events with higher magnitudes generates detectable deformations inland, especially in thrust fault zones. In addition, some inland earthquakes can also marginally contribute to a tsunami trigger some tens of kilometres away from the shore; dense GNSS networks would help to rapidly quantify a possible coupling

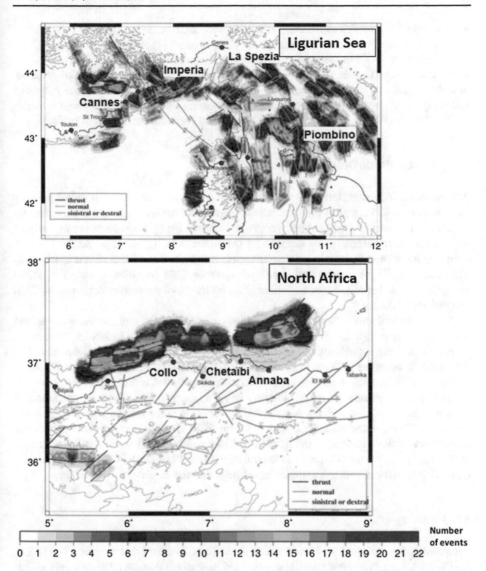

Fig. 16 Number of events mapped for the Ligurian Sea (top) and for Algeria (bottom), inducing a deformation higher than 10 cm (in absolute value) for magnitude up to 6.9 for Liguria, and to 7.5 for Algeria. Unit faults are 25 km long and 20 km, and the associated unit tsunami scenarios consider a 1 m slip. For the Ligurian Sea, 226 unit faults are considered and used as rupture zone triggering tsunamigenic earthquakes of magnitudes of 6.0, 6.3, 6.5, 6.7 and 6.9. For the North Algerian area, 142 unit faults are taken into account for the magnitudes of 6.5, 6.7, 6.9, 7.2 and 7.5. Some coupling of two, three or eight contiguous segments can be done on the offshore area in order to achieve magnitudes beyond 6.9 (Gailler et al. 2013), increasing the total number of scenarios to 462 along the Algerian margin. The tsunami simulation code is used to calculate the initial deformation for each scenario on the GEBCO grid (GEBCO 2014), at 30 s resolution

offshore. Having had a modern real-time dense GNSS network on the North African margin, the tsunami trigger following the 1980 El Asnam (Algeria) earthquake could have been estimated in a few minutes after the rupture (Roger et al. 2011). Together with an accurate numerical forecast, it could have contributed to estimate the range of the expected impact in Spain (significant, but less than 50 cm in most harbours).

Although most of the tsunamigenic deformation occurs in submarine parts, it is reasonable to foresee that denser networks will be available in the future with lower detection thresholds, and could increasingly contribute to rapid imagery of a rupture extension following an earthquake in these regions. Areas described previously, with large tsunamis expected, such as Japan and New Zealand, are already very innovative in that respect.

6.5 Rapid Landslide Detection

As previously indicated, landslide detection should rely on a dense geophysical network close to any area exposed to possible failure. Significant achievement has been made, for instance in the area of the Stromboli volcano which was hit by tsunami waves following a rock avalanche triggered by the eruption of December 2002 (Tinti et al. 2005). The volcano was monitored through a seismic network, and a tilt and GNSS network originally deployed in the 1990s, which had been updated from 2003 on (Bonaccorso et al. 2009) and greatly contributed to real-time information for the Civil Protection Authorities (CPA) (Puglisi et al. 2005).

The available datasets were statistically interpreted into thresholds so as to define and send alert messages to decision makers (Carlà et al. 2016). Thanks to InSAR, time series processing and the availability of free SAR data with a 6- to 12-day revisit time offered by Sentinel-1, it would be possible to predict catastrophic slope failures (Carlà et al. 2019a).

However, in general, landslide characterization is achieved a few days after its occurrence. For example, satellite optical imagery offers a complementary view of the landslide volume induced, both in the case of a subaerial event, and when images are available a few days later. This was the case, for instance, for the 2017 cliff collapse in Greenland, when tsunami waves were generated and flooded a village nearby (Paris et al. 2019). In the case of essentially submarine collapses, as in Krakatau 2018 (Paris et al. 2020), only high-resolution differential bathymetric mapping could help, and this is very rarely available.

6.6 Altimetry

Future altimetry missions with swath acquisition (e.g., international project SWOT, Surface Water Ocean Topography) will have reduced instrumental noise (about 1 cm in amplitude) and will ensure acquisition on a larger area than along single profiles. In the case of a scenario in the Mediterranean Sea, a tsunami following a large earthquake in Algeria (magnitude 6.8) would produce a signal of about 5–10 cm offshore, while the mesoscale signal, though in different wavelengths, is about 5 times larger in amplitude (Fig. 17). Accurate signal processing may allow the recovery of the tsunami signal in such future missions that have repetition times of only a few days (or less, with a suitable constellation).

The perspective of using such future swath altimetry not restricted to acquisition along a single profile will increase the number of records made across the oceans when a tsunami occurs. This, however, remains limited to tracks crossing the tsunami waves by chance, and, unless a dedicated constellation of such satellites is developed, with or without geostationary platforms that are able to watch a larger basin, this very informative data is not ready to integrate a full operational chain delivering messages within a few minutes after the tsunami triggering.

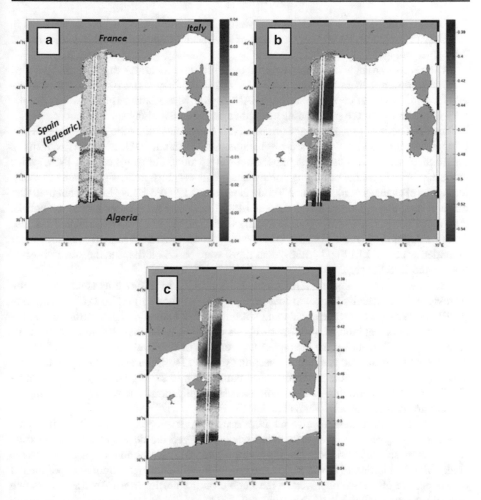

Fig. 17 Modelling of altimetry signals in the Western Mediterranean, in the case of a tsunami initiated in northern Algeria after a magnitude 6.8 earthquake (colour scale in metres). a) Model of the tsunami propagation with instrumental noise, b) model of the mesoscale dynamics showing larger wavelengths, and c) the sum of both contributions. The tsunami can be detected from the summed signal; its amplitude is about 20% smaller but with higher frequency content

6.7 Towards Real-Time Sea-Level Monitoring and Forecasting Using GNSS Buoys

Sea-level data available in real time throughout the warning is essential to monitor the propagation pattern during several hours, and it also integrates more and more numerical models on the fly, which in turn contributes to a better forecasting of remote tsunami impacts (Titov et al. 2005; Reymond et al. 2012; Clément and Reymond 2015; Jamelot and Reymond 2015). The offshore BPR (González et al. 2005), such as DART buoys, are very valuable as they can estimate the tsunami amplitude soon after their triggering by the earthquake. They are used to refine models and to compute real-time forecasts for specific sites.

This is, however, possible only when sea-level monitoring is available early to be incorporated, compared and/or assimilated (Yang et al. 2019) in the numerical approaches. In that case, warning products gradually evolve to numerically-based forecast mapping, as in the USA (Titov 2009; Percival et al. 2018) or in French Polynesia (Jamelot and Reymond 2015).

Such sensors are now integrated in three out of four regional warning systems, but they require costly investment to maintain the devices in the long term. No such offshore tsunami sensor has been deployed in the NEAM region where land-based networks are not very dense on any of the surrounding coastlines, in particular along seismogenic zones.

The development of a dense network of GNSS buoys thus offers the possibility of delivering timely information on the sea-level status. On 11 March 2011, at 14:46 (local time), the Tohoku tsunami had its largest impact and run-up along the Sanriku coast. JMA, which is the authority in charge of issuing domestic tsunami warnings in Japan, was able to monitor sea-level changes at more than 170 tide gauges and 12 GPS buoys located about 20 km off the coast. The initial warning issued at 14:49 (local time) was based on the earthquake magnitude computed in JMA (Ozaki 2012), which proved to be underestimated because of the saturation of most Honshu broadband seismic sensors. Tsunami warning major upgrades issued at 15:14 and 15:30 (local time) were based on the tsunami wave observations at the GPS buoys.

After the experience of the 2011 event, JMA planned to improve its tsunami warning (Ozaki 2012), especially thanks to real-time algorithms such as Precise Point Positioning (PPP), the use of accelerometer data and with a network extending farther from the coastline. The implementation of an array of buoys deployed more than 100 km off the coast (Kato et al. 2018) would allow the rapid incorporation of sea-level data to refine the source model used for near-field tsunami forecasting systems (Tsushima et al. 2014). Such GNSS buoys could also contribute to multi-hazard monitoring, to study seafloor geodesy, atmospheric, and ionospheric disturbances with possible applications for weather forecasting, in addition to tsunami warning (Kato et al. 2018).

In the near field too, innovative methods are being developed, such as those in Japan, where a very significant investment effort has encouraged the development of submarine cables composed of multi-technology sensors (Ohta 2012; Kawamoto et al. 2017; Inouye et al. 2019) in addition to GNSS buoys and conventional tide gauges. Similar experimental networks have also been deployed in the Pacific on the northwestern American coastline (e.g., the Neptune experiment, Thomson et al. 2011).

In conclusion, having GNSS buoys close to the coast would help to characterize the shoaling effect that is different depending on each specific bay or harbour. The parameters obtained from the analysis of continuous monitoring can then be put into operational numerical models relying on approximate amplification laws and providing forecasts in due time (Gailler et al. 2018).

6.8 The Promising Use of GNSS-TEC Data in Real Time

While GNSS data are practically already integrating early warning processes to retrieve the rupture pattern and the sea-level data (GNSS buoys), the TEC variations that are now often detected following large earthquakes and tsunamis are not yet operational. However, when detected through GNSS or altimetry data, they offer an interesting complementary constraint to characterize an ongoing tsunami, or even earlier, to estimate the possible efficiency of the tsunamigenic coupling. The number of events recorded through this technique is very tiny and the comprehensive analysis of the accuracy of the tsunami heights

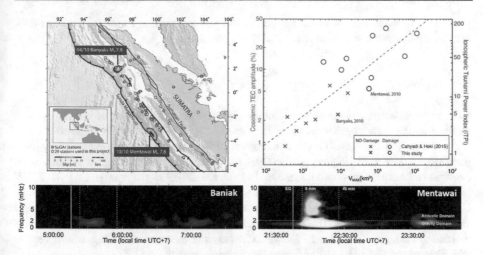

Fig. 18 Bottom: power density spectrum of the GNSS-TEC signals observed for Banyak (left) and Mentawai (right) events using the SuGAr (Sumatran GPS Array) network in Sumatra. The perturbations are 5% (left) and 15% (right) stronger than the mean background level. Top left: map showing epicentral areas of the April 2020 Mw 7.8 Banyak and October 2010 Mw 7.8 Mentawai earthquakes, and the 29 SuGAr stations used for the study (yellow dots). Top right: correlation Vmax versus TEC amplitude/TECI. The V_{max} is obtained from the product of the maximum seafloor uplift and the rupture area (A), it corresponds to the maximum volume of displaced water. Figures adapted from Manta et al. (2020)

is not well established. To integrate early warning procedures, further check of all these aspects should be performed and guaranteed.

A recent study presented the case of the twin events of Banyak and Mentawai in Sumatra, in 2010 (Manta et al. 2020) (Fig. 18): both $M = 7.8$ earthquakes were generated in the Sumatra mega thrust, and triggered a tsunami alert that was later cancelled. Unfortunately, the tsunami risk was underestimated in the case of the Mentawai event, and the resulting tsunami caused serious damage and more than 400 fatalities. The GNSS-TEC signature in the ionosphere of the AGW_{epi} highlights the important Mentawai uplift (causing the tsunami), compared to the Banyak event (Fig. 18 bottom).

A new TEC index (TECI) was introduced to estimate the energy transferred through the atmosphere (instead of the simple TEC amplitude used by previous authors) to better estimate the uplift and source extent (Fig. 18 top right). The link between the maximum volume of displaced water (Vmax) and the TECI/Amplitude-TEC improves the reliability of ionospheric monitoring for tsunami risk estimation. Indeed, the TEC observations were previously only correlated with the magnitude of the event (Cahyadi and Heki 2015) or the coseismic uplift (Astafyeva et al. 2013), and not with the Vmax directly related to the tsunami genesis. The fact that the Mentawai event was characterized as a slow earthquake amplifying the tsunami generation (a *tsunami earthquake*) (Newman et al. 2011) thus seems to be detectable through the TEC picture.

Other developments to rapidly locate the initial sea surface motion and estimate its amplitude accurately include TEC onset detection (Liu et al. 2010; Astafyeva et al. 2011) and rapid modelling of the near-field coseismic ionospheric pulse (Rolland et al. 2013; Mikesell et al. 2019).

Therefore, the double GNSS-derived observation of ground motion and TEC could be gradually included in future TWSs once the accuracy of the measurement is fully checked. The advantage of the double use of GNSS/TEC ionospheric monitoring during tsunami genesis and ground motion is their synergy: the first is sensitive to the vertical sea floor displacement and the second to the slip at the source area.

6.9 Airglow

Ground-based airglow cameras also could help to monitor the arrival of tsunami waves, as soon as the wave characteristics of the detected signals can be fully determined in a few minutes. The successful observations of the IGW_{tsuna} by airglow cameras already mentioned open new technological perspectives in tsunami remote sensing by ground-based cameras and cameras on-board a satellite. Indeed, the CNES satellite project IONOGLOW is currently exploring the possibility of an airglow camera as a payload on a commercial GEO satellite to cover the entire Pacific Ocean and to complete the coverage of the airglow cameras on a global scale. This would overcome the limitation of the airglow cameras, which are currently restricted to localized, night-time measurements, but would require real- time processing to be useful during the warning process.

7 Conclusions and Perspectives

The striking development of satellite platforms since the middle of the 20th century has offered a revolutionary perspective on Earth imagery and global communications. The strong impulse given by the growing use of data acquired or transmitted by satellites has already produced key pictures and interpretations of physical processes at work on our planet, and also of the hazards it is exposed to. For tsunami hazards as well, they contribute to a better knowledge of the processes, to better preparedness, and they will undoubtedly become more and more incorporated in the monitoring and warning procedures of the future.

The 26 December 2004 Mw 9.1 Sumatra earthquake and resultant tsunami were devastating, claiming over 250,000 human lives in South and Southeast Asia. Following this disaster, the global awareness of tsunami exposure in these basins triggered the expansion of TWS for all regions (Bernard and Titov 2015). Meanwhile, since the 1990s, the use of massive electronic communication on a global scale has opened a new era to exchange data, warnings, procedures and scientific studies with large volumes of data involved. This was made possible through high-rate communications via the Internet, and using submarine cables and communication satellites, especially those in the Global Telecommunication System (GTS) from the World Meteorological Organization. Every single stage of a tsunami can thus now be monitored, thanks to data and communications taking advantage of satellites.

To summarize the various space data described in this paper, Table 1 presents their level of applicability regarding the different goals to be addressed. The fundamental role played by the satellite missions dedicated to transmission is underlined. Since the 1970s, satellites have been transmitting many types of geophysical and oceanographic data, in particular the sea-level data from Bottom Pressure Recorders (BPRs) (using Meteosat, Inmarsat, GTS, Iridium). Since the 1980s, high sampling rate seismic and geophysical data has been sent by Very Small Aperture Terminals (VSATs) from everywhere in the world. All these data are used either for source characterization and/or tsunami wave detection and measurement.

Table 1 Synthesis of the satellite missions described in this paper

Satellite mission	Source information	Early detection of tsunami	Sea-level measurement	Coastal impact	Prevention plan	Emergency management	Operational warning and forecast current/future
Altimetry	−	−	+	−	−	−	−/−
Optical imagery	++	−	−	+++	++	+	−/−
InSAR	+++	−	−	++	+	−	−/−
GNSS geodesy	++	−	−	++	++	++	+/++
GNSS buoy	−	+++	+++	−	−	+++	+/+++
GNSS–TEC	++	+	+	−	−	−	−/++
GRACE, GOCE	+	−	+	−	−	−	−/+
Airglow	++	+	+	−	−	−	−/+
Alert and data transmission	+++	+++	+++	−	−	+++	+++/+++

The symbols reflect the efficiency for a given purpose, from low or non-applicable (−), to effective (+), very effective (++), and essential (+++)

For alert and warning communication, the downstream part of the warning system is currently using the Global Telecommunication system (GTS) implemented and managed by the World Meteorological Organization (WMO). For other applications, INMARSAT and Iridium are used, for example, by remote siren systems automatically triggered by satellite signals to reach remote communities and authorities who manage the information for alert and emergency management services. Cell broadcasts and alert systems to the population, such as developed in the federal emergency agency (FEMA) in the USA, are essentially relying on ground mobiles and antennas, but satellite operators participate in this emergency broadcasting as well.

Beyond this essential role of communication, satellites have played an increasing role in collecting the geophysical data. We have shown here that technologies for oceanography such as altimetry are not that efficient at monitoring tsunamis but, on the contrary, the most promising technique relies on ground-based receivers of GNSS signals. These offer an impressive capacity of detecting ground motions following an earthquake, which can be inverted in terms of source pattern, of monitoring sea-level variations when the receivers are installed on buoys, and finally of monitoring the upward ionospheric propagation of the tsunami signal through TEC signatures. Other applications such as GNSS-R remain to be confirmed with actual measures. The outstanding development of GNSS networks since the 1990s, with a growing number of real-time processing methods, offers a promising way of incorporating such data for operational purposes.

While many sensors dedicated to geophysical studies are developed within infrastructures or global networks to characterize and to warn, a growing number of less reliable sensors, but on very dense networks, can also bring an original complement to robust means. This is the case for the GNSS data acquired on ships and aircraft from commercial fleets that could also provide valuable, dense information across the oceans when a tsunami occurs (e.g., UNESCAP 2019). Although less secured and robust, they could increase the volume of data describing tsunami propagation, either of the direct waveform or its ionospheric signature.

As for the downstream component to populations and local authorities, satellite imagery helps for post-disaster surveys and for risk prevention plans. Though without direct physical relationship to global climate change, tsunamis are also more of a threat to shorelines where the mean sea level undergoes a gradual rise, exposing some areas with a low altitude above sea level to possible violent sea hazards. The continuous population density increase in coastal areas also contributes to the general increase of the tsunami risk in many places around the global ocean. Satellite data contributes to multi-hazard studies useful to mitigate tsunami impacts.

Finally, it is quite a paradox that the seafloor processes, such as those involving a rapid coseismic deformation, a landslide failure and a tsunami trigger, usually require expensive deployments in the sea (submarine cables, seafloor pressure gauges, ship cruises, etc.) not always available rapidly when a crisis occurs (e.g., monitoring seafloor ground deformation near Mayotte since 2018, e.g., Cesca et al. 2020). Although the satellite platforms contribute more and more to tsunami monitoring, the remote sensing within the oceanic layer remains an expensive challenge that should also be continuously supported for the sake of tsunami science and warnings.

Acknowledgements H. Hébert thanks the support given by the International Space Science Institute (Bern, Switzerland) to propose this review paper, which was originally presented in a solicited presentation at the multidisciplinary workshop held in April 2019 in Bern. Part of the work on the TEC data interpretation was supported by French ANR Cattell IONONAMI (Grant ANR-05-CATT-0004) project, and the TEC inversion and Pamatai airglow monitoring was developed within the NRL TWIST project. The altimetric swaths in the Mediterranean were synthesized in the ANR RISKNAT MAREMOTI (Grant ANR-08-RISKNAT-05-01 project). The study on the design of a GNSS network in western Mediterranean areas has received funding from the EU FP7 project ASTARTE, Assessment, Strategy and Risk Reduction for Tsunamis in Europe grant No. 603839 (Project ASTARTE). The authors wish to thank two anonymous reviewers who helped to revise and improve the clarity of the paper, as well as the Guest Editor, Jérôme Benveniste, and the Editor in Chief, Michael Rycroft, for improving the English language.

References

Abadie S, Harris JC, Grilli ST, Fabre R (2012) Numerical modelling of tsunami waves generated by the flankcollapse of the Cumbre Vieja Volcano (La Palma, Canary Islands): tsunami source and near field effects. J Geophys Res Oceans 117:C05030. https://doi.org/10.1029/2011JC007646

Adriano B, Xia J, Baier G, Yokoya N, Koshimura S (2019) Multi-source data fusion based on ensemble learning for rapid building damage mapping during the 2018 Sulawesi Earthquake and Tsunami in Palu, Indonesia. Remote Sens 11:886. https://doi.org/10.3390/rs11070886

Ajo-Franklin JB, Dou S, Lindsey NJ et al (2019) Distributed acoustic sensing using dark fiber for near-surface characterization and broadband seismic event detection. Sci Rep 9:1328. https://doi.org/10.1038/s41598-018-36675-8

Ammon C, Kanamori H, Lay TA (2008) A great earthquake doublet and seismic stress transfer cycle in the central Kuril islands. Nature 451:561–565. https://doi.org/10.1038/nature06521

Angove M, Arcas D, Bailey R, Carrasco P, Coetzee D, Fry B, Gledhill K, Harada S, von Hillebrandt-Andrade C, Kong L, McCreery C, McCurrach S-J, Miao Y, Sakya AE, Schindelé F (2019) Ocean observations required to minimize uncertainty in global Tsunami forecasts, warnings, and emergency response. Front Mar Sci 6:350. https://doi.org/10.3389/fmars.2019.00350

Artru J, Farges T, Lognonné P (2004) Acoustic waves generated from seismic surface waves: propagation properties determined from Doppler sounding observation and normal-mode modelling. Geophys J Int 158:1067–1077. https://doi.org/10.1111/j.1365-246X.2004.02377.x

Artru J, Ducic V, Kanamori H, Lognonné P, Murakami M (2005) Ionospheric detection of gravity waves induced by tsunamis. Geophys J Int 160(3):840–848. https://doi.org/10.1111/j.1365-246X.2005.02552.x

Assier-Rzadkiewicz S, Mariotti C, Heinrich P (1997) Numerical simulation of submarine landslides and their hydraulic effects. J Waterway Port Coast Ocean Eng 123(4):149–157. https://doi.org/10.1061/(ASCE)0733-950X(1997)123:4(149)

Astafyeva E (2019) Ionospheric detection of natural hazards. Rev Geophys. https://doi.org/10.1029/2019RG000668

Astafyeva E, Lognonné P, Rolland L (2011) First ionosphere images for the seismic slip of the Tohoku-oki earthquake. Geophys Res Lett 38:L22104. https://doi.org/10.1029/2011GL049623

Astafyeva E, Shalimov S, Olshanskaya E, Lognonné P (2013) Ionospheric response to earthquakes of different magnitudes: larger quakes perturb the ionosphere stronger and longer. Geophys Res Lett 40(9):1675–1681. https://doi.org/10.1002/grl.50398

Bai Y, Gao C, Singh S, Koch M, Adriano B, Mas E, Koshimura S (2018) A framework of rapid regional tsunami damage recognition from post-event TerraSAR-X imagery using deep neural networks. IEEE Geosci Remote S 15(1):43–47. https://doi.org/10.1109/lgrs.2017.2772349

Barrientos S, National Seismological Center (CSN) team (2018) The seismic network of Chile. Seismological Res Lett 89(2A):467–474. https://doi.org/10.1785/0220160195

Bernard E (2001) Tsunami: reduction of impacts through three key actions (TROIKA). In: Proceedings of the international tsunami symposium 2001 (ITS 2001), Session 1-1, Seattle, WA, 7–10 August 2001, pp 247–262. https://www.pmel.noaa.gov/pubs/docs/ITS2001/1-01_Bernard.pdf. Accessed Aug 2020

Bernard E, Titov V (2015) Evolution of tsunami warning systems and products. Philos Trans R Soc A 373:20140371. https://doi.org/10.1098/rsta.2014.0371

Bessette-Kirton E, Allstadt K, Godt J, Pursley J (2017) Preliminary analysis of satellite imagery and seismic observations of the Nuugaatsiaq Landslide and Tsunami, Greenland, U.S. Geological Survey—Landslide Hazards Program. https://www.usgs.gov/natural-hazards/landslide-hazards/science/preliminary-analysis-satellite-imagery-and-seismic. Last accessed Aug 2020

Bessis J-L, Béquignon J, Mahmood A (2004) The international charter "space and major disasters" initiative. Acta Astronaut 54(3):183–190. https://doi.org/10.1016/S0094-5765(02)00297-7

Bilitza D, Rawer K (1996) International reference ionosphere. In: Dieminger W, Hartmann G, Leitinger R (eds) The upper atmosphere—data analysis and interpretation. Springer, Berlin, pp 735–772

Blanc E, Ceranna L, Hauchecorne A et al (2018) Toward an improved representation of middle atmospheric dynamics thanks to the ARISE Project. Surv Geophys 39:171–225. https://doi.org/10.1007/s10712-017-9444-0

Bonaccorso A, Gambino S, Mattia M, Guglielmino F, Puglisi G, Boschi E (2009) Insight on recent Stromboli eruption inferred from terrestrial and satellite ground deformation measurements. J Volcanol Geoth Res 18:172–181. https://doi.org/10.1016/j.jvolgeores.2009.01.007

Borrero JC, Synolakis CE, Fritz HM (2006) Northern Sumatra field surveys after the December 2004 great Sumamtra earthquake and Indian Ocean tsunami. Earthq Spectra 22(S3):S93–S104. https://doi.org/10.1193/1.2206793

Broutman D, Eckermann SD, Drob DP (2014) The partial reflection of tsunami-generated gravity waves. J Atmos Sci 71:3416–3426. https://doi.org/10.1175/JAS-D-13-0309.1

Bürgmann R, Chadwell D (2014) Seafloor geodesy. Annu Rev Earth Planet Sci 42:509–534. https://doi.org/10.1146/annurev-earth-060313-054953

Cahyadi MN, Heki K (2015) Coseismic ionospheric disturbance of the large strike-slip earthquakes in North Sumatra in 2012: Mw dependence of the disturbance amplitudes. Geophys J Int 200:116–129. https://doi.org/10.1093/gji/ggu343

Caplan-Auerbach J, Fox CG, Duennebier FK (2001) Hydroacoustic detection of submarine landslides on Kilauea Volcano. Geophys Res Lett 28:1811–1813. https://doi.org/10.1029/2000GL012545

Carlà T, Intrieri E, Di Traglia F, Casagli N (2016) A statistical-based approach for determining the intensity of unrest phases at Stromboli volcano (Southern Italy) using one-step-ahead forecasts of displacement time series. Nat Hazard 84:669–683. https://doi.org/10.1007/s11069-016-2451-5

Carlà T, Intrieri E, Raspini F, Bardi F, Farina P, Ferretti A, Colombo D, Novali F, Casagli N (2019a) Perspectives on the prediction of catastrophic slope failures from satellite InSAR. Sci Rep 9:14137. https://doi.org/10.1038/s41598-019-50792-y

Carlà T, Tofani V, Lombardi L, Raspini F, Bianchini S, Bertolo D, Thuegaz P, Casagli N (2019b) Combination of GNSS, satellite InSAR, and GBInSAR remote sensing monitoring to improve the understanding of a large landslide in high alpine environment. Geomorphology 335:62–75. https://doi.org/10.1016/j.geomorph.2019.03.014

Cesca S, Letort J, Razafindrakoto HNT et al (2020) Drainage of a deep magma reservoir near Mayotte inferred from seismicity and deformation. Nat Geosci 13:87–93. https://doi.org/10.1038/s41561-019-0505-5

Chao WA, Wu TR, Ma KF, Kuo YT, Wu YM, Zhao L, Chung MJ, Wu H, Tsai YL (2018) The large Greenland landslideof 2017: Was a tsunami warning possible? Seismol Res Lett 89:1335–1344. https://doi.org/10.1785/0220170160

Chen P, Liew SC, Kwoh LK (2006) Tsunami damage mapping and assessment in Sumatra using remote sensing and GIS techniques. In: Proceedings of IEEE international conference on geoscience and remote sensing symposium, 2006 July 31–August 4, Denver, CO, pp 297–300. https://doi.org/10.1109/IGARSS.2006.81

Clément J, Reymond D (2015) New tsunami forecast tools for the French Polynesia Tsunami warning system. Pure appl Geophys 172:791–804. https://doi.org/10.1007/s00024-014-0888-6

Coïsson P, Occhipinti G, Lognonné P, Rolland LM (2011) Tsunami signature in the ionosphere: the innovative role of OTH radar. Radio Sci. https://doi.org/10.1029/2010RS004603

Coïsson P, Lognonné P, Walwer D, Rolland LM (2015) First tsunami gravity wave detection in ionospheric radio occultation data. Earth Space Sci 2:125–133. https://doi.org/10.1002/2014EA000054

Duputel Z, Rivera L, Kanamori H, Hayes GP, Hirsorn B, Weinstein S (2011) Real-time W phase inversions during the 2011 Tohoku-oki earthquake. Earth Planets Space 63:535–539. https://doi.org/10.5047/eps.2011.05.032

Elliott JR, Walters RJ, Wright TJ (2016) The role of space-based observation in understanding and responding to active tectonics and earthquakes. Nat Commun 7:13844. https://doi.org/10.1038/ncomms13844

Endo Y, Adriano B, Ma Koshimura S (2018) New insights into multiclass damage classification of tsunami-induced building damage from SAR images. Remote Sens 10:2059. https://doi.org/10.3390/rs10122059

Fritz HM, Hager WH, Minor H-E (2004) Near field characteristic of landslide generated impulse waves. J Waterway Port C Div 130:287–302. https://doi.org/10.1061/(ASCE)0733-950X(2004)130:6(287)

Gaffet S, Guglielmi Y, Cappa F, Pambrun C, Monfret T, Amitrano D (2010) Use of the simultaneous seismic, GPS and meteorological monitoring for the characterization of a large unstable mountain slope in the southern French Alps. Geophys J Int 182(3):1395–1410. https://doi.org/10.1111/j.1365-246X.2010.04683.x

Gailler A, Hébert H, Loevenbruck A, Hernandez B (2013) Simulation systems for tsunami wave propagation forecasting within the French tsunami warning center. Nat Hazard Earth Sys 13:2465–2482. https://doi.org/10.5194/nhess-13-2465-2013

Gailler A, Hébert H, Schindelé F, Reymond D (2018) Coastal amplification laws for the french tsunami warning center: numerical modeling and fast estimate of tsunami wave heights along the French Riviera. Pure appl Geophys 175:1429. https://doi.org/10.1007/s00024-017-1713-9

Galvan D, Komjathy A, Hickey MP, Mannucci AJ (2011) The 2009 Samoa and 2010 tsunamis as observed in the ionosphere using GPS total electron content. J Geophys Res-Space 116(6):A06318. https://doi.org/10.1029/2010JA016204

Garcia RF, Doornbos E, Bruinsma S, Hébert H (2014) Atmospheric gravity waves due to the Tohoku-Oki tsunami observed in the thermosphere by GOCE. J Geophys Res Atmos 119:4498–4506. https://doi.org/10.1002/2013JD021120

Glimsdal S, Pedersen GK, Harbitz CB, Løvholt F (2013) Dispersion of tsunamis: Does it really matter? Nat Hazards Earth Syst Sci 13:1507–1526. https://doi.org/10.5194/nhess-13-1507-2013

González FI, Bernard EN, Meinig C, Eble MC, Mojfeld HO, Stalin S (2005) The NTHMP tsunameter network. Nat Hazards 35:25. https://doi.org/10.1007/s11069-004-2402-4

Grandin R, Klein E, Métois M, Vigny C (2016) Three-dimensional displacement field of the 2015 Mw 8.3 Illapel earthquake (Chile) from across- and along-track Sentinel-1 TOPS interferometry. Geophys Res Lett 43:2552–2561. https://doi.org/10.1002/2016GL067954

Grawe MA, Makela JJ (2015) The ionospheric responses to the 2011 Tohoku, 2012 Haida Gwaii, and 2010 Chile tsunamis: effects of tsunami orientation and observation geometry. Earth Space Sci 2:472–483. https://doi.org/10.1002/2015EA000132

Grawe MA, Makela JJ (2017) Observation of tsunami-generated ionospheric signatures over Hawaii caused by the 16 September 2015 Illapel earthquake. J Geophys Res Space Phys 122:1128–1136. https://doi.org/10.1002/2016JA023228

Guérin C, Binet R, Pierrot-Deseilligny M (2014) Automatic detection of elevation changes by differential DSM analysis: application to urban areas. IEEE J Sel Top Appl 7(10):4020–4037. https://doi.org/10.1109/JSTARS.2014.2300509

Gupta HK (2005) Mega-tsunami of 26th December, 2004: Indian initiative for early warning system and mitigation of oceanogenic hazards. Episodes 28(1):2–5

Gupta HK (2008) India's initiative in mitigating tsunami and storm surge hazard. J Earthq Tsunami 2(4):287–295

Gupta HK, Gahalaut VK (2015) Seismotectonics and large earthquake generation in the Himalayan region. Gondwana Res 25(1):204–213. https://doi.org/10.1016/j.gr.2012.11.006

Hébert H, Heinrich P, Schindelé F, Piatanesi A (2001) Far-field simulation of tsunami propagation in the Pacific Ocean: impact on the Marquesas Islands (French Polynesia). J Geophys Res 106(C5):9161–9177. https://doi.org/10.1029/2000JC000552

Hébert H, Sladen A, Schindelé F (2007) Numerical modeling of the Great 2004 Indian Ocean Tsunami: focus on the Mascarene Islands. Bull Seismol Soc Am 97(1A):S208–S222. https://doi.org/10.1785/0120050611

Hébert H, Burg P-E, Binet R, Lavigne F, Allgeyer S, Schindelé F (2012) The 2006 July 17 Java (Indonesia) tsunami from satellite imagery and numerical modelling: a single or complex source? Geophys J Int 191(3):1255–1271. https://doi.org/10.1111/j.1365-246X.2012.05666.x

Heinrich P, Piatanesi A, Hébert H (2001) Numerical modelling of tsunami generation and propagation from submarine slumps: the 1998 Papua New Guinea event. Geophys J Int 145(1):97–111. https://doi.org/10.1111/j.1365-246X.2001.00336.x

Hickey MP, Schubert G, Walterscheid RL (2009) Propagation of tsunami-driven gravity waves into the thermosphere and ionosphere. J Geophys Res 114:A08304. https://doi.org/10.1029/2009JA014105

Hickey MP, Walterscheid RL, Schubert G (2010a) Wave mean flow interactions in the thermosphere induced by a major tsunami. J Geophys Res 115:A09309. https://doi.org/10.1029/2009JA014927

Hickey MP, Schubert G, Walterscheid RL (2010b) Atmospheric airglow fluctuations due to a tsunami driven gravity wave disturbance. J Geophys Res 115:A06308. https://doi.org/10.1029/2009JA014977

Howe BM et al (2019) SMART cables for observing the global ocean: science and implementation. Front Mar Sci 6:24. https://doi.org/10.3389/fmars.2019.00424

Huba JD, Drob DP, Wu T-W, Makela JJ (2015) Modeling the ionospheric impact of tsunami-driven gravity waves with SAMI3: conjugate effects. Geophys Res Lett 42:5719–5726. https://doi.org/10.1002/2015GL064871

Hupe P, Ceranna L, Le Pichon A (2019) How can the international monitoring system infrasound network contribute to gravity wave measurements? Atmosphere 10(7):399. https://doi.org/10.3390/atmos10070399

Inouye M, Tanioka Y, Yamanaka Y (2019) Method for near-real time estimation of tsunami sources using ocean bottom pressure sensor network (S-Net). Geosciences 9:310. https://doi.org/10.3390/geosciences9070310

IOC (Intergovernmental Oceanographic Commission) (1965) International coordination group for the tsunami warning system in the Pacific (ICG/ITSU), IOC 4th Session, Resolution IV-6

IOC (Intergovernmental Oceanographic Commission) (2005a) Intergovernmental coordination group for the Indian Ocean Tsunami warning and mitigation system, IOC 23rd Session, Resolution XXIII-12

IOC (Intergovernmental Oceanographic Commission) (2005b) Establishment of an intergovernmental coordination group for tsunami and other coastal hazards warning system for the Caribbean and adjacent regions, IOC 23rd Session, Resolution XXIII-13

IOC (Intergovernmental Oceanographic Commission) (2005c) Intergovernmental coordination group for a tsunami early warning system in the North-eastern Atlantic and the mediterranean and connected seas (ICG/NEAMTWS), IOC 23rd Session, Resolution XXIII-14

Iyyappan M, Usha T, Ramakrishnan SS et al (2018) Evaluation of tsunami inundation using synthetic aperture radar (SAR) data and numerical modeling. Nat Hazards 92:1419. https://doi.org/10.1007/s11069-018-3257-4

Jamelot A, Reymond D (2015) New Tsunami forecast tools for the French Polynesia tsunami warning system part II: numerical modelling and tsunami height estimation. Pure appl Geophys 172:805. https://doi.org/10.1007/s00024-014-0997-2

Jamelot A, Gailler A, Heinrich P, Vallage A, Champenois J (2019) Tsunami Simulations of the Sulawesi Mw 7.5 event: comparison of seismic sources issued from a Tsunami warning context versus post-event finite source. Pure appl Geophys 176:3351–3376. https://doi.org/10.1007/s00024-019-02274-5

Ji Y, Sri Sumantyo JT, Chua MY, Waqar MM (2018) Earthquake/Tsunami damage assessment for urban areas using post-event PolSAR data. Remote Sens 10:1088. https://doi.org/10.3390/rs10071088

Jin S (2019) GNSS atmospheric seismology: theory, observations and modelling. Springer, Singapore. https://doi.org/10.1007/978-981-10-3178-6

Jin S, Occhipinti G, Jin R (2015) GNSS ionospheric seismology: recent observation evidences and characteristics. Earth Sci Rev 147:54–64. https://doi.org/10.1016/j.earscirev.2015.05.003

JMA, Japanese Meteorological Agency (2013) Lessons learned from the tsunami disaster caused by the 2011 Great East Japan Earthquake and improvements in JMA's tsunami warning system. http://www.data.jma.go.jp/svd/eqev/data/en/tsunami/LessonsLearned_Improvements_brochure.pdf. Last accessed August 2020

Kanamori H (1972) Mechanism of tsunami earthquakes. Phys Earth Planet Int 6(5):346–359. https://doi.org/10.1016/0031-9201(72)90058-1

Kanamori H, Rivera L (2008) Source inversion of W phase: speeding up seismic tsunami warning. Geophys J Int 175(1):222–238. https://doi.org/10.1111/j.1365-246X.2008.03887.x

Kato T, Terada Y, Nagai T, Koshimura S (2010) Tsunami monitoring system using GPS buoy: present status and outlook. IEEE Int Geosci Remote Se. https://doi.org/10.1109/IGARSS.2010.5654449

Kato T, Terada Y, Tadokoro K, Kinugasa N, Futamura A, Toyoshima M, Yamamoto S, Ishii M, Tsugawa T, Nishioka M, Takizaka K, Shoji Y, Seka H (2018) Development of GNSS buoy for a synthetic geohazard monitoring system. J Disaster Res 13(3):460–471. https://doi.org/10.20965/jdr.2018.p0460

Kawai H, Satoh M, Miyata M, Kobayashi T (2012) 2010 Chilean Tsunami observed by NOWPHAS GPS buoys, seabed wave gauges and coastal tide gauges. Int J Offshore Polar Eng 22(3):177–185

Kawamoto S, Ohta Y, Hiyama Y, Todoriki M, Nishimura T, Furuya T, Sato Y, Yahagi T, Miyagawa K (2017) REGARD: a new GNSS-based real-time finite fault modeling system for GEONET. J Geophys Res 122:1324–1349. https://doi.org/10.1002/2016JB013485

Kerr RA (2005) Model shows islands muted tsunami after latest Indonesian Quake. Science 308(5720):341. https://doi.org/10.1126/science.308.5720.341a

Kherani EA, Lognonné P, Hébert H, Rolland L, Astafyeva E, Occhipinti G, Coïsson P, Walwer D, de Paula ER (2012) Modelling of the total electronic content and magnetic field anomalies generated by the 2011 Tohoku-Oki tsunami and associated acoustic-gravity waves. Geophys J Int 191:1049–1066. https://doi.org/10.1111/j.1365-246X.2012.05617.x

Kherroubi A, Déverchère J, Yelles A, Mercier de Lépinay B, Domzig A, Cattaneo A, Bracène R, Gaullier V, Graindorge D (2009) Recent and active deformation pattern off the easternmost Algerian margin, Western Mediterranean Sea: new evidence for contractional tectonic reactivation. Mar Geol 261(1–4):17–32. https://doi.org/10.1016/j.margeo.2008.05.016

Kong LSL, Dunbar PK, Arcos N (eds) (2015) Pacific Tsunami warning system: a half century of protecting the Pacific (1965-2015). International Tsunami Information Center, Government Printing Office, 188 pp

 Springer

Labbé M, Donnadieu C, Daubord C, Hébert H (2012) Refined numerical modeling of the 1979 tsunami in Nice (French Riviera): comparison with coastal data. J Geophys Res-Earth 117:F01008. https://doi.org/10.1029/2011JF001964

Lavigne F, Paris R, Grancher D, Wassmer P, Brunstein D, Vautier F, Leone F, Flohic F, De Coster B, Gunawan T, Gomez C, Setiawan A, Cahyadi R, Fachrizal A (2009) Reconstruction of tsunami inland propagation on December 26, 2004 in Banda Aceh, Indonesia, through field investigations. Pure appl Geophys 13(166):259–281. https://doi.org/10.1007/s00024-008-0431-8

Lay T, Kanamori H, Ammon CJ, Nettles M, Ward SN, Aster RC, Beck SL, Bilek SL, Brudzinski MR, Butler R, DeShon HR, Ekström G, Satake K, Sipkin S (2005) The Great Sumatra-Andaman Earthquake of 26 December 2004. Science 308:1127–1133. https://doi.org/10.1126/science.1112250

Lay T, Ammon C, Kanamori H, Rivera L, Koper KD, Hutko AR (2010) The 2009 Samoa-Tonga great earthquake triggered doublet. Nature 466:964–968. https://doi.org/10.1038/nature09214

Le Pichon A, Herry P, Mialle P, Vergoz J, Brachet N, Garcés M, Drob D, Ceranna L (2005) Infrasound associated with 2004–2005 large Sumatra earthquakes and tsunami. Geophys Res Lett 32:L19802. https://doi.org/10.1029/2005GL023893

Leone F, Lavigne F, Paris R, Denain J-C, Vinet F (2010) A spatial analysis of the December 26th, 2004 tsunami-induced damages: lessons learned for a better risk assessment integrating buildings vulnerability. Appl Geogr 31(1):363–375. https://doi.org/10.1016/j.apgeog.2010.07.009

Li J, Zhang H, Hou P, Fu B, Zheng G (2016) Mapping the bathymetry of shallow coastal water using single-frame fine-resolution optical remote sensing imagery. Acta Oceanol Sin 35(1):60–66. https://doi.org/10.1007/s13131-016-0797-x

Liu JY, Tsai HF, Lin CH, Kamogawa M, Chen YI, Lin CH et al (2010) Coseismic ionospheric disturbances triggered by the Chi-Chi earthquake. J Geophys Res Space Phys 115(A8):A08303. https://doi.org/10.1029/2009JA014943

Lognonné P (2009) Seismic waves from atmospheric sources and atmospheric/ionospheric signatures of seismic waves, Chapter 10. In: Le Pichon A (ed) Infrasound monitoring for atmospheric studies, Springer, New-York. https://doi.org/10.1007/978-1-4020-9508-5_10

Lognonné P, Clévédé E, Kanamori H (1998) Normal mode summation of seismograms and barograms in an spherical Earth with realistic atmosphere. Geophys J Int 135:388–406. https://doi.org/10.1046/j.1365-246X.1998.00665.x

Lognonné P, Artru J, Garcia R, Crespon F, Ducic V, Jeansou E, Occhipinti G, Helbert J, Moreaux G, Godet P-E (2006) Ground based GPS tomography of ionospheric post-seismic signal. Planet Space Sci 54:528–540. https://doi.org/10.1016/j.pss.2005.10.021

Løvholt F, Pedersen G, Harbitz CB, Glimsdal S, Kim J (2015) On the characteristics of landslide tsunamis. Philos Trans R Soc A 373:20140376. https://doi.org/10.1098/rsta.2014.0376

Lynett PJ et al (2017) Inter-model analysis of tsunami-induced coastal currents. Ocean Model 114:14–32. https://doi.org/10.1016/j.ocemod.2017.04.003

Ma S, Tao Z, Yang X, Yu Y, Zhou X, Li Z (2014) Bathymetry retrieval from hyperspectral remote sensing data in optical-shallow water. IEEE T Geosci Remote 52(2):1205–1212. https://doi.org/10.1109/TGRS.2013.2248372

Mader CL, Curtis G (1991) Modeling Hilo, Hawaii tsunami inundation. Sci Tsunami Hazard 9:85–94

Mai C-L, Kiang J-F (2009) Modeling of ionospheric perturbation by 2004 Sumatra tsunami. Radio Sci. https://doi.org/10.1029/2008RS004060

Makela JJ, Lognonné P, Hébert H, Gehrels T, Rolland L, Allgeyer S, Kherani A, Occhipinti G, Astafyeva E, Coïsson P, Loevenbruck A, Clévédé E, Kelley MC, Lamouroux J (2011) Imaging and modeling the ionospheric airglow response over Hawaii to the tsunami generated by the Tohoku earthquake of 11 March 2011. Geophys Res Lett. https://doi.org/10.1029/2011GL047860

Mangeney A, Heinrich P, Roche R, Boudon G, Cheminée J (2000) Modeling of debris avalanche and generated water waves: application to real and potential events in Montserrat. Phys Chem Earth Pt A 25(9–11):741–745. https://doi.org/10.1016/S1464-1895(00)00115-0

Mannucci AJ, Wilson BD, Yuan DN, Ho CH, Lindqwister UJ, Runge TF (1998) A global mapping technique for GPS-derived ionospheric total electron content measurements. Radio Sci 33:565–582. https://doi.org/10.1029/97RS02707

Manta F, Occhipinti G, Feng L, Hill EM (2020) Rapid identification of tsunami earthquakes using GPS ionospheric sounding. Sci Rep 10:11054. https://doi.org/10.1038/s41598-020-68097-w

Martin-Neira M (1993) A passive reflectometry and interferometry system (PARIS): application to ocean altimetry. ESA J 17:331–355. https://doi.org/10.1109/36.898676

Martin-Neira M, Colmenarejo P, Ruffini G, Serra C (2002) Altimetry precision of 1 cm over a pond using the wide-lane carrier phase of GPS reflected signals. Can J Remote Sens 28:394–403. https://doi.org/10.5589/m02-039

Marty J (2019) The IMS infrasound network: current status and technological developments. In: Le Pichon A, Blanc E, Hauchecorne A (eds) Infrasound monitoring for atmospheric studies. Springer, Cham. https://doi.org/10.1007/978-3-319-75140-5_1

Meilianda E, Pradhan B, Syamsidik S, Comfort L, Alfian D, Juanda R, Syahreza S, Munadi K (2019) Assessment of post-tsunami disaster land use/land cover change and potential impact of future sea-level rise to low-lying coastal areas: a case study of Banda Aceh coast of Indonesia. Int J Disaster Risk Reduct 41:101292. https://doi.org/10.1016/j.ijdrr.2019.101292

Melgar D, Hayes GP (2019) Characterizing large earthquakes before rupture is complete. Sci Adv. https://doi.org/10.1126/sciadv.aav2032

Meng X, Komjathy A, Verkhoglyadova OP, Yang Y-M, Deng Y, Mannucci AJ (2015) A new physics-based modeling approach for tsunami-ionosphere coupling. Geophys Res Lett 42:4736–4744. https://doi.org/10.1002/2015GL064610

Meng X, Verkhoglyadova OP, Komjathy A, Savastano G, Mannucci AJ (2018) Physics-based modeling of earthquake-induced ionospheric disturbances. J Geophys Res Space Phys 123:8021–8038. https://doi.org/10.1029/2018JA025253

Mikesell T, Rolland L, Lee R, Zedek F, Coïsson P, Dessa J-X (2019) IonoSeis: a package to model coseismic ionospheric disturbances. Atmosphere 10(8):443. https://doi.org/10.3390/atmos10080443

Milburn HB, Nakamura AI, González FI (1996) Real-time tsunami reporting from the deep ocean. In: Proceedings of the oceans 96 MTS/IEEE conference, 23–26 September 1996, Fort Lauderdale, FL, pp 390–394. https://doi.org/10.1109/oceans.1996.572778

Mohammed F, Fritz HM (2012) Physical modeling of tsunamis generated by three-dimensional deformable granular landslides. J Geophys Res 117:C11015. https://doi.org/10.1029/2011JC007850

Moreno M, Rosenau M, Oncken O (2010) 2010 Maule earthquake slip correlates with pre-seismic locking of Andean subduction zone. Nature 467:198–202. https://doi.org/10.1038/nature09349

Nayak S, Srinivas Kumar T (2008a) Indian Tsunami warning system. In: The international archives of the photogrammetry. Remote sensing and spatial information sciences XXXVII, Part B4, Beijing, pp 1501–1506

Nayak S, Srinivasa Kumar T (2008b) The first tsunami warning centre in the Indian ocean. In: Risk wise, Tudor Rose Publishers, UK, pp 175–177

NCEI (National Centers for Environmental Informations) (2017) Tsunami sources 1610 B.C. to A.D. 2017 from earthquakes, volcanic eruptions, landslides, and other causes, International Tsunami Information Center. http://itic.ioc-unesco.org/images/stories/awareness_and_education/map_posters/2017_tsu_poster_20180313_a2_low_res.pdf

Newman AV, Hayes G, Wei Y, Convers J (2011) The 25 October 2010 Mentawai tsunami earthquake, from real-time discriminants, finite-fault rupture, and tsunami excitation. Geophys Res Lett 38:L05302. https://doi.org/10.1029/2010GL046498

NOWPHAS, Nationwide Ocean Wave information network for Ports and HArbourS. https://nowphas.mlit.go.jp/. Last accessed Aug 2020

Occhipinti G (2015) The seismology of the planet mongo. In: Morra G, Yuen DA, King D, Lee S, Stein S (eds) Subduction dynamics. https://doi.org/10.1002/9781118888865.ch9

Occhipinti G, Lognonné P, Kherani EA, Hébert H (2006) Three-dimensional waveform modeling of ionospheric signature induced by the 2004 Sumatra tsunami. Geophys Res Lett. https://doi.org/10.1029/2006GL026865

Occhipinti G, Kherani EA, Lognonné P (2008) Geomagnetic dependence of ionospheric disturbances induced by tsunamigenic internal gravity waves. Geophys J Int 173(3):753–765. https://doi.org/10.1111/j.1365-246X.2008.03760.x

Occhipinti G, Dorey P, Farges T, Lognonné P (2010) Nostradamus: the radar that wanted to be a seismometer. Geophys Res Lett 37:L18104. https://doi.org/10.1029/2010GL044009

Occhipinti G, Coïsson P, Makela JJ, Allgeyer S, Kherani A, Hébert H, Lognonné P (2011) Three-dimensional numerical modeling of tsunami-related internal gravity waves in the Hawaiian atmosphere. Earth Planets Space 63(7):847–851. https://doi.org/10.5047/eps.2011.06.051

Occhipinti G, Rolland L, Lognonné P, Watada S (2013) From Sumatra 2004 to Tohoku-Oki 2011: the systematic GPS detection of the ionospheric signature induced by tsunamigenic earthquakes. J Geophys Res Space 118:3626–3636. https://doi.org/10.1002/jgra.50322

Occhipinti G, Aden-Antoniow F, Bablet A et al (2018) Surface waves magnitude estimation from ionospheric signature of Rayleigh waves measured by Doppler sounder and OTH radar. Sci Rep 8:1555. https://doi.org/10.1038/s41598-018-19305-1

Ohta Y (2012) Quasi real-time fault model estimation for near-field tsunami forecasting based on RTK-GPS analysis: application to the 2011 Tohoku-Oki earthquake (Mw9.0). J Geophys Res. https://doi.org/10.1029/2011JB008750

Ohta Y, Inoue T, Koshimura S, Kawamoto S, Hino R (2018) Role of real-time GNSS in near-field tsunami forecasting. J Disaster Res 13:453–459. https://doi.org/10.20965/jdr.2018.p0453

Okada Y (1985) Surface deformation due to shear and tensile faults in a half-space. B Seismol Soc Am 75(4):1135–1154

Okal EA, Piatanesi A, Heinrich P (1999) Tsunami detection by satellite altimetry. J Geophys Res 104:599–615. https://doi.org/10.1029/1998JB000018

Ozaki T (2012) JMA's Tsunami warning for the 2011 Great Tohoku earthquake and Tsunami improvement plan. J Disaster Res 7:439–445. https://doi.org/10.20965/jdr.2012.p0439

Ozawa S, Nishimura T, Suito H, Kobayashi T, Tobita M, Imakiire T (2011) Coseismic and postseismic slip of the 2011 magnitude-9 Tohoku-Oki earthquake. Nature 475:373–376. https://doi.org/10.1038/nature10227

Paris A, Okal EA, Guérin C, Heinrich P, Schindelé F, Hébert H (2019) Numerical modeling of the June 17, 2017 landslide and Tsunami events in Karrat Fjord, West Greenland. Pure appl Geophys 176(7):3035–3057. https://doi.org/10.1007/s00024-019-02123-5

Paris A, Heinrich P, Paris R, Abadie S (2020) The December 22, 2018 Anak Krakatau, Indonesia, landslide and tsunami: preliminary modeling resultsThe December 22, 2018 Anak Krakatau, Indonesia, landslide and tsunami: preliminary modeling results. Pure appl Geophys 177:571–590. https://doi.org/10.1007/s00024-019-02394-y

Percival DB, Denbo DW, Gica E, Huang PY, Mofjeld HO, Spillane MC, Titov VV (2018) Evaluating the effectiveness of DART® Buoy networks based on forecast accuracy. Pure appl Geophys 175:1445. https://doi.org/10.1007/s00024-018-1824-y

Polcari M, Albano M, Montuori A, Bignami C, Tolomei C, Pezzo G, Falcone S, La Piana C, Doumaz F, Salvi S, Stramondo S (2018) InSAR monitoring of italian coastline revealing natural and anthropogenic ground deformation phenomena and future perspectives. Sustainability 10(9):3152. https://doi.org/10.3390/su10093152

Poupardin A, de Michele M, Raucoules D, Idier D (2014) Water depth inversion from satellite dataset. IEEE Geosci Remote Sens Symp 54(4):2329–2342. https://doi.org/10.1109/IGARSS.2014.6946924

Poupardin A, Heinrich P, Frère A, Imbert D, Hébert H, Flouzat M (2017) The 1979 submarine landslide-generated tsunami in Mururoa, French Polynesia. Pure Appl Geoph 174(8):3293–3311. https://doi.org/10.1007/s00024-016-1464-z

Puglisi G, Bonaccorso A, Mattia M, Aloisi M, Bonforte A, Campisi O, Cantarero M, Falzone G, Puglisi B, Rossi M (2005) New integrated geodetic monitoring system at Stromboli volcano (Italy). Eng Geol 79(1–2):13–31. https://doi.org/10.1016/j.enggeo.2004.10.013

Raharja R, Gunawan E, Meilano I et al (2016) Long aseismic slip duration of the 2006 Java tsunami earthquake based on GPS data. Earthq Sci 29:291–298. https://doi.org/10.1007/s11589-016-0167-y

Rakoto V, Lognonné P, Rolland L (2017) Tsunami modeling with solid earth-ocean-atmosphere coupled normal modes. Geophys J Int 211(2):1119–1138. https://doi.org/10.1093/gji/ggx322

Rakoto V, Lognonné P, Rolland L, Coïsson P (2018) Tsunami wave height estimation from GPS-derived ionospheric data. J Geophys Res-Space 123:4329–4348. https://doi.org/10.1002/2017JA024654

Reymond D, Okal EA, Hébert H, Bourdet M (2012) Rapid forecast of tsunami wave heights from a database of pre-computed simulations, and application during the 2011 Tohoku tsunami in French Polynesia. Geophys Res Lett 39:L11603. https://doi.org/10.1029/2012GL051640

Rius A, Aparicio JM, Cardellach E, Martín-Neira M, Chapron B (2002) Sea surface state measured using GPS reflected signals. Geophys Res Lett 29:2122–2125. https://doi.org/10.1029/2002GL015524

Roch J, Duperray P, Schindelé F (2016) Very fast characterization of fault parameters through W-phase centroid inversion in the context of tsunami warning. Pure appl Geophys 173:3881–3893. https://doi.org/10.1007/s00024-016-1258-3

Roger J, Hébert H, Ruegg J-C, Briole P (2011) The El Asnam October 10th, 1980 inland earthquake: a new hypothesis of tsunami generation. Geophys J Int 185:1135–1146. https://doi.org/10.1111/j.1365-246X.2011.05003.x

Rolland LM, Occhipinti G, Lognonné P, Loevenbruck A (2010) Ionospheric gravity waves detected offshore Hawaii after tsunamis. Geophys Res Lett 37:L17101. https://doi.org/10.1029/2010GL044479

Rolland LM, Lognonné P, Astafyeva E, Kherani EA, Kobayashi N, Mann M, Munekane H (2011) The resonant response of the ionosphere imaged after the 2011 off the Pacific coast of Tohoku Earthquake. Earth Planets Space 63(7):853–857. https://doi.org/10.5047/eps.2011.06.020

Rolland LM, Vergnolle M, Nocquet J-M, Sladen A, Dessa J-X, Tavakoli F et al (2013) Discriminating the tectonic and non-tectonic contributions in the ionospheric signature of the 2011, Mw 7.1, dip-slip Van earthquake, Eastern Turkey. Geophys Res Lett 40(11):2518–2522. https://doi.org/10.1002/grl.50544

Ruffini G, Soulat F, Caparrini M, Germain O, Martín-Neira M (2004) The Eddy experiment: accurate GNSS-R ocean altimetry from low altitude aircraft. Geophys Res Lett 31:L12306. https://doi.org/10.1029/2004GL019994

Salameh E, Frappart F, Almar R, Baptista P, Heygster G, Lubac B, Raucoules D, Almeida LP, Bergsma EWJ, Capo S, de Michele M, Idier D, Li Z, Marieu V, Poupardin A, Silva PA, Turki I, Laignel B (2019) Monitoring beach topography and nearshore bathymetry using spaceborne remote sensing: a review. Remote Sens 11:2212. https://doi.org/10.3390/rs11192212

Satake K (1988) Effects of bathymetry on tsunami propagation: application of ray tracing to tsunamis. Pure Appl Geoph. https://doi.org/10.1007/BF00876912

Savastano G, Komjathy A, Verkhoglyadova O, Mazzoni A, Crespi M, Wei Y, Mannucci AJ (2017) Real-time detection of tsunami ionospheric disturbances with a stand-alone GNSS receiver: a preliminary feasibility demonstration. Sci Rep 7:46607. https://doi.org/10.1038/srep46607

Sladen A, Hébert H (2008) On the use of satellite altimetry to infer the earthquake rupture characteristics: application to the 2004 Sumatra event. Geophys J Int 172(2):707–714. https://doi.org/10.1111/j.1365-246X.2007.03669.x

Sladen A, Rivet D, Ampuero JP et al (2019) Distributed sensing of earthquakes and ocean-solid Earth interactions on seafloor telecom cables. Nat Commun 10:5777. https://doi.org/10.1038/s41467-019-13793-z

Smith WHF (1993) On the accuracy of digital bathymetric data. J Geophys Res 98(B6):9591–9603. https://doi.org/10.1029/93JB00716

Smith WHF, Sandwell DT (1997) Global seafloor topography from satellite altimetry and ship depth soundings. Science 277:1956–1962. https://doi.org/10.1126/science.277.5334.1956

Smith SM, Martinis CR, Baumgardner J, Mendillo M (2015) All-sky imaging of transglobal thermospheric gravity waves generated by the March 2011 Tohoku earthquake. J Geophys Res Space Phys 120:10992–10999. https://doi.org/10.1002/2015JA021638

Song YT, Fukumori I, Shum CK, Yi Y (2012) Merging tsunamis of the 2011 Tohoku-Oki earthquake detected over the open ocean. Geophys Res Lett 39:L05606. https://doi.org/10.1029/2011GL050767

Stosius R, Beyerle G, Helm A, Hoechner A, Wickert J (2010) Simulation of space-borne tsunami detection using GNSS-Reflectometry applied to tsunamis in the Indian Ocean. Nat Hazards Earth Syst 10:1359–1372. https://doi.org/10.5194/nhess-10-1359-2010

Stosius R, Beyerle A, Hoechner A, Wickert J, Lauterjung J (2011) The impact on tsunami detection from space using GNSS-reflectometry when combining GPS with GLONASS and Galileo. Adv Space Res 47:843–853. https://doi.org/10.1016/j.asr.2010.09.022

Synolakis CE, Bernard EN (2006) Tsunami science before and beyond Boxing Day 2004. Philos Trans R Soc A 364:2231–2265. https://doi.org/10.1098/rsta.2006.1824

Terada, Y, Kato T, Nagai T, Koshimura S, Imada N, Sakaue H, Tadokoro K (2015) Recent developments of GPS tsunami meter for a far offshore observations. In: Hashimoto M (ed) International symposium on geodesy for earthquake and natural hazards (GENAH). International Association of Geodesy Symposia, Springer, Cham. https://doi.org/10.1007/1345_2015_151

Thomson R, Fine I, Rabinovich A, Mihály S, Davis E, Heesemann M, Krassovski M (2011) Observation of the 2009 Samoa tsunami by the NEPTUNE-Canada cabled observatory: test data for an operational regional tsunami forecast model. Geophys Res Lett 38:L11701. https://doi.org/10.1029/2011GL046728

Tinti S, Manucci A, Pagnoni G, Armigliato A, Zaniboni F (2005) The 30 December 2002 landslide-induced tsunamis in Stromboli: sequence of the events reconstructed from the eyewitness accounts. Nat Hazard Earth Sys 5:763–775. https://doi.org/10.5194/nhess-5-763-2005

Titov VV (2009) Tsunami forecasting. In: Bernard EN, Robinson AR (eds) The sea, vol 15. Harvard University Press, Cambridge, pp 371–400

Titov VV, González FI (1997) Implementation and testing of the method of splitting tsunami (MOST) model. Technical report NOAA Tech. Memo. ERL PMEL-112 (PB98-122773) NOAA/Pacific Marine Environmental Laboratory Seattle, WA

Titov VV, González FI, Bernard EN, Eble MC, Mofjeld HO, Newman JC, Venturato AJ (2005) Real-time tsunami forecasting: challenges and solutions. Nat Hazard 35(1):41–58. https://doi.org/10.1007/s11069-004-2403-3

Tsushima H, Hino R, Ohta Y, Iinuma T, Miura S (2014) tFISH/RAPiD: rapid improvement of near-field tsunami forecasting based on offshore tsunami data byincorporating onshore GNSS data. Geophys Res Lett 41:3390–3397. https://doi.org/10.1002/2014GL059863

Uma Devi E, Sunanda V, Ajay Kumar B, Patanjali Kumar Ch, Srinivasa Kumar T (2016) Real-time earthquake monitoring at the Indian Tsunami Early Warning System for tsunami advisories in the Indian Ocean. Int J Ocean Clim Syst 7(1):20–26. https://doi.org/10.1177/1759313115623164

UNESCAP (UN Economic and Social Commission for Asia and the Pacific) (2019) Roundtable: maritime sector strategies to augment Tsunami monitoring with economic, safety and environmental co-benefits. Roundtable, 22–23 Aug 2019. https://www.unescap.org/sites/default/files/Roundtable%20Outcome%20Report%20APril%202020.pdf. Accessed August 2020

Vadas SL, Makela JJ, Nicolls MJ, Milliff RF (2015) Excitation of gravity waves by ocean surface wave packets: upward propagation and reconstruction of the thermospheric gravity wave field. J Geophys Res Space Phys 120:9748–9780. https://doi.org/10.1002/2015JA021430

Vallée M, Ampuero J-P, Juhel K, Bernard P, Montagner J-P, Barsuglia M (2017) Observations and modeling of the elastogravity signals preceding direct elasticity waves. Science 358(6367):1164–1168. https://doi.org/10.1126/science.aao0746

Walker JG (1984) Satellite constellations. J Br Interplanet Soc 37:559–571

Wang R, Parolai S, Ge M, Jin M, Walter TR, Zschau J (2013) The 2011 Mw 9.0 Tohoku Earthquake: comparison of GPS and strong-motion data. Bull Seismol Soc Am 103:1336–1347. https://doi.org/10.1785/0120110264

WMO (2018) Manual on the global telecommunication system, annex III to the WMO technical regulations, 2015 Edition, updated in 2018, 200 pp

Yamamoto N, Hirata K, Aoi S, Suzuki W, Nakamura H, Kunugi T (2016) Rapid estimation of tsunami source centroid location using a dense offshore observation network. Geophys Res Lett 43:4263–4269. https://doi.org/10.1002/2016GL068169

Yan X, Sun Y, Yu T, Liu J-Y, Qi Y, Xia C et al (2018) Stratosphere perturbed by the 2011 Mw90 Tohoku earthquake. Geophys Res Lett 45(10):10050–10056. https://doi.org/10.1029/2018GL079046

Yang Y-M, Verkhoglyadova O, Mlynczak MG, Mannucci AJ, Meng X, Langley RB, Hunt LA (2017) Satellite-based observations of tsunami-induced mesosphere airglow perturbation. Geophys Res Lett 44:522–532. https://doi.org/10.1002/2016GL070764

Yang Y, Dunham EM, Barnier G, Almquist M (2019) Tsunami wavefield reconstruction and forecasting using the ensemble Kalman filter. Geophys Res Lett 46:853–860. https://doi.org/10.1029/2018GL080644

Ye L, Kanamori H, Rivera L, Lay T, Zhou Y, Sianipar D, Satake K (2020) The 22 December 2018 tsunami from flank collapse of Anak Krakatau volcano during eruption. Sci Adv. https://doi.org/10.1126/sciadv.aaz1377

Yu Y, Wang W, Hickey MP (2017) Ionospheric signatures of gravity waves produced by the 2004 Sumatra and 2011 Tohoku tsunamis: a modeling study. J Geophys Res Space Phys 122:1146–1162. https://doi.org/10.1002/2016JA023116

Publisher's Note Springer Nature remains neutral with regard to jurisdictional claims in published maps and institutional affiliations.

Affiliations

H. Hébert[1] · G. Occhipinti[2] · F. Schindelé[1] · A. Gailler[1] · B. Pinel-Puysségur[1] · H. K. Gupta[3] · L. Rolland[4] · P. Lognonné[2] · F. Lavigne[5] · E. Meilianda[6] · S. Chapkanski[5] · F. Crespon[7] · A. Paris[1] · P. Heinrich[1] · A. Monnier[1] · A. Jamelot[8] · D. Reymond[8]

[1]   CEA, DAM, DIF, 91297 Arpajon Cedex, France

[2]   Université de Paris, Institut de Physique du Globe de Paris, CNRS, 75005 Paris, France

[3]   CSIR-National Geophysical Research Institute, Uppal Road, Hyderabad 500007, India

[4]   OCA, CNRS, IRD, Géoazur, Université Côte d'Azur, 06560 Valbonne, France

[5]   Laboratoire de Géographie Physique UMR 8591, Université Paris 1 Panthéon-Sorbonne, 1 Place Aristide Briand, 92195 Meudon, France

[6]   Tsunami and Disaster Mitigation Research Center (TDMRC), Syiah Kuala University, Jalan Prof. Dr. Ibrahim Hasan, Ulee Lheue, Kecamatan Meuraxa, Banda Aceh 23232, Indonesia

[7]   Noveltis, 31672 Labège, France

[8]   LDG/Pamatai, Papeete, Tahiti 98713, French Polynesia

Surveys in Geophysics (2020) 41:1583–1609
https://doi.org/10.1007/s10712-020-09599-0

Air Pollution and Sea Pollution Seen from Space

Camille Viatte[1] · Cathy Clerbaux[1] · Christophe Maes[2] · Pierre Daniel[3] · René Garello[4] · Sarah Safieddine[1] · Fabrice Ardhuin[2]

Received: 19 November 2019 / Accepted: 19 May 2020 / Published online: 11 June 2020
© The Author(s) 2020

Abstract

Air pollution and sea pollution are both impacting human health and all the natural environments on Earth. These complex interactions in the biosphere are becoming better known and understood. Major progress has been made in recent past years for understanding their societal and environmental impacts, thanks to remote sensors placed aboard satellites. This paper describes the state of the art of what is known about air pollution and focuses on specific aspects of marine pollution, which all benefit from the improved knowledge of the small-scale eddy field in the oceans. Examples of recent findings are shown, based on the global observing system (both remote and in situ) with standardized protocols for monitoring emerging environmental threats at the global scale.

Keywords Pollution · Atmosphere · Ocean · Space-borne instrument

1 Introduction

According to the World Health Organization (WHO), air pollution currently kills more than 4 million people per year (more than cholesterol), and projections indicate that 6.5 million people will be affected in 2050. Pollution levels are very unevenly distributed across the globe (Asia being the most affected), and each individual is affected differently by pollution episodes. On average, an adult inhales about 10 L of air every minute and up to 100 L during sports activities. Children, seniors, and people with cardiovascular or respiratory problems are the most affected. It is estimated that air pollution contributes to 7.6% of all deaths mainly due to cardiovascular and respiratory diseases (World Health Organization 2017). Air pollution is ranked as the fifth highest mortality risk factor associated with

✉ Camille Viatte
  camille.viatte@latmos.ipsl.fr

[1] LATMOS/IPSL, Sorbonne Université, UVSQ, CNRS, BC 102, 4 place Jussieu,
     75252 Paris Cedex 05, France

[2] LOPS, Univ. Brest CNRS/IFREMER/IRD/UBO, 29280 Brest, Plouzané, France

[3] Météo-France, Direction des Opérations pour la Prévision, Département Prévision Marine et
     Océanographique, 42 Av. Coriolis, BP 45712, 31057 Toulouse Cedex 1, France

[4] UMR Lab-STICC 6285 - IMT Atlantique Bretagne-Pays de la Loire, Technopôle Brest-Iroise,
     Campus de Brest, CS 83818, 29238 Brest cedex 03, France

Surveys in Geophysics (2020) 41:1583–1609

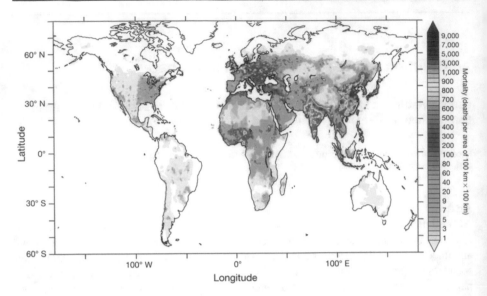

Fig. 1 Mortality related to outdoor pollution (in 2010). The color scale represents the number of deaths over areas of 100×100 km, as an annual average, for fine particulate matter ($PM_{2.5}$) and ozone. White areas are areas where concentrations are below the limits of health impact. Figure from Lelieveld et al. (2015)

147 million years of healthy life lost (Health Effect Institute 2019). For instance, pollution accounts for 800,000 additional deaths per year in Europe (Fig. 1).

Covering more than 70% of our planet, oceans contain the Earth's most valuable natural resources but are also the end point for so much of the pollution from societies and civilization generated on land. The types of ocean pollution are vast, ranging from carbon emissions, to choking marine debris, to leaking oil, to noise, and to simply introducing harmful or toxic substances and wastes into the natural environment. The majority of the garbage that enters the ocean each year is made out of plastic. Because of its properties, this type of pollution is not biodegradable and persists for long periods of time, making plastic pollution one of the most ubiquitous and pressing contemporary threats to the worldwide coastal and marine environments. Due to its potential impact through the food chain on human health, the WHO calls in early 2019 for a further assessment of microplastics in the whole environment. Despite the increasing body of evidence and the literature that microplastic pollution is accumulating in the oceans, the real diffusion and pathways in the oceans remain uncertain and long-term life cycle of such pollution is unknown.

This paper is organized as follows: Sect. 1 focuses on air pollution, starting with a few highlights on the main pollutants and their evolution with time. A general description on how to monitor pollutants from space is then provided, with illustrations of four case studies involving different pollutants. Section 2 is dedicated to sea pollution, starting with an introduction, and followed by a description of three major issues: marine accident and safety, oil spill pollution, and the problem of plastics as marine debris.

2 Air Pollution

2.1 Main Atmospheric Pollutants and Trends

A pollutant is a chemical compound that affects health, not to be confused with a greenhouse gas that affects the climate. They can be emitted directly from the surface (primary pollutants) and/or formed through chemical reactions in the atmosphere (secondary pollutants). Depending on the chemical composition, number, size, and shape of the gases and particles, effects on humans, animals, plants, microorganisms and the environment may vary. The health impacts of each pollutant are difficult to estimate, and it is only relatively recently that documented and quantified correlations between particulate matter levels and health have been established, although it has long been known that life expectancy in polluted countries is lower (Brunekreef 1997).

Air quality standards generally define and set the norms of the following air pollutants: sulfur dioxide (SO_2), nitrogen dioxide (NO_2), carbon monoxide (CO), ozone (O_3), ammonia (NH_3) and particulate matter (with a diameter less than 10 and 2.5 micrometers in size, PM_{10} and $PM_{2.5}$, respectively). The sources of these pollutants are various and mainly linked to human activities: transport, heating, industrial activity, agriculture, etc.

- Sulfur dioxide (SO_2): is primarily produced from the burning of fossil fuels from power plants, industry, automobiles, planes and ships. When combined with water in the atmosphere, it forms sulfuric acid that is the main component of acid rain. Exposure to SO_2 affects the respiratory system and causes irritation of the eyes.
- Nitrogen dioxide (NO_2): is an odorous gas mainly emitted by industrial and traffic sources and is an important constituent of particulate matter and a precursor of ozone. These affect health through respiratory diseases and are linked to premature mortality.
- Carbon monoxide (CO): is a colorless and odorless gas that is toxic and causes death at high concentrations. In the atmosphere, CO is produced by incomplete combustion of carbon-containing fuels, such as gasoline, natural gas, oil, coal, and wood.
- Ozone (O_3): is a secondary pollutant produced when CO, methane, or other volatile organic compounds (VOCs) are oxidized in the presence of nitrogen oxides (NO_x) and sunlight. It is a major component of photochemical smog and leads to breathing problems, asthma, reduced lung function and respiratory diseases.
- Ammonia (NH_3): is a reactive pollutant for which emissions are at least four times higher than in the pre-industrial era leading to environmental damages such as acidification and eutrophication. It is mainly emitted from the agriculture sector including nitrogen fertilizers and livestock manure. It is also a precursor of particulate matter.
- Particulate matter (PM): can be composed of many constituents such as sulfates, nitrates, ammonia, black carbon, and mineral dust. They are emitted by traffic (both diesel and petrol), by industry of energy production as well as other industrial activities (building, mining, manufacture of cement). Size of PM varies, from a diameter of less than 10 microns (PM_{10}) to 2.5 microns ($PM_{2.5}$) and less. PM poses major risks to health, as they are capable of penetrating lungs and entering the bloodstream.

Depending on the country, pollutant emissions are more or less significant (Elguindi et al. 2020), and air quality is regulated or not; currently, it is regulated in about fifty countries, but with variable population alert thresholds. Trends in the emissions depend strongly on national implemented mitigation strategies (Fig. 2). In most industrial countries (where

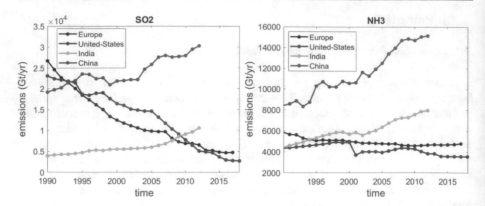

Fig. 2 Trends of emissions of SO_2 (left panel) and NH_3 (right panel) over Europe[a] (black), USA[b] (blue), India[c] (green), and China[c] (magenta). *Sources* [a]European Environment Agency (EEA) (2018), [b]United States Environmental Protection Agency (EPA), [c]Crippa et al. (2018)

coal is not used anymore), the pollutants that often exceed the harmful thresholds are ozone, ammonia and particles (PM_{10} and $PM_{2.5}$). Ammonia (related to fertilizers and livestock) also remains difficult to control.

China and India are the two most polluted countries. India is currently implementing strategies for effective air pollution control (Apte and Pant 2019) especially by mitigating emissions from household sources (Chowdhury et al. 2019). While China's ambient air is still very polluted, the remarkable improvements seen in recent years (Fig. 3) highlight the potential for air quality management efforts to rapidly and substantially improve air quality both in China and around the world.

2.2 How to Monitor Pollution from Space?

Instrumental remote sensing techniques dedicated to the monitoring of atmospheric gases concentrations have greatly improved in the past few decades. They now allow the probing into the lowest layers of the troposphere and contribute to air quality assessment. For atmospheric composition, satellite observational tools are mainly in the form of

Fig. 3 Trends of emissions of SO_2 (black), NO_x (in Gg/yr of NO_2, blue), PM_{10} (red), and NH_3 (green) over China resulting from aggressive pollution control. Figure adapted from Zheng et al. (2018)

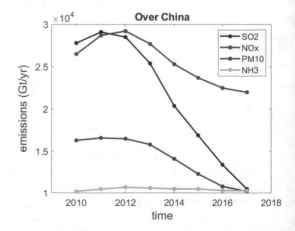

spectrometers with high spectral resolution ranging from the thermal infrared (IR) to the ultraviolet (UV) spectral range. Combined with ground-based measurements and atmospheric models, space observations are now essential and used in most scientific studies and many atmospheric applications to improve our knowledge about the physical and chemical processes from the global to the local scale.

Back in the 1980s, global space-borne observations of trace gases from space were dedicated to water vapor and ozone measurements. For instance, the total ozone mapping spectrometer (TOMS) instruments were able to characterize the ozone hole formation over Antarctica and monitor its evolution, specifically in response to the implementation of the Montreal Protocol in 1987 to ban ozone depleting substances. With the improvement of measurement techniques and retrieval algorithms of species concentration, it is now possible to provide quantitative information about many trace gases and pollutants concentration at regional or even local scales and with an increasingly fine vertical resolution. The latest achievements even allow sounding down to the Earth's surface for some reactive species.

To observe atmospheric constituents, passive remote sensing instruments from satellites measure atmospheric spectra resulting from the interaction between radiation (solar or emitted by the Earth or the atmosphere) and molecules. The exploitation of these signals containing specific features or signatures of the different molecules allows retrieving their concentrations at different altitudes (for strong absorbers) or their total column concentration (integrated along the vertical, for weak absorbers). Each molecular absorption/emission line (in the IR) or cross section (in the UV) in a spectrum has a characteristic signature: The position indicates the identity of the molecule, and the intensity infers its atmospheric concentration. To properly retrieve atmospheric concentrations from any raw satellite spectrum, other parameters are required, such as characteristics of the instrument (detector, optics, etc.), spectroscopic data, auxiliary data such as temperature or air mass factors, and an atmospheric radiative transfer algorithm.

There are many satellite instruments at the moment able to measure atmospheric pollution. The state-of-the-art instruments using the UV–visible and thermal infrared radiation are TROPOMI (TROPOspheric Monitoring Instrument, Veefkind et al., 2012) and IASI (Infrared Atmospheric Sounding Interferometer, Clerbaux et al., 2009), respectively. These missions have shown their ability to accurately monitor concentrations and variability of key gases involved in atmospheric pollution. TROPOMI is a multispectral imaging spectrometer which has ultraviolet and visible spectral (UV–Vis—270–500 nm), near infrared (675–775 nm) and infrared (2305–2385 nm) bands (Fig. 4). It was launched in October 2017 by the European Space Agency (ESA) to monitor ground-level pollutants such as nitrogen dioxide, ozone, formaldehyde (HCHO), and sulfur dioxide with an accuracy and spatial resolution (3.5×7 km^2) never achieved before (see https://sentinels.copernicus.eu/web/sentinel/missions/sentinel-5p).

In the infrared part of the spectrum, the IASI instrument performs complementary measurements of pollutants from numerous (8641) channels in the thermal infrared spectral range with more than 31 molecules detected in its spectra (Fig. 5). Three identical IASI instruments were built by Centre National d'Etudes Spatiales (CNES) and European Organization for the Exploitation of Meteorological Satellites (EUMETSAT) for three successive launches: Metop-A on October 19, 2006, Metop-B on September 17, 2012, and Metop-C on November 7, 2018. All these measurements are analyzed and distributed in near real time to deliver gas concentration maps around the world less than 3 h after the measurement is taken.

The complementarity of multi-platform measurements, on board satellites or in situ from the ground, allows a better observation of the chemical composition of the

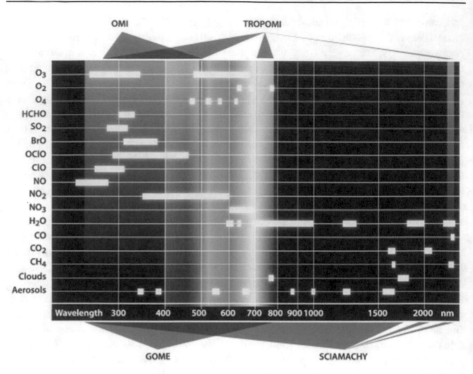

Fig. 4 Spectral ranges for TROPOMI and the heritage instruments: OMI (Ozone Monitoring Instrument), SCIAMACHY (SCanning Imaging Absorption SpectroMeter for Atmospheric CHartographY) and GOME (Global Ozone Monitoring Experiment). Figure from Veefkind et al. (2012)

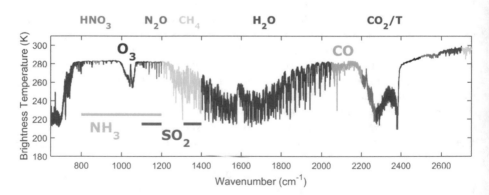

Fig. 5 Example of an IASI spectrum containing absorption features of different species important for air pollution and climate monitoring

atmosphere. In situ measurements are more accurate at local scale, while satellite measurements provide better spatial coverage, with information on the transport of pollution plumes. In addition, satellite instruments can monitor in remote areas with very limited coverage by ground-based measurement networks. These satellite observations are therefore important and are used to constrain emission inventories, validate our knowledge about physical and chemical processes of the atmosphere, and improve the prediction of pollution peaks when assimilated into dedicated atmospheric models.

Both the infrared and ultraviolet observations provide the user community with valuable information to carry out research on pollution processes across the globe. On the operational side, routine pollution forecasts are provided by the Copernicus Atmospheric Composition Monitoring (CAMS; https://atmosphere.copernicus.eu/ service: data from a dozen of instruments on board satellites are assimilated into regional and global models, to deliver air quality information for the main pollutants).

2.3 A Few Illustrations of Atmospheric Pollution

2.3.1 Case study 1: Mapping Nitrogen Dioxide (NO₂) from Space at the City Scale

Nitrogen dioxide and nitrogen oxide (NO) together are usually referred to as nitrogen oxides ($NO_x = NO + NO_2$). They are important pollutants emitted by combustion processes in the troposphere. In the presence of sunlight, a photochemical cycle involving ozone converts very rapidly NO into NO_2 making it a robust measure for nitrogen oxides concentrations. With a relatively short lifetime in the troposphere (few hours), the NO_2 will remain relatively close to its source, making the NO_x sources well detectable from space, whereas NO is disappearing too rapidly to be measured from space.

With TROPOMI's high spatial resolution, pollution from individual power plants, other industrial complexes, and major highways can be observed. Figure 6 shows hot spots of NO_2 pollution over the urban regions of UK, the Netherlands, the Pô Valley and the Ruhr area in the monthly average tropospheric NO_2 for April 2018 over Europe derived from TROPOMI data (van Geffen et al. 2019). This is very valuable in improving our knowledge

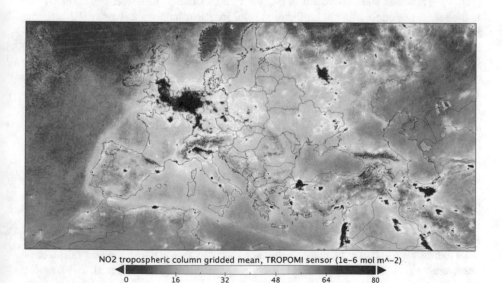

NO2 tropospheric column gridded mean, TROPOMI sensor (1e–6 mol m^–2)

| 0 | 16 | 32 | 48 | 64 | 80 |

Fig. 6 Monthly average distribution of tropospheric NO_2 columns for April 2018 over Europe based on TROPOMI data, derived with processor version 1.2.0. Figure from van Geffen et al. (2019)

on how different sectors contribute to the overall emission of nitrogen oxides all over the world.

2.3.2 Case Study 2: Regional Atmospheric Ammonia Variability over EUROPE and Link with Local PM Formation over the Paris Urban Area

With respect to primary pollutants, such as sulfur dioxide (SO_2) or nitrogen oxides (NO_x), in most countries no mitigation policies have been implemented to reduce NH_3 emissions despite the importance of this gas in the formation of secondary aerosols, visibility degradation and/or acidification and eutrophication. Despite this, our knowledge of (1) spatiotemporal distributions of NH_3 sources at the regional scale, and (2) the role of NH_3 in the $PM2_{2.5}$ formation is very limited (mainly because there are very few systematic measurements), and emission inventories are uncertain.

Recently, IASI satellite observations were used to map for the first time the distribution of NH_3 concentrations to the nearest kilometer (Van Damme et al. 2018). More than 400 NH_3 point and extended sources have been identified worldwide, and this study showed that emissions from known sources are largely underestimated by emission inventories.

The Paris megacity (which is one of the most populated cities in the European Union with 10.5 million inhabitants) experiences frequent particulate matter pollution episodes in springtime (March–April). At this time of the year, major parts of of the particles consist of ammonium sulfate and nitrate which are formed from ammonia released during fertilizer spreading practices and transported from the surrounding areas to Paris and affect local air quality. There is still limited knowledge on the emission sources around Paris, both in terms of magnitude and seasonality.

Using IASI space-borne NH_3 observation records of 10-years (2008–2017) a regional pattern of NH_3 variabilities (seasonal and interannual) was derived (Viatte et al. 2020). Three main regions of high NH_3 occurring between March and August were identified (Fig. 7). Those hot spots are found in (A) the French Champagne-Ardennes region which is the leader region for mineral fertilization used for sugar industry in France (Ramanantenasoa et al. 2018); (B) the northern part of the domain known for its animal farming (Scarlat et al. 2018); and (C) in the Pays de la Loire region also areas of livestock farming (Robinson et al. 2014). It has been found that air masses originated from rich NH_3 areas, mainly over Belgium and the Netherlands, increase the observed NH_3 total columns measured by IASI over the urban area of Paris. The latter study also suggests that meteorological parameters are the limiting factors and that low temperature, thin boundary layer, coupled with almost no precipitation and wind coming from the northeast, all favor the $PM_{2.5}$ formation with the presence of atmospheric NH_3 in the Paris urban region.

2.3.3 Case Study 3: Summertime Tropospheric Ozone over the Mediterranean Region

Tropospheric ozone is a greenhouse gas, air pollutant, and a primary source of the hydroxyl radical (OH), the most important oxidant in the atmosphere (Chameides and Walker 1973). Previous observations and studies have shown that tropospheric O_3 over the Mediterranean exhibits a significant increase during summertime, especially in the east of the basin (Kouvarakis et al. 2000; Im et al. 2011; Richards et al. 2013; Safieddine et al. 2014, among others).

Ozone is a secondary pollutant which is not directly emitted in the troposphere but is formed when its precursors (i.e., specific chemical compounds such as carbon monoxide,

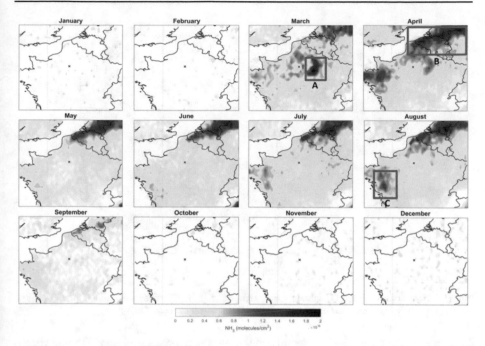

Fig. 7 Seasonal variability of NH$_3$ derived from 10 years of IASI measurements. The blue cross represents Paris. Figure from Viatte et al. (2020)

methane and nitrogen oxides) react with other chemical compounds, in the presence of solar radiation. Meteorological conditions such as frequent clear-sky conditions and high exposure to solar radiation in summer enhance the formation of photochemical O$_3$ due to the availability of its precursors. Locally, the eastern part of the basin is surrounded by megacities such as Cairo, Istanbul, and Athens that are large sources of local anthropogenic emissions. The geographical location of the basin makes it a receptor for anthropogenic pollution from Europe both in the boundary layer and the mid-troposphere. Moreover, numerous studies have also shown that the eastern part of the basin is associated with strong summer anti-cyclonic activity leading to downward transport of upper tropospheric ozone into the lower troposphere (Zanis et al. 2014; Kalabokas et al. 2015). Summertime stratospheric intrusions are also common events above this region influencing both the upper and mid-troposphere (Safieddine et al. 2014; Akritidis et al. 2016). Figure 8 shows the tropospheric [0-8] km ozone column above Europe and the Mediterranean basin. The east side of the basin exhibits the highest ozone concentration as seen by the IASI instrument.

2.3.4 Case study 4: CO Pollution Across Europe Linked with the Hurricane Ophelia

With the remnants of hurricane Ophelia reaching the British Isles in October 2017, a rare pollution event crossed Europe, with transport of dust from the Sahara and of thick smoke from disastrous wildfires in northern Portugal/Spain. On October 15–16, both dust and smoke were rapidly advected by Ophelia-energized strong southerly winds into the Bay of Biscay, northwestern France (Brittany) and across the UK. On October 17, residents of

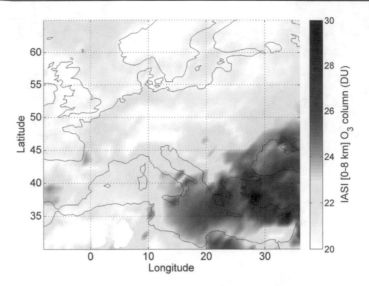

Fig. 8 Tropospheric [0–8] km O_3 column during summer 2013. High values are recorded in the Mediterranean, especially to the east part of the basin. Figure from Safieddine et al. (2014)

Belgium and the Netherlands woke up under a surprising orange-red Sun with low visibility enduring for most of the day (Osborne et al. 2019).

Two unusual events were intersecting: a tropical storm and high-intensity late-season fires. We now benefit from a range of observations above Europe allowing us to track the transport of a pollution plume. Key observations on the spreading of this event are provided by IASI. From its radiance, data near real-time global scale maps of dust, carbon monoxide (Fig. 9) and ammonia (two proxies for fire plumes) were derived. The highly polluted plumes could be followed for a full week using IASI CO data, as they spread over large parts of northern and eastern Europe, and into Asia (Clerbaux 2018).

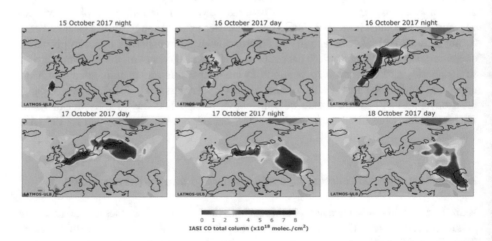

Fig. 9 CO plume associated with fires transported from Spain/Portugal in October 2017. Each subplot is representing the CO total column measured at each IASI overpass, in the morning and in the afternoon. Figure from Clerbaux (2018)

3 Sea Pollution

3.1 Marine Accident and Safety

Of increasing concern in parallel to air pollution are the marine ecosystem degradation and the sea pollution. More than 3 billion people depend crucially on the oceans for their livelihoods. Associated with their presence, it has been reported that litter contaminates marine habitats from the poles to the equator and from the coastal borders to the deep environments. Sea pollution is far from being uniform and that depends on our knowledge of the oceans themselves. Monitoring and forecasting of accidental marine pollution, for which oil spills are an important factor, depend on access to quality information on ocean circulation. It is one of the most important applications of operational oceanography.

Over the past decades, the accuracy of weather and oceanographic forecasts has improved, but millions of dollars of goods and thousands of lives are still lost at sea each year due to extreme weather and oceanographic conditions. In the marine environment, vessels of all sizes are exposed and vulnerable to the elements. High winds, high waves, fog, thunderstorms, sea ice and freezing spray make shipping a high-risk business. Nevertheless, the ocean and seas are a sustainable transport route for the global economy—a "blue economy" estimated at 84% of world trade (The International Union of Marine Insurance Stats report 2018: https://iumi.com). The value of the global ocean economy is estimated at 3–6 trillion USD per year (United Nations Conference on Trade and Development, 2019: https://unctad.org). In addition, ferries carry more than a quarter of the world's population each year (InterFerry, 2019, see https://interferry.com/ferry-industry-facts/). Maritime incidents endanger lives and property on board and can also cause environmental disasters.

The International Convention for the Safety of Life at Sea (SOLAS), which was created after the Titanic tragedy in 1912 and maintained by the International Maritime Organization (IMO), sets out various standards for the safety, security and operation of ships. IMO and the World Meteorological Organization (WMO), which are specialized agencies of the United Nations, ensure that consistent weather-related marine safety information is available globally for ships at sea. WMO provides a coordination structure that establishes protocols to ensure that coastal countries provide standardized weather information, forecasts and warnings to ensure the safety of people and goods at sea. This includes the requirement to issue explicit warnings for tropical cyclones and for gale and storm wind speeds. Other services include warnings for other severe conditions such as abnormal or devastating waves, reduced visibility due to fog and ice accumulation. Services are prepared and disseminated by National Meteorological Services according to areas of responsibility called Meteorological Areas (METAREA) as shown in Fig. 10. METAREAs are closely aligned with navigation zones, which are used to provide navigational information and warnings to ships at sea. Weather-related maritime safety information is transmitted to ships at sea by countries through the WMO Maritime Broadcast System, which supports the IMO Global Maritime Distress and Safety System (GMDSS).

These warning and forecasting services are widely used by all mariners, safety and security organizations and economic sectors that make informed decisions using marine weather information to improve the safety of people and goods, environmental safety and socio-economic benefits in marine and coastal environments. Numerical weather, wave and ocean prediction models provide marine forecasters with indications that underpin the forecasting process. Weather and ocean forecasting models are run two

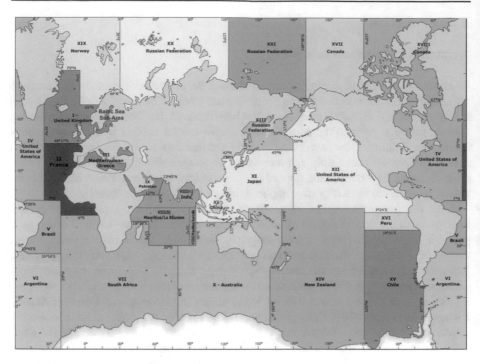

Fig. 10 Global map of the METAREAs with issuing services

to four times a day using the data collected to make future forecasts. These models are based on supercomputers that perform calculations based on complex systems of differential equations that approximate the physical processes of the atmosphere, the oceans and sea ice. Experienced marine forecasters evaluate all available observations and numerical forecasts to ensure that the atmosphere and the ocean are represented as accurately as possible in the initial state of the model and to determine where the greatest model uncertainties lie on a given day. Forecasters then prepare and disseminate their marine weather forecasts, warn of hazardous weather conditions as required and send their forecasts for broadcast on the Internet, radio or satellite.

As a result of decades of coordination with universities, research centers, research infrastructures and private companies, operational oceanographic services, such as the Copernicus Marine Environment Monitoring Service (CMEMS: http://marine.coper nicus.eu), have been set up. The combination of advanced observations and forecasts offers new downstream service opportunities to meet the needs of heavily populated coastal areas and climate change. Over the next decade, the challenge will be to maintain ocean observations, monitor small-scale variability, resolve sub-basin/seasonal and interannual variability in circulation, understand and correct model biases, and improve model data integration and overall forecasting for uncertainty estimation. Better knowledge and understanding of the level of variability will enable subsequent assessment of the impacts and mitigation of human activities and climate change on biodiversity and ecosystems, which will support environmental assessments and decisions. Other challenges include the extension of value-added scientific products to socially relevant downstream services and engagement with communities to develop initiatives that will contribute, inter alia, to the United Nations Decade of Ocean Sciences for Sustainable Development, thereby helping to bridge the gap between science and policy.

Forecasting services can play an important role in both incident decision making and the design of response services. Monitoring, forecasting and, to some extent, detection of marine pollution depend mainly on reliable and timely access to environmental data, observations and predictions products. These products provide an overview of the current and future state of meteorological and oceanographic conditions. They can also be used to drive pollutant fate prediction models, either directly or by providing boundary conditions for high-resolution nested models. Access to large geophysical datasets that are interoperable with regional and sub-regional observation and modeling systems, using standard service formats and specifications such as those provided by CMEMS, is an undeniable achievement.

Oil spill prediction is generally performed using a numerical model to predict the advection and weathering of oil in the sea (examples will be discussed in the next section). Although the treatment of these processes can vary considerably from one model to another, they all critically depend on geophysical forcing to determine the fate of the oil spill, particularly its movement implying currents and winds. Oil spill models consider different forcing fields from data on wind, ocean currents, wave energy, Stokes drift, air temperature, water temperature and salinity, turbulent kinetic energy, depending on the settings used by the particular model. For the prediction of the movement of oil slicks on the high seas, ocean circulation data has the greatest potential for improvement, mainly because ocean forecasts are less mature than weather and wave forecasts. Operational ocean forecasting systems offer the promise of improved forecast accuracy through the assimilation of available ocean observations. The geographical scope of these systems extends from basin to global scale, facilitating global oil spill modeling capabilities. It should be noted at this point that oil spill modeling systems can use ocean forecast data in two ways: as direct forcing to the oil drift model and as boundary conditions to higher-resolution local ocean models which, in turn, provide forcing data to the oil drift model.

The forecasting systems used to assist search-and-rescue operations are very similar to those used for oil spills. Speed of response is an essential criterion for finding people lost at sea. The size of the search area grows rapidly over time, and it is important to have accurate environmental data (winds and currents) to reduce this area and quickly find the target you are being looked for. Futch and Allen (2019) state that 60% of SAR incidents under the US Coast Guard areas of responsibility are outside areas where high-resolution wind and current data are available, and only global forecasts are available. Finding solutions that reduce the size of these areas of intervention is therefore crucial.

3.1.1 The Example of the Grande America Accident

The Grande America accident is a recent example of the use of ocean current data from several models. On March 10, 2019, a fire broke out on board the Italian merchant ship Grande America. The ship carries 365 containers, 45 of which contain dangerous goods and 2000 vehicles (cars, trucks, trailers, construction machinery) in its car decks. It sank on March 12 at a depth of 4600 m, 350 km off the French coast, with about 2200 tonnes of bunker fuel on board. More than 1000 tonnes of heavy fuel oil were released into the marine environment on the day the ship sank, followed by 35 days of continuous leaks before the breaches in the hull were filled by an underwater robot. A technical committee bringing together experts from Cedre, Météo-France, IFREMER and SHOM assessed the drift forecasts provided by the MOTHY oil spill drift model (Daniel 1996). The Drift Committee was responsible for providing the maritime authorities with consistent and relevant information on oil drift, observations and forecasts on a daily basis. MOTHY was used daily during the aerial surveillance and recovery period using oceanic forcing fields provided by CMEMS. Significant discrepancies were found between ocean models. Finally, the drift predictions closest to the observations were obtained with the currents provided by the ocean model of lower resolution (Fig. 11). This shows the importance of having several ocean models and the difficulty of using high-resolution ocean forecasts that generate many eddies whose precise location is sometimes difficult to model.

This example highlights the main challenge for data providers in the future, namely improving the accuracy of current forecasts. Therefore, international collaboration should continue to consolidate work on validation measurements and model comparisons to ensure that a minimum set of measurements is implemented internationally. In particular, it is necessary to include spatially explicit estimates of uncertainty, both for forcing data and for the results of the oil spill model. From this point of view, ensemble forecasting is a very promising research area for quantifying the uncertainties inherent in drift prediction at sea.

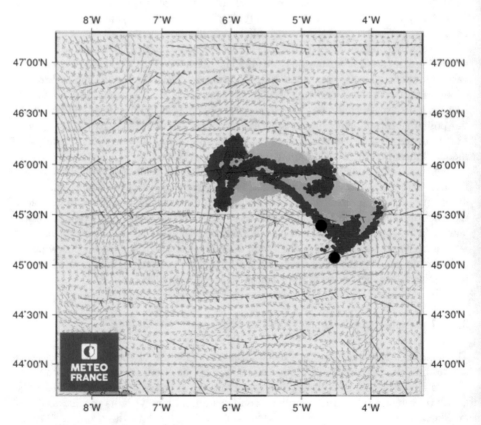

Fig. 11 MOTHY output from a simulated 10-day leak from the Grande America wreck (dark green star near 46° N–6° W). Black disks are aerial observations. Particle drift: in green without ocean current data, in red using ocean current data from the operational Iberian Biscay Irish (IBI) Ocean Analysis and Forecasting system at 1/36 degree, in blue using ocean current data from the operational Mercator global ocean analysis and forecast system at 1/12 degree

It should be further studied. Products should be delivered to users efficiently and should be provided with adequate spatial and temporal resolution. The interaction between users and production must be an important criterion for future developments.

3.2 Oil Spill Pollution

The end of the last century and the beginning of this new one have been plagued by major environmental catastrophes in terms of ocean oil pollution. Amoco Cadiz (1978), Exxon Valdez (1989), Sea Empress (1996), Erika (1999), Prestige (2002) and Tasman Spirit (2003) are but examples of oil tanker accidents. Nevertheless, these tanker accidents account for only a maximum of 2–3% of the whole oil pollutions worldwide. Another small percentage (2–3%) is due to platform accidents (Deep Water Horizon is a good example) leaving the rest to wild discharges (see Fig. 12 extracted from Pavlakis et al. 2001).

Detecting and tracking oil spills were necessary, and rapidly, the best and most efficient means of observation and analysis were recognized through synthetic-aperture radar (SAR or imaging radar) measurements. Indeed, on board satellite or aircraft, SAR is an adapted tool for monitoring the surface of the ocean and therefore for detecting sea surface pollution. Due to microwave properties, SAR measurements do not depend on weather or cloud cover, and slicks have significant backscattering behaviors that can be highlighted by SAR systems. Imaging radars measure mainly the surface roughness and because slicks are producing a strong impact on short waves they modify seawater viscosity and dampened backscatter returns. Nevertheless, automatic spill recognition is not an easy task as the presence of extended dark manifestations in the image, due to occurrences other than spills but comparable to them, yields similar SAR signatures.

These extended dark patches are due to other phenomena also modifying the sea surface conditions: wind, sea state, sea temperature, rain showers, and currents (Girard-Ardhuin et al. 2003). An experienced image interpreter is essential to sort out irrelevant disturbances linked to natural phenomena. Furthermore, specific SAR parameters, using different SAR modes, are to be considered and provide more flexibility for the analysis: radar wave polarization, wavelength, and incidence angle for example, but also several other

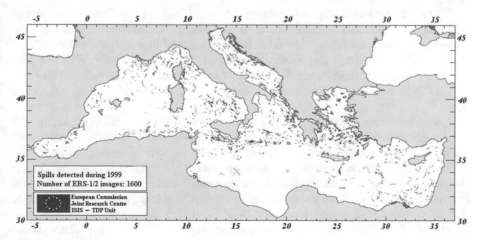

Fig. 12 Fingerprints of illicit vessel discharges detected on ERS-1 and ERS-2 SAR images, during 1999 within the Mediterranean Sea. Figure from Pavlakis et al. (2001)

parameters such as satellite or aircraft flight direction, waves and wind directions. Then, many of the possible confusions arising from automatic recognition can be smoothed by integrating this information surplus, helping to discriminate between different sources of sea surface disturbances.

In order to go beyond oil spill detection and tracking, and for setting an efficient prevention system, one must also detect and recognize the ships responsible for the oil discharges. For that purpose, synergetic approaches have been used, mixing radar imaging inputs with met information and mandatory ship identification systems (AIS—automatic identification system) mainly. This effort is quite important as the vast majority of these pollutions occur near the coastal zones where 80% of the world population lives.

3.2.1 Effect of Slicks on the Ocean Surface and Radar Imaging Measurements

Slicks have the effect of damping the waves at the surface (Alpers and Hühnerfuss 1988, 1989; Solberg et al. 1999). When a slick covers the surface, wind has a lesser effect, and the amplitude of wave crest/trough decreases, implying a surface stress gradient, with opposite strength to this alternated motion. Therefore, a film at the ocean surface implies an energy decrease with distribution to high and short waves, corresponding to the important surface stress decrease. Because the radar is only sensitive to the surface roughness, several other modifying aspects need to be considered: waves, sea properties (diffusion, mixing, turbulence, air–sea exchanges, biology, etc.), surface currents, eddies, internal waves, oceanography, meteorology, atmospheric dynamics, etc., in order to understand the all influences on slick dissipation.

The SAR technology is one of the most suitable (see for instance Girard-Ardhuin et al. 2003, among many others) instruments for oil slick measurements since it does not depend on weather conditions (clouds) nor sunshine, which allows showing illegal discharges that most frequently appear during night; it can also survey storms areas, where accident risks are increased. Added to local aircraft tracking, SAR is helpful for synoptic oil spill monitoring. Its spatial coverage can be as wide as 500×500 km, with high-resolution images of about 10 m, or in very high-resolution mode down to a meter-scale resolution with a much smaller coverage (20×20 km).

Figure 13a shows an example of a large extent zone North West of Spain, on the path of the Finisterre TSS (traffic separation scheme) used by the oil tankers going back and forth between Rotterdam and Athens. The radar image is acquired by Radarsat-2 on 03/25/2016 at 06:41 UTC. This image was received, processed and analyzed by the VIGISAT 24/7 operational center in Brest, dedicated to the near real-time maritime surveillance services. Figure 13b shows, in the enlargement, not only the slick but several ships as well (white dots).

Surface roughness, linked by gravity–capillarity waves, is dampened by slicks, and the radar backscattering level is therefore decreased, yielding dark patches with weak backscattering in comparison with the surrounding regions in radar image, with at least 3 dB contrasts. From a synthesis of experimental previous studies, the best parameters to use to detect slicks are a function of radar configuration, slick nature and meteorological and oceanic conditions. SAR parameters are particularly of importance as their selection may affect the oil slick detection capabilities. Per construction, one can choose the radar wavelength, the radar polarization or the incidence angle of the radar wave. Other parameters strongly affecting the end results are conjunctural, such as the angle between flight direction and wind direction and, of course, the nature of the slick itself. The influence of

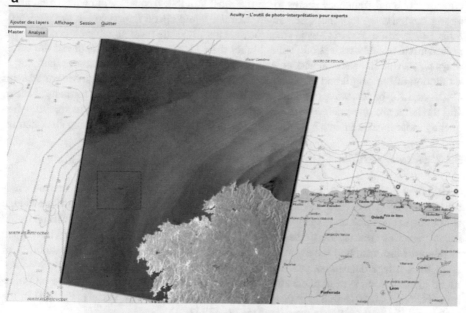

Fig. 13 a Radar image showing an oil discharge (blue square), **b** zoom in on the oil slick (Radarsat2@2016)

meteorological and oceanic conditions has been discussed before and is entering into the category of additional parameters allowing some possible discrimination (Girard-Ardhuin et al. 2003).

For two decades (1980–2000), several experiments were conducted with a clear choice for higher microwave frequencies (C- to Ku-bands) over longer wavelength (S- or L-band), and an ability to show about 5 dB contrast for a slick made with "light" fuel, and 10–15 dB contrast for a "heavy" fuel one. Wind speed being an external parameter to consider, C-band frequency seems to be the most suitable frequency allowing to measure strong contrasts up to wind speed about 13 m/s. Wind speed intervenes in the choice of radar polarization. Again C-band frequency seems to be the most suitable choice with VV polarization, notably with strong wind, higher than 11 m/s. Finally, the nature of the slicks plays also a role in detection and, potentially, identification. Indeed, backscatter damping is a function of the origin and of the slick viscosity and elasticity properties. Measured with C-, X- and Ku-bands, wave damping is more important for oil slicks than for natural films, and the latter are often detectable only at low wind speed (< 5 m/s). Thus, using multi-frequency radar measurements is a good solution, when possible, for determining slick nature.

3.2.2 SAR and AIS Coupled Detection

SAR imagery provides a great potential in observing and monitoring the marine environment. Ship detection in SAR imagery is well advanced (Crisp 2004), and knowledge about vessels position and type yields a wide range of applications, such as maritime traffic safety, fisheries control, illegal discharges and border surveillance. Applications developed in this area are vast, and the evolution of SAR sensors capabilities and the availability of large SAR image dataset coupled with cooperative positioning data (such as AIS—automatic identification system ones) allows a much finer monitoring for man-made oil pollution (Pelich et al. 2015a). The AIS system allows the tracks of all vessels (Fig. 14) in the vicinity of the spill (AIS data from 01:00 to 06:47 UTC).

In order to identify the potential polluter, a series of steps must be applied. As too many candidates are present, there is a need to reduce the selection of potential ships. Figure 15a then shows the track of only three ships sailing southbound, in the vicinity of the spill. One AIS track is more likely to be the polluter. In order to assess the correct track, a drift algorithm is run for confirmation.

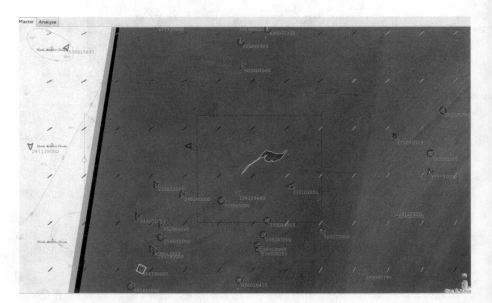

Fig. 14 Oil slick highlighted with added information on AIS ship positioning and wind speed and direction. Blue color: AIS positions are 6 h before the SAR image acquisition time. Red color: AIS positions at the SAR image acquisition time (courtesy by CLS)

Finally, to assess which track is more likely to point to the polluter, a drift model is applied on all three tracks with an estimation of "fake" pollutions, as if all three ships had polluted. From the AIS-based locations of vessels, and considering the wind and the current information, the displacement of the pollution at sea is computed for a specific time interval (typically 6 h). The pollution transport model is based on a 2D Lagrangian advection scheme which linearly combines wind and current effects, with an additional diffuse scheme (Pelich et al. 2015b; Longépé et al. 2015). Figure 15 shows that only one AIS track is consistent with the oil spill shape.

In summary, ocean oil pollutions are mainly coming from illegal discharges and one of the best tools for detecting them is the imaging radar on board satellites. But SAR instruments cannot, alone, achieve the level of recognition expected by organizations such as European Maritime Safety Agency (EMSA). More evidences are required, and we showed

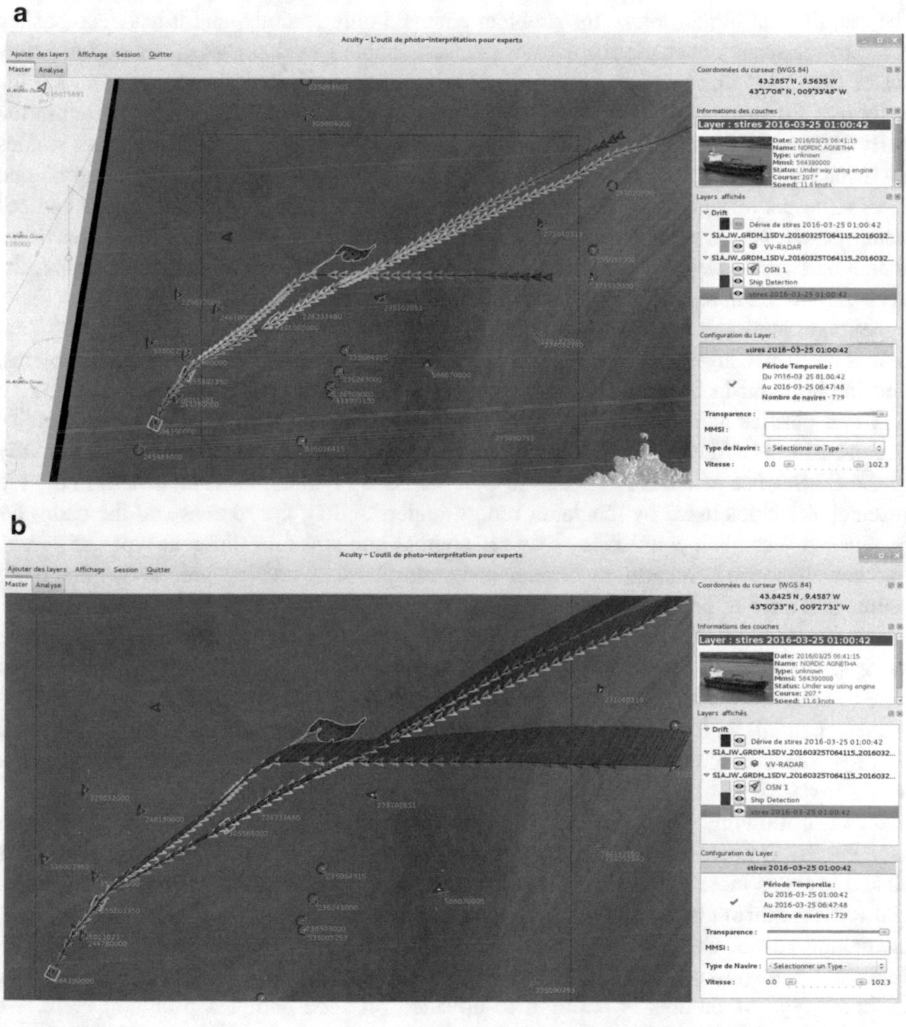

Fig. 15 **a** Tracks of three main candidates as polluters, **b** only one track is consistent with the oil spill (courtesy by CLS)

here that by adding informations directly derived from the SAR image (ship detection, wind speed and direction, image interpreters) coupled with external ones (met conditions, AIS positions, drift model), we are able to accumulate enough proofs for having a significant impact on the polluters, including prosecution (a set of evidences can be accepted as a proof).

3.3 The Problem of Plastics as Marine Debris

World plastic production has increased exponentially in recent decades, from 2.3 million of tonnes (Mt) in 1950 to 162 million in 1993 to 448 million by 2015. As a consequence of their high durability (and associated low cost), substantial quantities of end-of-life plastics are accumulating in the environment. Presently, the part of unmanaged debris and plastics that end up in the World Ocean is estimated between 8 and 15 Mt per year. Nevertheless, the real dimension of the problem emerged only recently, and it has been pushed forward the scene under the auspices of the Sustainable Development Goals on life below water (SDG 14) that goals to "conserve and sustainably use the oceans, seas and marine resources for sustainable development." Before that step, scientific research started in the early 1970s with the report of plastics on the seafloor and at sea and with numerous studies focusing on their impact on a variety of marine animals (Ryan 2015). Marine debris and litter represent nowadays a growing environmental problem and plastic, the largest component of litter, is now widely reported within the marine and as well as all natural environments on Earth. The problem is so huge that it represents threats to the environment, the economy, and human well-being on a global scale (Nielsen et al. 2019). Despite its many benefits on land and extensive usage inside our societies, the plastics and other artificial debris are increasingly being recognized as the source of severe environmental problems, and in particular in the oceans. Among the numerous aspects of the plastic problem, the fact that plastics and debris could travel across large distances, typically at the scale of ocean basins, is at the heart of the research priorities addressed by the scientific community to the G7 Science Ministers at Berlin in October 2015 (Williamson et al. 2016). In the top research priorities listed by this latter report, understanding the sources and the pathways in order to "establish connections between sources and sinks for different types of debris, and how these are influenced by oceanic and atmospheric dynamics" are the two first basic points that need to be solved to define the extent of the problem. More or less, all the different international reports dedicated to the plastic problem at global scales addressed the same starting points in addition to the fact that present ocean circulation models are not able to accurately simulate drift of debris because of its complex hydrodynamics (Maximenko et al. 2019). It is also important to note that, despite some progress made in spectral and hyperspectal remote sensing technologies (Garaba et al. 2018; Goddijn-Murphy and Williamson 2019), observations from space remain limited.

In the following, a simple Lagrangian assessment of dispersion from modeled surface current trajectories at the global ocean scales is addressed. The reason for that is linked to focus more on the physics of the oceanic environment rather than on the material itself requiring to specify, in such a case, different parameters like its density, size, shape and so on. Considering the problem of microplastic marine particles, Khatmullina and Chubarenko (2019) recently conclude that further integration or coupling with different type of models is required to advance on such complex problem. Here, we consider that plastics and marine floating debris represent one unique tracer that provides the opportunity to learn more about the physics and dynamics of the oceans across multiple scales. The evolution of a tracer in any turbulent flows is given by the paradigm of Eckart (1948): At first, during the stirring phase, the variance of the scalar gradient is increased, and later, during the mixing phase the molecular diffusion dominates and the strong gradient disappears (homogeneous final state). However, the "stirring and mixing" problem in a stratified ocean relies on the physics that need to be parametrized in ocean models and great challenging open problems remain at all levels, from very fundamental to highly applied aspects (Müller and Garrett 2002).

For centuries, surface currents were inferred from bottles and drifting objects but, since the 1980s, the World Climate Research Program initiated the global array of surface drifters (with anchors drogue set at 15 m depth). Accumulation of observations and continuous improvement leads to a quite precise time-mean circulation resolving details such as the cross-stream structure of western boundary currents, recirculation cells, and zonally elongated mid-ocean striations (Laurindo et al. 2017). If the averaged seasonal climatology is quite pertinent, daily and near real-time observations are only based on 1200–1300 active buoys on global scales representing approximately $\sim 80\%$ of $5° \times 5°$ coverage within the 60 °N–60 °S band. On the other hand, satellite remote sensing proposes surface geostrophic currents derived from the altimetry since the 1990s, and with the addition of a wind-driven Ekman current, one estimate of surface current could be delivered (i.e., Sudre et al. 2013). If the variability of large-scale currents (200-km wavelength and 15-day period) is well depicted by satellite altimetry and vector winds from remote sensing, it leaves important observation gaps that will require other technology such as the understanding of Doppler properties of radar backscatter from the ocean (Ardhuin et al. 2019a, b). Representing another important tool in climate services, ocean reanalyses combine ocean models, atmospheric forcing fluxes, and observations using data assimilation to give a four-dimensional description of the ocean (i.e., Storto and Masina 2016). Based on these estimates, the dispersion dimension of the problem should be addressed from a Lagrangian point of view.

Complementing the traditional Eulerian approach, Lagrangian ocean methodology is a powerful way to analyze the ocean circulation or dynamics. The purpose is based on large sets of virtual (passive or fictive) particles whose displacements (trajectories) are integrated within the 3D/2D and time evolving velocity field. Among the expectations, questions about pathways and flow connectivity can be addressed and it avoids the recourse to the classical use of passive tracers such as geotrace gases. In a simplified framework, the analyses include the evaluation of the ocean currents, the determination of the initial positions of the particle and the computation of trajectories under physical assumptions, and finally their post-treatment. Some concrete examples that have been studied during the recent couple of years include: the tracking of the origins of plastic bottles across the Coral Sea (Maes and Blanke 2015), the discovering of "exit routes" near the core of the subtropical convergence zones of the Pacific Ocean (Maes et al. 2016), the determination of the mesoscale features characterizing the dynamics of the Coral Sea with its impact on water masses and circulation (Rousselet et al. 2016), the potential connection between the South Indian and the South Pacific subtropical convergence zones revealing that large-scale convergence zones should not be seen as "closed systems" (Maes et al. 2018), and more recently, the important role of the Stokes drift in the surface dispersion of floating debris in the South Indian Ocean (Dobler et al. 2019). Based on these different studies and other consistent results reported in the literature, we were able to identity these important findings: We need to know the time variability of surface currents to adequately address the long distance connectivity and pathways, we need to consider high-resolution horizontal currents (mesoscale at least) to estimate the time transfer on specific pathways, and finally, we need to evaluate more accurately the emission scenario (and the sinks) of surface floating litter to access the distributions at global scales.

Regarding the latter aforementioned point, additional experiments were performed using two different scenarios of particle release: The first one is based on one instantaneous release over the full ocean domain of the model (i.e., one particle on each ocean grid point at the initial timestep), whereas the second one is based on a permanent release along the coastlines (i.e., the ocean grid point adjacent to the land mask), the distribution being based on the population density as used by van Sebille et al. (2015). Not surprisingly, both experiments exhibit the presence of the subtropical convergence zones mainly due to the ocean dynamics (Fig. 16), which results in accumulation in the absence of vertical mixing and potential sinks. Nevertheless, the experiment with continuous release along the coastlines shows much more complex structures and higher concentrations in some areas including the different marginal Seas of the western Pacific Ocean, the Indonesian Seas, the Mediterranean Sea and the Bay of Bengal as well as the South Indian Ocean. It is also clear that some places within the tropical band such as in the Eastern Pacific Ocean near the Galapagos Islands or in the Gulf of Guinea are also characterized by some convergence of particles (Fig. 16b). Such places could not be identified in the first experiment (Fig. 16a) due to the divergent nature of the ocean dynamics, whereas they could be seen as semi-permanent places in the second case. To determine which one of these spatial distributions represents the actual concentration of marine debris or plastic at sea is quite impossible, due to the absence of in situ observations, and they are probably not realistic either. It is important

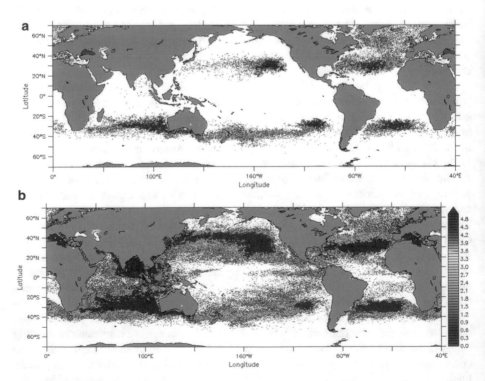

Fig. 16 Number of particles per one fourth degree cell resulting from the initially instantaneous experiment (**a**) and from the permanent release along coastlines assuming a lifetime of 20 years for the particles (**b**) after 29 years of dispersion

to remember that most of the present studies put their focus on the surface layer (and so, on floating material), whereas the vertical distribution of floating plastic depends not only on the particle's buoyancy, but also on the dynamic pressure due to vertical movements of ocean water. Understanding of the vertical flow in the ocean is challenging because it is induced by several processes acting at different temporal and spatial scales, not always fully resolved by present ocean models. Secondly, most of the present studies have assumed a steady, non-changing ocean circulation. However, low-frequency variations, such as the El Niño Southern Oscillation (ENSO) and the Pacific Decadal Oscillation (PDO), modify the ocean circulation and modulate the different physical processes. Trends associated with climate change can also amplify some of the dispersion processes at work in the future. It is not clear how these processes and the coupling with the internal variability of the ocean will or not affect the dispersion of marine debris and plastics at the global scales.

4 Conclusions

Atmospheric pollution is one of the greatest environmental disasters today with more than 90% of the population living in places exceeding WHO air quality guidelines. Air pollution is a complex mixture of many constituents that can be transported far from its source. It is a cross-border problem that affects each individual differently. Concentrations of key pollutants need to be known, quantified, and evaluated against atmospheric models as the first step toward crucial scientific understanding and a common awareness to address global environmental problems. Satellite instruments are powerful tools to monitor the atmospheric composition and its evolution globally, and the Copernicus CAMS infrastructure allows us to predict air quality for the next few days based on these observations. With technological advancements, space-borne observations are expected to be even more refined in space and time and hence help attributing pollutant's sources and mitigating air pollution.

Recourse to short-term forecasts to modeling oil spill drift or trajectory for search-and-rescue support is quite recent in operational oceanography but essential to be efficient. Such processes and systems depend strongly on our understanding of the intrinsic variability in ocean dynamics. Thanks to the development of the technologies of satellite remote sensing, which began in the 1970s, tremendous advances have been achieved on large-scale oceanographic phenomena that lead to significant contributions to ocean and climate research for instance. However, satellite observations also emphasize the importance of the strongly energetic mesoscale turbulent eddy field in all the oceans involving energetic scale interactions that are not completely understood so far. Refining our understanding of ocean motions (and ultimately the coupling between the atmosphere, ocean and surface waves) as well as our capability of observations of current velocities in the top few meters of the ocean is crucial to provide the framework to relevant national and intergovernmental organizations to tackle the different types of sea pollution at local and global scales.

4.1 Perspectives

In the coming decades, we expect that satellite measurements will achieve a very high resolution in space as well as a more frequent revisit in time to track pollution on hourly basis within one kilometer or less (atmosphere) and a few meters' (ocean) footprints. At the same time, many in situ sensors will provide additional information allowing better estimation of vertical information (atmosphere) and drifts due to currents and surface winds (ocean). These new datasets will boost research on pollutants variability and transport and afterward increase the accuracy of atmospheric and ocean models.

The large increase in data collection amount will be challenging for the processing systems, based on data science new developments (machine/deep learning and artificial intelligence) which will be in the near future massively used.

It means that researchers in atmospheric and marine sciences and technology will be able to capture pollution episodes (such as rush-hour traffic for instance) and detect whether pollution within a given region was generated on the spot or transported over from other places.

Acknowledgements The authors would like to acknowledge the International Space Science Institute (ISSI) in Bern (Switzerland) for organizing inspiring workshops and their help during this book's conception. A special mention is addressed to Anny Cazenave and the other organizers for their availability and time.

References

Akritidis D, Pozzer A, Zanis P, Tyrlis E, Škerlak B, Sprenger M, Lelieveld J (2016) On the role of tropopause folds in summertime tropospheric ozone over the eastern Mediterranean and the Middle East. Atmos Chem Phys 16:14025–14039. https://doi.org/10.5194/acp-16-14025-2016

Alpers W, Hühnerfuss H (1988) Radar signatures of oil films floating on the sea surface and the Marangoni effect. J Geophys Res 93(C4):3642–3648

Alpers W, Hühnerfuss H (1989) The damping of ocean waves by surface films: a new look at an old problem. J Geophys Res 94(C5):6251–6265

Apte JS, Pant P (2019) Toward cleaner air for a billion Indians. PNAS 116(22):10614–10616. https://doi.org/10.1073/pnas1905458116

Ardhuin F, Brandt P, Gaultier L, Donlon C, Battaglia A, Boy F, Casal T, Chapron B, Collard F, Cravatte S, Delouis J-M, De Witte E, Dibarboure G, Engen G, Johnsen H, Lique C, Lopez-Dekker P, Maes C, Martin A, Marié L, Menemenlis D, Nouguier F, Peureux C, Rampal P, Ressler G, Rio M-H, Rommen B, Shutler JD, Suess M, Tsamados M, Ubelmann C, van Sebille E, van den Oever M, Stammer D (2019a) SKIM a candidate satellite mission exploring global ocean currents and waves. Front Mar Sci. https://doi.org/10.3389/fmars.2019.00209

Ardhuin F, Chapron B, Maes C, Romeiser R, Gommenginger C, Cravatte S, Morrow R, Donlon C, Bourassa M (2019b) Satellite Doppler observations for the motions of the oceans. Bull Am Meteorol Soc 100:ES215–ES219

Brunekreef B (1997) Air pollution and life expectancy: is there a relation? Occup Environ Med 54(11):781–784. https://doi.org/10.1136/oem.54.11.781

Chameides W, Walker JCG (1973) A photochemical theory of tropospheric ozone. J Geophys Res 78:8751–8760

Chowdhury S, Dey S, Guttikunda S, Pillarisetti A, Smith KR, Di Girolamo L (2019) Indian annual ambient air quality standard is achievable by completely mitigating emissions from household sources. Proc Natl Acad Sci. https://doi.org/10.1073/pnas.1900888116

Clerbaux C (2018) Ciel jaune et soleil rouge: l'ouragan Ophélia décrypté. The conversation. https://theconversation.com/ciel-jaune-et-soleil-louragan-ophelia-decrypte-86894. Accessed 30 Oct 2019

Clerbaux C, Boynard A, Clarisse L, George M, Hadji-Lazaro J, Herbin H, Hurtmans D, Pommier M, Razavi A, Turquety S, Wespes C, Coheur P-F (2009) Monitoring of atmospheric composition using the thermal infrared IASI/MetOp sounder. Atmos Chem Phys 9:6041–6054. https://doi.org/10.5194/acp-9-6041-2009

Crippa M, Guizzardi D, Muntean M, Schaaf E, Dentener F, van Aardenne JA, Monni S, Doering U, Olivier JGJ, Pagliari V, Janssens-Maenhout G (2018) Gridded emissions of air pollutants for the period 1970–2012 within EDGAR v4.3.2. Earth Syst Sci Data 10:1987–2013. https://doi.org/10.5194/essd-10-1987-2018

Crisp DJ (2004) The state-of-the-art in ship detection in synthetic aperture radar imagery. Intelligence, Surveillance and Reconnaissance Division Information Sciences Laboratory, Edinburgh, South Australia, Australia, Technical Report, May 2004

Daniel P (1996) Operational forecasting of oil spill drift at METEO-FRANCE. Spill Sci Technol Bull 3(1/2):53–64

Dobler D, Huck T, Maes C, Grima N, Blanke B, Martinez E, Ardhuin F (2019) Large impact of Stokes drift on the fate of surface floating debris in the South Indian Basin. Mar Pollut Bull 148:202–209

Eckart C (1948) An analysis of the stirring and mixing processes in incompressible fluids. J Mar Res 7:265–275

Elguindi N, Granier C, Stravakou T, Darras S, Bauwens M, Cao H, Chen C, Denier van der Gon HAC, Dubovik O, Fu TM, Henze D, Jiang Z, Kuenen JJP, Kurokawa J, Liousse C, Miyazaki K, Muller JF, Qu Z, Sekou K, Solmon F, Zheng B (2020) Analysis of recent anthropogenic surface emissions from bottom-up inventories and top-down estimates: are future emission scenarios valid for the recent past? Earth Space Sci Open Arch. https://doi.org/10.1002/essoar.10502317.1

European Environment Agency (EEA) (2018) https://www.eea.europa.eu/data-and-maps/data/national-emissions-reported-to-the-convention-on-long-range-transboundary-air-pollution-lrtap-convention-13#tab-european-data. Accessed 30 Oct 2019

Futch V, Allen A (2019) Search and rescue applications: on the need to improve ocean observing data systems in offshore or remote locations. Front Mar Sci 6:301. https://doi.org/10.3389/fmars.2019.00301

Garaba SP, Aitken J, Slat B, Dierssen HM, Lebreton L, Zielinski O, Reisser J (2018) Sensing ocean plastics with an airborne hyperspectral shortwave infrared imager. Environ Sci Technol 52(20):11699–11707. https://doi.org/10.1021/acs.est.8b02855

Girard-Ardhuin F, Mercier G, Garello R (2003) Oil slick detection by SAR imagery: potential and limitation. In: IEEE/MTS proceedings of the marine technology and ocean science conference OCEANS2003, San Diego, CA, USA, vol 1, pp 164–169, 22–26 September 2003

Goddijn-Murphy L, Williamson B (2019) On thermal infrared remote sensing of plastic pollution in natural waters. Remote Sens 11:2159. https://doi.org/10.3390/rs11182159

Health Effects Institute State of Global Air (2019) Special report, Boston MA, Health Effects Institute, ISSN 2578-687

Im U, Markakis K, Poupkou A, Melas D, Unal A, Gerasopoulos E, Daskalakis N, Kindap T, Kanakidou M (2011) The impact of temperature changes on summer time ozone and its precursors in the Eastern Mediterranean. Atmos Chem Phys 11:3847–3864. https://doi.org/10.5194/acp-11-3847-2011

Kalabokas PD, Thouret V, Cammas J-P, Volz-Thomas A, Boulanger D, Repapis CC (2015) The geographical distribution of meteorological parameters associated with high and low summer ozone levels in the lower troposphere and the boundary layer over the Eastern Mediterranean (Cairo case). Tellus B Chem Phys Meteorol 67:27853. https://doi.org/10.3402/tellusb.v67.27853

Khatmullina L, Chubarenko I (2019) Transport of marine microplastic particles: why is it so difficult to predict? Anthropocene Coasts 2:293–305. https://doi.org/10.1139/anc-2018-0024

Kouvarakis G, Tsigaridis K, Kanakidou M, Mihalopoulos N (2000) Temporal variations of surface regional background ozone over Crete Island in the southeast Mediterranean. J Geophys Res 105:4399–4407

Laurindo LC, Mariano A, Lumpkin R (2017) An improved surface velocity climatology for the global ocean from drifter observations. Deep Sea Res Part I 124:73–92. https://doi.org/10.1016/j.dsr.2017.04.009

Lelieveld J, Evans JS, Fnais M, Giannadaki D, Pozzer A (2015) The contribution of outdoor air pollution sources to premature mortality on a global scale. Nature 525:367–371. https://doi.org/10.1038/nature15371

Longépé NA, Mouche M, Goacolou N, Granier L, Carrere J-Y, Le Bras P, Lozac'h P, Besnard S (2015) Polluter identification with spaceborne radar imagery AIS and forward drift modeling. Mar Pollut Bull 101:826–833

Maes C, Blanke B (2015) Tracking the origins of plastic debris across the Coral Sea: a case study from the Ouvéa Island New Caledonia. Mar Pollut Bull 97:160–168

Maes C, Blanke B, Martinez E (2016) Origin and fate of surface drift in the oceanic convergence zones of the eastern Pacific. Geophys Res Lett 43:3398–3405. https://doi.org/10.1002/2016GL068217

Maes C, Grima N, Blanke B, Martinez E, Paviet-Salomon T, Huck T (2018) A surface "superconvergence" pathway connecting the South Indian Ocean to the subtropical South Pacific gyre. Geophys Res Lett 45:1915–1922. https://doi.org/10.1002/2017GL076366

Maximenko N, Corradi P, Law KL, Van Sebille E, Garaba SP, Lampitt RS, Galgani F, Martinez-Vicente V, Goddijn-Murphy L, Veiga JM, Thompson RC, Maes C, Moller D, Löscher CR, Addamo AM, Lamson MR, Centurioni LR, Posth NR, Lumpkin R, Vinci M, Martins AM, Pieper CD, Isobe A, Hanke G, Edwards M, Chubarenko IP, Rodriguez E, Aliani S, Arias M, Asner GP, Brosich A, Carlton JT, Chao Y, Cook A-M, Cundy AB, Galloway TS, Giorgetti A, Goni GJ, Guichoux Y, Haram LE, Hardesty BD, Holdsworth N, Lebreton L, Leslie HA, Macadam-Somer I, Mace T, Manuel M, Marsh R, Martinez E, Mayor DJ, Le Moigne M, Molina Jack ME, Mowlem MC, Obbard RW, Pabortsava K, Robberson B, Rotaru A-E, Ruiz GM, Spedicato MT, Thiel M, Turra A, Wilcox C (2019) Toward the integrated marine debris observing system. Front Mar Sci 6:447. https://doi.org/10.3389/fmars.2019.00447

Müller P, Garrett C (2002) From stirring to mixing in a stratified ocean. Oceanography 15:12–19

Nielsen TD, Hasselbalch J, Holmberg K, Stripple J (2019) Politics and the plastic crisis: a review throughout the plastic life cycle. WIREs Energy Environ. https://doi.org/10.1002/wene.360

Osborne M, Malavelle FF, Adam M, Buxmann J, Sugier J, Marenco F, Haywood J (2019) Saharan dust and biomass burning aerosols during ex-hurricane Ophelia: observations from the new UK lidar and sun-photometer network. Atmos Chem Phys 19:3557–3578. https://doi.org/10.5194/acp-19-3557-2019

Pavlakis P, Tarchi D, Sieber AJ (2001) On the monitoring of illicit vessel discharges using space-borne SAR remote sensing—a reconnaissance study in the Mediterranean Sea. Ann Telecommun 56(11–12):700–718

Pelich N, Longépé N, Mercier G, Hajduch G, Garello R (2015a) Performance evaluation of Sentinel-1 data in SAR ship detection. In: 2015 IEEE international geoscience and remote sensing symposium (IGARSS), pp 2103–2106

Pelich N, Longépé N, Mercier G, Hajduch G, Garello R (2015b) AIS-based evaluation of target detectors and SAR sensors characteristics for maritime surveillance. IEEE J Sel Top Appl Earth Obs Remote Sens 8(8):3892–3901

Ramanantenasoa MMJ, Gilliot J-M, Mignolet C, Bedos C, Mathias E, Eglin T, Makowski D, Génermont S (2018) A new framework to estimate spatio-temporal ammonia emissions due to nitrogen fertilization in France. Sci Total Environ 645:205–219. https://doi.org/10.1016/j.scitotenv.2018.06.202

Richards NAD, Arnold SR, Chipperfield MP, Miles G, Rap A, Siddans R, Monks SA, Hollaway MJ (2013) The Mediterranean summertime ozone maximum: global emission sensitivities and radiative impacts. Atmos Chem Phys 13:2331–2345. https://doi.org/10.5194/acp-13-2331-2013

Robinson TP, Wint GRW, Conchedda G, Van Boeckel TP, Ercoli V, Palamara E, Cinardi G, D'Aietti L, Hay S, Gilbert M (2014) Mapping the global distribution of livestock. PLoS ONE 9(5):e96084. https://doi.org/10.1371/journal.pone.0096084

Rousselet LA, Doglioli M, Maes C, Blanke B, Petrenko AA (2016) Impacts of mesoscale activity on the water masses and circulation in the Coral Sea. J Geophys Res Oceans 121:7277–7289. https://doi.org/10.1002/2016jc011861

Ryan PG (2015) A brief history of marine litter research. In: Bergmann M, Gutow L, Klages M (eds) Marine anthropogenic litter. Springer, Cham, pp 1–25. https://doi.org/10.1007/978-3-319-.16510-3_1

Safieddine S, Boynard A, Coheur P-F, Hurtmans D, Pfister G, Quennehen B, Thomas JL, Raut J-C, Law KS, Klimont Z, Hadji-Lazaro J, George M, Clerbaux C (2014) Summertime tropospheric ozone assessment over the Mediterranean region using the thermal infrared IASI/MetOp sounder and the WRF-Chem model. Atmos Chem Phys 14:10119–10131. https://doi.org/10.5194/acp-14-10119-2014

Scarlat N, Fahl F, Dallemand J-F, Monforti F, Motola V (2018) A spatial analysis of biogas potential from manure in Europe. Renew Sustain Energy Rev 94:915–930. https://doi.org/10.1016/j.rser.2018.06.035

Solberg AH, Storvik SG, Solberg R, Volden E (1999) Automatic detection of oil spills in ERS SAR images. IEEE Trans Geosci Remote Sens 37(4):1916–1924

Storto A, Masina S (2016) C-GLORSv5 an improved multipurpose global ocean eddy-permitting physical reanalysis. Earth Syst Sci Data 8(2):679–696. https://doi.org/10.5194/essd-8-679-2016

Sudre J, Maes C, Garcon V (2013) On the global estimates of geostrophic and Ekman surface currents. Limnol Oceanogr Fluids Environ 3(1):1–20. https://doi.org/10.1215/21573689-2071927

United States Environmental Protection Agency (EPA) https://www.epa.gov/air-emissions-inventories/air-pollutant-emissions-trends-data. Accessed 30 Oct 2019

Van Damme M, Clarisse L, Whitburn S, Hadji-Lazaro J, Hurtmans D, Clerbaux C, Coheur P-F (2018) Industrial and agricultural ammonia point sources exposed. Nature 564:99–103. https://doi.org/10.1038/s41586-018-0747-1

van Geffen JHGM, Eskes HJ, Boersma KF, Maasakkers JD, Veefkind JP (2019) TROPOMI ATBD of the total and tropospheric NO2 data products Tech Rep S5P-KNMI-L2-0005-RP Koninklijk Nederlands Meteorologisch Instituut (KNMI). https://sentinels.copernicus.eu/documents/247904/2476257/Sentinel-5P-TROPOMI-ATBD-NO2-data-products, CI-7430-ATBD, issue 1.4.0

van Sebille E, Wilcox C, Lebreton L, Maximenko N, Hardesty BD, Van Franeker JA (2015) A global inventory of small floating plastic debris. Environ Res Lett 10(12):124006. https://doi.org/10.1088/1748-9326/10/12/124006

Veefkind JP, Aben EAA, McMullan K, Forster H, de Vries J, Otter G et al (2012) TROPOMI on the ESA Sentinel-5 Precursor: a GMES mission for global observations of the atmospheric composition for climate air quality and ozone layer applications. Remote Sens Environ 120:70–83. https://doi.org/10.1016/j.rse.2011.09.027

Viatte C, Wang T, Van Damme M, Dammers E, Meleux F, Clarisse L, Shephard MW, Whitburn S, Coheur PF, Cady-Pereira KE, Clerbaux C (2020) Atmospheric ammonia variability and link with PM formation: a case study over the Paris area. Atmos Chem Phys. https://doi.org/10.5194/acp-20-577-2020

Williamson P, Smythe-Wright D, Burkill P (2016) A non-governmental scientific perspective on seven marine research issues of G7 interest. Future of the ocean and its seas. ICSU-IAPSO-IUGG-SCOR, Paris

World Health Organization (2017) https://www.who.int/airpollution/ambient/about/en/. Accessed 30 Oct 2019

Zanis P, Hadjinicolaou P, Pozzer A, Tyrlis E, Dafka S, Mihalopoulos N, Lelieveld J (2014) Summertime free-tropospheric ozone pool over the eastern Mediterranean/Middle East. Atmos Chem Phys 14:115–132. https://doi.org/10.5194/acp-14-115-2014

Zheng B, Tong D, Li M, Liu F, Hong C, Geng G, Li H, Li X, Peng L, Qi J, Yan L, Zhang Y, Zhao H, Zheng Y, He K, Zhang Q (2018) Trends in China's anthropogenic emissions since 2010 as the consequence of clean air actions. Atmos Chem Phys 18:14095–14111. https://doi.org/10.5194/acp-18-14095-2018

Publisher's Note Springer Nature remains neutral with regard to jurisdictional claims in published maps and institutional affiliations.

Surveys in Geophysics (2020) 41:1611–1627
https://doi.org/10.1007/s10712-020-09598-1

Geomagnetic Field Processes and Their Implications for Space Weather

Mioara Mandea[1] · Aude Chambodut[2]

Received: 6 November 2019 / Accepted: 15 May 2020 / Published online: 14 June 2020
© Springer Nature B.V. 2020

Abstract

Understanding the magnetic environment of our planet and the geomagnetic field changes in time and space is a very important issue for assessing the Sun–Earth interactions. All changes in solar activity impact the delicate balance between influences of interplanetary magnetic field and of geomagnetic field. The most dynamic events eventually result in disturbances in the magnitude and direction of the Earth's magnetic field and therefore impact our planet and its magnetosphere as a whole. The dynamics of the ionosphere and thermosphere during magnetic storms and substorms involves the heating, expansion, and composition changes at high latitudes, but also the surface-level response in terms of geomagnetically induced currents and other geomagnetic and geoelectric disturbances. Here, we provide a short overview of the current knowledge of the Earth's magnetic field, its present shape and the way it responds to external forces. The main aim of the paper is not to present the complexity of the space weather processes, but rather to bring the attention of the geohazard community to the possible dramatic effects of space weather events. For this, the paper highlights some societal implications of space weather on our increasingly technology-dependent society, including some possible effects of geomagnetically induced currents, the disruption of satellite communications and navigation, and risks of radiation damage both in space and in aviation.

Keywords Geomagnetic field · Space weather · Geoeffectivity · Geomagnetic risks · Hazards

1 Introduction

A decade ago, the Eyjafjallajökull Icelandic volcano eruption (April 2010) and the resulting ash cloud demonstrated that the whole world can suffer as a consequence of natural events. One year after this event, the Tohoko Japanese earthquake and the related tsunami (March 2011) showed just how devastating a natural event could be. Another year later, a

---

✉ Mioara Mandea
   mioara.mandea@cnes.fr

[1]   CNES - Centre National d'Etudes Spatiales, 2 Place Maurice Quentin, 75039 Paris, France

[2]   Institut de Physique du Globe de Strasbourg, UMR 7516, Université de Strasbourg/EOST, CNRS, 5 rue René Descartes, 67084 Strasbourg Cedex, France

solar active region gave rise to a powerful coronal mass ejection (July 2012), with an initial speed of some 2500 ± 500 km s^{-1} (Baker et al. 2013). The solar eruption was directed away from Earth, but what would have happened if this powerful event had been Earthward directed? To answer such a question we need first to better understand the Sun–Earth magnetic environment, but also the intrinsic structure and variations of the core magnetic field.

Space weather is a very complex concept and involves variations in the Sun, solar wind, magnetosphere, ionosphere, and thermosphere. Like the terrestrial weather, space weather is pervasive and its understanding is a challenge. Indeed, for space weather, as for weather-like variations, there are different types of variations: on some days, space weather is more severe than on others. Slight magnetic storms are rather common events; extremely large ones occur very rarely, perhaps once every century or two (Yokoyama and Kamide 1997; Riley 2012; Moriña et al. 2019). Moreover, space weather exhibits a climatology which varies over timescales ranging from a day to the 11-year solar cycle and even longer (Lockwood 2012).

To understand the Earth's complicated magnetic environment, the origins of the geomagnetic field need to be well known and well described. The Earth's magnetic field has its sources inside the Earth (internal contributions) and outside it (external contributions). The predominant internal source is the core field (also called "main field"), originating in the external fluid core, and the lithospheric field (also called "crustal field"), caused by magnetic minerals in the crust and, to a lesser extent, the upper mantle. The external sources originate in the ionosphere, the magnetosphere and also from electrical currents coupling the ionosphere and magnetosphere (named "field aligned currents", or FAC). These external sources induce secondary fields in the Earth.

An important player in the space weather is the core magnetic field, which acts like a shield to the solar wind that the Sun continually emits. However, currently the core field is changing dramatically (Mandea and Purucker 2018), and we are not yet able to deeply evaluate how these changes can increase or attenuate geoeffectivity, i.e. the ability to generate geomagnetic disturbances and more prolonged geomagnetic storms.

In recent decades, considerable progress has been made in solar physics and in understanding processes operating in the Earth's magnetosphere and ionosphere. The concept of space weather has developed as a scientific discipline and, increasingly, as a need to assess the vulnerabilities and to gauge the resilience of today's infrastructures built by humans on the ground and at satellite altitudes. A crucial requirement of space weather is to understand the magnetic terrestrial environment and to quantify the possibility of predictability of its variations.

To take into consideration these aspects, the remainder of this paper is organised as follows. Section 2 summarises the Earth's magnetic field contributions as measured on recent observational platforms. The following section highlights some particular features of their temporal variations. In Sect. 4, some aspects related to the geomagnetic risks and impacts are given, and finally Sect. 5 draws some conclusions and maps out future directions.

2 The Geomagnetic Field Sources

2.1 Internal Sources

The Earth's magnetic field is a dynamic system and varies on a wide range of timescales from seconds to hundreds of millions of years.

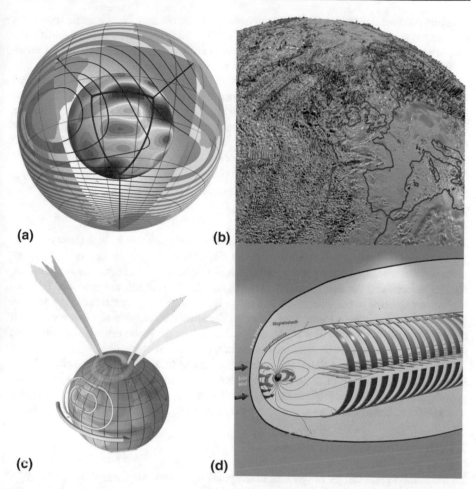

Fig. 1 Sketches of various sources of the geomagnetic field. **a** Contour lines of the dominating core field intensity at Earth's surface and downwards continued (radial field component) to the core-mantle boundary (CMB). The South Atlantic weak field anomaly is seen as a minimum in the intensity at the surface, caused by a large patch or reverse (blue) magnetic flux at the CMB. **b** The lithospheric field caused by magnetised rocks and tectonic structures from the World Digital Magnetic Anomaly Map. In the Atlantic region (left side of panel) the striped normal and reverse magnetic field pattern is seen. **c** The ionospheric Sq (orange) and equatorial electrojet (red) currents, with field aligned currents (yellow) linking high latitude ionospheric currents to the more distant magnetospheric current systems. **d** The magnetosphere shielding our planet against the solar wind, with Chapman-Ferraro currents on the dayside, tail current on the night side and the ring current circling the Earth at a few Earth's radii distance (all in pink). Magnetic field lines are stylised in blue. ©Korte and Mandea (2019)

The internal sources of the geomagnetic field include those originating in the core and lithosphere of the Earth (see Fig. 1a, b).

The core field (Hulot et al. 2015) is dominated by fields generated from a self-sustaining dynamo in the Earth's fluid outer core. This creates around 95% of the magnetic field strength at the Earth's surface. The measured magnetic field averages a strength on the order of 50 µT, varying however between some 20 and 60 µT. The strongest scalar magnetic fields originating from the Earth's core are found in the Earth's polar regions. The area just offshore Antarctica, in the ocean between Australia and the Antarctic, possesses the largest scalar magnetic fields, averaging 66 µT. The core field exhibits large Earth-fixed spatial scales and varies on timescales of months to millennia (known as the secular variation), even millions of years when reversal processes are considered.

Another internal source is the lithospheric field (Purucker and Whaler 2015), generated in rocks containing minerals carrying the magnetisation and situated below the Curie temperature, generally in the upper 5–30 km of the Earth's surface. Globally, this contribution is much smaller at around 20 nT; however, locally it can be much larger. The lithospheric field covers a wide range of spatial scales that reflect the geological setting of the crust. It changes on timescales of millions of years except at sources such as active metamorphic processes, volcanic regions or along mid-ocean ridges.

2.2 External Sources

The external sources are created by solar-terrestrial interactions and include those originating in the ionosphere, those which couple the magnetosphere and ionosphere, and those associated with more distant regions of the magnetosphere, for example the ring current, the magnetotail and the magnetopause (see Fig. 1c, d). These fields are produced by electric current systems in the ionosphere and magnetosphere. Strong electric currents are also generated and flow down to the ionosphere in the auroral zones where intense east-west currents, named the auroral electrojets, are produced. The magnetic disturbances observed at high latitudes are due to the magnetic fields of these auroral electrojets.

The magnitude of these fields at the surface varies in local time (LT), in space, and in response to solar forcing. The external fields have magnitudes of a few pT to 100 nT on geomagnetically quiet days, but can change rapidly within minutes to thousands of nT, for example from the impact of an interplanetary coronal mass ejection (CME) upon the Earth.

2.3 Recent Observational Platforms

To understand these ranges of spatial and temporal variations, highly accurate and precise measurements of the near-Earth geomagnetic field from space are needed. They began with NASA's MAGSAT satellite which observed for 6 months from 1979 to 1980. This was followed in the first decade of the twenty-first century by the Ørsted[1], CHAMP[2] and SAC-C[3] magnetic field satellites.

Swarm[4], the 3-satellite ESA constellation, is the most recent and most advanced "geomagnetic observatory" in space (Fig. 2). The three satellites were launched on 22 November 2013, with a 4-year nominal mission, after a 3-month commissioning phase. In November 2017, the mission was granted a 4-year extension to 2021.

One of the novelties of the mission is not only the constellation concept, but also the new absolute scalar magnetometer: an optically pumped helium magnetometer which provides absolute scalar measurements of the magnetic field with high accuracy and stability for the calibration of the vector field magnetometer (Leger et al. 2009). The mission

---

[1] http://www.space.dtu.dk/english/Research/Projects/Project-descriptions/Oersted.

[2] http://www.gfz-potsdam.de/champ/.

[3] http://www.space.dtu.dk/english/Research/Scientific_data_and_models/Magnetic_Satellites.

[4] https://earth.esa.int/web/guest/missions/esa-operational-eo-missions/swarm.

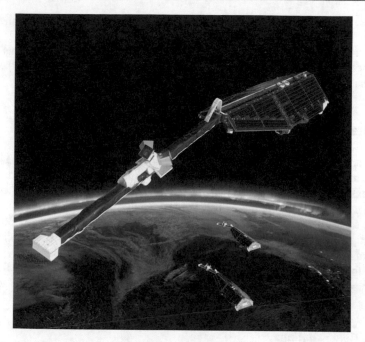

Fig. 2 The Swarm constellation above polar aurora in progress ©ESA

was designed to derive the first global representation of the geomagnetic field variations on timescales from an hour to several years, addressing the crucial problem of source separation.

Swarm satellites remain at a low altitude, below 500 km, whereas other satellites are at much higher altitudes, up to several tens of Earth radii. These missions have a different scientific purpose as they have the capability of making highly precise and accurate magnetic field measurements directly within the magnetosphere or ionosphere, as Cluster, MMS and RBSP.

Cluster II[5] is an ESA mission launched in 2000 and consists of four satellites free flying in a tetrahedral formation around the Earth. The satellites provide information in three dimensions about how the solar wind affects the magnetosphere. Each satellite has the same set of instruments for measuring the plasma environment around the Earth, including magnetic and electric fields, ions and electrons. The relative positions and distances between satellites can be modified to answer various scientific goals.

As Cluster mission, the Magnetospheric Multiscale Mission (MMS)[6] is a NASA mission to study the Earth's magnetosphere, using four identical spacecraft flying in a tetrahedral formation. As for Cluster's case, by acquiring data simultaneously at multiple points in space, MMS allows scientists to differentiate between spatial variations and temporal evolution of the magnetospheric plasma processes.

The Radiation Belt Storm Probes (RBSP)[7] is a mission of NASA's Living With a Star program. The two spacecrafts are devoted to study the Van Allen radiation belts, providing

---

[5] http://sci.esa.int/cluster/.

[6] https://mms.gsfc.nasa.gov/.

[7] http://www.nasa.gov/mission_pages/rbsp/mission/.

insight into their physical dynamics and inputs to understand changes in this critical region of space.

Measuring the Earth's magnetic field at a wide range of altitudes represents a unique possibility to get a full picture of the field variations, by taking into account the complementary nature of these observational platforms.

3 Magnetic Field Variations

Here, we turn to an important part of this review, dedicated to the magnetic field variations and some of their specific features. The geomagnetic field is in a permanent state of change on timescales from seconds to hundreds of millions of years. The shortest timescales are linked to processes arising in the ionosphere and magnetosphere, and their coupling, and cover the scope of space weather. Temporal variations over timescales of months to decades mostly reflect changes in the Earth's core, and the very long ones, from centuries to millions of years reveal different aspects of core processes and space climate.

In the following, we list the main characteristics of the internal and external magnetic fields temporal variations and the need to consider them for space climate and space weather.

3.1 Core Field Variations

The geometry of the core field directly impacts the geoeffectivity. Moreover, from an operational point of view, some specific features of the secular variation may affect space weather tools and models, as the secular variation coefficients are incorporated as inputs (e.g., the geomagnetic coordinates are calculated from reference field models—such as the International Geomagnetic Reference Field (IGRF)[8]). We thus present in the following relevant internal magnetic field variations.

3.1.1 Dipole Moment Decay

The dipole moment has decreased by nearly 6% per century since intensity magnetic field measurements exist, and by about 30% over the past 2000 yrs according to archeomagnetic measurements and models. The magnetosphere, ionosphere and ground geomagnetic field perturbations respond as the geomagnetic dipole moment changes. The implications of this rapid change are critical, as a strongly reduced dipole moment allows access of solar protons into the Earth's atmosphere even at midlatitudes, with implications to the end users of space weather information.

Figure 3 shows the axial component of the dipole field as computed from the IGRF field model. The general decreasing rate is of some 16 nT yr^{-1}. However, at some epochs the field has very different rate fluctuations. For example around 1980, it diminished twice as fast as nowadays. If the same rate of decay is extrapolated in time, the axial dipole would reach zero values in less than 2000 yrs.

Aubert (2015) investigated the axial dipole decay and the place where the field is minimum at the Earth's surface, for the next 100 years. It appears that the South

---

[8] https://www.ngdc.noaa.gov/IAGA/vmod/igrf.html.

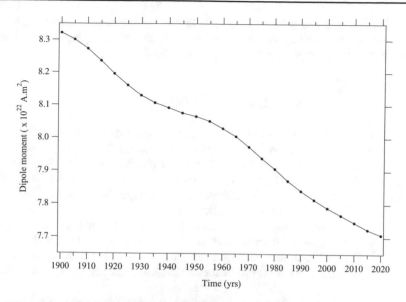

Fig. 3 Geomagnetic dipole moment as computed from the IGRF-13 model

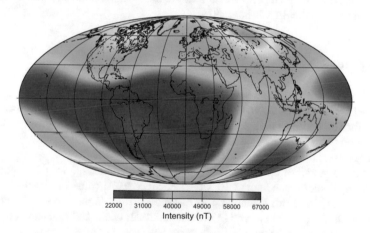

Fig. 4 Magnetic field intensity at the Earth' surface calculated from the core field model IGRF-13 for epoch 2020.5

Atlantic/South America Anomaly (SAA, see Fig. 4) behaviour could be connected to the general decrease of the dipolar field and to the significant increase of the non-dipolar field in the Southern Atlantic region.

A question arises: What may be the future evolution and the implications of this significant dipole moment decay? To offer possible answers it is crucial to understand the physical process responsible for the dipole moment decay. Fluid motion in the outer core transports magnetic field lines, converting kinetic energy into magnetic energy. This ensures the maintenance of the magnetic field against Ohmic dissipation. Considering the free decay of the dipole, if the fluid motions ceased, estimates of some 55,000 yrs are needed to reach a zero level of the axial dipole magnitude. This time interval is more than one order of magnitude larger that the value of 2000 yrs, based on recent observations. An explanation could be found in different scenarios for the fluid motions within the core proposed by Finlay et al. (2016).

3.1.2 South Atlantic Anomaly/South America Anomaly

One of the striking characteristics of the geomagnetic field was known as the South Atlantic Anomaly and now known as the South America Anomaly (SAA), where the total field intensity is unusually low and the fluxes of energetic protons and electrons trapped in the geomagnetic field are unusually high (see purple area on Fig. 4). In this area, the field intensity reaches less of 1/3 of the field strength at comparable latitudes. This regional weakening in field intensity allows energetic particles and cosmic rays to penetrate much deeper into the magnetosphere and even atmosphere than in other regions, resulting in significant space weather effects in space, such as satellite outages (Heirtzler et al. 2002; Willis et al. 2016) or even on the ground, to radio communications or induced currents in pipelines and transmission lines (Pulkkinen et al. 2012; Boteler and Pirjola 2014) (see next section).

The weakness of the field intensity in the SAA is caused by a patch of opposite magnetic flux compared to the dipole direction at the core-mantle boundary (stressing that the main contribution to the SAA is internal).

3.1.3 Magnetic Poles

The positions of the north and south magnetic poles gradually change. Considering the *gufm* model (Jackson et al. 2000) and more recent models, such as IGRF-13, locations and velocities for the two magnetic poles can be computed. Such computations indicate that both magnetic poles have changed their positions; however, the rate of movement for each pole is very different.

During the first half of the twentieth century the north and south magnetic poles velocities are comparable, around some 10 km yr^{-1}. Since 1970 (time of a well-documented geomagnetic jerk) the north magnetic pole has moved from Canada towards Siberia, with velocities reaching some 60 km yr^{-1} (Newitt et al. 2002; Olsen and Mandea 2007) and the south pole towards Australia, with a much lower velocity of around 5 km yr^{-1}.

Figure 5 shows the north and south magnetic and geomagnetic poles positions as obtained from IGRF-13 model. The north magnetic pole velocity has entered into a deceleration phase, decreasing from about 50 km yr^{-1} in 2015 to $\sim$ 45 km yr^{-1} in 2020.

The rapid change in the north magnetic pole positions can be thought as a signature of a sudden change in the behaviour of the field. However, in the other hemisphere the south magnetic pole has kept nearly the same velocity rate (Mandea and Dormy 2003). The location of the north magnetic pole appears to be governed by two large-scale patches of magnetic field, one beneath Canada and one beneath Siberia, and could be linked to a high-speed jet of liquid iron beneath Canada (Livermore et al. 2016). The fast accelerating/ decelerating of the north magnetic field has a large significance for space weather, considering the dynamic polar processes.

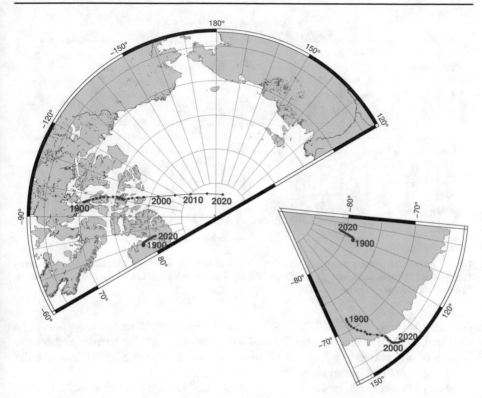

Fig. 5 North (left) and south (right) magnetic (red) and geomagnetic (blue) poles positions obtained from IGRF-13 model. All symbols are at 5-yr interval

3.2 Some Specific External Field Features

Processes originating from the Sun produce rapid variations, with periods from seconds to days. Electric currents within the ionosphere and magnetosphere produce magnetic fields, which are combined with the Earth's own magnetic field. An important role is played by the magnetosphere which acts as a shield against the solar wind coming from the Sun. Severe space weather events compress Earth's magnetic shield, releasing enough power to blind satellites, disrupt radio signals and impact human-made infrastructure on the Earth's surface plunging entire cities into electrical blackouts.

3.2.1 Solar Cycles

Solar cycles are depicted by the number of sunspots on the Sun's surface. The present period, at the time of the edition of this article (mid-2020), is a solar minimum between the 24th and the 25th Solar Cycles. Solar Cycle 25 Prediction Panel experts coordinated by NOAA/NASA—preliminary forecast for Solar Cycle 25 (5 April 2019), considering that the Solar Cycle 25 may have a slow start, but is anticipated to peak with solar maximum

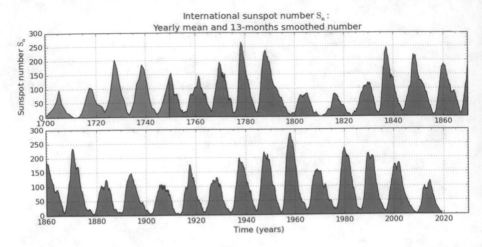

Fig. 6 Yearly mean sunspot number (black) until 1749 and monthly 13-month smoothed sunspot number (blue) from 1749 until nowadays. (Source: WDC SILSO-Sunspot Index and Long-term Solar Observations, Royal Observatory of Belgium, Brussels)

occurring between 2023 and 2026[9]. A low sunspot range of 95 to 130 (very similar in shape to Cycle 24) is also expected. Figure 6 shows the long-running international sunspot number (yearly values) as made available by the World Data Center SILSO (Sunspot Index and Long-term Solar Observations).[10]

Sunspot number is a dataset used in an increasing number of studies linked to Solar Activity and Global Change in the Earth's environment. This long-time series is a valuable complement to scattered historical magnetic measurements to infer space climate/space weather over a few centuries.

3.2.2 Quiet-Day Geomagnetic Variation

The variations are related to the Earth–Sun relation. The Sun illuminates and heats the day-side of the Earth, and Sun's radiations also affect the ionosphere causing convection. This convection implies a motion of the charged particles through the geomagnetic field creating a dynamo action. These daily or diurnal variations, commonly also named as "Solar Quiet" (Sq), are caused by electric currents of several $\mu A\ m^{-2}$ flowing on the sunlit side of the E-region, at about 90–150 km heights. Recently, new advances in our understanding of the geomagnetic daily variations are provided by Yamazaki and Maute (2016), a basis for observations and existing theories to describe them.

3.2.3 Geomagnetic Storms and Substorms

The geomagnetic storms are caused by the interaction of the solar wind with the Earth's magnetic field. These phenomena can be driven by fast solar wind originating from the solar coronal holes (so-called co-rotating interactive regions, CIRs) or by coronal mass

---

[9] https://www.weather.gov/news/190504-sun-activity-in-solar-cycle.

[10] http://sidc.be/silso/home; http://doi.org/10.17616/R38322.

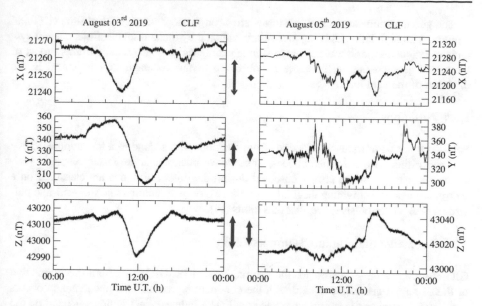

Fig. 7 Magnetograms for a quiet day (left, 03 August 2019) and a disturbed day (right, 05 August 2019), as recorded at Chambon la Forêt observatory (X—North component, Y—East component and Z—vertical component). Red arrows, representing 20 nT, allow an easier comparison of vertical axis

ejections (CMEs). The geoeffectiveness of a CME is related to the southward-pointing component of the interplanetary magnetic field (IMF). The continuing interaction between the solar wind and the magnetosphere increases the number of charged particles trapped within the magnetosphere. During disturbed times, charged particles are guided down the magnetic field lines into the upper atmosphere creating the polar aurora. A recent review on the geomagnetic storms from an historical perspective to the modern view is provided by Lakhina and Tsurutani (2016).

The substorms are related to the intensification of the ionospheric auroral electrojet current system and drive the field aligned currents (FAC) which connect the magnetosphere into the ionosphere at high latitude.

Geomagnetic storms are often characterised by an initial unexpected sharp increase in the northern (X) magnetic component of ground measurements at magnetic observatories, called a sudden storm commencement (SSC). The largest geomagnetic storms tend to occur near the maximum and during the declining phase of a solar cycle. The substorms occurrence is much more frequent than storms (about 1500 substorms occur per year).

The daily magnetic field variation gives an important indication about the status of the day, i.e. a magnetically "quiet" or "disturbed" day. An example is given in Fig. 7, the choice being to show close days in August 2019, the 3rd, a quiet day, and the 5th, a moderate geomagnetic storm. This kind of information is important in distinguishing the status of the Earth's magnetic environment.

The Carrington solar event in 1859, the largest recorded solar magnetic event, has been associated with external field changes of more than −1500 nT at the Colaba Indian magnetic observatory (Cliver and Dietrich 2013). The solar cycle of this event (the solar cycle 10) was a weak cycle, typical of the nineteenth century. The cycle during which occurred the Carrington event was well below the nowadays average solar cycle. These specific conditions may allow such extreme events.

The geomagnetic storms and substorms are strongly dependent on local time (LT) and latitude, with high latitudes being particularly affected by the auroral electrojet current systems or magnetospheric waves. Due to simple geometric reasons (zonal currents), most of the above geomagnetic disturbances are observed in the geomagnetic component in the direction of the field lines.

3.2.4 Pulsations

Pulsations are wave trains with periods from fractions of a second to a few minutes. They are seen riding on the magnetic traces. These wave phenomena are faster magnetic variations and can be almost always observed during daytime. Pulsations are classified on a morphological basis dividing them into two classes: regular, continuous, sinusoid-like pulsations (Pc) and impulsive, irregular pulsations (Pi).

3.2.5 Geomagnetically Induced Currents (GIC)

GIC represent a large interest for the geomagnetic risks. GIC arise from the interaction of the disturbed geomagnetic field with the conductive ground. This phenomenon occurs with every geomagnetic storm, large and small. The induced electric fields depend on the local geological conditions (the electrical conductivity of the ground) and magnetic latitude, mainly.

The Earth's electrical conductivity is a complex parameter, as also is a geomagnetic storm, so the geoelectric field conditions are highly variable in time and space. Generally, the geoelectric field peaks appear during the largest fluctuations, such as at the beginning of a geomagnetic storm, but high-frequency components at later times during a storm can also lead to a large impact.

4 Which Magnetic Risks?

4.1 Hazard and Risk Factors

One of the largest solar storms ever seen struck the Earth in 1859 (see Sect. 2). It then spawned power surges that set the telegraphs on fire, while observers of the tropical islands were able to admire polar auroras, as in Cuba and Hawaii. What would be the consequences of a geomagnetic storm on our technology-dependent civilisation?

To understand the hazard and risk factors related to the geomagnetic field variations, it is crucial to consider the Earth's system as a whole (see Sect. 3). Let us consider a well-known example linked to the induced geoelectric field that represents the natural hazard for GIC. The important risk factors to be considered in this specific case are summarised in the following.

4.1.1 The Core Magnetic Field

The magnetic field produced in the Earth's core plays an important role in space weather hazard. Indeed, the core field strength and geometry impose the shape of the magnetosphere surrounding the Earth and act as a shield which protects our Planet against penetration of highly energetic cosmic and solar particles. For example, the evolution of the

SAA particle flux can be seen as the result of two main effects, on the one hand the secular variation of core magnetic field and on the other hand the modulation of the density of the inner radiation belts during a solar cycle. Domingos et al. (2017) analysis allows to well recover the westward drift rate, as well as the latitudinal and longitudinal solar cycle oscillations, important parameters to forecast the impact of the SAA on space weather.

4.1.2 The Magnetic Latitude

The magnetic latitude, defined as a function of the magnetic poles positions, plays an important role on the GIC hazard. In polar regions, the electric current systems are enhanced during geomagnetic storms and can expand southwards with effects at lower magnetic latitudes. The effects of large geomagnetic storm are reasonably well understood and modelled (Baker et al. 2019); however, it is still debated on how far from the pole these ionospheric current systems can move during the extreme events. Recently, Rogers et al. (2020) show how the likelihood estimates for the extreme horizontal geomagnetic field fluctuations depend on the magnetic latitude, besides some other parameters.

4.1.3 The Lithosphere Electrical Conductivity

The electrical conductivuty of the lithosphere has a large impact on GIC hazard. The variations of the electrical conductivity are important in coastal regions, due to the large difference in conductivity between continental and oceanic crusts, as well as in areas with complex geology of heterogeneous facies. GIC may reach very high values for several minutes ($> 1\ V\,km^{-1}$), at high latitudes, during disturbed conditions. When strong magnetic variations reach low resistance man-made structures, as most of fluid transport infrastructures (e.g., pipelines, power lines) or transport and communication frameworks (e.g., railways), they induce important DC currents that propagate into systems causing important damages.

4.2 Space Weather Impacts

Unlike some other natural hazards such as meteorological conditions, flooding or disease, space weather has no direct impact on humans. The early impacts of space weather (e.g., on the electric telegraph, on telephones and HF radio communications) affected systems used only occasionally in everyday life. The wider societal impact of space weather was limited until the latter part of the twentieth century. The increased susceptibility of technological society to extreme space weather events has driven research in this area.

O'Hare et al. (2019) shows that about 2610 years ago, a huge solar storm hit the Earth, detected by the analysis of ice cores in Greenland. Such an event, comparable to that which occurred in 774–775 AD, would cause a cataclysm in our modern civilisation, as suggested by the "Carrington" magnetic storm of 1859.

Recently, Chapman et al. (2020) analysed a catalogue of geomagnetic field changes going back to 1868, using the *aa* index. They found that severe storms, with possible implications on our modern life, occurred in 42 of the last 150 years, while the most extreme storms, with possible significant damage and disruption occurred in 6 of those years, or once every 25 years. The top 10 years with the most severe geomagnetic activity from 1868 to nowadays are the following: 1921, 1938, 2003, 1946, 1989, 1882, 1941, 1909, 1960 and 1958 (in order of decreasing intensity of an event).

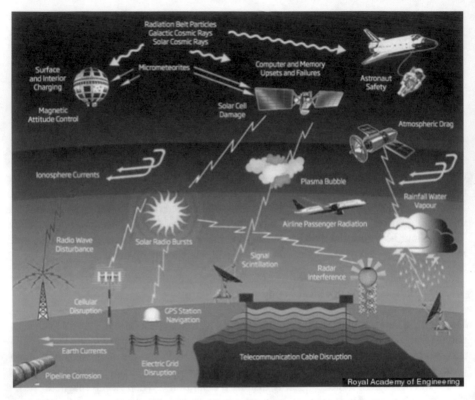

Fig. 8 Space weather impacts (©Royal Academy of Engineering)

The event recorded in May 1921, with disturbances on the US telegraph services, also brought northern and southern auroras, visible at far lower latitudes than usual, with one observatory claiming to detect the southern lights from the Samoa island. The August 1972 storm has detonated sea mines and the one of March 1989 produced huge perturbations of a Canadian power grid.

The severity of the Carrington magnetic storm is not uncommon. According to a report published by the National Academy of Sciences (2008), an exceptional solar event as it is recorded every hundred years could have an economic impact equal to the hurricane "Katrina" times 20! A solar storm of an intensity equivalent to those that have already occurred in the last 200 years could deprive entire regions of electricity for several months (Hapgood 2012). The author quotes a 2009 American study that estimates that a giant blackout could cost $ 2 trillion in the USA alone, because of repairs that would require 4 to 10 years of work.

Figure 8 summarises the space weather effects on different human facilities. Some of the main space weather impacts are summarised in the following. The interested reader is referred to Hapgood (2019) for more details.

4.2.1 Impacts on the Human-made Infrastructures at the Earth's Surface

All lines made of electrically conductive materials may be affected, such as pipelines, copper long distance communication cables, railways, etc. In particular for power grids, the risk of space weather disruption to electricity supplies is a consequence of additional electric currents that are injected into power transmission networks. The impact on power grids creates two primary risks: power outages and damage to grid systems. An example of a dramatic effect linked to a geomagnetic storms is the well-known collapse of Hydro-Quebec's electricity transmission system. Some six million people were left without power for 9 h. The geomagnetic storm causing this event on 13 March 1989 was a severe one, with extremely intense auroras seen as far south as Texas.

4.2.2 Impacts on the Satellite Mission Design, Lifetime, and Navigation Technology

During geomagnetic events, Low Earth Orbit satellites (< 1000 km altitude) may experience an increased air drag force that causes a reduction of satellite altitude. Such processes lead to disturbance of all nominal orbit parameters for constellation control, collision avoidance, re-entry prediction, attitude dynamics. The Global Navigation Satellite Systems (GNSS), at an altitude of about 20,000 km, give accurate measurements of position anywhere on Earth and also the current time with a high accuracy. A wide range of applications is based on GNSS information, from all forms of navigation to synchronisation of activities across networked infrastructures including communications, broadcasting and finance. The GNSS is vulnerable to space weather through the need to correct for the ionospheric delay of signals and through signal disruption by ionospheric scintillation.

4.2.3 Impacts on Spacecraft Physical Integrity

All spacecrafts are exposed to energetic particle radiation (electrons, protons and heavier ions). The fluxes of these particles vary considerably in time and space and lead to a range of effects. It is important also to note the radiation damage to the structure of materials, a major constraint on spacecraft lifetime via the needed power to keep all sub-systems in full running mode.

4.2.4 Impacts on Manned Air- and Spacecraft

During quiet space weather conditions, the radiations from space contribute at some 8% of the radiation observed at the surface, being the primordial contribution up to 3 km altitude. The aircraft cruising at some 10 km altitude flies in a radiation environment, not mentioning higher altitudes vehicles with the International Spatial Station and manned space flights. During severe solar events the contribution of radiation from space can be greatly enhanced.

Finally, let us note that the colossal coronal mass ejection of July 2012 missed Earth, but would have been of Carrington scale if we were in its path. So, devastating magnetic storms could be much more frequent than previously thought, and the new technological modern world needs to take this into account.

5 Conclusions and Future Directions

Continuous global observations of multiple parameters are needed to better understand and predict the closely coupled aspects of Earth's space weather system. An important message that emerges from this contribution is the need for a very strong interplay between the production and analysis of observational data, the development, validation and use of advanced models of the different types, to understand disturbances of the Sun–Earth's system propagating from above (classical space weather), from the coupled Earth's system (terrestrial and solar magnetic fields long-term variations and global electric circuit) and from below (seismic and volcano activity, injection of greenhouse gases into the atmosphere, atmospheric waves produced by solar heating and weather systems). Thus, different sources of hazard are complex, and to address all of them is out of the scope of this paper. Here, we mainly offer a tour of the Earth's magnetic field variations and how they are essential for space weather. We also point out some extreme events and their different consequences.

That relevant processes for the variability of the geomagnetic field need to be better described by a comprehensive separation and understanding of its internal and external sources. With respect of core field, two of its first-order features are of prime importance for space weather—the rapid decay of the dipole field, and the expansion and the decreasing intensity of the SAA—because they accentuate the impact of space weather events.

In the framework of studying space weather impacts on man-made infrastructure a special attention needs to be paid to fast and sometimes violent dynamic changes in any of the magnetospheric and ionospheric current systems. A better knowledge of phenomena in these regions will improve space weather forecasts, with immediate applications, as the correction of GNSS signals in real time. It is clear that the impact of a severe space weather event on the modern technology requests vigilance and new mitigation strategies. The information about conditions in near-Earth's space together with a full understanding of Earth's deep interior is important to make progress towards a complete understanding of Solar-Terrestrial coupling and its effects on humankind and human societies.

References

Aubert J (2015) Geomagnetic forecasts driven by thermal wind dynamics in the earth's core. Geophys J Int. https://doi.org/10.1093/gji/ggv394

Baker DN, Li X, Pulkkinen A, Ngwira CM, Mays ML, Galvin AB, Simunac KDC (2013) A major solar eruptive event in July 2012: defining extreme space weather scenarios. Space Weather. https://doi.org/10.1002/swe.20097

Baker DN, Balogh A, Gombosi TI, Koskinen HEJ, Veronig A, von Steiger R (2019) The scientific foundation of space weather. Space sciences series of ISSI. Springer, Dordrecht

Boteler DH, Pirjola RJ (2014) Comparison of methods for modelling geomagnetically induced currents. Ann Geophys. https://doi.org/10.5194/angeo-32-1177-2014

Chapman SC, Horne RB, Watkins NW (2020) Using the aa index over the last 14 solar cycles to characterize extreme geomagnetic activity. Geophys Res Lett. https://doi.org/10.1029/2019GL086524

Cliver EW, Dietrich WF (2013) The 1859 space weather event revisited: limits of extreme activity. J Space Weather Space Clim 3:31. https://doi.org/10.1051/swsc/2013053

Domingos J, Jault D, Pais MA, Mandea M (2017) The south atlantic anomaly throughout the solar cycle. Earth Planet Sci Lett 473:154–163. https://doi.org/10.1016/j.epsl.2017.06.004

Finlay CC, Aubert J, Gillet N (2016) Gyre-driven decay of the Earth's magnetic dipole. Nat Commun 7:10422. https://doi.org/10.1038/ncomms10422

Hapgood M (2012) Prepare for the coming space weather storm. Nature. https://doi.org/10.1038/484311a

Hapgood M (2019) Technological impacts of space weather. In: Mandea M, Korte M, Yau A, Petrovsky E (eds) Geomagnetism, aeronomy and space weather: a journey from the earth's core to the sun

Heirtzler JR, Allen JH, Wilkinson DC (2002) Ever-present South Atlantic Anomaly damages spacecraft. EOS Trans Am Geophys Union 83(15):165–172

Hulot G, Olsen N, Sabaka TJ, Fournier A (2015) The present and future geomagnetic field. In: Schubert G (ed) Treatise on geophysics (2nd edn), pp 33–78. https://doi.org/10.1016/B978-0-444-53802-4.00096-8

Jackson A, Jonkers ART, Walker MR (2000) Four centuries of geomagnetic secular variation from historical records. Philos Trans R Soc Lond A 358:957–990

Korte M, Mandea M (2019) Geomagnetism: from alexander von humboldt to current challenges. Geochem Geophys Geosyst 20(8):3801–3820. https://doi.org/10.1029/2019GC008324

Lakhina GS, Tsurutani BT (2016) Geomagnetic storms: historical perspective to modern view. Geosci Lett. https://doi.org/10.1186/s40562-016-0037-4

Leger JM, Bertrand F, Jager T, Le Prado M, Fratter I, Lalaurie JC (2009) Swarm absolute scalar and vector magnetometer based on helium 4 optical pumping. In: Proceedings of the Eurosensors XXIII conference, pp 634–637. https://doi.org/10.1016/j.proche.2009.07.158

Livermore P, Finlay C, Hollerbach R (2016) An accelerating high-latitude jet in Earth's core. In: SEDI meeting abstracts

Lockwood M (2012) Solar influence on global and regional climates. Surv Geophys. https://doi.org/10.1007/s10712-012-9181-3

Mandea M, Dormy E (2003) Asymmetric behaviour of magnetic dip poles. Earth Planets Space 55:153–157

Mandea M, Purucker M (2018) The varying core magnetic field from a space weather perspective. Space Sci Rev 214:11. https://doi.org/10.1007/s11214-017-0443-8

Moriña D, Serra I, Puig P, Corral Á (2019) Probability estimation of a carrington-like geomagnetic storm. Sci Rep 9(1):2393

Newitt LR, Mandea M, McKee LA, Orgeval J-J (2002) Recent acceleration of the North Magnetic Pole linked to magnetic jerks. EOS Trans Am Geophys Union 83(35):381–389

O'Hare P, Mekhaldi F, Adolphi F, Raisbeck G, Aldahan A, Anderberg E, Beer J, Christl M, Fahrni S, Synal H-A, Park J, Possnert G, Southon J, Bard E (2019) ASTER Team, R. Muscheler, Multiradionuclide evidence for an extreme solar proton event around 2,610 b.p. (660 bc). Proc. Natl. Acad. Sci. USA, pp 5961–5966 . https://doi.org/10.1073/pnas.1815725116

Olsen N, Mandea M (2007) Will the magnetic north pole wind up in Siberia? EOS Transactions American Geophysical Union

Pulkkinen A, Bernabeu E, Eichner J, Beggan C, Thomson AWP (2012) Generation of 100-year geomagnetically induced current scenarios. Space Weather. https://doi.org/10.1029/2011SW000750

Purucker M, Whaler K (2015) Crustal magnetism. In: Schubert G (ed) Treatise on geophysics, 2nd ed, pp 185–218

Riley P (2012) On the probability of occurrence of extreme space weather events. Space Weather 10(2):1–12

Rogers Neil C, Wild James A, Eastoe Emma F, Gjerloev Jesper W, Thomson Alan WP (2020) A global climatological model of extreme geomagnetic field fluctuations. J Space Weather Space Clim. https://doi.org/10.1051/swsc/2020008

Willis P, Heflin MB, Haines BJ, Bar-Sever YE, Bertiger WI, Mandea M (2016) Is the jason-2 doris oscillator also affected by the South Atlantic anomaly? Adv Space Res. https://doi.org/10.1016/j.asr.2016.09.015

Yamazaki Y, Maute A (2016) Sq and eej—a review on the daily variation of the geomagnetic field caused by ionospheric dynamo currents. Space Sci Rev. https://doi.org/10.1007/s11214-016-0282-z

Yokoyama N, Kamide Y (1997) Statistical nature of geomagnetic storms. J Geophys Res Space Phys 102(A7):14215–14222

Publisher's Note Springer Nature remains neutral with regard to jurisdictional claims in published maps and institutional affiliations.

Printed in the United States
by Baker & Taylor Publisher Services